The Humongous Book of

Calculus Problems

Translated for People who don't Speak Math!!

W. Michael Kelley

ALPHA

A member of Penguin Random House LLC

ALPHA BOOKS

Published by Penguin Random House LLC

Penguin Random House LLC, 375 Hudson Street, New York, New York 10014, USA • Penguin Random House LLC (Canada), 90 Eglinton Avenue East, Suite 700, Toronto, Ontario M4P 2Y3, Canada (a division of Pearson Penguin Canada Inc.) • Penguin Books Ltd., 80 Strand, London WC2R 0RL, England • Penguin Ireland, 25 St. Stephen's Green, Dublin 2, Ireland (a division of Penguin Books Ltd.) • Penguin Random House LLC (Australia), 250 Camberwell Road, Camberwell, Victoria 3124, Australia (a division of Pearson Australia Group Pty. Ltd.) • Penguin Books India Pvt. Ltd., 11 Community Centre, Panchsheel Park, New Delhi—110 017, India • Penguin Random House LLC (NZ), 67 Apollo Drive, Rosedale, North Shore, Auckland 1311, New Zealand (a division of Pearson New Zealand Ltd.) • Penguin Books (South Africa) (Pty.) Ltd., 24 Sturdee Avenue, Rosebank, Johannesburg 2196, South Africa • Penguin Books Ltd., Registered Offices: 80 Strand, London WC2R 0RL, England

International Standard Book Number: 978-1-59257-512-1
Library of Congress Catalog Card Number: 2006930703

19 21

021-189413-January2007

Interpretation of the printing code: The rightmost number of the first series of numbers is the year of the book's printing; the rightmost number of the second series of numbers is the number of the book's printing. For example, a printing code of 06-1 shows that the first printing occurred in 2006.

Printed in the United States of America

Note: This publication contains the opinions and ideas of its author. It is intended to provide helpful and informative material on the subject matter covered. It is sold with the understanding that the author and publisher are not engaged in rendering professional services in the book. If the reader requires personal assistance or advice, a competent professional should be consulted.

The author and publisher specifically disclaim any responsibility for any liability, loss, or risk, personal or otherwise, which is incurred as a consequence, directly or indirectly, of the use and application of any of the contents of this book.

Most Alpha books are available at special quantity discounts for bulk purchases for sales promotions, premiums, fund-raising, or educational use. Special books, or book excerpts, can also be created to fit specific needs.

For details, write: Special Markets, Alpha Books, 375 Hudson Street, New York, NY 10014.

Contents

Introduction

Are you in a calculus class? Yes? Then you NEED this book. Here's why:

Fact #1: The best way to learn calculus is by working out calculus problems.

There's no denying it. If you could figure this class out just by reading the textbook or taking good notes in class, everybody would pass with flying colors. Unfortunately, the harsh truth is that you have to buckle down and work problems out until your fingers are numb.

Fact #2: Most textbooks only tell you WHAT the answers to their practice problems are but not HOW to do them!

Sure your textbook may have 175 problems for every topic, but most of them only give you the answers. That means if you don't get the answer right you're totally screwed! Knowing you're wrong is no help at all if you don't know WHY you're wrong. Math textbooks sit on a huge throne like the Great and Terrible Oz and say, "Nope, try again," and we do. Over and over. And we keep getting the problem wrong. What a delightful way to learn! (Let's not even get into why they only tell you the answers to the odd problems. Does that mean the book's actual AUTHOR didn't even feel like working out the even ones?)

Fact #3: Even when math books try to show you the steps for a problem, they do a lousy job.

Math people love to skip steps. You'll be following along fine with an explanation and then all of a sudden BAM, you're lost. You'll think to yourself, "How did they do that?" or "Where the heck did that 42 come from? It wasn't there in the last step!" Why do almost all of these books assume that in order to work out a problem on page 200, you'd better know pages 1 through 199 like the back of your hand? You don't want to spend the rest of your life on homework! You just want to know why you keep getting a negative number when you're calculating the minimum cost of building a pool whose length is four times the sum of its depth plus the rate at which the water is leaking out of a train that left Chicago at 4:00 a.m. traveling due west at the same speed carbon is decaying.

Fact #4: Reading lists of facts is fun for a while, but then it gets old. Let's cut to the chase.

Just about every single kind of calculus problem you could possibly run into is in here—after all, this book is HUMONGOUS! If a thousand problems aren't enough, then you've got some kind of crazy math hunger, my friend, and I'd seek professional help. This practice book was good at first, but to make it GREAT, I went through and worked out all the problems and took notes in the margins when I thought something was confusing or needed a little more explanation. I also drew little skulls next to the hardest problems, so you'd know not to freak out if they were too challenging. After all, if you're working on a problem and you're totally stumped, isn't it better to know that the problem is SUPPOSED to be hard? It's reassuring, at least for me.

All of my notes are off to the side like this and point to the parts of the book I'm trying to explain.

I think you'll be pleasantly surprised by how detailed the answer explanations are, and I hope you'll find my little notes helpful along the way. Call me crazy, but I think that people who WANT to learn calculus and are willing to spend the time drilling their way through practice problems should actually be able to figure the problems out and learn as they go, but that's just my 2¢.

Good luck and make sure to come visit my website at www.calculus-help.com. If you feel so inclined, drop me an email and give me your 2¢. (Not literally, though—real pennies clog up the Internet pipes.)

—Mike Kelley

Dedication

This book is for my family, who'd love and support me whether I wrote ridiculously long math books or not. For my wife, Lisa, whose grip on sanity is firm when mine starts to buckle, I couldn't possibly love you more than I do. For my swashbuckling pirate son, Nick, who I hope will continue to end the majority of his sentences with "me hardies" even when he's not 3 anymore. And for my beautiful twin girls Erin and Sara, who just said their first word: "shoes." I have a sinking feeling I'll be hearing that word a lot in the not-too-distant future.

Special thanks to Mike Sanders, who helped turn my idea about a marked-up book of calculus problems into a reality, and to my editors Sue Strickland and Ginny Munroe, who work hard to keep me from looking silly.

This book is in memory of Joe, who passed from us in 2006. Back when I wrote The Complete Idiot's Guide to Calculus, Joe told me (in a thick Long Island former trucker accent) that it'd be a "home run." With sincerity unmatched by anyone I have ever met, his simple words of encouragement meant so much to me as a struggling new author. Thanks, Joe. You were right.

Chapter I
LINEAR EQUATIONS AND INEQUALITIES
Problems containing x to the first power

A proper and rigorous understanding of linear equations and their standard forms, linear segments and the associated algorithms, systems of multiple linear equations, and linear inequalities is an essential prerequisite for the study of calculus. Though the majority of calculus students are familiar with the topics in this chapter, mere familiarity is insufficient. In order to succeed in the more advanced topics of the chapters that follow, student mastery of these foundational skills and concepts must be ensured.

Points and lines are the most basic geometric concepts, so you'll need to understand how they are related before you can move on to more complex functions and their graphs. You'll need to know how to create equations of lines, graph lines in the coordinate plane, and even find the lengths and midpoints of line segments. You'll also need to know what to do with expressions containing <, >, ≤, and ≥ signs. Once you've got that down pat, you'll review how to find solutions to systems of equations and inequalities (where you're working with more than one equation or inequality at a time).

Linear Geometry
Creating, graphing, and measuring lines and line segments

1.1 Solve the equation: $3x - (x - 7) = 4x - 5$.

Distribute –1 through the parentheses and combine like terms.
$$3x - x + 7 = 4x - 5$$
$$2x + 7 = 4x - 5$$

Subtract $4x$ and 7 from both sides of the equation to separate the variable and constant terms.

$$
\begin{array}{rcrcr}
2x & + 7 & = & 4x & - 5 \\
-4x & - 7 & & -4x & - 7 \\
\hline
-2x & & = & & - 12
\end{array}
$$

Divide both sides by –2 to get the solution.
$$\frac{-2x}{-2} = \frac{-12}{-2}$$
$$x = 6$$

1.2 Calculate the slope, m, of the line $4x - 3y = 9$.

> Slope-intercept form of a line is $y = mx + b$, where m is the slope of the line and b is the y-intercept.

Solve the equation for y in order to rewrite it in slope-intercept form.
$$-3y = -4x + 9$$
$$y = \frac{4}{3}x - 3$$

The slope of the line is the coefficient of x: $m = \frac{4}{3}$.

1.3 Prove that the slope of a line in standard form, $Ax + By = C$, is $-\frac{A}{B}$.

> The equation's in standard form if it has: (1) No fractions, (2) Only x- and y-terms on the left side, (3) Just the constant on the right side, and (4) A positive x-coefficient.

Write the equation in slope-intercept form by solving it for y.
$$Ax + By = C$$
$$By = -Ax + C$$
$$y = -\frac{A}{B}x + \frac{C}{B}$$

The coefficient of x is the slope of the line: $m = -\frac{A}{B}$.

1.4 Rewrite the linear equation $3x - 4\left(x - \frac{2}{3}y\right) = \frac{4}{5}x - (7y + 3)$ in standard form.

Distribute the constants and combine like terms.
$$3x - 4x + \frac{8}{3}y = \frac{4}{5}x - 7y - 3$$
$$-x + \frac{8}{3}y = \frac{4}{5}x - 7y - 3$$

Multiply by 15, the least common denominator, to eliminate fractions.

$$-15x + 40y = 12x - 105y - 45$$

Separate the variable and constant terms.

$$-27x + 145y = -45$$
$$27x - 145y = 45$$

> Multiply the entire equation by −1 so that the x-coefficient is positive. (It's a requirement of standard form.)

1.5 Write the equation of the line passing through the points (−3,−8) and (−6,2) in slope-intercept form.

Calculate the slope of the line.

$$m = \frac{y_2 - y_1}{x_2 - x_1} = \frac{2 - (-8)}{-6 - (-3)} = \frac{10}{-3} = -\frac{10}{3}$$

Substitute the slope into the slope-intercept formula ($y = mx + b$) for m, replace x and y using one of the coordinate pairs, and solve for b.

> Think of the point (−3,−8) as (x_1, y_1) and (−6,2) as (x_2, y_2), so $x_1 = -3$, $y_1 = -8$, $x_2 = -6$, and $y_2 = 2$.

$$y = mx + b$$
$$-8 = -\frac{10}{3}(-3) + b$$
$$-8 = 10 + b$$
$$b = -18$$

Substitute m and b into the slope-intercept formula.

$$y = mx + b$$
$$y = -\frac{10}{3}x - 18$$

1.6 Calculate the x- and y-intercepts of $3x - 4y = -6$ and use them to graph the line.

To calculate the x-intercept, substitute 0 for y and solve for x. Similarly, substitute 0 for x to calculate the y-intercept.

$$3(0) - 4y = -6 \qquad 3x - 4(0) = -6$$
$$-4y = -6 \qquad\qquad 3x = -6$$
$$y = \frac{3}{2} \qquad\qquad\quad x = -2$$

Therefore, the graph of $3x - 4y = -6$ intersects the x-axis at (−2,0) and the y-axis at $\left(0, \frac{3}{2}\right)$, as illustrated by Figure 1-1.

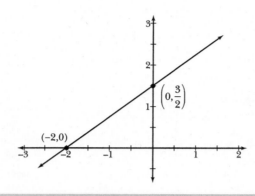

Figure 1-1

The graph of 3x – 4y = –6 with its x- and y-intercepts identified.

The point-slope formula creates an equation based on the slope of the line, m, and a point on the line, (x_1, y_1). Don't plug anything in for the x and y that don't have little numbers next to them.

1.7   Assume that line p contains the point (–3,1) and is parallel to $x – 4y = 1$. Write the equation of p in slope-intercept form.

Calculate the slope of $x – 4y = 1$ using the method of Problem 1.3.

$$m = -\frac{A}{B} = -\frac{1}{-4} = \frac{1}{4}$$

Plug this slope and the coordinates $(x_1, y_1) = (-3,1)$ into the point-slope formula.

$$y - y_1 = m(x - x_1)$$

$$y - 1 = \frac{1}{4}(x - (-3))$$

$$y - 1 = \frac{1}{4}x + \frac{3}{4}$$

Solve for y to express the equation in slope-intercept form.

$$y = \frac{1}{4}x + \frac{7}{4}$$

Note: Problems 1.8 - 1.10 refer to parallelogram ABCD in Figure 1-2.

The midpoint of a line segment with endpoints (x_1, y_1) and (x_2, y_2) is $\left(\frac{x_1 + x_2}{2}, \frac{y_1 + y_2}{2}\right)$. In other words, the x-coordinate of a segment's midpoint is the average of the x-coordinates of its endpoints. The y-coordinate works the same way.

1.8.   According to a basic Euclidean geometry theorem, the diagonals of a parallelogram bisect each other. Demonstrate this theorem for parallelogram *ABCD*.

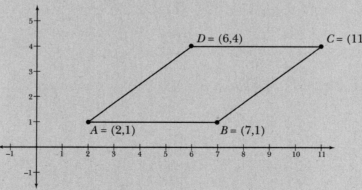

Figure 1-2

Parallelogram ABCD.

Calculate the midpoints of $\overline{AC}$ and $\overline{BD}$; the diagonals bisect one another if and only if those midpoints are equal.

Midpoint of $\overline{AC}$:   Midpoint of $\overline{BD}$:

$$\left(\frac{2+11}{2},\frac{1+4}{2}\right)=\left(\frac{13}{2},\frac{5}{2}\right) \qquad \left(\frac{7+6}{2},\frac{1+4}{2}\right)=\left(\frac{13}{2},\frac{5}{2}\right)$$

Note: Problems 1.8 - 1.10 refer to parallelogram ABCD in Figure 1-2.

1.9 Prove that *ABCD* is a rhombus by verifying that its sides are congruent.

Apply the distance formula four times, once for each side.

The distance between the points (x_1,y_1) and (x_2,y_2) is $\sqrt{(x_2-x_1)^2+(y_2-y_1)^2}$.

$$AB = \sqrt{(7-2)^2+(1-1)^2} \qquad BC = \sqrt{(11-7)^2+(4-1)^2}$$
$$= \sqrt{25+0} \qquad\qquad\quad = \sqrt{16+9}$$
$$= 5 \qquad\qquad\qquad\quad = 5$$

$$CD = \sqrt{(6-11)^2+(4-4)^2} \qquad AD = \sqrt{(6-2)^2+(4-1)^2}$$
$$= \sqrt{25+0} \qquad\qquad\qquad = \sqrt{16+9}$$
$$= 5 \qquad\qquad\qquad\qquad = 5$$

Note: Problems 1.8 - 1.10 refer to parallelogram ABCD in Figure 1-2.

1.10 Prove that *ABCD* is a rhombus by verifying that its diagonals are perpendicular to one another.

Calculate the slopes of the diagonals using the slope formula from Problem 1.5.

$$\text{Slope of } \overline{AC}: \ m_1 = \frac{4-1}{11-2} \qquad \text{Slope of } \overline{BD}: \ m_2 = \frac{4-1}{6-7}$$
$$m_1 = \frac{1}{3} \qquad\qquad\qquad\qquad m_2 = \frac{3}{-1}$$

Parallel lines have equal slopes. The slopes of perpendicular lines are reciprocals of one another and have opposite signs.

The diagonals are negative reciprocals, so the line segments are perpendicular.

Linear Inequalities and Interval Notation
Goodbye equal sign, hello parentheses and brackets

1.11 Write the expression $x \geq -4$ using interval notation.

An interval is defined by the two values that bound an inequality statement, the lower followed by the upper bound. You must indicate whether or not each endpoint is included in the interval. (A bracket next to an endpoint signifies inclusion, and a parenthesis indicates exclusion.)

Any number greater than or equal to −4 makes this statement true; −4 is the lower bound and must be included. The upper bound is infinity. Therefore, $x \geq -4$ is written $[-4,\infty)$.

Always use parentheses next to ∞ when writing intervals. You can't "include" something that's not a finite, real number.

1.12 Write the expression $x < 10$ using interval notation.

The upper bound is 10 and should be excluded (since 10 is not less than 10). Any number less than 10 makes this statement true; there are infinitely many such values in the negative direction, so the lower bound is $-\infty$. Therefore, the inequality statement is written $(-\infty, 10)$.

1.13 Write the expression $6 \geq x > -1$ using interval notation.

The lower bound must always precede the upper bound, regardless of how the expression is written: $(-1, 6]$.

1.14 Write the solution to the inequality using interval notation: $4x - 2 > x + 13$.

Separate variables and constants, then divide by the coefficient of x.

$$4x - x > 13 + 2$$
$$3x > 15$$
$$x > 5$$

Write the solution in interval notation: $(5, \infty)$.

1.15 Write the solution to the inequality using interval notation: $3(2x - 1) - 5 \leq 10x + 19$.

Distribute the constant, combine like terms, and isolate x on the left side of the equation.

$$6x - 3 - 5 \leq 10x + 19$$
$$6x - 8 \leq 10x + 19$$
$$-4x \leq 27$$

Dividing by a negative constant fundamentally changes the inequality: $x \geq -\dfrac{27}{4}$.

Write the solution in interval notation: $\left[-\dfrac{27}{4}, \infty\right)$.

1.16 Graph the inequality: $-2 \leq x < 3$.

Rewrite the inequality as an interval: $[-2, 3)$. To graph the interval on a number line, place a dot at each boundary (closed dots for included boundaries and open dots for excluded boundaries). All values between those boundaries belong to the interval, so darken the number line between the dots, as illustrated by Figure 1-3.

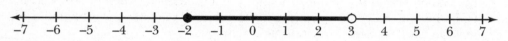

Figure 1-3 *The graph of $-2 \leq x < 3$ includes the interval boundary $x = -2$, but excludes $x = 3$.*

RULE OF THUMB: Use a bracket if the inequality symbol next to the number is ≤ or ≥—otherwise use a parenthesis. Always use parentheses next to ∞ and −∞.

If you multiply or divide both sides of an inequality by a negative number, reverse the inequality sign. In this case, ≤ becomes ≥.

Some textbooks use brackets instead of closed dots and parentheses instead of open dots on the number line.

1.17 Graph the inequality: $x > -1$.

There is no upper bound for the interval $(-1, \infty)$, but all values greater than -1 satisfy the inequality. Therefore, shade all numbers greater than -1 on the number line, as illustrated in Figure 1-4.

Figure 1-4   *The graph of $x > -1$ excludes the lower boundary, $x = -1$.*

1.18 Solve and graph the inequality: $-7 \le 1 - 2x < 11$.

Isolate $-2x$ in the middle of the compound inequality by subtracting 1 from each expression. Next, divide each expression by -2 to isolate x, reversing the inequality signs as you do.

$$-7 - 1 \le -2x < 11 - 1$$
$$\frac{-8}{-2} \ge \frac{-2x}{-2} > \frac{10}{-2}$$
$$4 \ge x > -5$$

The graph of the solution, $(-5, 4]$, is illustrated in Figure 1-5.

Figure 1-5   *The graph of $-7 \le 1 - 2x < 11$ includes $x = 4$ and excludes $x = -5$.*

1.19 Graph the inequality: $y < -\dfrac{1}{3}x + 2$.

This inequality contains two variables, x and y, so it must be graphed on the coordinate plane. Note that the inequality is solved for y and (apart from the inequality sign) looks like a linear equation in slope-intercept form. The linear inequality has y-intercept $(0, 2)$ and slope $-\dfrac{1}{3}$.

Stick the negative sign on top: $\dfrac{-1}{3}$. Starting at the y-intercept, go down 1 unit, right 3 units, and mark the point. Connect the dots to graph the line.

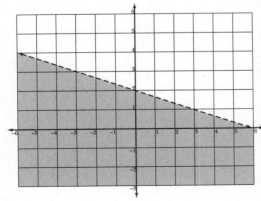

Figure 1-6

The graph of $y < -\dfrac{1}{3}x + 2$ is dotted rather than solid, because it is excluded from the solution (like an open dot indicates exclusion from an inequality graph on a number line).

> By shading the region, you're saying, "All of these points, not just the one I tested, make the inequality true."

The dotted graph separates the coordinate plane into two regions (one above and one below the line). To determine which region represents the solution, choose a point (x,y) from one of the regions and substitute the values into the inequality. If the resulting statement is true, shade the region that contains that point. If not, shade the other region.

1.20 Solve the equation: $2x - y \leq 4$.

Solve the inequality for y.

$$-y \leq -2x + 4$$
$$y \geq 2x - 4$$

> If you don't feel like testing points to figure out where to shade, solve the equation for y and use this rule of thumb: Shade above the line for "greater than" and below the line for "less than."

This graph is solid (not dotted) because the line itself belongs to the solution. Shade the region above the line, as illustrated in Figure 1-7.

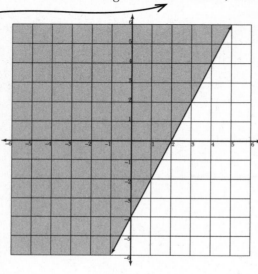

Figure 1-7

All of the ordered pairs above the line are valid solutions to the inequality $2x - y \leq 4$.

Absolute Value Equations and Inequalities
Solve two things for the price of one

1.21 Solve the equation: $|3x - 7| = 8$.

In order for this statement to be valid, the absolute value expression must either equal 8 (since $|8| = 8$) or –8 (since $|-8| = 8$).

$$
\begin{array}{ll}
3x - 7 = 8 & \qquad 3x - 7 = -8 \\
3x = 15 & \qquad 3x = -1 \\
x = 5 & \qquad x = -\dfrac{1}{3}
\end{array}
$$

> You're not asked to graph the solution because a number line with dots at $-\frac{1}{3}$ and at 5 isn't very interesting.

The solution is $x = -\dfrac{1}{3}$ or $x = 5$.

1.22 Solve the equation: $1 - 2|x + 6| = -4$.

Isolate the absolute value expression on the left side of the equation.

$$-2|x+6| = -5$$

$$|x+6| = \frac{5}{2}$$

Apply the technique described in Problem 1.21.

> Remove the absolute value bars and create two equations—one with a positive right side and one with a negative right side.

$$x+6 = \frac{5}{2} \qquad\qquad x+6 = -\frac{5}{2}$$

$$x = \frac{5}{2} - \frac{12}{2} \qquad\qquad x = -\frac{5}{2} - \frac{12}{2}$$

$$x = -\frac{7}{2} \qquad\qquad x = -\frac{17}{2}$$

1.23 Solve the equation: $9 - 3|x+2| = 15$.

Isolate the absolute value expression on the left side of the equation.

$$-3|x+2| = 6$$

$$|x+2| = -2$$

This equation has no solution.

> Absolute values always produce a positive number, so there's no way something in absolute values can equal –2.

1.24 Solve the inequality: $|x-5| < 1$.

The solution to the absolute value inequality $|x+a| < b$, where a and b are real numbers (and $b > 0$), is equivalent to the solution of the compound inequality $-b < x+a < b$.

$$-1 < x-5 < 1$$

Solve the inequality using the method described in Problem 1.18.

$$-1+5 < x < 1+5$$

$$4 < x < 6$$

The solution, in interval notation, is $(4,6)$.

> Drop the inequality bars, stick a matching inequality sign on the left, and then put the opposite of the right side on the left side.

1.25 Graph the solution to the inequality: $2|x-7| - 5 \le -1$.

Isolate the absolute value expression on the left side of the inequality.

$$2|x-7| \le -1+5$$

$$\frac{2|x-7|}{2} \le \frac{4}{2}$$

$$|x-7| \le 2$$

Create a compound inequality (as explained in Problem 1.24) and solve.

$$-2 \qquad\ \le x-7\ \le 2$$

$$-2+7 \ \le x \qquad\ \le 2+7$$

$$5 \qquad\ \le x \qquad\ \le 9$$

The solution, $[5,9]$, is graphed in Figure 1-8.

Chapter One — Linear Equations and Inequalities

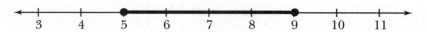

Figure 1-8 The solution graph of $2|x-7|-5 \leq -1$ is a closed interval because both endpoints (x = 5 and x = 9) are included.

1.26 Solve the inequality: $|2x+5| \geq 3$.

Rewrite an inequality of form $|ax+b| \geq c$ as two new inequalities, $ax+b \geq c$ and $ax+b \leq -c$, and solve. The union of the solutions is equivalent to the solution of the original inequality.

$$2x+5 \geq 3 \qquad 2x+5 \leq -3$$
$$2x \geq -2 \quad \text{or} \quad 2x \leq -8$$
$$x \geq -1 \qquad \qquad x \leq -4$$

The solution, in interval notation, is $(\infty, -4]$ or $[-1, \infty)$. The word "or" does not imply that either interval by itself is an acceptable answer, but rather that both intervals together (and therefore all values from both intervals) constitute the solution.

1.27 Solve the inequality and graph the solution: $2-3|x+1| < -5$.

Isolate the absolute value expression on the left side of the inequality.

$$-3|x+1| < -7$$
$$|x+1| > \frac{7}{3}$$

Dividing by −3 reverses the inequality sign; apply the solution method outlined in Problem 1.26.

$$x+1 > \frac{7}{3} \qquad x+1 < -\frac{7}{3}$$
$$\text{or}$$
$$x > \frac{4}{3} \qquad \qquad x < -\frac{10}{3}$$

The solution is $\left(-\infty, -\frac{10}{3}\right)$ or $\left(\frac{4}{3}, \infty\right)$. Graph both intervals on the same number line to generate the graph of $2-3|x+1| < -5$, as illustrated by Figure 1-9.

To solve a > or ≥ absolute value inequality, set up two inequalities without absolute value bars. One will match the original inequality. The other looks the same on the left side, but the number on the right is negative and the inequality sign is reversed.

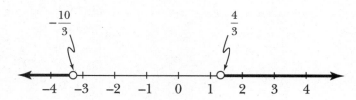

Figure 1-9 All real numbers satisfy the inequality $2-3|x+1| < -5$, except those on the interval $\left[-\frac{10}{3}, \frac{4}{3}\right]$.

10 The Humongous Book of Calculus Problems

Systems of Equations and Inequalities
Find a common solution shared by multiple equations or inequalities

1.28 Solve the following system of equations using the substitution method.

$$\begin{cases} -8x + 2y = -5 \\ 2x - y = 1 \end{cases}$$

Solve the second equation for y and substitute its value into the first equation.

$$-8x + 2(2x - 1) = -5$$
$$-8x + 4x - 2 = -5$$
$$-4x = -3$$
$$x = \frac{3}{4}$$

> Solving $2x - y = 1$ for y gives you $y = 2x - 1$. This allows you to replace y in the other equation with $2x - 1$.

Substitute this x-value into the equation solved for y at the start of the problem.

$$y = 2x - 1 = 2\left(\frac{3}{4}\right) - 1 = \frac{6}{4} - 1 = \frac{1}{2}$$

The coordinate pair (x,y) is the solution: $\left(\frac{3}{4}, \frac{1}{2}\right)$.

> $\left(\frac{3}{4}, \frac{1}{2}\right)$ is also the point where the graphs of the lines $-8x + 2y = -5$ and $2x - y = 1$ intersect.

1.29 Solve the following system of equations using the elimination method.

$$\begin{cases} 2x - 5y = -11 \\ 3x + 13y = 4 \end{cases}$$

To eliminate x from the system, multiply the first equation by -3, multiply the second equation by 2, and then add the equations together.

$$\begin{array}{rrcr} -6x & +15y & = & 33 \\ 6x & +26y & = & 8 \\ \hline & 41y & = & 41 \\ & y & = & 1 \end{array}$$

> Another option is to eliminate y by multiplying the top equation by 13 and the bottom equation by 5.

Substitute $y = 1$ into either of the original equations and solve for x.

$$2x - 5y = -11$$
$$2x - 5(1) = -11$$
$$2x = -11 + 5$$
$$x = -\frac{6}{2} = -3$$

The point $(-3,1)$ is the solution to the system of equations.

1.30 Solve the system of equations: $\begin{cases} x - 6y = 24 \\ \dfrac{1}{3}x - 2y = 8 \end{cases}$

Use the substitution technique, as the first equation is easily solved for x: $x = 6y + 24$.

$$\frac{1}{3}(6y + 24) - 2y = 8$$

$$2y + 8 - 2y = 8$$

$$8 = 8$$

The end result is a true statement ($8 = 8$), but no variables remain. This indicates that the equations of the system are multiples of each other (dividing the first equation by 2 results in the second equation); the system is therefore *dependent* and possesses infinitely many solutions.

> It's "easily solved for x" because the coefficient of x is 1. Not needing to divide by an x-coefficient means one less fraction to deal with.

> Even though the equations in "dependent" systems look different, their graphs are exactly the same. No matter what x you plug into the equations, you'll get the same y, so they intersect at infinitely many points and have infinitely many common solutions.

> This is the slope shortcut formula from problem 1.3.

1.31 Determine the real number value of k in the system of equations below that makes the system indeterminate.

$$\begin{cases} x - 6y = -13 \\ 4x - ky = 1 \end{cases}$$

An indeterminate system of equations has no solution. Consider this geometric explanation: If the solution to a system of equations is the point(s) at which the graphs of its equations intersect, then an indeterminate system has no solution because the graphs of the linear equations do not intersect. The slope of the first line is $-\dfrac{1}{-6} = \dfrac{1}{6}$, and the slope of the second line is $-\dfrac{4}{-k} = \dfrac{4}{k}$. Set the slopes equal to create parallel lines, and solve for k.

$$\frac{4}{k} = \frac{1}{6}$$

Cross multiply to solve the proportion.

$$4 \cdot 6 = k \cdot 1$$

$$24 = k$$

1.32 Graph the solution to the below system of inequalities.

$$\begin{cases} y < 3 \\ x \geq -4 \\ y > \dfrac{2}{3}x - 1 \end{cases}$$

Graph the inequalities on the same coordinate plane, as illustrated by Figure 1-10. The region of the plane upon which the shaded solutions of all three inequality graphs overlap is the solution to the system.

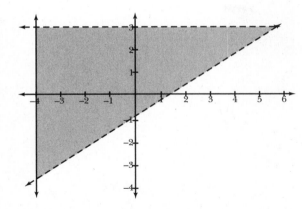

Figure 1-10

The solution to a system of inequalities is a two-dimensional shaded region. Note that y = 3 is a horizontal line 3 units above the x-axis and x = −4 is a vertical line four units left of the y-axis.

1.33 Solve the following system of equations.

$$\begin{cases} 3x + 2y - z = 0 \\ 5x - y - 8z = 9 \\ x + 4y - 3z = -22 \end{cases}$$

Solve the first equation for z: $z = 3x + 2y$. Use this expression to replace z in the other two equations of the system.

$$5x - y - 8z = 9$$
$$5x - y - 8(3x + 2y) = 9$$
$$5x - y - 24x - 16y = 9$$
$$-19x - 17y = 9$$

$$x + 4y - 3z = -22$$
$$x + 4y - 3(3x + 2y) = -22$$
$$x + 4y - 9x - 6y = -22$$
$$-8x - 2y = -22$$
$$-4x - y = -11$$

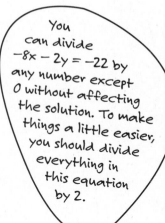

You can divide $-8x - 2y = -22$ by any number except 0 without affecting the solution. To make things a little easier, you should divide everything in this equation by 2.

You're left with a system of two equations in two variables:

$$\begin{cases} -19x - 17y = 9 \\ -4x - y = -11 \end{cases}$$

Solve this system using substitution. (Solve the second equation for y to get $y = -4x + 11$ and substitute that expression into the other equation.)

$$-19x - 17(-4x + 11) = 9$$
$$-19x + 68x - 187 = 9$$
$$49x = 196$$
$$x = 4$$

Plug $x = 4$ into the equation you previously solved for y.

$$y = -4x + 11$$
$$y = -4(4) + 11$$
$$y = -5$$

Plug $x = 4$ and $y = -5$ into the equation previously solved for z.

$$z = 3x + 2y$$
$$z = 3(4) + 2(-5)$$
$$z = 2$$

The solution to the system of equations is $(x, y, z) = (4, -5, 2)$.

Chapter 2
POLYNOMIALS
Because you can't have exponents of 1 forever

The usefulness of the exponential properties and algorithms taught in an elementary algebra course does not expire at the final exam. In fact, they are essential to the study of polynomials, one of the building blocks of calculus. Polynomials, though simple in structure, can possess quite complex graphs, and are so versatile that even the most fundamental differentiation and integration techniques (the power rules for differentiation and integration) are defined in terms of polynomials. This chapter affords you the opportunity to review polynomials and properties of the exponents they contain.

Exponents will play a key role later in calculus. If you're skeptical, consider this: The two most important calculus topics (derivatives and integrals) are actually tied to exponents. The derivative of a polynomial is always one degree (exponent) lower than the polynomial you started with. It works the other way as well—the exponents of an integral are always one greater than the original polynomial. Since exponents and polynomials play such an important role in the upcoming chapters, make sure you know how to manipulate them by working through all the problems here.

Exponential and Radical Expressions
Powers and square roots

2.1 Simplify the expression: $x^2y^3x^4y^6x^9y^8$.

> The commutative property is just a fancy way to say that you can add or multiply things in any order, and it won't change the result, so $4 + 5 = 5 + 4 = 9$ and $(2)(9) = (9)(2) = 18$.

Apply the commutative property of multiplication to rewrite the expression by grouping like variables together.

$$x^2x^4x^9y^3y^6y^8$$

Recall the algebraic axiom concerning the product of exponential expressions with equivalent bases: $x^ax^b = x^{a+b}$.

$$x^{2+4+9}y^{3+6+8} = x^{15}y^{17}$$

2.2 Simplify the expression: $\dfrac{w^{10}x^6y^3z^8}{w^2x^4y^7z^{10}}$.

> In other words, to multiply things that have the same base, add the exponents together: $w^3w^{10} = w^{3+10} = w^{13}$.

If the product of two exponential expressions with the same base requires you to add the exponents (as demonstrated in Problem 2.1), then the quotient is equal to the difference of the exponents

$$w^{10-2}x^{6-4}y^{3-7}z^{8-10} = w^8x^2y^{-4}z^{-2}$$

Rewrite the expression to eliminate the negative exponents.

$$\frac{w^8x^2}{y^4z^2}$$

2.3 Simplify the expression: $\left(\dfrac{4x}{7y^3}\right)^{-2}$.

> Something to a negative power should be moved to the opposite part of the fraction. Once it's moved, change the power back to positive.

The entire rational expression is raised to a negative exponent, so take the reciprocal of the fraction.

$$\left(\frac{7y^3}{4x}\right)^2$$

Square each factor. If a factor is already raised to a power, such as y^3, multiply that power by 2.

$$\frac{7^2y^{3(2)}}{4^2x^2} = \frac{49y^6}{16x^2}$$

2.4 Simplify the expression: $\dfrac{\left(x^{-2}y^6\right)^3}{\left(3xy^{-5}\right)^{-2}}$.

Raise the numerator to the third power and the denominator to the −2 power.

$$\frac{x^{(-2)(3)}y^{6(3)}}{3^{-2}x^{-2}y^{(-5)(-2)}} = \frac{x^{-6}y^{18}}{3^{-2}x^{-2}y^{10}}$$

Eliminate the negative exponents.

$$\frac{3^2x^2y^{18}}{x^6y^{10}} = \frac{9x^2y^{18}}{x^6y^{10}}$$

Simplify using the method outlined in Problem 2.2.

$$9x^{2-6}y^{18-10} = 9x^{-4}y^8 = \frac{9y^8}{x^4}$$

2.5   Prove that $x^0 = 1$.

The multiplicative inverse property of algebra states that any nonzero real number times its reciprocal equals 1. Therefore, if x^a is a real number and is multiplied by its reciprocal, the result must be 1.

$$x^a \cdot \frac{1}{x^a} = 1$$

Rewrite the fraction using a negative exponent.

$$(x^a)(x^{-a}) = 1$$

The expressions have the same base, so calculate the product by adding the exponents.

$$x^{a-a} = 1$$
$$x^0 = 1$$

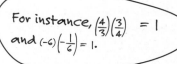

For instance, $\left(\frac{4}{3}\right)\left(\frac{3}{4}\right) = 1$ and $(-6)\left(-\frac{1}{6}\right) = 1$.

2.6   Simplify the expression: $\sqrt{72x^5 y^8}$.

Because the expression is a square root (with an unwritten, implied index of 2), rewrite the factors in terms of perfect squares. ←

$$\sqrt{36 \cdot 2 \cdot x^2 \cdot x^2 \cdot x \cdot y^4 \cdot y^4} = \sqrt{6^2 \cdot 2 \cdot \left(x^2\right)^2 \cdot x \cdot \left(y^4\right)^2}$$

Each perfect square can be removed from the radical: $6x^2 y^4 \sqrt{2x}$.

A **perfect square** is made up of something multiplied by itself. That means 36, x^4, and y^8 are perfect squares because $6^2 = 36$, $(x^2)^2 = x^4$, and $(y^4)^2 = y^8$.

2.7   Explain why the expression $\sqrt{x^2 y}$ should be simplified as $|x|\sqrt{y}$, rather than $x\sqrt{y}$.

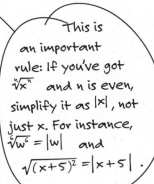

Consider the effect of these values in the original expression: $x = -2$, $y = 5$.

$$\sqrt{x^2 y} = \sqrt{(-2)^2 (5)} = \sqrt{20}$$

Simplify the result.

$$\sqrt{20} = \sqrt{4 \cdot 5} = \sqrt{2^2 \cdot 5} = 2\sqrt{5}$$

Now test the expressions $|x|\sqrt{y}$ and $x\sqrt{y}$ by substituting $x = -2$ and $y = 5$.

$$x\sqrt{y} = -2\sqrt{5} \qquad |x|\sqrt{y} = |-2|\sqrt{5} = 2\sqrt{5}$$

Notice that only $|x|\sqrt{y}$ gives the correct result.

This is an important rule: If you've got $\sqrt[n]{x^n}$ and n is even, simplify it as $|x|$, not just x. For instance, $\sqrt[6]{w^6} = |w|$ and $\sqrt{(x+5)^2} = |x+5|$.

2.8   Simplify the expression: $\sqrt[3]{-8x^7 y^5 z^2}$.

The index of the radical is 3, so rewrite the factors of the radicand as perfect cubes and remove those perfect cubes from beneath the radical sign.

$$\sqrt[3]{(-2)^3 \cdot \left(x^2\right)^3 \cdot x \cdot \left(y\right)^3 \cdot y^2 \cdot z^2} = (-2x^2 y)\sqrt[3]{xy^2 z^2}$$

There's no need to include absolute value bars (as discussed in Problem 2.7) because the index of the expression is odd.

The **radicand** is what you call what's underneath the radical.

2.9 Simplify the expression: $\sqrt{112xy} - \sqrt{28xy}$.

So you can't simplify $3\sqrt{x} + 8\sqrt{y}$, but you can simplify $2\sqrt{y} - 5\sqrt{y}$ to get $-3\sqrt{y}$.

Although $112xy - 28xy = 84xy$, the answer is not $\sqrt{84xy}$; such an answer violates numerous arithmetic rules. Terms containing radicals may be combined only via addition or subtraction if the corresponding radicands match exactly. Notice that simplifying the individual radicals in the expression produces such radicands.

$$\sqrt{112xy} - \sqrt{28xy} = \sqrt{16 \cdot 7 \cdot xy} - \sqrt{4 \cdot 7 \cdot xy}$$
$$= \sqrt{4^2 \cdot 7 \cdot xy} - \sqrt{2^2 \cdot 7 \cdot xy}$$
$$= 4\sqrt{7xy} - 2\sqrt{7xy}$$
$$= 2\sqrt{7xy}$$

2.10 Rewrite $250^{2/3}$ as a radical expression and simplify.

You can also rewrite $x^{a/b}$ $\sqrt[b]{x^a}$.

A quantity raised to a rational exponent can be rewritten as a radical expression: $x^{a/b} = \left(\sqrt[b]{x}\right)^a$.

$$250^{2/3} = \left(\sqrt[3]{250}\right)^2 = \left(\sqrt[3]{125 \cdot 2}\right)^2 = \left(\sqrt[3]{5^3 \cdot 2}\right)^2 = \left(5\sqrt[3]{2}\right)^2 = 25\sqrt[3]{4}$$

Operations on Polynomial Expressions

How to $+$, $-$, $\times$, and $\div$ polynomials

2.11 Simplify the expression: $3(x^2 - 5xy + 6y^2) - 5(x^2 + 4xy - 1)$.

Distribute 3 through the first set of parentheses and -5 through the second set.
$$3x^2 - 15xy + 18y^2 - 5x^2 - 20xy + 5$$

Combine like terms to simplify.

$$(3x^2 - 5x^2) + 18y^2 + (-15xy - 20xy) + 5 = -2x^2 - 35xy + 18y^2 + 5$$

The term $-35xy$ has a power of 2 since both of its variables are raised to the first power. If there's more than one variable in the term, add their powers together.

Convention dictates that variables should be written in decreasing order of exponential power. If two terms have the same power (in the above expression, x^2, y^2, and xy are raised to the second power), write the variables in alphabetical order.

2.12 Simplify the following expression.
$$-2x^2(y - 4) + x(x + 6) - 4(3x - y) + 7y^2(x + 1) + 6y(y - 9) - 3(y + 5x)$$

Distribute the constants.
$$-2x^2y + 8x^2 + x^2 + 6x - 12x + 4y + 7xy^2 + 7y^2 + 6y^2 - 54y - 3y - 15x$$

Combine like terms.
$$-2x^2y + 7xy^2 + 9x^2 + 13y^2 - 21x - 53y$$

2.13 Find the product and simplify: $(y - 1)(2y + 3)$.

Use the FOIL method to multiply pairs of binomials. "FOIL" is a technique requiring you to multiply pairs of terms and add all of the results; its name is an acronym describing those pairs: first (the first terms in each binomial, in this example, y and $2y$), outside (the terms at the outer edges of the product, y and 3), inside (the terms in the middle of the product, -1 and $2y$), and last (the last term in each binomial, -1 and 3).

$$y(2y) + y(3) + (-1)(2y) + (-1)(3)$$
$$= 2y^2 + 3y - 2y - 3$$
$$= 2y^2 + y - 3$$

> Any binomial squared follows the same pattern: $(x + y)^2 = x^2 + 2xy + y^2$. Be careful! $(x + y)^2$ does NOT equal $x^2 + y^2$. You've got to use the FOIL method—don't just square the terms individually.

2.14 Find the product and simplify: $(a - 3b)^2$.

Rewrite the squared binomial as a product and multiply using the FOIL method.

$$(a - 3b)(a - 3b)$$
$$= a \cdot a + a(-3b) + (-3b)(a) + (-3b)(-3b)$$
$$= a^2 - 3ab - 3ab + 9b^2$$
$$= a^2 - 6ab + 9b^2$$

2.15 Find the product and simplify: $(2x - y)(x + 5y - 1)$.

The FOIL method is useful only when multiplying exactly two binomials. In order to multiply two polynomials containing any number of terms, multiply each term in the left polynomial by each term in the right polynomial, one at a time, and add the results.

$$2x(x) + 2x(5y) + 2x(-1) + (-y)(x) + (-y)(5y) + (-y)(-1)$$
$$= 2x^2 + 10xy - 2x - xy - 5y^2 + y$$
$$= 2x^2 + 9xy - 5y^2 - 2x + y$$

> In other words, start with $2x$ and multiply it by every term in the right polynomial: $2x(x) = 2x^2$, $2x(5y) = 10xy$, and $2x(-1) = -2x$. Now repeat the process, this time with $-y$.

2.16 Evaluate the quotient using polynomial long division: $\dfrac{2x^3 - 5x^2 + 9x - 8}{x + 3}$.

Prepare the polynomials for long division: $x + 3$ is the divisor and $2x^3 - 5x^2 + 9x - 8$ the numerator is the dividend.

$$x + 3 \overline{)\,2x^3 - 5x^2 + 9x - 8}$$

> The **dividend** is inside the division symbol, and the **divisor** is outside.

What value, when multiplied by x (the first term of the divisor) results in $2x^3$, the first term of the dividend? The only such value is $2x^2$: $(x)(2x^2) = 2x^3$. Write that value above its like term in the dividend.

$$\begin{array}{r} 2x^2 \\ x + 3 \overline{)\,2x^3 - 5x^2 + 9x - 8} \end{array}$$

Multiply each term in the divisor by $2x^2$, and write the *opposite* of each result beneath its like term in the dividend.

$$\begin{array}{r} 2x^2 \\ x + 3 \overline{)\,2x^3 - 5x^2 + 9x - 8} \\ -2x^3 - 6x^2 \end{array}$$

Combine like terms. Then copy the third term of the divisor next to the results.

> At this point, $2x^3$ disappears. The entire purpose of the steps leading up to this moment was to find a way to add $2x^3$ to its opposite so that it goes away.

$$x+3 \overline{)\,2x^3 - 5x^2 + 9x - 8\,}$$
$$\begin{array}{r} 2x^2 \\ \underline{-2x^3 - 6x^2} \\ -11x^2 + 9x \end{array}$$

Repeat the process, this time identifying the value that (when multiplied by x) will result in the first term of the new expression: $-11x^2$. Place that value, $-11x$, above the dividend, multiply it by the terms of the divisor, record the *opposites* of those results, combine like terms, and copy the final term of the divisor (-8).

$$
\begin{array}{r}
2x^2 - 11x \\
x+3 \overline{)\,2x^3 - 5x^2 + 9x - 8\,} \\
\underline{-2x^3 - 6x^2} \\
-11x^2 + 9x \\
\underline{11x^2 + 33x} \\
42x - 8
\end{array}
$$

The process repeats once more, this time with the value 42 above the divisor, since $x \cdot 42 = 42x$.

> You can check the answer by multiplying the quotient and the divisor and then adding the remainder: $(2x^2 - 11x + 42)(x + 3) - 134$. You should get the dividend.

$$
\begin{array}{r}
2x^2 - 11x + 42 \\
x+3 \overline{)\,2x^3 - 5x^2 + 9x \quad\; - 8\,} \\
\underline{-2x^3 - 6x^2} \\
-11x^2 + 9x \\
\underline{11x^2 + 33x} \\
42x - 8 \\
\underline{-42x - 126} \\
-134
\end{array}
$$

The remainder is -134, and should be added to the quotient, $2x^2 - 11x + 42$, as the numerator of a fraction whose denominator is the divisor: $2x^2 - 11x + 42 - \dfrac{134}{x+3}$.

2.17 Evaluate the quotient using polynomial long division: $(x^4 + 6x - 2) \div (x^2 + 3)$.

> The numerator has no x term—it skips right from x^2 to 3. The denominator skips two terms, because it has no x^3 or x^2. Write those "missing" terms as $0x$, $0x^3$, and $0x^2$.

Use the method described in Problem 2.16, but when setting up the division problem, ensure that *every* power of x is included from the highest power to a constant. Use a coefficient of 0 for missing terms.

$$
\begin{array}{r}
x^2 + 0x - 3 \\
x^2 + 0x + 3 \overline{)\,x^4 + 0x^3 + 0x^2 + 6x - 2\,} \\
\underline{-x^4 - 0x^3 - 3x^2} \\
0x^3 - 3x^2 + 6x \\
\underline{0x^3 + 0x^2 + 0x} \\
-3x^2 + 6x - 2 \\
\underline{3x^2 + 0x + 9} \\
6x + 7
\end{array}
$$

The solution is $x^2 - 3 + \dfrac{6x+7}{x^2+3}$.

2.18 Calculate $(3x^2 + 10x - 8) \div (x + 4)$ using synthetic division.

List the coefficients of the dividend, and place the *opposite* of the divisor's constant in a box to the left. Leave some space below that and draw a horizontal line.

$$\boxed{-4} \quad 3 \quad 10 \quad -8$$

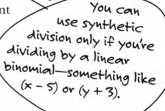

You can use synthetic division only if you're dividing by a linear binomial—something like (x – 5) or (y + 3).

Copy the first constant (3) below the horizontal line, multiply it by the boxed constant (−4), and record the result (−12) in the next column, below the number 10. Add the numbers in that column (10 − 12 = −2) and record the result beneath the line.

$$
\begin{array}{r|rrr}
\boxed{-4} & 3 & 10 & -8 \\
 & & -12 & \\
\hline
 & 3 & -2 &
\end{array}
$$

Repeat the process; multiply the new constant (−2) by the boxed constant (−4). Record the result (8) in the next column and, once again, add the numbers in that column.

$$
\begin{array}{r|rrr}
\boxed{-4} & 3 & 10 & -8 \\
 & & -12 & 8 \\
\hline
 & 3 & -2 & 0
\end{array}
$$

Since (x + 4) divided evenly into $3x^2 + 10x - 8$ (the remainder was 0), (x + 4) is a factor of $3x^2 + 10x - 8$.

The numbers below the horizontal line are the coefficients of the quotient. Note that the degree of the quotient is always one less than the degree of the dividend (so this quotient is linear), and the rightmost number below the horizontal line (in this case 0) is the remainder. The solution is $3x - 2$.

2.19 Calculate $(4x^3 - 11x^2 - 1) \div (x - 3)$ using synthetic division.

List the coefficients and the opposite of the divisor's constant as demonstrated in Problem 2.18, but notice that the dividend contains no x-term. Just like long division, synthetic division requires you to insert a 0 coefficient for missing terms.

$$
\begin{array}{r|rrrr}
\boxed{3} & 4 & -11 & 0 & -1 \\
 & & 12 & 3 & 9 \\
\hline
 & 4 & 1 & 3 & 8
\end{array}
$$

The quotient is $4x^2 + x + 3 + \dfrac{8}{x - 3}$.

Factoring Polynomials
Reverse the multiplication process

2.20 Factor the expression: $18x^2y^5 - 9xy^3$.

Both terms can be divided evenly by 9, x, and y^3 (i.e. there will be no remainder). Therefore, the greatest common factor is $9xy^3$. Factor the expression by writing the greatest common factor followed by the quotients of each term divided by that factor.

$$9xy^3 \left(\frac{18x^2y^5}{9xy^3} - \frac{9xy^3}{9xy^3} \right)$$

$$= 9xy^3 \left(2xy^2 - 1 \right)$$

2.21 Factor the expression: $21x^5y^9z^6 - 15x^4y^2z^{11} + 36x^8y^3z$.

The greatest common factor is $3x^4y^2z$. Use that value to factor the expression, applying the method outlined in Problem 2.20.

$$3x^4y^2z \left(\frac{21x^5y^9z^6}{3x^4y^2z} - \frac{15x^4y^2z^{11}}{3x^4y^2z} + \frac{36x^8y^3z}{3x^4y^2z} \right)$$

$$= 3x^4y^2z \left(7xy^7z^5 - 5z^{10} + 12x^4y \right)$$

2.22 Factor the expression: $x^2 + 13x + 40$.

> *In other words, find two numbers that equal the x-coefficient when they're added, and equal the constant when they're multiplied.*

To factor a trinomial of the form $x^2 + ax + b$, find two numbers k_1 and k_2 such that $k_1 + k_2 = a$ and $k_1 \cdot k_2 = b$. The factors of the polynomial will be $(x + k_1)$ and $(x + k_2)$. In this case, $k_1 = 5$ and $k_2 = 8$, since $5 + 8 = 13$ and $5(8) = 40$, so the factored form of the polynomial is $(x + 5)(x + 8)$.

2.23 Factor the expression $x^2 - 7x - 18$.

Determine k_1 and k_2 as directed in Problem 2.22. Note that the constant (-18) is negative so k_1 and k_2 have opposite signs. (The product of two numbers with the same sign is always positive.) Additionally, the x-coefficient is -7, which means the larger of the two numbers must be negative: $k_1 = -9$ and $k_2 = 2$, as $-9 + 2 = -7$ and $(-9)(2) = -18$. The factored form of the polynomial is $(x - 9)(x + 2)$.

2.24 Factor the expression: $x^2 - 49$.

> *You can also factor the difference of perfect cubes: $x^3 - y^3 = (x - y)(x^2 + xy + y^2)$.*

This expression is the difference of perfect squares ($x^2 = x \cdot x$ and $49 = 7 \cdot 7$). Note that $a^2 - b^2$ is factored $(a + b)(a - b)$, so the factored form of the polynomial is $(x + 7)(x - 7)$.

2.25 Factor the expression: $8a^3 + 125b^3$.

This expression is the sum of perfect cubes: $8a^3 = (2a)^3$ and $125b^3 = (5b)^3$. Much like a difference of perfect squares follows a distinct factoring pattern, so does the sum of perfect cubes: $x^3 + y^3 = (x + y)(x^2 - xy + y^2)$. To apply the formula, set $x = 2a$ and $y = 5b$.

$$x^3 + y^3 = (x + y)(x^2 - xy + y^2)$$

$$(2a)^3 + (5b)^3 = (2a + 5b)\left((2a)^2 - (2a)(5b) + (5b)^2\right)$$

$$8a^3 + 125b^3 = (2a + 5b)\left(4a^2 - 10ab + 25b^2\right)$$

2.26 Factor the expression: $4x^3 - 20x^2 - 3x + 15$.

A polynomial containing four terms from which no common denominator can be extracted is often factored by grouping. Use parentheses to split the polynomial into the sum of two binomials, one containing the first two terms and one containing the remaining two terms.

$$(4x^3 - 20x^2) + (-3x + 15)$$ ⟵

Factor the greatest common factor out of each quantity.

$$= 4x^2(x - 5) - 3(x - 5)$$

Both terms now have a common factor: $(x - 5)$. Factor out that binomial; the first term is left with $4x^2$ and the second term is left with -3.

$$= (x - 5)(4x^2 - 3)$$

> The two groups are always added. Even though the coefficient of x is –3, keep that negative sign inside the second set of parentheses.

2.27 Factor the expression by decomposition: $6x^2 + 7x - 24$.

The coefficient of the x^2-term in this trinomial does not equal 1, so the method described in Problems 2.22 and 2.23 is invalid and you should factor by decomposition. To factor the expression $ax^2 + bx + c$ (when $a \neq 1$), you once again seek two constants (k_1 and k_2), but in this case those numbers meet slightly different criteria.

$$k_1 + k_2 = b \text{ and } (k_1)(k_2) = ac$$ ⟵

With some experimentation you'll determine that $k_1 = 16$ and $k_2 = -9$. Replace the x-coefficient with $k_1 + k_2$.

$$6x^2 + 7x - 24$$
$$= 6x^2 + (16 - 9)x - 24$$
$$= 6x^2 + 16x - 9x - 24$$

Factor by grouping, as explained in Problem 2.26.

$$= (6x^2 + 16x) + (-9x - 24)$$
$$= 2x(3x + 8) - 3(3x + 8)$$
$$= (3x + 8)(2x - 3)$$

> k_1 and k_2 still have to add up to the x-coefficient, but now their product equals the constant times the x^2-coefficient. In this problem, $k_1 + k_2 = 7$ and $(k_1)(k_2) = 6(-24) = -144$.

Solving Quadratic Equations
Equations with a highest exponent of 2

2.28 Solve the equation by factoring: $4x^2 + 4x = 15$.

Subtract 15 from both sides of the equation, so that the polynomial equals 0.

$$4x^2 + 4x - 15 = 0$$

Factor by decomposition, as explained in Problem 2.27.

$$4x^2 + (10 - 6)x - 15 = 0$$
$$4x^2 + 10x - 6x - 15 = 0$$
$$2x(2x + 5) - 3(2x + 5) = 0$$
$$(2x - 3)(2x + 5) = 0$$

Set each factor equal to 0, creating two separate equations to be solved.

$$2x - 3 = 0 \qquad 2x + 5 = 0$$
$$2x = 3 \quad \text{or} \quad 2x = -5$$
$$x = \frac{3}{2} \qquad x = -\frac{5}{2}$$

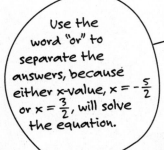

Use the word "or" to separate the answers, because either x-value, $x = -\frac{5}{2}$ or $x = \frac{3}{2}$, will solve the equation.

The solution is $x = -\frac{5}{2}$ or $x = \frac{3}{2}$.

2.29 Solve the equation using the quadratic formula: $5x^2 = 3x - 6$.

Begin by setting the equation equal to 0.

$$5x^2 - 3x + 6 = 0$$

Apply the quadratic formula to solve the equation.

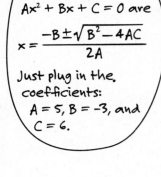

The solutions of the equation $Ax^2 + Bx + C = 0$ are

$$x = \frac{-B \pm \sqrt{B^2 - 4AC}}{2A}$$

Just plug in the coefficients: $A = 5$, $B = -3$, and $C = 6$.

$$x = \frac{-(-3) \pm \sqrt{(-3)^2 - 4(5)(6)}}{2(5)}$$
$$= \frac{3 \pm \sqrt{9 - 120}}{10}$$
$$= \frac{3 \pm \sqrt{-111}}{10}$$

In order to simplify a square root with a negative radicand, recall that $i = \sqrt{-1}$.

$$\frac{3 \pm \sqrt{-111}}{10} = \frac{3 \pm \sqrt{(-1)(111)}}{10} = \frac{3 \pm i\sqrt{111}}{10}$$

Therefore, there are two imaginary solutions to the quadratic equation: $\frac{3 + i\sqrt{111}}{10}$ or $\frac{3 - i\sqrt{111}}{10}$.

2.30 Find the value of k such that the quadratic equation $6(x^2 + 2x) = k(2x - 1) + 5x^2$ has the double root $x = 3$.

Simplify and set the equation equal to 0.

$$6x^2 + 12x = 2kx - k + 5x^2$$
$$(6x^2 - 5x^2) + (12x - 2kx) + k = 0$$

Factor x^2 out of the first quantity and factor x out of the second.

$$x^2(6 - 5) + x(12 - 2k) + k = 0$$
$$x^2 + x(12 - 2k) + k = 0$$

A quadratic equation has a double root when the same root appears twice. For example, $(x - 3)(x - 3) = 0$ has matching solutions: $x = 3$ or $x = 3$.

Notice that $(x - 3)(x - 3) = 0$ is another quadratic equation whose solution is the double root $x = 3$. Set the equivalent expressions equal to one another.

$$x^2 + x(12 - 2k) + k = (x - 3)(x - 3)$$
$$x^2 + x(12 - 2k) + k = x^2 - 6x + 9$$

Quadratics are equal only if the coefficients of their corresponding terms are equal. The constant of the left quadratic is k, whereas the constant of the right quadratic is 9. Therefore, $k = 9$. (Substituting $k = 9$ into $(12 - 2k)$ results in -6, so the coefficients of the x-terms in both polynomials will be equal as well.)

2.31 Solve the equation by completing the square: $x^2 - 14x + 3 = 0$.

Unlike the methods of factoring and the quadratic formula, you should not set the equation equal to 0, but instead, isolate the constant on the right side of the equation.

$$x^2 - 14x = -3$$

Calculate one half of the x-coefficient and square it: $(-14) \div 2 = -7$, and $(-7)^2 = 49$. Add the result to both sides of the equation.

$$x^2 - 14x + 49 = -3 + 49$$

This creates a perfect square on the left side of the equation.

$$(x - 7)(x - 7) = 46$$
$$(x - 7)^2 = 46$$

Take the square root of both sides of the equation and solve for x.

$$\sqrt{(x - 7)^2} = \pm\sqrt{46}$$
$$x - 7 = \pm\sqrt{46}$$
$$x = 7 \pm \sqrt{46}$$

2.32 Solve the equation by completing the square: $4x^2 - 20x + 7 = 0$.

Note that you *cannot* complete the square unless the coefficient of the x^2-term is 1. Although that was true in Problem 2.31, it is not true here. Divide everything by the coefficient of x^2, and then use the method described in Problem 2.31.

$$\frac{4x^2}{4} - \frac{20x}{4} + \frac{7}{4} = \frac{0}{4}$$

$$x^2 - 5x + \frac{7}{4} = 0$$

$$x^2 - 5x = -\frac{7}{4}$$

$$x^2 - 5x + \frac{25}{4} = -\frac{7}{4} + \frac{25}{4}$$

$$\left(x - \frac{5}{2}\right)^2 = \frac{18}{4}$$

$$\sqrt{\left(x - \frac{5}{2}\right)^2} = \pm\sqrt{\frac{9}{2}}$$

$$x = \frac{5}{2} \pm \frac{3}{\sqrt{2}}$$

> $\frac{25}{4}$ is half of the x-coefficient squared: $-5 \div 2 = -\frac{5}{2}$ and $\left(-\frac{5}{2}\right)^2 = \frac{25}{4}$.

> Rationalizing a fraction eliminates the radicals from its denominator. Multiply the top and bottom of the fraction by the radical you're trying to get rid of: $\frac{3}{\sqrt{2}} \cdot \frac{\sqrt{2}}{\sqrt{2}} = \frac{3\sqrt{2}}{\sqrt{4}} = \frac{3\sqrt{2}}{2}$

Rationalize the expression to get the equivalent solution $x = \dfrac{5}{2} \pm \dfrac{3\sqrt{2}}{2}$.

Chapter 3
RATIONAL EXPRESSIONS
Fractions, fractions, and more fractions

During a course of review, rational expressions are the natural successors of polynomial expressions for myriad reasons. For one, the most common rational expressions are merely polynomial quotients. Additionally, rational expressions (like polynomial expressions) are restricted under the binary operations of addition and subtraction, but face far fewer restrictions if products or quotients are calculated. It is, therefore, logical to investigate how the skills already applied to linear and polynomial expressions (in Chapter 2) extend to rational expressions.

A rational expression, like a rational number, is just a fraction. Of course, you probably don't have a whole lot of love in your heart for fractions—they can be pretty hard to work with, especially when it comes to adding and subtracting. Most of the time, finding the common denominator is the most difficult thing about fractions, and unfortunately, it's usually the first thing you have to do. (After all, if you can't add two things together, why even bother learning how to multiply or divide them?) This chapter will get you back up to speed with fractions, help you solve complicated fractional equations, and end with fractional inequalities (which don't work anything like the inequalities from Chapter 1).

Adding and Subtracting Rational Expressions

Remember the least common denominator?

3.1 Simplify the expression: $\dfrac{5}{y} - \dfrac{3}{x^2}$.

> In this problem, the unique factors are x and y. The highest power of x is 2 and the highest power of y is 1, so the LCD is x^2y.

Begin by calculating the least common denominator (abbreviated LCD), the term that includes all unique factors of each denominator raised to the highest power reached by each individually. In this problem, the LCD is x^2y.

Multiply the numerator *and* denominator of $\dfrac{5}{y}$ by x^2, so that the resulting denominator is the LCD (x^2y). Similarly, multiply the numerator and denominator of $\dfrac{3}{x^2}$ by y.

$$\frac{5}{y}\left(\frac{x^2}{x^2}\right) + \frac{3}{x^2}\left(\frac{y}{y}\right) = \frac{5x^2}{x^2y} + \frac{3y}{x^2y}$$

Now that the denominators are equal, add the numerators.

$$= \frac{5x^2 + 3y}{x^2y}$$

3.2 Simplify the expression: $\dfrac{x^2}{18y} - \dfrac{7y}{3z} + \dfrac{5z^3}{12x}$.

> To figure out the LCD, factor the denominators: $18y = 3^2 \cdot 2 \cdot y$, $12x = 2^2 \cdot 3 \cdot x$, and $3z = 3 \cdot z$. Multiply the unique factors raised to their highest powers: $2^2 \cdot 3^2 \cdot x^1 \cdot y^1 \cdot z^1$.

Notice that the least common denominator is $36xyz$. Rewrite each fraction in an equivalent form that contains the LCD.

$$\frac{x^2}{18y}\left(\frac{2xz}{2xz}\right) - \frac{7y}{3z}\left(\frac{12xy}{12xy}\right) + \frac{5z^3}{12x}\left(\frac{3yz}{3yz}\right)$$

$$= \frac{2x^3z}{36xyz} - \frac{84xy^2}{36xyz} + \frac{15yz^4}{36xyz}$$

$$= \frac{2x^3z - 84xy^2 + 15yz^4}{36xyz}$$

3.3 Simplify the expression: $\dfrac{7}{10xy^3} - 3$.

> To find out what you should multiply a fraction by to get the LCD, divide the LCD by the fraction's denominator. For example, take $\dfrac{7y}{3z}$. Dividing the LCD by $3z$ gives you $\dfrac{36xyz}{3z} = \dfrac{36}{3} \cdot \dfrac{xy\cancel{z}}{\cancel{z}} = 12xy$.

Every integer has an implied denominator of 1, but that denominator can be ignored when determining the LCD. By default, then, the LCD of this expression is $10xy^3$—the left term already has this denominator, so it does not need to be modified.

$$\frac{7}{10xy^3} - \frac{3}{1} = \frac{7}{10xy^3} - \frac{3}{1}\left(\frac{10xy^3}{10xy^3}\right)$$

$$= \frac{7}{10xy^3} - \frac{30xy^3}{10xy^3}$$

$$= \frac{7 - 30xy^3}{10xy^3}$$

3.4 Simplify the expression: $\dfrac{w+1}{14w^4} - \dfrac{5}{w^3 - 3w^2}$.

Factor the denominators.

$$\frac{w+1}{2\cdot 7\cdot w^4} - \frac{5}{w^2(w-3)}$$

The LCD is $2\cdot 7\cdot w^4 \cdot (w-3) = 14(w^4)(w-3)$. Manipulate the factored versions of the fractions so that they share the LCD.

$$= \frac{w+1}{14\cdot w^4}\left(\frac{w-3}{w-3}\right) - \frac{5}{w^2(w-3)}\left(\frac{14w^2}{14w^2}\right)$$

$$= \frac{(w+1)(w-3)}{14w^4(w-3)} - \frac{70w^2}{14w^4(w-3)}$$

$$= \frac{(w^2 - 2w - 3) - 70w^2}{14w^4(w-3)}$$

$$= \frac{-69w^2 - 2w - 3}{14w^4(w-3)}$$

3.5 Simplify the expression: $\dfrac{x-2}{5x^2 - 45} - \dfrac{2x-3}{x^2 - x - 12}$.

Factor the denominators.

$$\frac{x-2}{5(x+3)(x-3)} - \frac{2x-3}{(x-4)(x+3)}$$

The LCD is $5(x+3)(x-3)(x-4)$.

$$= \frac{x-2}{5(x+3)(x-3)}\left(\frac{x-4}{x-4}\right) - \frac{2x-3}{(x-4)(x+3)}\left[\frac{5(x-3)}{5(x-3)}\right]$$

$$= \frac{(x-2)(x-4)}{5(x+3)(x-3)(x-4)} - \frac{(2x-3)(5x-15)}{5(x+3)(x-3)(x-4)}$$

$$= \frac{(x^2 - 6x + 8) - (10x^2 - 45x + 45)}{5(x+3)(x-3)(x-4)}$$

$$= \frac{-9x^2 + 39x - 37}{5(x+3)(x-3)(x-4)}$$

> Factoring $5x^2 - 45$ takes two steps. First, factor out the greatest common factor: $5(x^2 - 9)$. Now factor $x^2 - 9$, the difference of perfect squares: $5(x + 3)(x - 3)$.

3.6 Explain why the process used to simplify the following expression is incorrect.

$$\frac{x^2 - 4}{4} = \frac{x^2 - \cancel{4}}{\cancel{4}} = \frac{x^2 - 1}{1} = x^2 - 1$$

The fraction $\frac{x^2 - 4}{4}$ can be rewritten as two fractions with a common denominator.

$$\frac{x^2 - 4}{4} = \frac{x^2}{4} - \frac{4}{4} = \frac{x^2}{4} - 1$$

Therefore, properly reducing the fraction $\frac{x^2 - 4}{4}$ to simplest terms results in $\frac{x^2}{4} - 1$, not $x^2 - 1$.

3.7 Simplify the expression: $\dfrac{6x^2y^5 + 2xy^8}{10x^3y^3}$.

Factor the numerator.

$$\frac{2xy^5\left(3x + y^3\right)}{10x^3y^3}$$

For the moment, ignore the quantity $(3x + y^3)$ and reduce the rational expression $\frac{2xy^5}{10x^3y^3}$.

$$\frac{2xy^5\left(3x + y^3\right)}{10x^3y^3} = \frac{2}{10} \cdot \frac{x}{x^3} \cdot \frac{y^5}{y^3} \cdot \frac{\left(3x + y^3\right)}{1}$$

$$= \frac{1}{5} \cdot \frac{1}{x^2} \cdot \frac{y^2}{1} \cdot \frac{\left(3x + y^3\right)}{1}$$

$$= \frac{y^2\left(3x + y^3\right)}{5x^2}$$

According to the exponential rules from Chapter 2, you can subtract exponents when equal bases are divided:
$\frac{2}{10}x^{1-3}y^{5-3} = \frac{1}{5}x^{-2}y^2 = \frac{y^2}{5x^2}$.

3.8 Simplify the expression: $\dfrac{64x^3 - y^3}{4x^2 - 33xy + 8y^2}$.

Factor the numerator and denominator. Note that the numerator is the difference of perfect cubes.

$$\frac{\left(4x - y\right)\left(16x^2 + 4xy + y^2\right)}{\left(4x - y\right)\left(x - 8y\right)}$$

Eliminate the common factor of $(4x - y)$.

$$\frac{\left(4x - y\right)\left(16x^2 + 4xy + y^2\right)}{\left(4x - y\right)\left(x - 8y\right)} = \frac{16x^2 + 4xy + y^2}{x - 8y}$$

Neither the numerator nor the denominator can be factored further.

To factor the difference of perfect cubes, remember the formula $a^3 - b^3 = a^2 + ab + b^2$. In this problem, $a = 4x$ and $b = y$. To factor $4x^2 - 33xy + 8y^2$, use factoring by decomposition (like in Problem 2.27).

Multiplying and Dividing Rational Expressions
Multiplying = easy, dividing = almost as easy

3.9 Calculate the product: $\left(\dfrac{3xz^3}{y^2}\right)\left(\dfrac{7x^4y^3}{9z^8}\right)$.

The numerator of the product is the product of the numerators, and the denominator of the product is the product of the denominators.

$$\frac{\left(3xz^3\right)\left(7x^4y^3\right)}{\left(y^2\right)\left(9z^8\right)} = \frac{21x^5y^3z^3}{9y^2z^8}$$

Notice that $\frac{21}{9} = \frac{7 \cdot 3}{3 \cdot 3} = \frac{7}{3}$. To reduce the variable portion of the expression, recall that $\frac{x^a}{x^b} = x^{a-b}$.

$$\frac{7}{3}x^5y^{3-2}z^{3-8} = \frac{7}{3}x^5yz^{-5} = \frac{7x^5y}{3z^5}$$

3.10 Calculate the product: $\left(\dfrac{33x^2y^5}{10w^3}\right)\left(\dfrac{5w^7}{11x^9y^2}\right)$.

Before you multiply the numerators together, it's helpful to reduce the coefficients. Any factor in either numerator can be reduced using any factor in either denominator: $\frac{33}{11}=\frac{3}{1}$ and $\frac{5}{10}=\frac{1}{2}$.

$$\frac{3}{1}\cdot\frac{1}{2}\cdot\frac{w^7x^2y^5}{w^3x^9y^2}=\frac{3}{2}w^{7-3}x^{2-9}y^{5-2}=\frac{3}{2}w^4x^{-7}y^3=\frac{3w^4y^3}{2x^7}$$

3.11 Calculate the product: $\left(\dfrac{3x^2}{z}\right)^3\cdot\left(\dfrac{6y^3}{z^2}\right)^{-2}$.

Before multiplying, raise each fraction to the power indicated.

$$\left(\frac{3^3x^{2\cdot3}}{z^{1\cdot3}}\right)\left(\frac{6^{1(-2)}y^{3(-2)}}{z^{2(-2)}}\right)$$

$$=\left(\frac{27x^6}{z^3}\right)\left(\frac{z^4}{6^2y^6}\right)$$

$$=\frac{27x^6z^4}{36y^6z^3}$$

$$=\frac{3x^6z}{4y^6}$$

> You could also interpret the negative exponent as a reciprocal. In other words, rewrite $\left(\frac{6y^3}{z^2}\right)^{-2}$ as $\left(\frac{z^2}{6y^3}\right)^2$ so you don't have to mess with lots of negative exponents later.

3.12 Calculate the product: $\left(\dfrac{x-1}{4xy^2}\right)\left(\dfrac{2x-5}{x+6}\right)$.

Apply the FOIL method when multiplying the numerators.

$$\frac{(x-1)(2x-5)}{4xy^2(x+6)}=\frac{2x^2-7x+5}{4x^2y^2+24xy^2}$$

3.13 Calculate the quotient: $\dfrac{3x}{4}\div\dfrac{6x-9}{10}$.

> In other words, "dividing by x" means the same as "multiplying by the reciprocal of x." So $\frac{a}{b}\div\frac{c}{d}$ really equals $\left(\frac{a}{b}\right)\left(\frac{d}{c}\right)$.

The quotient $\frac{a}{b}\div\frac{c}{d}$ (where b and d are nonzero real numbers) is equivalent to $\frac{a}{b}\cdot\left(\frac{c}{d}\right)^{-1}$.

$$\left(\frac{3x}{4}\right)\left(\frac{10}{6x-9}\right)=\frac{30x}{24x-36}$$

To reduce the fraction, factor the denominator.

$$\frac{30x}{12(2x-3)}=\frac{\cancel{6}\cdot5\cdot x}{\cancel{6}\cdot2\cdot(2x-3)}$$

$$=\frac{5x}{2(2x-3)}$$

3.14 Calculate the following quotient and write the answer in simplest form.

$$\frac{3x^2 - 11x + 10}{x^2 - 49} \div \frac{3x^2 - 2x - 5}{x^2 - 6x - 7}$$

Factor the polynomials and rewrite the quotient as a product (as explained in Problem 3.13).

$$\frac{(3x-5)(x-2)}{(x+7)(x-7)} \div \frac{(3x-5)(x+1)}{(x-7)(x+1)} = \frac{(3x-5)(x-2)}{(x+7)(x-7)} \cdot \frac{(x-7)(x+1)}{(3x-5)(x+1)}$$

$$= \frac{(3x-5)(x-2)(x-7)(x+1)}{(x+7)(x-7)(3x-5)(x+1)}$$

Eliminate any binomial factors that appear in both the numerator and denominator.

$$\frac{\cancel{(3x-5)}(x-2)\cancel{(x-7)}\cancel{(x+1)}}{(x+7)\cancel{(x-7)}\cancel{(3x-5)}\cancel{(x+1)}} = \frac{x-2}{x+7}$$

3.15 Calculate the following quotient and write the answer in simplest form.

$$\frac{x^3 + 8}{(x+8)^3} \div \frac{8x + 16}{x^2 + 16x + 64}$$

Notice that $x^3 + 8$ is a perfect cube (which should be factored according to the formula in Problem 2.25) and $x^2 + 16x + 64$ is a perfect square (because it has two equivalent factors).

$$\frac{(x+2)(x^2 - 2x + 4)}{(x+8)^3} \div \frac{8(x+2)}{(x+8)^2}$$

Convert this quotient into a product and simplify.

$$= \frac{(x+2)(x^2 - 2x + 4)}{(x+8)^3} \cdot \frac{(x+8)^2}{8(x+2)}$$

$$= \frac{\cancel{(x+2)}(x^2 - 2x + 4)\cancel{(x+8)}\cancel{(x+8)}}{\cancel{(x+8)}\cancel{(x+8)}(x+8)(8)\cancel{(x+2)}}$$

$$= \frac{x^2 - 2x + 4}{8x + 64}$$

3.16 Write the following expression in simplest form.

$$-\frac{x}{x-2} + \frac{16x^2}{x+4} \div \frac{4x^2 + 10x}{4x^2 + 13x - 12}$$

Division comes before addition, according to the order of operations.

Begin by calculating the quotient. Rewrite the quotient as a multiplication problem and factor.

$$= -\frac{x}{x-2} + \frac{16x^2}{x+4} \cdot \frac{4x^2 + 13x - 12}{4x^2 + 10x}$$

$$= -\frac{x}{x-2} + \frac{16x^2}{x+4} \cdot \frac{(4x-3)(x+4)}{2x(2x+5)}$$

$$= -\frac{x}{x-2} + \frac{8 \cdot \cancel{2} \cdot x \cdot x \cdot (4x-3)\,\cancel{(x+4)}}{\cancel{2}\,x\,\cancel{(x+4)}\,(2x+5)}$$

$$= -\frac{x}{x-2} + \frac{8x(4x-3)}{2x+5}$$

Calculate the sum.

> Don't forget that you need to use the least common denominator: $(x-2)(2x+5)$.

$$\left(-\frac{x}{x-2}\right)\left(\frac{2x+5}{2x+5}\right) + \left[\frac{8x(4x-3)}{2x+5}\right]\left(\frac{x-2}{x-2}\right)$$

$$= \frac{-x(2x+5) + \left(32x^2 - 24x\right)(x-2)}{(x-2)(2x+5)}$$

$$= \frac{\left(-2x^2 - 5x\right) + \left(32x^3 - 64x^2 - 24x^2 + 48x\right)}{2x^2 + x - 10}$$

$$= \frac{32x^3 - 90x^2 + 43x}{2x^2 + x - 10}$$

Solving Rational Equations

Here comes cross multiplication

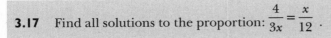

3.17 Find all solutions to the proportion: $\dfrac{4}{3x} = \dfrac{x}{12}$.

Cross multiply and solve for x.

> Multiply the left numerator times the right denominator and set it equal to the left denominator multiplied by the right numerator: $\frac{a}{b} = \frac{c}{d}$ becomes $ad = bc$.

$$4(12) = 3x(x)$$
$$48 = 3x^2$$
$$16 = x^2$$
$$\pm 4 = x$$

Therefore, $x = -4$ or 4.

3.18 Find all solutions to the proportion: $\dfrac{2x-7}{x+6} = \dfrac{2x-1}{x-2}$.

Cross multiply and solve for x.

$$(2x-7)(x-2) = (x+6)(2x-1)$$
$$2x^2 - 11x + 14 = 2x^2 + 11x - 6$$
$$2x^2 - 2x^2 - 11x - 11x = -14 - 6$$
$$-22x = -20$$
$$x = \frac{20}{22} = \frac{10}{11}$$

3.19 Find all solutions to the equation: $\dfrac{2}{x+3} - \dfrac{x-4}{4} = 0$.

Transform the equation into a proportion by adding $\frac{x-4}{4}$ to both sides of the equation.

$$\frac{2}{x+3} = \frac{x-4}{4}$$

Cross multiply to eliminate the fractions.

$$2(4) = (x+3)(x-4)$$
$$8 = x^2 - x - 12$$

Apply the factoring method to solve the quadratic equation.

$$x^2 - x - 20 = 0$$
$$(x-5)(x+4) = 0$$
$$x = -4 \text{ or } 5$$

> Here's how you get those answers:
>
> $$x = \frac{-7 \pm \sqrt{49 - 4(1)(3)}}{2}$$
> $$= \frac{-7 \pm \sqrt{49 - 12}}{2}$$
> $$= \frac{-7 \pm \sqrt{37}}{2}$$

3.20 Find all solutions to the equation: $\dfrac{x^2 + 2x + 3}{x} + 5 = 0$.

Create a proportion and cross multiply.

$$\frac{x^2 + 2x + 3}{x} = \frac{-5}{1}$$
$$x^2 + 2x + 3 = -5x$$
$$x^2 + 7x + 3 = 0$$

According to the quadratic formula, the solutions are $x = \dfrac{-7 - \sqrt{37}}{2}$ and $\dfrac{-7 + \sqrt{37}}{2}$. Simplify the left side of the equation.

3.21 Find all solutions to the equation: $\dfrac{4}{2x+1} - 3 = \dfrac{x+5}{x}$.

> Add $\frac{4}{2x+1}$ to -3 together using the common denominator $2x + 1$ to get one fraction on the left side of the equation, so you'll be able to cross multiply.

$$\frac{4}{2x+1} - \frac{3}{1}\left(\frac{2x+1}{2x+1}\right) = \frac{x+5}{x}$$
$$\frac{4 - 3(2x+1)}{2x+1} = \frac{x+5}{x}$$
$$\frac{4 - 6x - 3}{2x+1} = \frac{x+5}{x}$$
$$\frac{-6x+1}{2x+1} = \frac{x+5}{x}$$

Cross multiply and solve using the quadratic formula.

$$(-6x+1)(x) = (2x+1)(x+5)$$
$$-6x^2 + x = 2x^2 + 11x + 5$$
$$0 = 8x^2 + 10x + 5$$

The solution is $x = \dfrac{-5 \pm i\sqrt{15}}{8}$.

3.22 Solve the equation in Problem 3.21 again $\left(\dfrac{4}{2x+1} - 3 = \dfrac{x+5}{x}\right)$, but this time eliminate the fractions by multiplying the entire equation by its least common denominator. Verify that you get the same solutions.

The denominators of the terms, from left to right, are $(2x+1)$, 1, and x; the least common denominator is $x(2x+1)$. Multiply each term of the equation by that value.

$$\frac{x(2x+1)}{1}\left[\frac{4}{2x+1} - 3\right] = \frac{x(2x+1)}{1}\left(\frac{x+5}{x}\right)$$

$$\frac{4x\cancel{(2x+1)}}{\cancel{2x+1}} - \frac{3x(2x+1)}{1} = \frac{\cancel{x}(2x+1)(x+5)}{\cancel{x}}$$

$$4x - 6x^2 - 3x = 2x^2 + 11x + 5$$

$$0 = 8x^2 + 10x + 5$$

> Sometimes multiplying an entire equation by something containing an x adds extra, false answers. Make sure to check your answers in the original equation if you do this.

This matches the quadratic equation from Problem 3.21, so it will have identical solutions.

3.23 Find all solutions to the equation: $-\dfrac{5}{x^2 - 4x + 4} + \dfrac{3}{x^2} = \dfrac{1}{x^2 - 2x}$.

Factor the denominators.

$$-\frac{5}{(x-2)^2} + \frac{3}{x^2} = \frac{1}{x(x-2)}$$

The least common denominator of all three terms is $x^2(x-2)^2$. Multiply each term by the LCD to eliminate the fractions.

$$\frac{-5x^2\cancel{(x-2)^2}}{\cancel{(x-2)^2}} + \frac{3\cancel{x^2}(x-2)^2}{\cancel{x^2}} = \frac{x \cdot x \cdot \cancel{(x-2)}(x-2)}{\cancel{x}\cancel{(x-2)}}$$

$$-5x^2 + 3(x-2)^2 = x(x-2)$$

$$-5x^2 + 3\left(x^2 - 4x + 4\right) = x^2 - 2x$$

$$-5x^2 + 3x^2 - 12x + 12 = x^2 - 2x$$

$$-3x^2 - 10x + 12 = 0$$

According to the quadratic formula, the solutions are $x = -\dfrac{5 - \sqrt{61}}{3}$ and $-\dfrac{5 + \sqrt{61}}{3}$.

Polynomial and Rational Inequalities
Critical numbers break up your number line

3.24 Write the solution to the inequality using interval notation: $(x-3)(2x+1) < 0$.

Calculate the inequality's critical numbers.

> The x-values that make the polynomial equal 0.

$$\begin{array}{ccc} x - 3 = 0 & & 2x + 1 = 0 \\ x = 3 & \text{or} & x = -\dfrac{1}{2} \end{array}$$

Draw a number line with these points marked; they split the number line into three intervals: $\left(-\infty, -\frac{1}{2}\right)$, $\left(-\frac{1}{2}, 3\right)$, and $(3, \infty)$, as illustrated by Figure 3-1.

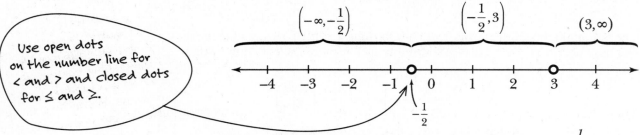

> Use open dots on the number line for < and > and closed dots for ≤ and ≥.

Figure 3-1   *The critical numbers of $(x-3)(2x+1) < 0$ are $x = -\frac{1}{2}$ and $x = 3$.*

The solution to the inequality may be any combination of those three intervals. To determine which belong to the solution, pick a value from each (called the "test value" of the interval) and plug them into the original inequality.

$\left(-\infty, -\frac{1}{2}\right)$ Test Value: $x = -1$	$\left(-\frac{1}{2}, 3\right)$ Test Value: $x = 0$	$(3, \infty)$ Test Value: $x = 5$
$(-1-3)(2(-1)+1) < 0$	$(0-3)(2(0)+1) < 0$	$(5-3)(2(5)+1) < 0$
$(-4)(-1) < 0$	$(-3)(1) < 0$	$(2)(11) < 0$
$5 < 0$	$-3 < 0$	$22 < 0$
False	*True*	False

> If the dots were closed on the graph, the solution would have been $\left[-\frac{1}{2}, 3\right]$.

Only values from the interval $\left(-\frac{1}{2}, 3\right)$ make the inequality true, so that is the final solution to the inequality. Note that $x = -\frac{1}{2}$ and $x = 3$ are excluded from the solution because the critical numbers are excluded from the graph in Figure 3-1.

3.25   Graph the solution to the inequality: $x^2 + x \geq 2$.

Move all the terms to the left side of the inequality and factor.

$$x^2 + x - 2 \geq 0$$
$$(x+2)(x-1) \geq 0$$

The critical numbers for this inequality are $x = -2$ and $x = 1$. Graph both on a number line using solid dots, as illustrated by Figure 3-2.

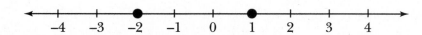

Figure 3-2   *The critical numbers of $x^2 + x \geq 2$ are $x = -2$ and $x = 1$.*

The number line is split into three intervals: $(-\infty, -2]$, $[-2, 1]$, and $[1, \infty)$. Choose one test value from each interval (such as $x = -5$, $x = 0$, and $x = 2$, respectively) and plug each into the inequality to determine the solution.

It is more expedient to substitute into the factored version of the polynomial, as the arithmetic is simpler.

Use brackets next to –2 and 1 since they're graphed as solid dots, but always use parentheses for ∞ no matter what dots are on any other numbers.

$(-\infty, -2]$	$[-2, 1]$	$[1, \infty)$
Test Value: $x = -5$	Test Value: $x = 0$	Test Value: $x = 2$
$(-5+2)(-5-1) \geq 0$	$(0+2)(0-1) \geq 0$	$(2+2)(2-1) \geq 0$
$(-3)(-6) \geq 0$	$(2)(-1) \geq 0$	$(4)(1) \geq 0$
$18 \geq 0$	$-2 \geq 0$	$4 \geq 0$
True	*False*	*True*

The solution is $(-\infty, -2]$ or $[1, \infty)$; graph both intervals on the same number line, as illustrated in Figure 3-3.

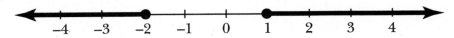

Figure 3-3 *The solution graph of $x^2 + x \geq 2$.*

3.26 Graph the solution to the inequality: $2x^2 - 3x - 8 < 0$.

Calculate the critical numbers via the quadratic formula: $x = \frac{3 \pm \sqrt{73}}{4}$. Use a calculator to determine approximate decimal values for the critical numbers: $\frac{3 - \sqrt{73}}{4} \approx -1.386$ and $\frac{3 + \sqrt{73}}{4} \approx 2.886$. Choose test values (such as $x = -2$, $x = 0$, and $x = 4$) to verify that the solution is $\left(\frac{3 - \sqrt{73}}{4}, \frac{3 + \sqrt{73}}{4} \right)$, as graphed in Figure 3-4.

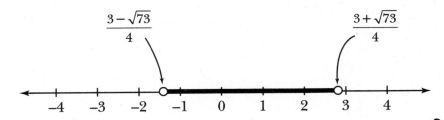

Figure 3-4 *The solution graph of $2x^2 - 3x - 8 < 0$.*

3.27 Graph the solution to the inequality: $16x^2 - 24x + 9 \leq 0$.

Factor the quadratic and note that this polynomial has a double root (because the polynomial is a perfect square).

$$(4x - 3)(4x - 3) \leq 0$$

The only critical number is $x = \frac{3}{4}$; graphing that value results in two possible interval solutions (as illustrated by Figure 3-5): $\left(-\infty, \frac{3}{4} \right]$ and $\left[\frac{3}{4}, \infty \right)$.

Basically, you're picking something less than $\frac{3}{4}$ for the left interval and something greater than $\frac{3}{4}$ for the right interval.

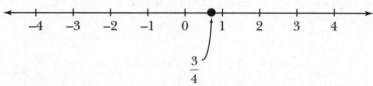

Figure 3-5 *The only critical number of $16x^2 - 24x + 9 \leq 0$ is $x = \dfrac{3}{4}$.*

Neither interval is a solution to the inequality; verify using the test values $x = 0$ and $x = 1$—both make the inequality false.

Test value $x = 0$	Test value $x = 1$
$(4(0) - 3)(4(0) - 3) \leq 0$	$(4(1) - 3)(4(1) - 3) \leq 0$
$(-3)(-3) \leq 0$	$(1)(1) \leq 0$
$9 \leq 0$	$1 \leq 0$
False	False

Therefore, the only solution to the inequality is the solid dot $x = \dfrac{3}{4}$ in Figure 3-5.

3.28 Write the solution to the inequality using interval notation: $\dfrac{x+1}{3x-1} \geq 0$.

> Basically, any x-value that makes the numerator or denominator of the fraction equal 0 is a critical number.

Critical numbers are x-values that cause an expression to equal 0 or x-values that cause an expression to be undefined. Whereas polynomial inequalities do not address the latter case, rational inequalities do.

Ensure that the rational expression alone appears on the left side of the inequality and that 0 appears on the right side. Set the numerator and denominator equal to 0 and solve the resulting equations.

$$x + 1 = 0 \qquad 3x - 1 = 0$$
$$x = -1 \quad \text{or} \quad x = \dfrac{1}{3}$$

Plot those critical numbers on the number line, as illustrated in Figure 3-6. Notice that $x = \dfrac{1}{3}$ is plotted with an open dot, even though the sign of the inequality is "≥". Any critical number generated by setting the denominator equal to 0 must be plotted as an open dot, regardless of the inequality sign. Therefore, the possible solution intervals are $(-\infty, -1]$, $\left[-1, \dfrac{1}{3}\right)$, and $\left(\dfrac{1}{3}, \infty\right)$.

> It can't be a solution, because plugging it into the inequality means dividing by 0, and that's not allowed.

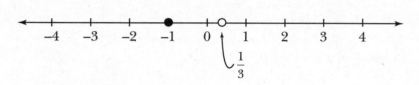

Figure 3-6 *The critical numbers of $\dfrac{x+1}{3x-1} \geq 0$ are $x = -1$ and $x = \dfrac{1}{3}$. Of those, only $x = -1$ can be part of the solution.*

Choose test values from those intervals (such as –3, 0, and 1, respectively) to verify that the solution is $(-\infty,-1]$ or $\left(\frac{1}{3},\infty\right)$.

$(-\infty,-1]$ Test Value: $x=-3$	$\left[-1,\frac{1}{3}\right)$ Test Value: $x=0$	$\left(\frac{1}{3},\infty\right)$ Test Value: $x=1$
$\dfrac{-3+1}{3(-3)-1}\geq 0$	$\dfrac{0+1}{3(0)-1}\geq 0$	$\dfrac{1+1}{3(1)-1}\geq 0$
$\dfrac{-2}{-10}\geq 0$	$\dfrac{1}{-1}\geq 0$	$\dfrac{2}{2}\geq 0$
$\dfrac{1}{5}\geq 0$	$-1\geq 0$	$1\geq 0$
True	*False*	*True*

A <u>union</u> makes one big thing by combining two smaller things. For example, the solution graph in Figure 3-7 is created by combining two inequality graphs.

The solution graph is the union of the graphs of the solution intervals, pictured in Figure 3-7.

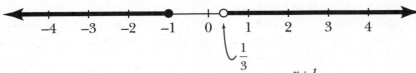

Figure 3-7 *The solution graph of $\frac{x+1}{3x-1}\geq 0$.*

3.29 Graph the solution to the inequality: $\dfrac{3x^2+17x+24}{3x^2+2x}\leq 0$.

Factor the numerator and denominator.

$$\frac{(3x+8)(x+3)}{x(3x+2)}\leq 0$$

Set both the numerator and denominator equal to 0 in order to calculate the critical numbers: $x=-3,-\frac{8}{3},-\frac{2}{3},$ and 0. Plot those values, keeping in mind that $x=0$ and $x=-\frac{2}{3}$ must be graphed with an open dot (because they make the expression undefined). Choose test values from each of the five resulting intervals and darken the intervals that satisfy the inequality, as illustrated by Figure 3-8.

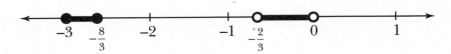

Figure 3-8 *The solution to the inequality $\dfrac{3x^2+17x+24}{3x^2+2x}\leq 0$ is $\left[-3,-\frac{8}{3}\right]$ or $\left(-\frac{2}{3},0\right)$.*

3.30 Write the solution to the inequality using interval notation: $\dfrac{x+2}{x-4} > 3$.

Make sure you've got 0 on the right side of the inequality and a single fraction on the left before you start looking for critical numbers.

Subtract 3 from both sides of the inequality and then identify critical numbers.

$$\frac{x+2}{x-4} - 3 > 0$$

Simplify the left side of the inequality using the least common denominator $x - 4$.

$$\frac{x+2}{x-4} - \frac{3}{1}\left(\frac{x-4}{x-4}\right) > 0$$

$$\frac{x+2}{x-4} - \frac{3x-12}{x-4} > 0$$

$$\frac{(x+2)-(3x-12)}{x-4} > 0$$

$$\frac{x-3x+2+12}{x-4} > 0$$

$$\frac{-2x+14}{x-4} > 0$$

The critical numbers of the inequality are $x = 4$ and $x = 7$, so the possible solution intervals are $(-\infty, 4)$, $(4, 7)$, and $(7, \infty)$. Use test values from each interval to determine that the solution is $(4, 7)$.

Chapter 4
FUNCTIONS

Now you'll start seeing f(x) all over the place

Whereas linear, polynomial, and rational equations are both useful and ubiquitous in advanced mathematics, concise definitions of mathematical relationships are essential once theory and proof are involved. That the function automatically assures us of predictable behavior is a given, but that you know how to manipulate such expressions is not, hence this chapter.

A function is an equation with certain benefits. They look a little different than the linear equations and polynomials from the preceding chapters, so they're easy to spot. Rather than equations that look like $y = 3x - 9$ and $y = 2x^2 + 7x - 10$, they look like $f(x) = 3x - 9$ and $g(x) = 2x^2 + 7x - 10$. (Technically, they don't have to start with $f(x)$ or $g(x)$ to qualify as a function, but most functions are written like that.) They carry this guarantee: for every input, you'll get one and only one output.

If you plug $x = 1$ into $f(x) = 3x - 9$, you'll always get the same thing: $f(1) = 3(1) - 9 = -6$. Most of calculus deals with functions exclusively, so they're well worth reviewing.

Combining Functions

Do the usual (+,−, ×, and ÷) or plug 'em into each other

> The domain of a function is the collection of all its inputs. The range of a function is the set containing all its outputs.

4.1 Is the relation s: {(−3,2), (4,1), (8,2)}, a function? Justify your answer.

Yes, s is a function because each member of the domain (−3, 4, and 8) corresponds to exactly one member of the range (2, 1, and 2). In other words, $s(-3) = 2$, $s(4) = 1$, and $s(8) = 2$. While it's true that $s(-3)$ and $s(8)$ both have the same output, that does not invalidate the function. Two inputs may have identical outputs, as long as each input corresponds *only* to that output and no other.

4.2 Find real number values for m and n such that the following relation is not a function.
$$p: \{(-4,-1), (-2,7), (0,3), (m,n), (10,8)\}$$

The relation p is not a function if any element of the domain has more than one corresponding range element. Therefore, if m is equal to any other member of the domain and n does not match that element's range element, p is not a function. Any of the four following possibilities are viable solutions to this problem: $m = -4$ and $n \neq -1$; $m = -2$ and $n \neq 7$; $m = 0$ and $n \neq 3$; and $m = 10$ and $n \neq 8$.

> There are an infinite number of answers. Here's one: m = 0, n = 5. That would mean p contains (0,3) and (0,5), so the input 0 corresponds to two different outputs, 3 and 5.

4.3 Find a real number value for c that ensures the following relation is one-to-one.
$$j: \{(-2,9), (-1,c), (0,-4), (1,-13), (2,-6)\}$$

As discussed in Problems 4.1-4.2, no function may have an input element that corresponds to two different output elements. If a function is one-to-one, the reverse is also true—each element of the range corresponds to exactly one domain element, so c may be any real number *except* −13, −6, −4, or 9. No matter what real number is substituted for c, j will be a function, but it will only be one-to-one if those four values are avoided.

4.4 Given the functions f: {(−1,10), (3,5), (10,−6)} and g: {(−1, −4), (3,0), (10, −5)}, calculate $(f+g)(10)$.

Note that $(f+g)(10) = f(10) + g(10)$.
$$(f+g)(10) = f(10) + g(10)$$
$$= (-6) + (-5)$$
$$= -11$$

4.5 Given the functions $f(x) = x^2 + 4$ and $g(x) = (x+4)^2$, calculate $(fg)(-2)$.

Note that $(fg)(-2) = f(-2) \cdot g(-2)$.
$$(fg)(-2) = f(-2) \cdot g(-2)$$
$$= \left((-2)^2 + 4\right) \cdot (-2 + 4)^2$$
$$= (4+4)(2)^2$$
$$= 8(4)$$
$$= 32$$

4.6   Given the functions $f(x) = x + 1$ and $g(x) = 6x^2 + 19x - 36$, determine the domain of $\left(\dfrac{f}{g}\right)(x)$.

The function $\left(\dfrac{f}{g}\right)(x) = \dfrac{f(x)}{g(x)} = \dfrac{x+1}{6x^2 + 19x - 36}$ is undefined whenever its denominator

equals 0. Set $g(x) = 0$ and solve to identify those values, which must be excluded

from the domain.

$$6x^2 + 19x - 36 = 0$$

Factor the polynomial. ←

$$(2x + 9)(3x - 4) = 0$$

$$2x + 9 = 0 \qquad\qquad 3x - 4 = 0$$

$$x = -\frac{9}{2} \qquad\qquad x = \frac{4}{3}$$

Here's how to factor by decomposition (the technique explained in Problem 2.27):
$$6x^2 + (27 - 8)x - 36$$
$$= 6x^2 + 27x - 8x - 36$$
$$= 3x(2x + 9) - 4(2x + 9)$$
$$= (2x + 9)(3x - 4)$$

The domain of $\left(\dfrac{f}{g}\right)(x)$ is all real numbers except $x = -\dfrac{9}{2}$ and $x = \dfrac{4}{3}$.

4.7   Given functions $f(x) = x^2$ and $g(x) = 2x + 5$, find $(f \circ g)(x)$ and $(g \circ f)(x)$.

The notation $(f \circ g)(x)$ is read "f composed with g of x," and is equivalent to $f(g(x))$. In other words, the function $g(x)$ should be substituted into $f(x)$. Similarly, $(g \circ f)(x) = g(f(x))$.

$$
\begin{aligned}
(f \circ g)(x) &= f(g(x)) \\
&= f(2x + 5) \\
&= (2x + 5)^2 \\
&= 4x^2 + 20x + 25
\end{aligned}
\qquad\qquad
\begin{aligned}
(g \circ f)(x) &= g(f(x)) \\
&= g(x^2) \\
&= 2(x^2) + 5 \\
&= 2x^2 + 5
\end{aligned}
$$

4.8   Given functions $f(x) = \sqrt{x}$ and $g(x) = x^2 - 12x + 36$, find $(f \circ g)(x)$ and $(g \circ f)(x)$.

$$
\begin{aligned}
(f \circ g)(x) &= f(g(x)) \\
&= f(x^2 - 12x + 36) \\
&= \sqrt{x^2 - 12x + 36} \\
&= \sqrt{(x - 6)^2} \\
&= |x - 6|
\end{aligned}
$$

$$
\begin{aligned}
(g \circ f)(x) &= g(f(x)) \\
&= g(\sqrt{x}) \\
&= (\sqrt{x})^2 - 12\sqrt{x} + 36 \\
&= x - 12\sqrt{x} + 36
\end{aligned}
$$

If you have something that looks like $\sqrt[n]{x^n}$ and n is even, then $\sqrt[n]{x^n} = |x|$.

4.9   Given the functions $f(x) = \dfrac{1}{x + 3}$, $g(x) = x - 2x^2$, and $h(x) = \sqrt[3]{x - 6}$, calculate $f(g(h(70)))$.

Start with h(70) and work your way from the inside out. In other words, plug what you get from h(70) into g(x), and then plug that result into f(x).

Evaluate the innermost function.

$$h(70) = \sqrt[3]{70 - 6} = \sqrt[3]{64} = \sqrt[3]{4^3} = 4$$

Substitute $h(70) = 4$ into the function: $f(g(h(70))) = f(g(4))$. Once again evaluate the innermost function, which is now $g(x)$.

$$g(4) = 4 - 2(4)^2 = 4 - 2(16) = -28$$

By substitution, $f(g(4)) = f(-28)$. Evaluate $f(-28)$ to complete the problem.

$$f(-28) = \frac{1}{-28 + 3} = \frac{1}{-25} = -\frac{1}{25}$$

Therefore, $f\left(g\left(h(70)\right)\right) = -\frac{1}{25}$.

4.10 Given the functions $f(x) = x^2$, $g(x) = \dfrac{1}{2x+1}$, and $h(x) = \dfrac{1-x}{2x}$, find $f(g(h(x)))$.

Substitute $h(x)$ into $g(x)$.

$$g(x) = \frac{1}{2x+1}$$

$$g(h(x)) = \frac{1}{2\left(\dfrac{1-x}{2x}\right)+1}$$

$$= \frac{1}{\dfrac{2}{1}\left(\dfrac{1-x}{2x}\right)+1}$$

$$= \frac{1}{\dfrac{1-x}{x}+1}$$

Fractions containing fractions are called complex. Add $\frac{1-x}{x}$ and 1 using the common denominator, x.

Simplify the complex fraction.

$$g(h(x)) = \frac{1}{\dfrac{1-x}{x}+1\left(\dfrac{x}{x}\right)}$$

$$= \frac{1}{\dfrac{(1-x)+x}{x}}$$

$$= \frac{1}{\dfrac{1}{x}}$$

Multiply the numerator and denominator of the complex fraction by the reciprocal of its denominator to simplify.

$$g(h(x)) = \frac{1}{\dfrac{1}{x}} \cdot \left(\dfrac{\dfrac{x}{1}}{\dfrac{x}{1}}\right)$$

$$g(h(x)) = \frac{x}{1} = x$$

Substitute $g(h(x)) = x$ into the expression: $f(g(h(x))) = f(x) = x^2$.

Graphing Function Transformations

Stretches, squishes, flips, and slides

Note: Problems 4.11-4.18 address transformations of the graph f(x) in Figure 4-1, so that you can more effectively juxtapose the effects of the transformations. It is not beneficial to determine the equations that generate the graph.

4.11 Graph $f(x) - 2$.

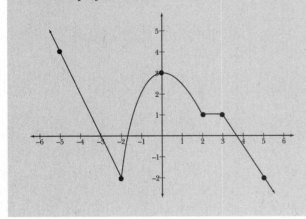

> Start over with the original graph of f(x) in each problem, but don't worry about what the actual function is. Just pay attention to where the points on the graph will go.

Figure 4-1

The graph of f(x), to be transformed in Problems 4.11–4.18.

When a constant is added to, or subtracted from, a function, it shifts the graph vertically. Adding a constant c moves each point on the graph up c units, and subtracting c moves the graph down c units. Therefore, each point on the graph of $f(x) - 2$ should be plotted two units below its corresponding point on the graph of $f(x)$, as illustrated by Figure 4-2.

> Moving the points of a graph down 2 units means you should subtract 2 from the y-coordinates of its points.

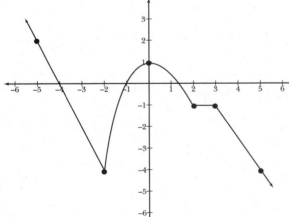

Figure 4-2

The graph of f(x) − 2 is the graph of f(x) moved down two units.

> Read that carefully. Adding inside the parentheses moves the function LEFT not right, and subtracting moves the function RIGHT.

Note: Problems 4.11-4.18 address transformations of the graph f(x) in Figure 4-1.

4.12 Graph $f(x - 3)$.

Notice that 3 is subtracted from x, whereas the constant was subtracted from $f(x)$ in Problem 4.11. This operation causes a horizontal shift on its graph. Note that subtracting a constant moves the graph to the right, and adding moves the graph left. Therefore, the graph of $f(x - 3)$ is simply the graph of $f(x)$ moved three units to the right, as illustrated in Figure 4-3.

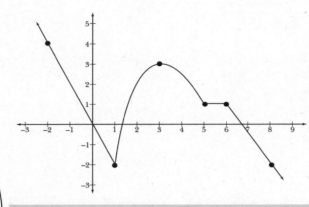

Figure 4-3

The graph of $f(x - 3)$ is the graph of $f(x)$ moved three units to the right.

The original $f(x)$ graph contains the points (−2,−2) and (3,1). Take the opposites of the y's to get the corresponding points on −f(x): (−2,2) and (3,−1).

Note: Problems 4.11-4.18 address transformations of the graph $f(x)$ in Figure 4-1.

4.13   Graph $-f(x)$.

Multiplying a function by −1 reflects its graph across the *x*-axis. If $f(x)$ contains the point (x,y), then $-f(x)$, graphed in Figure 4-4, contains the point $(x,-y)$.

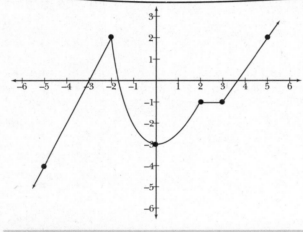

Figure 4-4

The graph of $-f(x)$ is the reflection of $f(x)$ across the x-axis.

Note: Problems 4.11-4.18 address transformations of the graph $f(x)$ in Figure 4-1.

4.14   Graph $f(-x)$.

Multiplying *x* by −1 reflects the graph of $f(x)$ across the *y*-axis; if $f(x)$ contains the point (x,y), then $f(-x)$ contains the point $(-x,y)$, as illustrated in Figure 4-5.

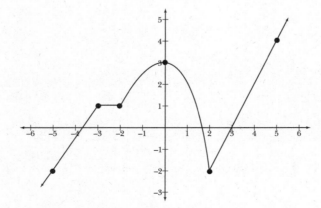

Figure 4-5

The graph of $f(-x)$ is the reflection of $f(x)$ across the y-axis.

Note: Problems 4.11-4.18 address transformations of the graph f(x) in Figure 4-1.

4.15 Graph $\frac{1}{2} f(x)$.

Multiplying a function by a constant affects the y-values of its coordinates. If the graph of $f(x)$ contains the point (x,y), then the graph of $c \cdot f(x)$ contains the point $(x, c \cdot y)$. In this case, each point on $\frac{1}{2} f(x)$ is half the distance from the x-axis as the corresponding point on graph of $f(x)$, as illustrated by Figure 4-6.

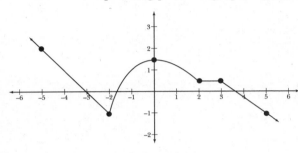

Multiply the y-values by $\frac{1}{2}$. Because f(x) contains the points (−5,4) and (3,1), $\frac{1}{2}$ f(x) will contain the points (−5,2) and $\left(3, \frac{1}{2}\right)$.

Figure 4-6
Points on the graph of $\frac{1}{2} f(x)$ are half as far from the x-axis as the corresponding points of f(x).

Note: Problems 4.11-4.18 address transformations of the graph f(x) in Figure 4-1.

4.16 Graph $f(2x)$.

Whereas multiplying $f(x)$ by a constant affects the distance of its points from the x-axis, stretching it away from (or compacting it toward) the x-axis (as explained in Problem 4.15), multiplying x by a constant inversely affects the distance of a function's coordinates from the y-axis. Although you might expect the coordinates of $f(cx)$ to be c times further from the y-axis than the corresponding points on $f(x)$, the points are actually $\frac{1}{c}$ times as far away, as illustrated by Figure 4-7.

The points on the graph of f(2x) have x-values that are $\frac{1}{2}$ as large as the original points on f(x). Because f(x) contained points (−5,4) and (2,1), f(2x) contains $\left(-\frac{5}{2}, 4\right)$ and (1,1).

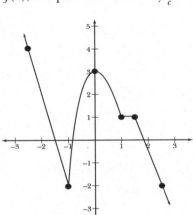

Figure 4-7
Points on the graph of f(2x) are half as far from the y-axis as the corresponding points of f(x).

Note: Problems 4.11-4.18 address transformations of the graph f(x) in Figure 4-1.

4.17 Graph $\left| f(x) \right|$.

By taking the absolute value of $f(x)$, you change all negative outputs into their opposites while leaving positive outputs unchanged. In other words, if $f(x)$ contains the point (x,y), then $\left| f(x) \right|$ contains the point $\left(x, \left|y\right|\right)$. Graphically, this means any portion of $f(x)$ below the x-axis is reflected above the x-axis, but the rest of the graph does not change.

Any coordinate that had a negative y-value now has a positive y-value: (−2,|−2|) = (−2,2) and (5,|−2|) = (5,2).

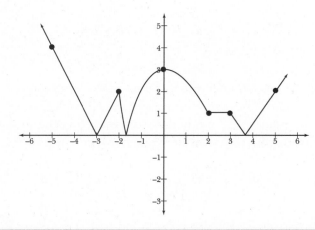

Figure 4-8

The graph of $|f(x)|$.

Note: Problems 4.11-4.18 address transformations of the graph f(x) in Figure 4-1.

4.18   Graph $f\left(|x|\right)$.

The $f\left(|x|\right)$ transformation affects coordinates with negative x-values. Consider this: $x = 1$ and $x = -1$ must have the same output for $f\left(|x|\right)$, since $f\left(|-1|\right) = f(1)$. In fact, every x in the domain of the function must output a value that matches its opposite; therefore, $f(x) = f(-x)$ *even if* $-x$ *does not belong to the domain of* $f(x)$.

Graphically speaking, if $f(x)$ contains point (x,y), and $x > 0$, then $f\left(|x|\right)$ contains the point $(-x,y)$. The net result: $f\left(|x|\right)$ is symmetric about the y-axis based on its positive domain, as illustrated by Figure 4-9. See Problem 4.21 for another example of this transformation for additional clarification.

To graph f(|x|) , completely erase the part of the graph that's left of the y-axis (but keep the y-intercept if there is one). Replace the entire left half of the coordinate axes with a reflection of its right side (sort of like you did in Problem 4.14— just leave the points in the first and fourth quadrants alone).

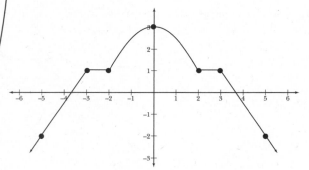

Figure 4-9

The graph of $f\left(|x|\right)$ *is symmetric about the y-axis*

4.19   Sketch the graph of $f(x) = (x+2)^2 - 1$ without a calculator by applying graphical transformations.

You need to know what the graph of x² looks like. If you don't, flip to the back of the book and check out Appendix A: Important Graphs and Transformations.

To transform the function $y = x^2$ (the dotted curve in Figure 4-10) into $f(x) = (x+2)^2 - 1$ (the solid curve in Figure 4-10), you must to add 2 to x (which shifts its graph left two units) and subtract 1 from $f(x)$ (which shifts the graph down one unit).

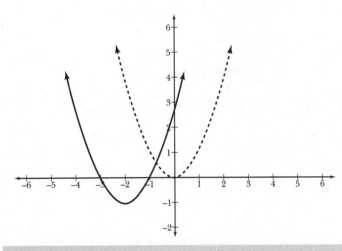

Figure 4-10

The graphs of y = x² (dotted) and f(x) = (x + 2)² − 1 (solid).

4.20 Sketch the graph of $g(x) = -\left|\dfrac{1}{3}x\right| + 4$.

There are three transformations that change $y = |x|$ (the dotted curve in Figure 4-11) into $g(x) = -\left|\dfrac{1}{3}x\right| + 4$ (the solid curve in Figure 4-11):

1. Multiplying x by $\frac{1}{3}$ stretches its graph horizontally along the x-axis by a factor of 3.
2. Multiplying by −1 reflects its graph across the x-axis
3. Adding 4 moves the graph up four units.

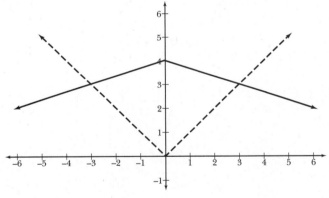

Figure 4-11

The graphs of y = |x| (dotted) and $g(x) = \left|-\dfrac{1}{3}x\right| + 4$ (solid).

4.21 Sketch the graph of $h(x) = 3\sqrt{|x|} - 4$ without a graphing calculator.

Three transformations are required to change $y = \sqrt{x}$ (the dotted curve in Figure 4-12) into $h(x) = 3\sqrt{|x|} - 4$ (the solid curve in Figure 4-12): the absolute value within the square root function replaces all points for which $x < 0$ (via the technique of Problem 4.18), multiplying the function by 3 stretches its graph vertically by a factor of 3, and subtracting 2 moves the graph down 2 units).

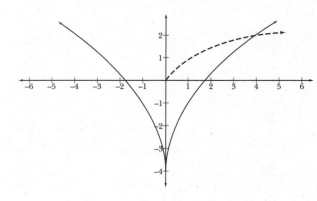

Figure 4-12
The graphs of $y = \sqrt{x}$ (dotted)
and $h(x) = 3\sqrt{|x|} - 4$ (solid).

Inverse Functions

Functions that cancel other functions out

4.22 Given the function $s : \left\{(-2,-1), \left(-1, \frac{3}{2}\right), (0,1), (1,4), (3,-9)\right\}$, define the inverse function, s^{-1}.

Any output of f(x) is an input for f⁻¹(x) and vice versa.

To generate the inverse of a function, reverse the coordinate pairs—if $s(x) = y$, then $s^{-1}(y) = x$.

$$s^{-1} : \left\{(-1,-2), \left(\frac{3}{2}, -1\right), (1,0), (4,1), (-9,3)\right\}$$

Rewrite s^{-1}, listing the domain elements from least to greatest.

$$s^{-1} : \left\{(-9,3), (-1,-2), (1,0), \left(\frac{3}{2}, -1\right), (4,1)\right\}$$

4.23 Given $f(x)$ graphed in Figure 4-13, sketch the graph of $f^{-1}(x)$.

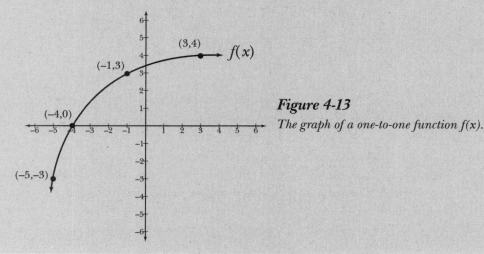

Reverse the x's and y's of the points on Figure 4-13 to get the new points (-3,-5), (0,-4), (3,-1), and (4,3) and connect the dots to get the graph of f⁻¹(x).

Figure 4-13
The graph of a one-to-one function f(x).

The graph of a function and its inverse are reflections of one another across the line $y = x$, as illustrated by Figure 4-14.

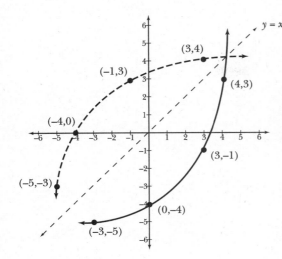

Figure 4-14

The dotted graph of a function f(x) and the solid graph of its inverse f⁻¹(x) are reflections of one another across the line y = x.

4.24 Given the function $g(x)$ graphed in Figure 4-15, explain why $g^{-1}(x)$ does not exist.

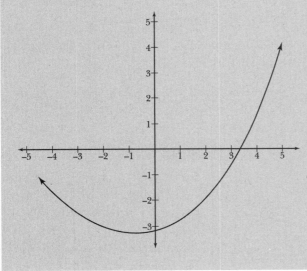

Figure 4-15

The graph of function g(x).

In order for a $g^{-1}(x)$ to exist, $g(x)$ must be one-to-one (as defined by Problem 4.3). However, $g(x)$ fails the horizontal line test, indicating that it is not one-to-one, and therefore does not have an inverse. According to the horizontal line test, any horizontal line drawn across a one-to-one function won't intersect that graph more than once. However, the horizontal lines $y = -2$ and $y = -3$ both intersect $g(x)$ twice.

4.25 Given $g(x) = \dfrac{x+4}{7}$, find $g^{-1}(x)$.

Rewrite $g(x)$ as y.

$$y = \frac{x+4}{7}$$

Reverse the x and y variables, substituting x for y and vice versa.

$$x = \frac{y+4}{7}$$

Solve for y.

Rewrite y as $g^{-1}(x)$.

$$7x = y + 4$$
$$y = 7x - 4$$

$$g^{-1}(x) = 7x - 4$$

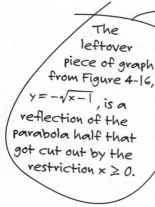

The graph of $f(x) = x^2 + 1$ fails the horizontal line test. That's why the problem includes $x \geq 0$, which says to forget about the half of the parabola to the left of the y-axis, because that way you'll pass the horizontal line test. (Look at the graph in 4-16.)

4.26 Given $f(x) = x^2 + 1$ $(x \geq 0)$, find $f^{-1}(x)$.

Rewrite $f(x)$ as y.

$$y = x^2 + 1$$

Reverse x and y.

$$x = y^2 + 1$$

Solve for y.

$$x - 1 = y^2$$
$$\pm\sqrt{x - 1} = y$$
$$\pm\sqrt{x - 1} = f^{-1}(x)$$

Note that $f^{-1}(x)$ doesn't equal *both* $\sqrt{x-1}$ and $-\sqrt{x-1}$; only one of those equations is a reflection of $f(x) = x^2 + 1$ $(x \neq 0)$ across the line $y = x$ (as illustrated by Figure 4-16): $f^{-1}(x) = \sqrt{x-1}$. Discard the negative radical.

The leftover piece of graph from Figure 4-16, $y = -\sqrt{x-1}$, is a reflection of the parabola half that got cut out by the restriction $x \geq 0$.

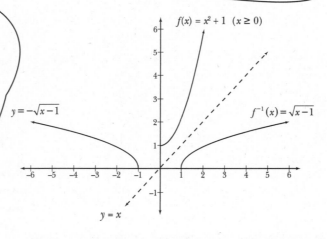

Figure 4-16

Notice that $f(x) = x^2 + 1$ and $f^{-1}(x) = \sqrt{x-1}$ are reflections of one another across the y-axis. The equation $y = -\sqrt{x-1}$, a byproduct of the inverse function creation process, is superfluous.

4.27 Verify that $g(x) = \dfrac{x+4}{7}$ and $g^{-1}(x) = 7x - 4$ (from Problem 4.25) are inverses by demonstrating that $g(g^{-1}(x)) = g^{-1}(g(x)) = x$.

Perform composition of functions using the method of Problems 4.7 and 4.8.

$$g\left(g^{-1}(x)\right) = g(7x - 4) \qquad g^{-1}\left(g(x)\right) = g^{-1}\left(\frac{x+4}{7}\right)$$
$$= \frac{(7x - 4) + 4}{7} \qquad\qquad = 7\left(\frac{x+4}{7}\right) - 4$$
$$= \frac{7x}{7} \qquad\qquad\qquad = x + 4 - 4$$
$$= x \qquad\qquad\qquad\qquad = x$$

4.28 Verify that $f(x) = x^2 + 1$ $(x \neq 0)$ and $f^{-1}(x) = \sqrt{x-1}$ (from Problem 4.26) are inverses by demonstrating that $f(f^{-1}(x)) = f^{-1}(f(x)) = x$.

$$f\left(f^{-1}(x)\right) = f\left(\sqrt{x-1}\right) \qquad f^{-1}\left(f(x)\right) = f^{-1}\left(x^2+1\right)$$
$$= \left(\sqrt{x-1}\right)^2 + 1 \qquad = \sqrt{\left(x^2+1\right)-1}$$
$$= x - 1 + 1 \qquad = \sqrt{x^2}$$
$$= x \qquad = x$$

> Technically, $\sqrt{x^2} = |x|$, but thanks to the $x \geq 0$ restriction on $f(x)$, x can't be negative and the absolute values aren't necessary.

Asymptotes of Rational Functions

Equations of the untouchable dotted line

4.29 Identify the vertical asymptote to the graph of $y = \dfrac{7}{x-3}$.

If substituting c into the rational expression causes the denominator to equal 0, then $x = c$ is a vertical asymptote of $f(x)$, as long as the numerator does not equal 0 as well. Set the denominator equal to 0 and solve.

$$x - 3 = 0$$
$$x = 3$$

Because $x = 3$ causes the denominator to equal 0 (and the numerator does not equal 0), $x = 3$ is a vertical asymptote of the function's graph.

> If c makes the top and the bottom of the fraction equal 0, it usually means there's a hole in the graph. In Chapter 10, you'll use limits to figure out where that hole is.

4.30 Identify the vertical asymptotes to the graph of $y = \dfrac{-3}{3x^2 - 23x - 36}$.

Set the denominator equal to 0 and solve the quadratic by factoring.

$$3x^2 - 23x - 36 = 0$$
$$(3x + 4)(x - 9) = 0$$
$$3x + 4 = 0 \qquad x - 9 = 0$$
$$3x = -4 \quad \text{or} \quad x = 9$$
$$x = -\frac{4}{3}$$

Both x-values cause the denominator (but not the numerator) to equal 0, so they are asymptotes to the graph: $x = -\frac{4}{3}$ and $x = 9$.

> If the numerator's degree is higher than the denominator's, the function won't have any horizontal asymptotes.

4.31 Identify the horizontal asymptote to the graph of $g(x) = \dfrac{x^2 - 2x + 4}{-2x^3 - 16}$.

Compare the degrees of the polynomials in the numerator and denominator to determine the equation of the horizontal asymptote (if it exists). In this case, the degree of the numerator is 2, and the degree of the denominator is 3. When the denominator's degree is greater than the numerator's degree, the function has horizontal asymptote $y = 0$.

4.32 Identify the horizontal asymptote to the graph of $h(x) = \dfrac{(2x-3)(x+6)}{(4x-5)(-3x+1)}$.

Calculate the products in the numerator and denominator.

$$h(x) = \frac{2x^2 + 9x - 18}{-12x^2 + 19x - 5}$$

> The coefficient of the x that's raised to the highest power.

When the degrees of the numerator and denominator are equal, the graph of the function will have horizontal asymptote $y = \frac{a}{b}$, where a and b are the leading coefficients of the numerator and denominator, respectively.

$$y = \frac{2}{-12} = -\frac{1}{6}$$

4.33 Determine the equations of all asymptotes to the graph of $f(x) = \dfrac{3x^2 - 13x + 4}{x^2 - 9}$.

The numerator and denominator have equal degrees, so the horizontal asymptote is equal to the quotient of their leading coefficients, as explained in Problem 4.32.

$$y = \frac{3}{1} = 3$$

> Slant asymptotes occur only when the degree of the numerator is one number greater than the degree of the denominator.

To determine the vertical asymptotes, factor the polynomials.

$$f(x) = \frac{(3x-1)(x-4)}{(x+3)(x-3)}$$

Set the factors of the denominator equal to 0 and solve to get $x = -3$ and $x = 3$. Neither of those values causes the numerator to equal 0 as well, so both represent vertical asymptotes. Therefore, the equations of the asymptotes to $f(x)$ are $x = -3$, $x = 3$, and $y = 3$.

4.34 Find the equation of the slant asymptote to the graph of $g(x) = \dfrac{2x^2 - 3x + 6}{x + 1}$.

The slant asymptote is the quotient of the rational function (omitting the remainder). The divisor is a linear binomial, so synthetic division (the method described in Problems 2.18 and 2.19) is the most efficient way to calculate the quotient.

> To get the equation of the slant asymptote, set "$y =$" the division result and completely ignore the remainder.

$$\begin{array}{r|rrr} \boxed{-1} & 2 & -3 & 6 \\ & & -2 & 5 \\ \hline & 2 & -5 & 11 \end{array}$$

The equation of the slant asymptote is the quotient: $y = 2x - 5$.

4.35 Identify asymptotes to the graph of $j(x) = \dfrac{5x^3 - 30x^2 - 4x + 24}{x^2 + 2x - 4}$.

Because the degree of the numerator is exactly one greater than the degree of the denominator, $j(x)$ has one slant asymptote and no horizontal asymptotes. (Note that a rational function may have only one slant asymptote *or* one horizontal asymptote, but may have multiple vertical asymptotes.

To determine the slant asymptote, use polynomial long division, as outlined in Problems 2.16 and 2.17.

$$
\begin{array}{r}
5x \quad -40 \\
x^2+2x-4\overline{)5x^3 \quad -30x^2 - \ 4x+ \ 24} \\
\underline{-5x^3 \ -10x^2+20x} \\
-40x^2+16x+ \ 24 \\
\underline{40x^2+80x-160} \\
96x-136
\end{array}
$$

The slant asymptote to $j(x)$ is $y = 5x - 40$. To determine the vertical asymptotes of $j(x)$, set the denominator equal to 0 and solve using the quadratic formula.

$$x^2+2x-4=0$$

$$
\begin{aligned}
x &= \frac{-2\pm\sqrt{2^2-4(1)(-4)}}{2\cdot 1} \\
&= \frac{-2\pm\sqrt{20}}{2} \\
&= \frac{-2\pm 2\sqrt{5}}{2} \\
&= \frac{\cancel{2}\left(-1\pm\sqrt{5}\right)}{\cancel{2}} \\
&= -1\pm\sqrt{5}
\end{aligned}
$$

The function $j(x)$ has three asymptotes: the slant asymptote $y = 5x - 40$ and the vertical asymptotes $x = -1 - \sqrt{5}$ and $x = -1 + \sqrt{5}$.

Chapter 5
LOGARITHMIC AND EXPONENTIAL FUNCTIONS

Functions like $\log_3 x$, $\ln x$, a^x, and e^x

Chapters 1–4 provide the opportunity to sharpen your skills in all matters concerning variable expressions raised to real number powers. This chapter begins by investigating the reverse, expressions containing real numbers raised to variable powers. Whereas exponential rules hold true despite this reversal, new techniques must be mastered in order to properly manipulate such expressions. Of course, one cannot discuss such an important function without exploring its inverse, the logarithmic function.

This chapter deals with exponential functions, which look more like 3^x than x^3. The problems feel very different when x is the exponent instead of the base. For one thing, you'll need some way of "canceling out" x exponents if you're ever going to solve for x, and that requires logarithmic functions. Good news and bad news: logarithmic functions (the inverses of exponential functions) really help you solve equations that have x in the exponent, but logarithms bring with them their own set of properties and rules.

Exploring Exponential and Logarithmic Functions

Harness all those powers

5.1 Graph the function $f(x) = 3^x$ without a graphing calculator.

An exponential function always equals 1 when you plug in $x = 0$. In Problem 5.1, you got $f(0) = 3^0 = 1$, but any number to the 0 power will equal 1.

Employ the most basic of graphic techniques: substitute consecutive values of x into $f(x)$ and plot the resulting coordinate pair.

x	$f(x)$
-1	$f(-1) = 3^{-1} = \dfrac{1}{3}$
0	$f(0) = 3^0 = 1$
1	$f(1) = 3^1 = 3$
2	$f(2) = 3^2 = 9$

Because $f(-1) = \dfrac{1}{3}$, the point $\left(-1, \dfrac{1}{3}\right)$ belongs on the graph. Similarly, the graph of $f(x)$ includes the points $(0,1)$, $(1,3)$, and $(2,9)$, as illustrated by Figure 5-1.

The more negative the input, the closer the output gets to 0. Even though $f(-5)$ is small, $f(-5) = 3^{-5} = \dfrac{1}{243}$, $f(-10)$ is a lot smaller: $f(-10) = 3^{-10} = \dfrac{1}{59,049}$.

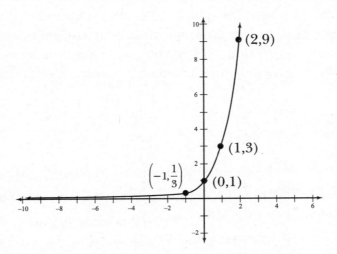

Figure 5-1

The graph of $f(x) = 3^x$.

Note that the x-axis is a horizontal asymptote of $f(x)$. Although negative values of x result in small $f(x)$ values, 3^x doesn't equal 0 for any x-value.

5.2 Identify the domain and range of the generic exponential function $g(x) = a^x$ (assuming a is a real number and $a > 1$).

Make sure to use a parenthesis, not a bracket, here. The function never actually equals 0, so you can't include 0 in the range.

Any real number can be substituted for x, but the positive real number a raised to any power (whether positive or negative) always results in a positive number. Therefore, the domain is $(-\infty, \infty)$ and the range is $(0, \infty)$.

5.3 Sketch the graph of $y = 2^{-x} - 1$ without a graphing calculator.

The graph of any exponential function $y = a^x$ will pass through the points $(0,1)$ and $(1,a)$ and have a horizontal asymptote of $y = 0$. Begin by graphing $y = 2^x$ as illustrated by the dotted curve in Figure 5-2.

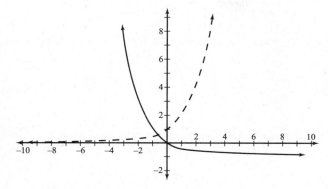

Figure 5-2

The graphs of y = 2ˣ (dotted) and
y = 2⁻ˣ − 1 (solid).

To graph $y = 2^{-x} - 1$, perform two translations on the graph of $y = 2^x$ (as explained in detail by Problems 4.11-4.21): Substituting $-x$ for x reflects the graph about the y-axis, and subtracting 1 moves the entire graph down one unit.

5.4 Determine the domain and range of the logarithmic function $h(x) = \log_3 x$.

Note that $h(x)$ is the inverse function of $f(x) = 3^x$, as defined in Problem 5.1. Therefore, the domain of $f(x)$ equals the range of $h(x)$ and vice versa. You can then conclude that the domain of $h(x)$ is $(0,\infty)$ and its range is $(-\infty,\infty)$.

5.5 Solve the equation: $\log_{10} x = 2$.

Rewrite the logarithmic equation as an exponential equation and solve for x.

$$10^2 = x$$
$$100 = x$$

> You can translate any log expression $\log_a x = b$ into $a^b = x$.

5.6 Solve the equation: $\log_4 x = -3$.

Rewrite in exponential form and solve for x.

$$4^{-3} = x$$
$$\frac{1}{4^3} = x$$
$$\frac{1}{64} = x$$

5.7 Solve the equation: $\log_5 625 = x$.

Rewrite the equation in exponential form: $5^x = 625$. Note that $5^4 = 625$, so $x = 4$.

5.8 Solve the equation: $\log_4 \dfrac{1}{8} = x$.

Rewrite as an exponential equation.

$$4^x = \frac{1}{8}$$

Express the fraction as a negative exponent.
$$4^x = 8^{-1}$$
Rewrite the equation using exponential expressions with equivalent bases ($4 = 2^2$ and $8 = 2^3$).
$$\left(2^2\right)^x = \left(2^3\right)^{-1}$$
$$2^{2x} = 2^{-3}$$

> You're allowed to drop the bases—if $a^x = a^y$, then $x = y$.

Two equivalent exponential expressions with equal bases must have equal exponents as well.
$$2x = -3$$
$$x = -\frac{3}{2}$$

5.9 Solve the equation: $\log_x 16 = 2$.

Rewrite as an exponential equation and solve for x.
$$x^2 = 16$$
$$x = \pm 4$$

> The base of a log has to be greater than 1, which is why $x = -4$ isn't a valid answer.

Only the solution $x = 4$ is valid; discard $x = -4$.

5.10 Solve the equation: $\log_x 81 = \frac{4}{3}$.

Rewrite in exponential form.
$$x^{4/3} = 81$$

To solve for x, raise both sides to the $\frac{3}{4}$ power.
$$\left(x^{4/3}\right)^{3/4} = 81^{3/4}$$
$$x^{\frac{4}{3}\cdot\frac{3}{4}} = \left(\sqrt[4]{81}\right)^3$$
$$x^{\frac{12}{12}} = \left(\sqrt[4]{3^4}\right)^3$$
$$x^1 = 3^3$$
$$x = 27$$

5.11 Graph the function $f(x) = \log_2 x$ without a graphing calculator.

The domain of $f(x)$, like the domain of $y = \log_3 x$ in Problem 5.4 or any other logarithm, consists only of positive numbers; do not substitute negative x-values into $f(x)$ as you plot points. The first column in the table of values below consists of the x-inputs, the second column substitutes x into $f(x)$, the third column is the equation expressed in exponential form, and the final column is the y-value that corresponds with x.

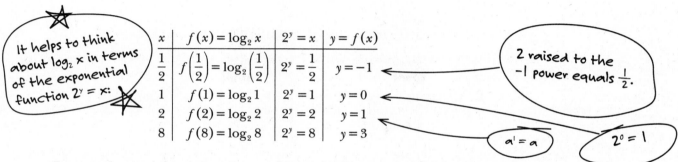

It helps to think about $\log_2 x$ in terms of the exponential function $2^y = x$:

x	$f(x) = \log_2 x$	$2^y = x$	$y = f(x)$
$\frac{1}{2}$	$f\left(\frac{1}{2}\right) = \log_2\left(\frac{1}{2}\right)$	$2^y = \frac{1}{2}$	$y = -1$
1	$f(1) = \log_2 1$	$2^y = 1$	$y = 0$
2	$f(2) = \log_2 2$	$2^y = 2$	$y = 1$
8	$f(8) = \log_2 8$	$2^y = 8$	$y = 3$

2 raised to the -1 power equals $\frac{1}{2}$.

$a^1 = a$

$2^0 = 1$

Because the x-axis is an asymptote to the graph of an exponential function, the y-axis is an asymptote to logarithmic graphs, as illustrated in Figure 5-3.

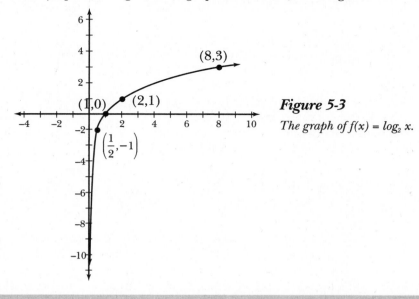

Figure 5-3

The graph of $f(x) = \log_2 x$.

5.12 Sketch the graph of $g(x) = -\log_3(x+4)$ without a graphing calculator.

The graph of $g(x)$ is simply the graph of $y = \log_3 x$ with two transformations applied to it. Adding 4 to x shifts the graph left four units, and multiplying $\log_3 x$ by -1 reflects the graph about the x-axis, as illustrated in Figure 5-4.

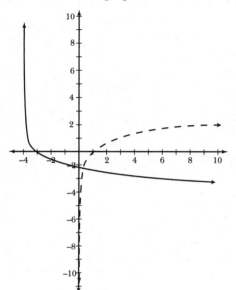

Figure 5-4

The graphs of $y = \log_3 x$ (dotted curve) and $g(x) = -\log_3(x+4)$ (solid curve).

Natural Exponential and Logarithmic Functions
Unwritten bases, bases with e, and change of base formula

The constant $e \approx 2.7182818284...$ is Euler's number, a long and irrational number like π that is (a) important, (b) way too long to memorize, and (c) programmed into your calculator.

5.13 Define the terms "common logarithm," "natural logarithm," and "natural exponential function"; indicate the notation used for each.

- The common logarithm has base 10 and is implied if no base is indicated: $\log x = \log_{10} x$.
- The natural logarithm has base e and is written $\ln x$: $\ln x = \log_e x$.
- The natural exponential function has base e and is written e^x.

5.14 Use a calculator to determine the value of log 19, accurate to five decimal places, and interpret the answer.

Read "ln 19" as "natural log of 19" or just read the letters: "L N of 19."

Note that $\log 19 = \log_{10} 19$, because unwritten logarithmic bases are understood to be 10. Calculating log 19 is the equivalent of solving the equation $\log_{10} 19 = x$, which can be rewritten in exponential form: $10^x = 19$. According to the calculator, $\log 19 \approx 1.27875$, which means $10^{1.27875} \approx 19$.

5.15 Use a calculator to determine the value of ln 19, accurate to five decimal places, and interpret the result.

Note that $\ln 19 = \log_e 19$, as any logarithmic expression written "ln" instead of "log" is a natural logarithm and has an implied base of e. Calculating ln 19 is the equivalent of solving the equation $\log_e 19 = x$, which can be rewritten in exponential form: $e^x = 19$. According to the calculator, $\ln 19 \approx 2.94444$, so $e^{2.94444} \approx 19$. Because Euler's number, e, is approximately equal to 2.7182818, you can also write $2.7182818^{2.94444} \approx 19$.

5.16 Evaluate $\log_3 25$, accurate to four decimal places, using the change of base formula and a calculator.

Most calculators can't compute logs that don't have a base of 10 or e. That's why the change of base formula is so handy. Just remember that the log of the base goes in the denominator.

The change of base formula allows you to rewrite the expression $\log_a b$ as $\frac{\log b}{\log a}$ or $\frac{\ln b}{\ln a}$.

$$\frac{\log 25}{\log 3} = \frac{\ln 25}{\ln 3}$$

$$\frac{1.3979400087}{0.4771212547} = \frac{3.2188758249}{1.0986122887}$$

$$2.92994704... = 2.92994704...$$

Therefore, $\log_3 25 \approx 2.9299$. Note that there is no need to calculate the value twice, as both calculation methods return the same value.

5.17 Evaluate $\log_6 10$, accurate to four decimal places, using the change of base formula and a calculator.

Apply the change of base formula.

$$\frac{\ln 10}{\ln 6} \approx \frac{2.302585093}{1.791759469} \approx 1.2851$$

5.18 Verify the solution to the equation in Problem 5.7 ($\log_5 625 = x$) using the change of base formula and a calculator.

Apply the change of base formula.

$$\frac{\log 625}{\log 5} = x$$

$$\frac{\log 625}{\log 5} = 4 \longleftarrow$$

5.19 Verify the solution to the equation in Problem 5.8 $\left(\log_4 \frac{1}{8} = x\right)$ using the change of base formula and a calculator.

Apply the change of base formula.

$$\frac{\ln \dfrac{1}{8}}{\ln 4} = -1.5 = -\frac{3}{2}$$

Properties of Logarithms
Expanding and squishing log expressions

5.20 Rewrite as a single logarithm: $\log_2 x + \log_2 5 + \log_2 y$.

The sum of logarithms with equal bases is equal to the logarithm of the product: $\log_a b + \log_a c = \log_a (bc)$.

$$\log_2 x + \log_2 5 + \log_2 y = \log_2 (5xy)$$

5.21 Rewrite as a single logarithm: $\log 3 - (\log 9 + \log x)$.

Rewrite the parenthetical expression, $\log 9 + \log x$, as $\log 9x$ (as explained in Problem 5.20). Combine those first, according to the order of operations.

$$\log 3 - (\log 9 + \log x) = \log 3 - (\log 9x)$$

If you write decimal approximations of log 625 and log 5 and then divide, you'll get something like this:

$$\frac{2.7958800173}{0.6989700043} \approx 4.000000000143.$$

The real answer is exactly 4—the fraction has some error because you didn't use all of the infinitely many decimal places. That's why you should type $\frac{\log 625}{\log 5}$ directly into your calculator, because you'll avoid rounding errors.

> If two logs added become one log that's multiplied, then it makes sense that two logs subtracted become one log that's divided—they're exact opposites of one another.

The difference of two logs with equal bases is equal to the logarithm of the quotient: $\log_a b - \log_a c = \log_a \dfrac{b}{c}$.

$$= \log \frac{3}{9x}$$

Reduce the fraction.

$$= \log \frac{1}{3x}$$

5.22 Rewrite as a single logarithm: $\log 7 + \log x - \log 3 + \log y - \log z$.

According to the order of operations, addition and subtraction should be performed from left to right. Begin by rewriting $\log 7 + \log x$ as a single logarithm.

$$\log 7 + \log x - \log 3 + \log y - \log z = \log 7x - \log 3 + \log y - \log z$$

Again manipulate the two leftmost terms of the expression; rewrite $\log 7x - \log 3$ as a quotient using the logarithmic property discussed in Problem 5.21.

$$= \log \frac{7x}{3} + \log y - \log z$$

> The answer is just all of the log values written in one fraction—all the positive logs (like log 7 and log y) end up in the numerator, and all the negative ones (–log 3 and –log z) end up in the denominator.

Adding $\log \dfrac{7x}{3}$ to $\log y$ results in a single log containing the product.

$$= \log \frac{7xy}{3} - \log z$$

Finally, divide $\dfrac{7xy}{3}$ by z, which is the equivalent of multiplying the fraction by the reciprocal of z.

$$= \log\left(\frac{7xy}{3} \div z\right) = \log\left(\frac{7xy}{3} \cdot \frac{1}{z}\right) = \log \frac{7xy}{3z}$$

5.23 Verify that $3 \ln x = \ln x^3$ and, based upon your proof, extrapolate and prove an equivalent conclusion for $\log_a x^n$ (if a and n are real numbers and $a > 1$).

Rewrite $3 \ln x$ as a sum.

$$3 \ln x = \ln x + \ln x + \ln x$$

> Just like $x + x + x = 3x$ or $w^2 + w^2 + w^2 = 3w^2$.

According to a logarithmic property (described in Problem 5.20), the sum of logarithms with the same base is equal to the logarithm of their product.

$$3 \ln x = \ln x + \ln x + \ln x = \ln(x \cdot x \cdot x) = \ln x^3$$

Therefore, $3 \ln x = x^3$. Expressed more generally, $n \log_a x = \log_a x^n$. As justification, recall that $n \log_a x$ can be rewritten as the sum of n terms (where each term is $\log_a x$).

$$n \log_a x = \underbrace{\log_a x + \log_a x + \log_a x + \cdots + \log_a x}_{\text{total of } n \text{ terms}}$$

Rewrite the sum of the logarithms as a single logarithm of their products.

$$\underbrace{\log_a x + \log_a x + \log_a x + \cdots + \log_a x}_{\text{total of } n \text{ terms}} = \log_a \left(\underbrace{(x)(x)(x)\cdots(x)}_{\substack{\text{there are } n \\ \text{factors of } x}} \right) = \log_a x^n$$

Therefore, $n \log_a x = \log_a x^n$.

5.24 Rewrite as a single logarithm: $5 \log x - 2 \log y + 4 \log (x - y)$.

Rewrite the coefficients of the logarithms as the exponents of their arguments, according to the logarithmic property $n \log_a x = \log_a x^n$, as explained in Problem 5.23.

$$5 \log x - 2 \log y + 4 \log (x - y) = \log x^5 - \log y^2 + \log (x - y)^4$$

The two leftmost logarithms are equivalent to a logarithmic quotient.

$$= \log \frac{x^5}{y^2} + \log (x - y)^4$$

The sum of logarithmic expressions with equal bases is equal to the logarithm of their product.

$$= \log \frac{x^5 (x - y)^4}{y^2}$$

5.25 Expand the logarithmic expression: $\ln xy^2$.

Expanding a logarithmic expression requires you to apply logarithmic properties in order to rewrite a single logarithmic expression as multiple logarithmic expressions. Because xy^2 is a product, rewrite the argument as the sum of two logarithms with equal bases.

$$\ln xy^2 = \ln x + \ln y^2$$

Recall that $\log x^a = a \log x$, so $\ln y^2 = 2 \ln y$.

$$= \ln x + 2 \ln y$$

5.26 Expand the logarithmic expression: $\log \frac{y^3}{5}$.

Recall that $\log \frac{a}{b} = \log a - \log b$.

$$\log \frac{y^3}{5} = \log y^3 - \log 5$$

Substitute $\log y^3 = 3 \log y$ into the expression.

$$= 3 \log y - \log 5$$

> The "argument" of a log is whatever's inside it—the argument of $7 \log_2 y^5$ is y^5.

> You can't rewrite $\log (x - y)^4$. You could change $\log x - \log y$ into $\log \frac{x}{y}$, but that works only when two separate logs are subtracted, not when you're subtracting in the same log.

5.27 Expand the logarithmic expression: $\log_8 \left[\dfrac{2x^2}{(x-y)^3} \right]$.

The logarithm of a quotient can be rewritten as the difference of two logarithms.

$$\log_8 2x^2 - \log_8 (x-y)^3$$

The leftmost logarithm contains a product, which can be rewritten as the sum of two logarithms.

$$= \log_8 2 + \log_8 x^2 - \log_8 (x-y)^3$$

Rewrite the exponents of the logarithmic arguments as the coefficients of their respective logarithms.

$$= \log_8 2 + 2 \log_8 x - 3 \log_8 (x-y)$$

Solving Exponential and Logarithmic Equations
Exponents and logs cancel each other out

> The expression $\log_4 4^3$ asks "4 raised to what power equals 4^3?" The answer must be 3!

5.28 Simplify the expression: $\log_4 4^3$.

Apply the logarithmic property $\log x^a = a \log x$, as described in Problem 5.23.

$$\log_4 4^3 = 3 \log_4 4$$

Apply the change of base formula to the logarithm.

$$3 \log_4 4 = 3 \left(\frac{\ln 4}{\ln 4} \right) = 3(1) = 3$$

This result demonstrates a fundamental logarithmic fact: $\log_a a^n = n$.

> Problems 5.28 and 5.29 show you that $\log_a a^x$ and $a^{\log_a x}$ both equal x. The only difference in Problems 5.30 and 5.31 is the base—e instead of a. Therefore, $\ln e^x$ and $e^{\ln x}$ also equal x.

5.29 Simplify the expression: $3^{\log_3 x}$.

This expression is the result of the composition of functions $f(g(x))$, where $f(x) = 3^x$ and $g(x) = \log_3 x$. Because $f(x)$ and $g(x)$ are inverse functions, they cancel one another out, leaving behind only the argument of the inner function: $3^{\log_3 x} = x$. Exponential and logarithmic functions with the same base ($y = a^x$ and $y = \log_a x$) are inverses of one another.

5.30 Simplify the expression: $\ln (e^4 \cdot e^{5x})$.

Multiply the natural logarithmic expressions within the parentheses.

$$\ln (e^4 \cdot e^{5x}) = \ln e^{5x+4}$$

Notice that the expression $\ln e^{5x+4}$ is a composition of inverse functions, because $\ln x$ and e^x have the same base. Therefore, the functions cancel one another out, leaving behind only the argument of the inner function.

$$\ln e^{5x+4} = 5x + 4$$

5.31 Simplify the expression: $e^{\ln x - \ln y}$.

Rewrite the exponent using logarithmic properties. (The difference of two logarithms with the same base is equal to a logarithmic quotient, as explained in Problem 5.21.)

$$e^{\ln x - \ln y} = e^{\ln (x/y)}$$

The natural logarithmic and exponential functions are inverses of one another, so only the argument of the inner functions remains when the functions are composed.

$$e^{\ln(x/y)} = \frac{x}{y}$$

5.32 Determine the exact solution to the equation: $2^x = 9$.

To eliminate the exponential function 2^x on the left side of the equation, apply its inverse function $\log_2 x$ to both sides of the equation.

$$\log_2 2^x = \log_2 9$$
$$x = \log_2 9$$

Although you could use the change of base formula to approximate $\log_2 9$, the problem specifically requests the *exact* answer. Therefore, you should not estimate the solution: $x = \log_2 9$.

Because $\log_2 9$ is an irrational number, its decimal places never terminate or repeat, so anything short of an infinitely long list of decimals isn't an EXACT answer.

5.33 Determine the exact solution to the equation: $2 - e^{5x} = -13$.

Subtract 2 from both sides of the equation to isolate the exponential function, then multiply the entire equation by -1.

$$-e^{5x} = -15$$
$$e^{5x} = 15$$

To eliminate the natural exponential function, take the natural logarithm of both sides of the equation.

$$\ln e^{5x} = \ln 15$$
$$5x = \ln 15$$
$$x = \frac{\ln 15}{5}$$

5.34 Determine the exact solution to the equation: $5 + \ln (x + 3) = 7$.

Isolate the logarithmic expression on the left side of the equation.

$$\ln (x + 3) = 2$$

Exponentiate the equation to eliminate the natural logarithm.

"Exponentiation" means turning both sides of the equation into exponential expressions. Because the base of ln x is e, both sides of the equation become exponents of e.

$$e^{\ln(x+3)} = e^2$$
$$x + 3 = e^2$$
$$x = e^2 - 3$$

5.35 Determine the exact solution to the equation: $\log_5 2x - \log_5 3 = \log_5 (4x - 19)$.

Apply logarithmic properties to rewrite $\log_5 2x - \log_5 3$ as a logarithmic quotient.

$$\log_5 \frac{2x}{3} = \log_5 (4x - 19)$$

Exponentiate the equation to eliminate the logarithms.

$$5^{\log_5 (2x/3)} = 5^{\log_5 (4x-19)}$$

Because 5^x and $\log_5 x$ are inverse functions, they cancel one another out, leaving behind only the arguments of the logarithms.

$$\frac{2x}{3} = 4x - 19$$

Solve for x.

$$2x = 3(4x - 19)$$
$$2x = 12x - 57$$
$$-10x = -57$$
$$x = \frac{57}{10}$$

> Before you exponentiate, make sure you have only logs on the left side of the equation and then combine them all into one log using the properties from Problems 5.20-5.24.

5.36 Determine the exact solution to the equation: $\log x + \log (x - 2) - 1 = 0$.

Move the constant term to the right side of the equation.

$$\log x + \log (x - 2) = 1$$

The sum of two logarithms with equal bases equals the logarithm of their product.

$$\log [x(x - 2)] = 1$$
$$\log (x^2 - 2x) = 1$$

Exponentiate the equation to eliminate the logarithmic function.

$$10^{\log(x^2 - 2x)} = 10^1$$
$$x^2 - 2x = 10$$
$$x^2 - 2x - 10 = 0$$

Solve using the quadratic formula.

$$x = \frac{-(-2) \pm \sqrt{(-2)^2 - 4(1)(-10)}}{2(1)}$$
$$= \frac{2 \pm \sqrt{44}}{2}$$
$$= \frac{2 \pm 2\sqrt{11}}{2}$$
$$= 1 \pm \sqrt{11}$$

> If you plug $x = 1 - \sqrt{11} \approx -2.32$ into the original equation, you get $\log(-2.32) + \log(-4.32) - 1 = 0$. However, you can only take the log of a positive number, so $x = 1 - \sqrt{11}$ isn't a valid solution.

Discard the invalid solution $x = 1 - \sqrt{11}$ (logarithmic functions have a domain of $x > 0$); the only valid solution to the equation is $x = 1 + \sqrt{11}$.

Chapter 6
CONIC SECTIONS

Parabolas, circles, ellipses, and hyperbolas

When a double-napped right circular cone is sliced by a plane, the perimeter of the resulting cross section will be a circle, an ellipse, a parabola, or a hyperbola. Hence, this family of four curves is known as the conic sections. Like their graphs share a similar origin, the equations of their standard forms share similar characteristics as well, though each has unique distinguishing features.

This chapter explores the nuances of the equations that generate the conic sections and investigates how the constants and variables in those equations affect their graphs.

Quadratic equations have graphs that are roughly u-shaped, that are called "parabolas." The most obvious feature of a quadratic equation is its degree—one of the variables is raised to the second power, and it's the highest exponent in the equation (like $y = x^2 - 7x + 2$). That's all well and good, but what happens when there's an x^2 AND a y^2 in the equation? Those graphs will be (a) circles, (b) ellipses, or (c) hyperbolas. In this chapter, you'll learn how to graph all four kinds of conic sections and how to manipulate an equation so it's easy to graph.

Parabolas

Graphs of quadratic equations

6.1 Write the equation of the parabola in standard form and identify its vertex: $y = x^2 + 6x - 4$.

Add 4 to both sides of the equation so that the right side contains only x-terms.

$$y + 4 = x^2 + 6x$$

Complete the square on the right side of the equation: take half of the x-coefficient $\left(\frac{1}{2} \cdot 6 = 3\right)$, square the result ($3^2 = 9$), and add that number (9) to both sides of the equation.

$$y + 4 + 9 = x^2 + 6x + 9$$
$$y + 13 = x^2 + 6x + 9$$

Factor the trinomial.

$$y + 13 = (x + 3)(x + 3)$$
$$y + 13 = (x + 3)^2$$

Solve for y.

$$y = (x + 3)^2 - 13$$

There is no constant outside the parentheses, so $a = 1$.

The standard form of a parabola is $y = a(x - h)^2 + k$, , so in this problem $a = 1$, $h = -3$, and $k = -13$. The vertex of a parabola in standard form is $(h,k) = (-3,-13)$.

Notice that the standard form includes "$-h$" so you should take the opposite of the number in the parentheses: $h = -3$, not 3.

6.2 Write the equation of the parabola in standard form and identify its vertex and axis of symmetry: $y = 2x^2 - 16x - 1$.

Move the constant to the left side of the equation.

$$y + 1 = 2x^2 - 16x$$

Complete the square on the right side of the equation. To do so, the coefficient of the x^2-term *must* be 1. Divide the entire equation by 2 (the x^2-coefficient) to accomplish this.

$$\frac{y}{2} + \frac{1}{2} = x^2 - 8x$$

The square of half the x-coefficient is 16: $\frac{1}{2}(-8) = -4$, and $(-4)^2 = 16$; add 16 to both sides of the equation.

$$\frac{y}{2} + \frac{1}{2} + 16 = x^2 - 8x + 16$$

Add the constants on the left side of the equation and factor the right side.

$$\frac{y}{2} + \frac{1}{2} + \frac{32}{2} = (x - 4)(x - 4)$$
$$\frac{y + 33}{2} = (x - 4)^2$$

Multiply the entire equation by 2 to eliminate the fraction; solve for y.

$$\left(\frac{\cancel{2}}{1}\right)\left(\frac{y+33}{\cancel{2}}\right)=\left(\frac{2}{1}\right)\left[\frac{(x-4)^2}{1}\right]$$

$$y+33=2(x-4)^2$$

$$y=2(x-4)^2-33$$

The equation is now in standard form, $y=a(x-h)^2+k$; therefore, $a=2$, $h=4$, and $k=-33$. The vertex of the parabola is $(h,k)=(4,-33)$, and the axis of symmetry is $x=4$.

> The axis of symmetry is a line that cuts right down the middle of a parabola through its vertex. When standard form contains x^2, the axis of symmetry is the vertical line $x=h$.

6.3 Write the equation of the parabola in standard form and identify its vertex: $-y^2+3x+5y-7=0$.

Notice that this equation (unlike Problems 6.1 and 6.2) contains a y^2-term rather than an x^2-term. Though the process is similar to those problems, the end result is an equation solved for x (not y) with standard form $x=a(y-k)^2+h$. Isolate the y-terms on the right side of the equation.

$$3x-7=y^2-5y$$

> The difference between the standard form of a y^2 parabola and an x^2 parabola is the position of the variables: x and y are switched, and so are h and k.

Complete the square.

$$3x-7+\frac{25}{4}=y^2-5y+\frac{25}{4}$$

$$\frac{12x-28}{4}+\frac{25}{4}=\left(y-\frac{5}{2}\right)\left(y-\frac{5}{2}\right)$$

$$\frac{12x-3}{4}=\left(y-\frac{5}{2}\right)^2$$

Multiply both sides of the equation by 4 in order to eliminate the fraction; solve for x.

$$\frac{\cancel{4}}{1}\left(\frac{12x-3}{\cancel{4}}\right)=4\left(y-\frac{5}{2}\right)^2$$

$$12x-3=4\left(y-\frac{5}{2}\right)^2$$

$$12x=4\left(y-\frac{5}{2}\right)^2+3$$

Multiply each term by $\frac{1}{12}$ to isolate x.

$$\left(\frac{1}{12}\right)12x=\left(\frac{1}{12}\right)(4)\left(y-\frac{5}{2}\right)^2+\left(\frac{1}{12}\right)3$$

$$x=\frac{4}{12}\left(y-\frac{5}{2}\right)^2+\frac{3}{12}$$

$$x=\frac{1}{3}\left(y-\frac{5}{2}\right)^2+\frac{1}{4}$$

> When a parabola in standard form is written in terms of y (like this one), k is the opposite of the number in parentheses and h is the constant. The vertex is still (h,k)—those coordinates are exactly the same as parabolas with x^2 in them.

The parabola is now in standard form, $x=a(y-k)^2+h$, with $a=\frac{1}{3}$, $h=\frac{1}{4}$, and $k=\frac{5}{2}$. The vertex of the parabola is $(h,k)=\left(\frac{1}{4},\frac{5}{2}\right)$.

6.4 Write the equation of the parabola in standard form and identify its vertex and axis of symmetry: $x - 9y^2 + 18y + 6 = 0$.

This equation contains y^2, so you should rewrite it in standard form $x = a(y - k)^2 + h$. Move all y-terms to the right side of the equation.

$$x + 6 = 9y^2 - 18y$$

Divide all of the terms by 9 to ensure the coefficient of y^2 is 1, so that you can complete the square. Solve for x.

$$\frac{x}{9} + \frac{6}{9} = \frac{9y^2}{9} - \frac{18y}{9}$$

$$\frac{x+6}{9} = y^2 - 2y$$

$$\frac{x+6}{9} + 1 = y^2 - 2y + 1$$

$$\frac{x+15}{9} = (y-1)^2$$

$$x + 15 = 9(y-1)^2$$

$$x = 9(y-1)^2 - 15$$

> You can double-check your algebra by expanding standard form to see if you get the equation you started with:
>
> $x = 9(y^2 - 2y + 1) - 15$
> $x = 9y^2 - 18y + 9 - 15$
> $x - 9y^2 + 18y + 6 = 0$

This is the standard form of the parabola $x = a(y - k)^2 + h$, where $a = 9$, $h = -15$, and $k = 1$, so the vertex is $(h,k) = (-15,1)$. A parabola written in terms of y has axis of symmetry $y = k$, so the axis of symmetry of this parabola is $y = 1$.

6.5 Graph the parabola defined in Problem 6.1: $y = x^2 + 6x - 4$.

> If you need to review function graph transformations, look back at Problem 4.19—it's very similar.

According to Problem 6.1, the standard form of the parabola is $y = (x + 3)^2 - 13$. Graph the parabola by applying two basic transformations to the graph of $y = x^2$. Adding 3 to x shifts the graph left 3 units, and subtracting 13 shifts the graph down 13 units.

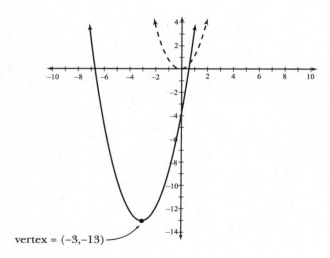

vertex = (−3,−13)

Figure 6-1

The dotted graph of $y = x^2$ and the solid graph of $y = (x + 3)^2 - 13$. All parabolas containing an x^2-term either open upward or downward, whereas parabolas written in terms of y either open left or right.

6.6   Graph the parabola defined in Problem 6.3: $-y^2 + 3x + 5y - 7 = 0$.

> A vertical line cannot intersect the graph of a function more than once.

According to Problem 6.3, the parabola has standard form $x = \frac{1}{3}\left(y - \frac{5}{2}\right)^2 + \frac{1}{4}$. The graphs of parabolas containing y^2-terms are not functions, because they fail the vertical line test. Thus, function transformations (such as those demonstrated in Problem 6.5) are not a reliable method to graph this equation. Instead, you should plot points; substitute a variety of values for y to get the corresponding x-values and graph the coordinates that result. Since the vertex is located at $\left(\frac{1}{4}, \frac{5}{2}\right)$, use y-values close to $\frac{5}{2}$.

$y = 0$	$y = 1$	$y = \dfrac{5}{2}$	$y = 3$	$y = 4$
$x = \frac{1}{3}\left(-\frac{5}{2}\right)^2 + \frac{1}{4}$ $= \frac{1}{3}\left(\frac{25}{4}\right) + \frac{1}{4} = \frac{7}{3}$	$x = \frac{1}{3}\left(-\frac{3}{2}\right)^2 + \frac{1}{4}$ $= \frac{1}{3}\left(\frac{9}{4}\right) + \frac{1}{4} = 1$	$x = \frac{1}{3}(0)^2 + \frac{1}{4} = \frac{1}{4}$	$x = \frac{1}{3}\left(\frac{1}{2}\right)^2 + \frac{1}{4}$ $= \frac{1}{3}\left(\frac{1}{4}\right) + \frac{1}{4} = \frac{1}{3}$	$x = \frac{1}{3}\left(\frac{3}{2}\right)^2 + \frac{1}{4}$ $= \frac{1}{3}\left(\frac{9}{4}\right) + \frac{1}{4} = 1$

According to this table of values, the points $\left(\frac{7}{3}, 0\right)$, $(1,1)$, $\left(\frac{1}{4}, \frac{5}{2}\right)$, $\left(\frac{1}{3}, 3\right)$, and $(1,4)$ belong on the graph, as illustrated by Figure 6-2.

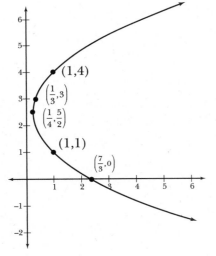

Figure 6-2

The graph of the parabola $-y^2 + 3x + 5y - 7 = 0$ opens to the right.

> A parabola is made up of all the points that are the same distance from a fixed point (called the FOCUS) and a fixed line (called the DIRECTRIX).

6.7   Identify the focus and directrix of the parabola: $2x^2 + 20x - 3y + 9 = 0$.

Rewrite the parabola in standard form. Divide the equation by 2, so that the coefficient of x^2 is 1, and complete the square.

$$\frac{2x^2}{2} + \frac{20x}{2} - \frac{3y}{2} + \frac{9}{2} = \frac{0}{2}$$

$$\frac{3y}{2} - \frac{9}{2} = x^2 + 10x$$

$$\frac{3y}{2} - \frac{9}{2} + 25 = x^2 + 10x + 25$$

$$\frac{3y}{2} - \frac{9}{2} + \frac{50}{2} = (x+5)(x+5)$$

$$\frac{3y + 41}{2} = (x+5)^2$$

$$3y + 41 = 2(x+5)^2$$

$$3y = 2(x+5)^2 - 41$$

$$y = \frac{2}{3}(x+5)^2 - \frac{41}{3}$$

Standard form of the parabola is $y = a(x - h)^2 + k$, so $a = \frac{2}{3}$, $h = -5$, and $k = -\frac{41}{3}$. In order to determine the focus of the parabola, first define the constant $c = \left|\frac{1}{4a}\right|$.

one divided by a fraction equals the reciprocal of the fraction you're dividing by.

$$c = \left| \frac{1}{\frac{4}{1}\left(\frac{2}{3}\right)} \right|$$

$$c = \left| \frac{1}{\left(\frac{8}{3}\right)} \right|$$

$$c = \frac{3}{8}$$

If a is negative, then the focus is (h, k − c) and the directrix is y = k + c.

It is important to note that $a = \frac{2}{3}$ is positive, because that means the parabola's focus is above its vertex and its directrix is below the vertex. Any such parabola has focus $(h, k + c)$ and directrix $y = k - c$. Substitute the values of h, k, and c into those formulas.

Focus $= (h, k + c)$

$$= \left(-5, -\frac{41}{3} + \frac{3}{8}\right)$$

$$= \left(-5, \frac{-328 + 9}{24}\right)$$

$$= \left(-5, -\frac{319}{24}\right)$$

Directrix: $y = k - c$

$$y = -\frac{41}{3} - \frac{3}{8}$$

$$y = -\frac{328}{24} - \frac{9}{24}$$

$$y = -\frac{337}{24}$$

6.8 Identify the focus and directrix of the parabola: $y^2 + 3y + 5x + 2 = 0$.

Rewrite the parabola in standard form $x = a(y - k)^2 + h$ by completing the square.

$$-5x - 2 = y^2 + 3y$$

$$-5x - 2 + \frac{9}{4} = y^2 + 3y + \frac{9}{4}$$

$$-5x - \frac{8}{4} + \frac{9}{4} = \left(y + \frac{3}{2}\right)\left(y + \frac{3}{2}\right)$$

$$-5x = \left(y + \frac{3}{2}\right)^2 - \frac{1}{4}$$

$$x = -\frac{1}{5}\left(y + \frac{3}{2}\right)^2 + \frac{1}{20}$$

> Half of the y-coefficient is $\frac{1}{2} \cdot \frac{3}{1} = \frac{3}{2}$, and $\left(\frac{3}{2}\right)^2 = \frac{9}{4}$, so add $\frac{9}{4}$ to both sides.

Therefore, $a = -\frac{1}{5}$, $h = \frac{1}{20}$, and $k = -\frac{3}{2}$; calculate c.

$$c = \left|\frac{1}{4a}\right| = \left|\frac{1}{\left(\frac{4}{1}\right)\left(-\frac{1}{5}\right)}\right| = \left|\frac{1}{\left(-\frac{4}{5}\right)}\right| = \left|-\frac{5}{4}\right| = \frac{5}{4}$$

> If $a > 0$ in a y^2 parabola, the focus is the point $(h + c, k)$ and the directrix is the line $x = h - c$.

When the parabola contains a y^2-term and $a < 0$, the focus is left of its vertex and the directrix is right of the vertex. Specifically, the focus is $(h - c, k)$ and the directrix is $x = h + c$.

Focus $= (h - c, k)$	Directrix: $x = h + c$
$= \left(\dfrac{1}{20} - \dfrac{5}{4}, -\dfrac{3}{2}\right)$	$x = \dfrac{1}{20} + \dfrac{5}{4}$
$= \left(\dfrac{1 - 25}{20}, -\dfrac{3}{2}\right)$	$x = \dfrac{1}{20} + \dfrac{25}{20}$
$= \left(-\dfrac{24}{20}, -\dfrac{3}{2}\right)$	$x = \dfrac{26}{20}$
$= \left(-\dfrac{6}{5}, -\dfrac{3}{2}\right)$	$x = \dfrac{13}{10}$

6.9 Write the equation of the parabola with focus $(-1, 3)$ and directrix $y = 1$ in standard form.

As noted in Problem 6.7, a parabola that contains an x^2-term has a horizontal directrix. (A parabola with a y^2-term has a vertical directrix, as demonstrated by Problem 6.8.) Therefore, you should use standard form $y = a(x - h)^2 + k$.

The vertex (h, k) of the parabola has the same x-value as the focus, so $h = -1$. Furthermore, it is equidistant from the directrix and the focus; therefore, k is the average of the y-coordinate of the focus and the constant in the directrix equation.

> Add the number in the directrix to the y-value of the focus and divide by 2.

$$k = \frac{3 + 1}{2} = 2$$

Note that c is the vertical distance between the vertex and the focus; calculate the absolute value of the difference of their y-values: $c = |3 - 2| = 1$. Calculate a using the formula from Problems 6.7 and 6.8, but omit the absolute value signs.

> The focus is $(-1,3)$ and the vertex is $(h,k) = (-1,2)$. Subtract the y-values to get c, and make sure c is positive! By the way, c is also the distance between the vertex and the directrix.

$$c = \frac{1}{4a}$$

$$1 = \frac{1}{4a}$$

$$4a = 1$$

$$a = \frac{1}{4}$$

You must now determine whether a is positive or negative. According to Problem 6.7, when a parabola's focus is above its vertex, $a > 0$. Therefore, $a = \frac{1}{4}$, $h = -1$, and $k = 2$.

$$y = a(x - h)^2 + k$$

$$y = \frac{1}{4}(x - (-1))^2 + 2$$

$$y = \frac{1}{4}(x + 1)^2 + 2$$

Circles

Center + radius = round shapes and easy problems

6.10 Graph the circle with center $(-3,0)$ and radius 2.

> The center of a circle, like the focus of a parabola, helps shape the graph but is not considered part of the graph.

Plot the center point on the coordinate plane, and then mark the points 2 units above, below, right of, and left of the center, as illustrated by Figure 6-3. Draw the graph of the circle through those four points.

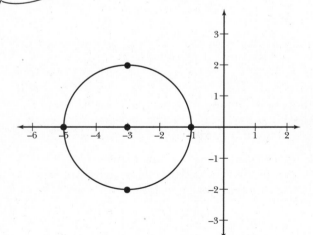

Figure 6-3

The graph of a circle with center $(-3,0)$ and radius 2.

6.11 If points $A = (-4,1)$ and $B = (6,-5)$ are the endpoints of line segment $\overline{AB}$, such that $\overline{AB}$ is a diameter of circle C, find the center and radius of C.

The center of the circle is located at the midpoint of the diameter, so apply the midpoint formula (as described in Problem 1.8).

$$\text{center} = \left(\frac{-4+6}{2}, \frac{1+(-5)}{2} \right) = \left(\frac{2}{2}, \frac{-4}{2} \right) = (1,-2)$$

The radius is half the length of the diameter. Calculate the diameter's length using the distance formula (as described in Problem 1.9) and divide the result by 2.

$$\text{radius} = \frac{AB}{2} = \frac{\sqrt{(6-(-4))^2 + (-5-1)^2}}{2} = \frac{\sqrt{10^2 + (-6)^2}}{2} = \frac{\sqrt{136}}{2} = \frac{2\sqrt{34}}{2} = \sqrt{34}$$

6.12 Identify the center and radius of the circle: $(x+9)^2 + (y-1)^2 = 16$.

The standard form of a circle is $(x-h)^2 + (y-k)^2 = r^2$, where (h,k) is the center of the circle and r is the radius. In this example, $(h,k) = (-9,1)$. The radius is equal to the square root of the constant on the right side of the equation.

$$r^2 = 16$$
$$r = 4$$

Though $r = -4$ is a valid solution to the equation $r^2 = 16$, it is not a valid radius; the radius of a circle must be positive.

> Pull the constants out of the squared binomials—just make sure you pull out the opposites of the numbers since the standard form contains $-h$ and $-k$.

6.13 Rewrite the equation of the circle in standard form: $x^2 + y^2 - 4x + 12y - 10 = 0$.

Write the x-terms in descending order of degree, leaving empty space before listing the y-terms. (You will use this space to complete the square in the next step.) Then write the y-terms in descending order of degree (leaving space before the equal sign). Move the constant by adding 10 to both sides of the equation.

$$x^2 - 4x \quad + y^2 + 12y \quad = 10$$

Complete the square *twice* (once for the x-terms and once for the y-terms), adding *both* constants to the right side of the equation to maintain equality.

$$x^2 - 4x + \mathbf{4} + y^2 + 12y + \mathbf{36} = 10 + \mathbf{4} + \mathbf{36}$$
$$(x-2)(x-2) + (y+6)(y+6) = 50$$
$$(x-2)^2 + (y+6)^2 = 50$$

> Half the x-coefficient is -2 and $(-2)^2 = 4$. Half the y-coefficient is 6 and $6^2 = 36$. Add 4 and 36 to the left side of the equation in the blank spots and add both of them to the right side also.

6.14 Rewrite the equation of the circle in standard form: $x^2 + y^2 + 7x - 2y + 6 = 0$.

Use the method of Problem 6.13 (completing the square twice simultaneously) to reach standard form.

$$x^2 + 7x + \frac{49}{4} + y^2 - 2y + 1 = -6 + \frac{49}{4} + 1$$

$$\left(x + \frac{7}{2}\right)^2 + \left(y - 1\right)^2 = \frac{29}{4}$$

Recall that the standard form of a circle is $(x - h)^2 + (y - k)^2 = r^2$; therefore, $h = -\frac{7}{2}$, $k = 1$, and $r = \sqrt{\frac{29}{4}} = \frac{\sqrt{29}}{2}$. The center of the circle is $\left(-\frac{7}{2}, 1\right)$, and the radius is $\frac{\sqrt{29}}{2}$.

6.15 Identify the center and radius of the circle: $2x^2 + 10x + 2y^2 - 7y + 12 = 0$.

In order to complete the square, the coefficients of the x^2- and y^2-terms must be 1. Note that they are equal, so divide all terms in the equation by the shared coefficient.

> The x^2- and y^2- coefficients of circles are always equal.

$$\frac{2x^2}{2} + \frac{10x}{2} + \frac{2y^2}{2} - \frac{7y}{2} + \frac{12}{2} = \frac{0}{2}$$

$$x^2 + 5x + y^2 - \frac{7}{2}y + 6 = 0$$

Rewrite the equation in standard form by completing the square once for the x-terms and again for the y-terms.

$$x^2 + 5x + \frac{25}{4} + y^2 - \frac{7}{2}y + \frac{49}{16} = -6 + \frac{25}{4} + \frac{49}{16}$$

$$\left(x + \frac{5}{2}\right)\left(x + \frac{5}{2}\right) + \left(y - \frac{7}{4}\right)\left(y - \frac{7}{4}\right) = -\frac{96}{16} + \frac{100}{16} + \frac{49}{16}$$

$$\left(x + \frac{5}{2}\right)^2 + \left(y - \frac{7}{4}\right)^2 = \frac{53}{16}$$

The center of the circle is $\left(-\frac{5}{2}, \frac{7}{4}\right)$ and the radius is $\sqrt{\frac{53}{16}} = \frac{\sqrt{53}}{4}$.

6.16 Graph the circle: $5x^2 - 30x + 5y^2 + 5y + 40 = 0$.

Rewrite the equation in standard form using the method outlined by Problem 6.15. Divide all terms of the equation by 5, the coefficient shared by the x^2- and y^2-terms.

$$x^2 - 6x + y^2 + y = -8$$

$$x^2 - 6x + 9 + y^2 + y + \frac{1}{4} = -8 + 9 + \frac{1}{4}$$

$$(x - 3)(x - 3) + \left(y + \frac{1}{2}\right)\left(y + \frac{1}{2}\right) = 1 + \frac{1}{4}$$

$$(x - 3)^2 + \left(y + \frac{1}{2}\right)^2 = \frac{5}{4}$$

The center of the circle is $\left(3, -\frac{1}{2}\right)$ and the radius is $\sqrt{\frac{5}{4}} = \frac{\sqrt{5}}{2}$. Use a calculator to find a decimal approximation of the radius $\left(\frac{\sqrt{5}}{2} \approx 1.118\right)$ and graph using the method outlined in Problem 6.10. The solution is illustrated in Figure 6-4.

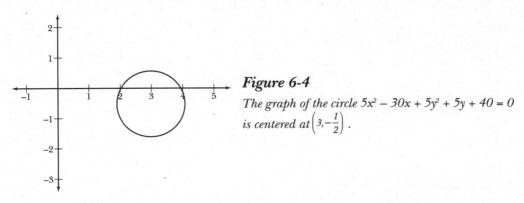

Figure 6-4

The graph of the circle $5x^2 - 30x + 5y^2 + 5y + 40 = 0$ is centered at $\left(3, -\frac{1}{2}\right)$.

Ellipses
Fancy word for "ovals"

Note: Problems 6.17–6.18 refer to the ellipse graphed in Figure 6-5.

6.17 Calculate the lengths of the major and minor axes of the ellipse and identify the vertices.

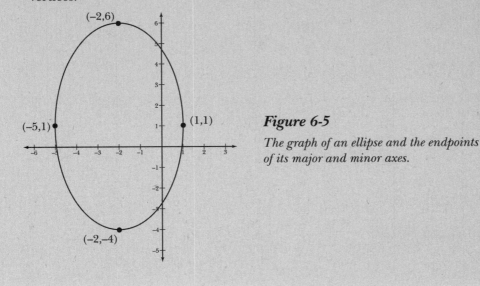

Figure 6-5

The graph of an ellipse and the endpoints of its major and minor axes.

The major axis of the ellipse has endpoints (–2,6) and (–2,–4). Its length, therefore, is the distance between those points. Note that endpoints are on the same vertical line ($x = -2$), so the distance between the points is the absolute value of the difference of the y-coordinates: $|-4 - 6| = 10$.

> **The endpoints of the minor axis are important, but they're not vertices—an ellipse has two vertices, not four.**

The endpoints of the minor axis are (1,1) and (–5,1); its length is $|-5-1| = 6$, the difference of the x-coordinates. Note that the major axis is always larger then the minor axis of an ellipse, and only the endpoints of the major axis are considered the vertices of the ellipse: (–2,6) and (–2,–4).

Note: Problems 6.17–6.18 refer to the ellipse graphed in Figure 6-5.

6.18   Write the equation of the ellipse in standard form.

The standard form of an ellipse with a vertical major axis is $\frac{(x-h)^2}{b^2} + \frac{(y-k)^2}{a^2} = 1$, where (h,k) is the center of the ellipse, a is half the length of the major axis, and b is half the length of the minor axis. Apply the midpoint formula to the endpoints of the major axis to determine the center of the ellipse.

$$\left(\frac{-2+(-2)}{2}, \frac{6+(-4)}{2}\right) = \left(\frac{-4}{2}, \frac{2}{2}\right) = (-2,1)$$

> **If the major axis is horizontal, a and b swap places: $\frac{(x-h)^2}{a^2} + \frac{(y-k)^2}{b^2} = 1$. Stick a underneath the variable (x or y) whose axis goes in the same direction as the major axis. In Problem 6.18, the major axis is vertical, so stick a under the y-variable since the y-axis is also vertical.**

Therefore, $h = -2$ and $k = 1$. (Note that the midpoint of the minor axis also marks the center of the ellipse—both midpoints are equal, because the center marks the point at which the axes bisect one another.) According to Problem 6.17, the length of the major axis is 10 and the length of the minor axis is 6, so $a = 10 \div 2 = 5$ and $b = 6 \div 2 = 3$. Plug h, k, a, and b into the standard form equation to generate the equation of the ellipse.

$$\frac{(x-h)^2}{b^2} + \frac{(y-k)^2}{a^2} = 1$$

$$\frac{(x-(-2))^2}{3^2} + \frac{(y-1)^2}{5^2} = 1$$

$$\frac{(x+2)^2}{9} + \frac{(y-1)^2}{25} = 1$$

Note: Problems 6.19–6.20 refer to the ellipse graphed in Figure 6-6.

6.19   Determine the values of a and b required to write the equation of the ellipse in standard form.

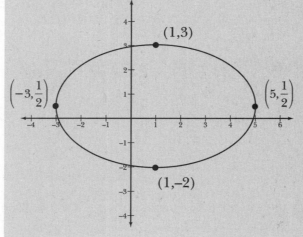

Figure 6-6

The endpoints of the major and minor axes of an ellipse.

The horizontal axis, connecting points $\left(-3, \frac{1}{2}\right)$ and $\left(5, \frac{1}{2}\right)$, is 8 units long; the vertical axis, connecting points (1,–2) and (1,3), is 5 units long. Since $8 > 5$, the horizontal axis is the major axis, and a is equal to half its length: $a = 8 \div 2 = 4$. Similarly, b is half the length of the remaining axis: $b = \frac{5}{2}$.

Note: Problems 6.19–6.20 refer to the ellipse graphed in Figure 6-6.

6.20 Write the equation of the ellipse in standard form.

Identify the center of the ellipse by calculating the midpoint of the major (or minor) axis.

$$\left(\frac{-3+5}{2}, \frac{1/2 + 1/2}{2} \right) = \left(\frac{2}{2}, \frac{1}{2} \right) = \left(1, \frac{1}{2} \right)$$

Therefore, $h = 1$ and $k = \frac{1}{2}$. According to Problem 6.19, $a = 4$ and $b = \frac{5}{2}$. Substitute these values into the standard form of an ellipse with a horizontal major axis.

$$\frac{(x-h)^2}{a^2} + \frac{(y-k)^2}{b^2} = 1$$

$$\frac{(x-1)^2}{4^2} + \frac{(y-1/2)^2}{(5/2)^2} = 1$$

$$\frac{(x-1)^2}{16} + \frac{(y-1/2)^2}{25/4} = 1$$

> Technically, a is the distance from the center to one of the vertices and b is the distance from the center to either endpoint of the minor axis.

6.21 Write the equation of the ellipse in standard form: $x^2 + 4y^2 - 8x + 24y + 36 = 0$.

Group the x- and y-terms together and move the constant to the right side of the equation.

$$x^2 - 8x + 4y^2 + 24y = -36$$

Reaching standard form requires you to complete the square twice, in a process similar to writing equations of circles in standard form (see Problems 6.13–6.16). However, the x^2- and y^2-coefficients of an elliptical equation are usually unequal, so rather than dividing by a shared coefficient (like in Problems 6.14 and 6.15), you factor out the leading coefficients. In this problem the coefficient of y^2 does not equal 1, so factor 4 out of the y-terms.

$$x^2 - 8x + 4\left(y^2 + 6y \right) = -36$$

> If the x^2- and y^2-coefficients of an ellipse are equal, their major and minor axes are the same length, and the ellipse is also a circle.

Complete the square for the x-terms, and work within the parentheses to complete the square for the y-terms.

$$x^2 - 8x + \mathbf{16} + 4\left(y^2 + 6y + \mathbf{9} \right) = -36 + \mathbf{16} + 4(\mathbf{9})$$

Although you add 9 to create the trinomial $y^2 + 6y + 9$, that expression is multiplied by 4, so you must add $4(9) = 36$ to the right side of the equation in order to maintain equality.

$$(x-4)^2 + 4(y+3)^2 = -36 + 52$$

$$(x-4)^2 + 4(y+3)^2 = 16$$

The right side of an elliptical equation in standard form equals 1, so divide the entire equation by 16 and reduce the resulting fractions.

$$\frac{(x-4)^2}{16} + \frac{4(y+3)^2}{16} = \frac{16}{16}$$

$$\frac{(x-4)^2}{16} + \frac{(y+3)^2}{4} = 1$$

6.22 Rewrite the ellipse in standard form and graph it: $9x^2 + y^2 - 90x - 4y + 220 = 0$.

Apply the method outlined in Problem 6.21: Regroup the variables, move the constant, factor the x^2-coefficient out of the x-terms, complete the square twice, and divide by the constant on the right side of the equation.

$$9x^2 - 90x + y^2 - 4y = -220$$

$$9\left(x^2 - 10x + \mathbf{25}\right) + y^2 - 4y + \mathbf{4} = -220 + \mathbf{225} + \mathbf{4}$$

$$9(x-5)^2 + (y-2)^2 = 9$$

$$\frac{9(x-5)^2}{9} + \frac{(y-2)^2}{9} = \frac{9}{9}$$

$$\frac{(x-5)^2}{1} + \frac{(y-2)^2}{9} = 1$$

It looks like you're adding 25 to the left side of the equation, but you're actually adding 9(25) = 225.

Compare the denominators. Because $9 > 1$, $a^2 = 9$ and $b^2 = 1$ (a^2 is always the larger denominator). The ellipse has a vertical major axis because a^2 appears below the y binomial, so apply the standard form equation $\frac{(x-h)^2}{b^2} + \frac{(y-k)^2}{a^2} = 1$, where $h = 5$, $k = 2$, $a = \sqrt{9} = 3$, and $b = \sqrt{1} = 1$.

To graph the ellipse, first plot the center: $(h,k) = (5,2)$. Plot the vertices by marking points three units above and below the center (since $a = 3$ and the major axis is vertical). Finally, plot the endpoints of the minor axes, which are one unit left and right of the center. As illustrated by Figure 6-7, the graph of the ellipse passes through all four endpoints.

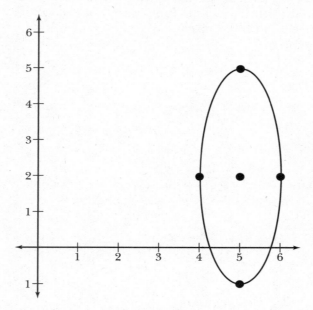

Figure 6-7

The graph of the ellipse
$9x^2 + y^2 - 90x - 4y + 220 = 0$, has center
(5,2), a vertical major axis 6 units in
length, a horizontal minor axis 2 units
in length.

6.23 Write the equation of the ellipse in standard form and identify the lengths of its major and minor axes: $x^2 + 16y^2 + 32y + 12 = 0$.

Apply the technique described in Problem 6.21, but note that you do not have to complete the square for the x^2-term.

$$x^2 + 16y^2 + 32y = -12$$
$$x^2 + 16\left(y^2 + 2y + \mathbf{1}\right) = -12 + 16(\mathbf{1})$$
$$x^2 + 16(y+1)^2 = 4$$
$$\frac{x^2}{4} + \frac{16(y+1)^2}{4} = \frac{4}{4}$$
$$\frac{x^2}{4} + \frac{4(y+1)^2}{1} = 1$$

> Dividing is the same as multiplying by a reciprocal (see Problem 3.13), so it makes sense that multiplying by 4 is the same as dividing by $\frac{1}{4}$.

This equation is not yet in standard form because $(y+1)^2$ has a coefficient of 4 and neither squared term has a coefficient in standard form. To remedy this, write the reciprocal of the coefficient $\left(\frac{1}{4}\right)$ in the denominator of the fraction that contains it.

$$\frac{x^2}{4} + \frac{(y+1)^2}{1/4} = 1$$

In this equation, $h = 0$, $k = -1$, $a = \sqrt{4} = 2$, and $b = \sqrt{\frac{1}{4}} = \frac{1}{2}$. The major axis has length $2a = 2(2) = 4$, and the minor axis has length $2b = 2\left(\frac{1}{2}\right) = 1$.

6.24 Find the coordinates of the foci of the ellipse $\dfrac{(x-6)^2}{25} + \dfrac{(y+11)^2}{16} = 1$.

Note that $a^2 = 25$ and $b^2 = 16$; substitute these values into the formula $c = \sqrt{a^2 - b^2}$ to determine the distance between the center of the ellipse and either of its foci.

$$c = \sqrt{25 - 16}$$
$$c = \sqrt{9}$$
$$c = 3$$

> A parabola has one focus, but an ellipse has two foci on its major axis. The distance from the center to either focus is $c = \sqrt{a^2 - b^2}$.

The center of the ellipse is $(h,k) = (6,-11)$ and its major axis is horizontal, so the foci are 3 units left and right of the center.

$$(h-c,k) \text{ and } (h+c,k)$$
$$(6-3,-11) \text{ and } (6+3,-11)$$
$$(3,-11) \text{ and } (9,-11)$$

6.25 Calculate the eccentricity of the ellipse from Problem 6.24:

$$\frac{(x-6)^2}{25} + \frac{(y+11)^2}{16} = 1.$$

The eccentricity, e, of an ellipse is a number between 0 and 1 that describes the "ovalness" of an ellipse. When e is close to 0, an ellipse looks a lot like a circle.

The eccentricity of an ellipse is calculated according to the formula $e = \dfrac{c}{a}$, where $c = \sqrt{a^2 - b^2}$. According to Problem 6.24, $c = 3$ and $a = 5$.

$$e = \frac{c}{a} = \frac{3}{5}$$

6.26 Calculate the eccentricity of the ellipse accurate to three decimal places:

$$\frac{(x+7)^2}{6} + \frac{(y-5)^2}{18} = 1.$$

Substitute, $a^2 = 18$ and $b^2 = 6$ into $c = \sqrt{a^2 - b^2}$.

$$c = \sqrt{18 - 6}$$
$$c = \sqrt{12}$$
$$c = 2\sqrt{3}$$

Apply the eccentricity formula defined in Problem 6.25.

$$e = \frac{c}{a} = \frac{2\sqrt{3}}{\sqrt{18}} = \frac{2\sqrt{3}}{3\sqrt{2}}\left(\frac{\sqrt{2}}{\sqrt{2}}\right) = \frac{2\sqrt{6}}{6} = \frac{\sqrt{6}}{3}$$

Use a calculator or other computer computational tool to calculate the decimal equivalent: $e \approx 0.816$.

6.27 Prove that an ellipse with eccentricity 0 is a circle.

If the eccentricity of an ellipse equals 0, then $e = \dfrac{c}{a} = 0$. In order for this fraction to equal 0, its numerator must equal 0, so assume that $c = \sqrt{a^2 - b^2} = 0$. Solve the equation for a and disregard negative values for a and b. (Because they represent distance in the coordinate plane, a and b must be positive real numbers.)

$$\left(\sqrt{a^2 - b^2}\right)^2 = (0)^2$$
$$a^2 - b^2 = 0$$
$$a^2 = b^2$$
$$\sqrt{a^2} = \sqrt{b^2}$$
$$a = b$$

If a and b are equal, then the major and minor axes have equal lengths and the endpoints of the axes (as well as all other points on the ellipse) are equidistant from the center; that distance is the radius of the circle.

Hyperbolas

Two-armed parabola-looking things

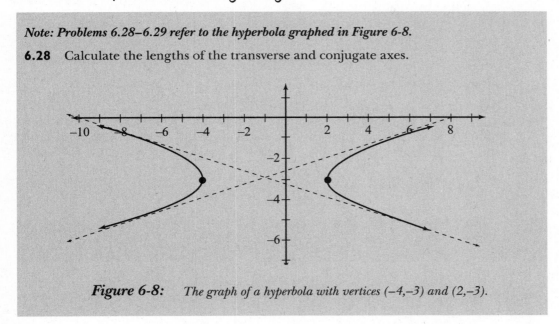

Note: Problems 6.28–6.29 refer to the hyperbola graphed in Figure 6-8.

6.28 Calculate the lengths of the transverse and conjugate axes.

Figure 6-8: *The graph of a hyperbola with vertices (−4,−3) and (2,−3).*

Extend vertical lines from the vertices to the asymptotes of the graph. Make note of the four resulting intersection points and use two horizontal lines to connect them. The end result is the rectangle pictured in Figure 6-9.

Use dotted lines to make the rectangle (like you do with asymptotes) so it's not mistaken as part of the actual graph of the hyperbola.

conjugate axis

transverse axis

Figure 6-9: *The transverse and conjugate axes of a hyperbola are perpendicular to, and bisect one another, at the center of the hyperbola, much like the major and minor axes of an ellipse. Their endpoints lie at the midpoints of the dotted rectangle.*

The transverse axis of a hyperbola is the segment connecting the vertices and has the same length as the horizontal sides of the dotted rectangle; the transverse axis in this problem is 6 units long. The conjugate axis is the vertical segment passing through the center of the hyperbola, which is 2 units in length.

Note: Problems 6.28–6.29 refer to the hyperbola graphed in Figure 6-8.

6.29 Write the equation of the hyperbola in standard form.

A hyperbola with a horizontal transverse axis has standard form $\frac{(x-h)^2}{a^2} - \frac{(y-k)^2}{b^2} = 1$, where a is half the length of the transverse axis, b is half the length of the conjugate axis, and (h,k) is the center of the hyperbola.

The center is the midpoint of the transverse axis, the midpoint of the conjugate axis, and the intersection point of the asymptotes, so $(h,k) = (-1,-3)$. According to Problem 6.28, the transverse axis has length 6 and the conjugate axis has length 2, so $a = 6 \div 2 = 3$ and $b = 2 \div 2 = 1$. Substitute the values into the standard form equation.

$$\frac{(x-(-1))^2}{3^2} - \frac{(y-(-3))^2}{1^2} = 1$$

$$\frac{(x+1)^2}{9} - \frac{(y+3)^2}{1} = 1$$

> **Rule of thumb:**
> The transverse axis is always parallel to the axis of the variable in the positive fraction. In problem 6.29, x is in the positive fraction so the transverse axis is parallel to the x-axis.

Note: Problems 6.30–6.31 refer to the hyperbola graphed in Figure 6-10.

6.30 Determine the values of a and b required to write the equation of the hyperbola in standard form.

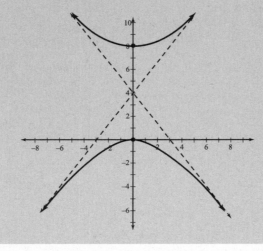

Figure 6-10

A hyperbola with vertices (0,0) and (0,8).

Draw a rectangle whose sides are congruent and parallel to the transverse and conjugate axes, as illustrated by Figure 6-11.

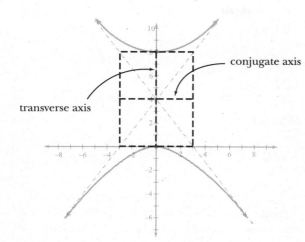

conjugate axis

transverse axis

Figure 6-11

The endpoints of the transverse and conjugate axes are the midpoints of the rectangle's sides.

Be careful: a is NOT always greater than b in equations of hyperbolas (like in equations of ellipses). Instead, a is always half the transverse axis (the one that contains the vertices).

According to Figure 6-11, the transverse axis is 8 units long, so $a = 8 \div 2 = 4$; the conjugate axis is 6 units long, so $b = 6 \div 2 = 3$.

Note: Problems 6.30–6.31 refer to the hyperbola graphed in Figure 6-10.

6.31 Write the equation of the hyperbola in standard form.

Because the transverse axis is vertical, the standard form of the hyperbola is $\dfrac{(y-k)^2}{a^2} - \dfrac{(x-h)^2}{b^2} = 1$. The midpoint of the transverse axis is the center of the hyperbola (as is the midpoint of the conjugate axis), so apply the midpoint formula to calculate h and k.

$$\left(\frac{0+0}{2}, \frac{0+8}{2}\right) = (0,4) = (h,k)$$

According to Problem 6.30, $a = 4$ and $b = 3$, so substitute those values into the standard form equation.

$$\frac{(y-4)^2}{4^2} - \frac{(x-0)^2}{3^2} = 1$$

$$\frac{(y-4)^2}{16} - \frac{x^2}{9} = 1$$

Either x^2 or y^2 will be negative in the equation of a hyperbola, because without that negative term, you wouldn't be able to get the necessary negative sign in standard form.

Note: Problems 6.32–6.33 refer to the equation $36x^2 - 25y^2 + 72x + 100y - 964 = 0$.

6.32 Rewrite the equation of the hyperbola in standard form and graph it.

Complete the square twice, using a technique similar to Problems 6.21–6.23. The only difference arises when dealing with the negative y^2—make sure to factor a negative constant out of the y-terms instead of a positive constant.

$$36x^2 + 72x - 25y^2 + 100y = 964$$

$$36\left(x^2 + 2x + 1\right) - 25\left(y^2 - 4y + 4\right) = 964 + 36(1) - 25(4)$$

$$36(x+1)^2 - 25(y-2)^2 = 900$$

$$\frac{36(x+1)^2}{900} - \frac{25(y-2)^2}{900} = \frac{900}{900}$$

Reduce the fractions: $\dfrac{36}{900} = \dfrac{1}{25}$ and $\dfrac{25}{900} = \dfrac{1}{36}$.

$$\frac{(x+1)^2}{25} - \frac{(y-2)^2}{36} = 1$$

The hyperbola is now in standard form.

$$\frac{(x-h)^2}{a^2} - \frac{(y-k)^2}{b^2} = 1$$

> Look back at my rule of thumb for Problem 6.29 if you don't know why this is true.

Therefore, $h = -1$, $k = 2$, $a = \sqrt{25} = 5$, and $b = \sqrt{36} = 6$. Plot the center point and then mark points 5 units left and right of it to plot the vertices. (The vertices are located left and right of the center—instead of above and below it—because the positive fraction contains x, so the hyperbola has a horizontal transverse axis.) Finally, plot the points 6 units above and below each vertex and draw a rectangle that passes through all of the points (except the center), as illustrated by Figure 6-12.

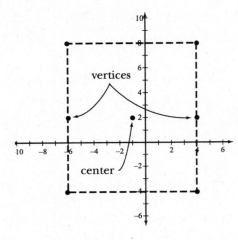

Figure 6-12

The hyperbola centered at (−1,2) has vertices (−6,2) and (4,2).

Extend the diagonals of the rectangle through its corners to draw the asymptotes of the hyperbola. Graph the branches of the hyperbola so that they pass through the vertices and quickly approach, but do not intersect, the asymptote lines. Because the transverse axis is horizontal, this hyperbola will open left and right (instead of up and down) as illustrated in Figure 6-13.

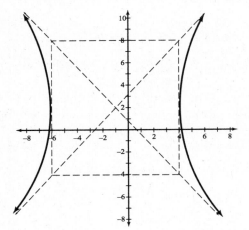

Figure 6-13

The graph of the hyperbola $36x^2 - 25y^2 + 72x + 100y - 964 = 0$.

Note: Problems 6.32–6.33 refer to the equation $36x^2 - 25y^2 + 72x + 100y - 964 = 0$.

6.33 Find the equations of the asymptotes to the hyperbola.

A hyperbola has two asymptotes, one with a positive slope and one with a negative slope. Notice that one asymptote passes through the center $(-1,2)$ and through the upper right-hand corner of the rectangle $(4,8)$; calculate its slope (using the method explained in Problem 1.5).

$$m = \frac{8-2}{4-(-1)} = \frac{6}{5}$$

Apply the point-slope formula using the slope and the coordinates of the center to write the equation of the line. (Substituting the coordinate pair $(x_1, y_1) = (4,8)$ into point-slope form results in the same linear equation.)

$$y - y_1 = m(x - x_1)$$
$$y - 2 = \frac{6}{5}(x - (-1))$$
$$y = \frac{6}{5}x + \frac{6}{5} + 2$$
$$y = \frac{6}{5}x + \frac{16}{5}$$

The asymptotes of hyperbolas have opposite slopes, so substitute $m = -\frac{6}{5}$ and $(x_1, y_1) = (-1,2)$ into the point-slope formula to generate the equation of the other asymptote.

> Both asymptotes pass through the center, so you can use it as (x_1, y_1) in both point-slope equations.

$$y - y_1 = m(x - x_1)$$
$$y - 2 = -\frac{6}{5}(x - (-1))$$
$$y = -\frac{6}{5}x - \frac{6}{5} + 2$$
$$y = -\frac{6}{5}x + \frac{4}{5}$$

6.34 Find the coordinates of the foci for the hyperbola: $\dfrac{(y-7)^2}{16} - \dfrac{(x+3)^2}{28} = 1$.

A hyperbola has two foci, which are located on the transverse axis c units further away from the center than the closest vertex (such that $c = \sqrt{a^2 + b^2}$). Note that the foci of an ellipse are *closer* to the center than the vertices, and thus "inside" the ellipse; however, the foci of a hyperbola are *farther away* from the center than the vertices, and are therefore "inside" the branches of the hyperbola.

The positive fraction of the standard form equation contains y, so the transverse axis is vertical (parallel to the y-axis); and each focus is a distance of c units away from the center of the ellipse along the transverse axis.

> In ellipses, $c = \sqrt{a^2 - b^2}$, but in hyperbolas, $c = \sqrt{a^2 + b^2}$.

Note that $a^2 = 16$ and $b^2 = 28$; calculate c.

$$c = \sqrt{a^2 + b^2}$$
$$c = \sqrt{16 + 28}$$
$$c = \sqrt{44}$$
$$c = 2\sqrt{11}$$

The center of the ellipse is $(h, k) = (-3, 7)$. The major axis is vertical, so the foci are located at the points $(h, k - c) = \left(-3, 7 - 2\sqrt{11}\right)$ and $(h, k + c) = \left(-3, 7 + 2\sqrt{11}\right)$.

Chapter 7
FUNDAMENTALS OF TRIGONOMETRY
Inject sine, cosine, and tangent into the mix

The study of calculus combines the arithmetic rigor of algebraic concepts with geometric and trigonometric postulates, theorems, and measurement techniques to expand, more accurately define, and generalize skills that are historically restrictive and unadaptable. Reacquainting yourself with trigonometric concepts must precede such a laudable endeavor.

Lots of calculus problems require prior knowledge of trigonometry. Not only do you have to know what sine, cosine, and tangent are, but you need to memorize some of their values (using the unit circle), understand their domains and ranges, and manipulate trig expressions using identities and theorems.

Trig helps you move beyond very strict rules in geometry—without trig, it's hard to find the area of a triangle unless you know its base and height. Besides, trig expressions are a lot more interesting (read: complicated) than linear and polynomial expressions, so they are often better tools to model real-life situations.

Basically, calculus is rife with trig, so make sure you understand the material in this and the next two chapters, because without this background, you'll struggle mightily in later chapters once calculus hits in full force.

Measuring Angles
Radians, degrees, and revolutions

7.1   Which is larger, m∠A = 15° or m∠B = 15 radians? Justify your answer.

> *Just like Fahrenheit and Celsius are two ways to measure a temperature.*

Radians and degrees represent two different ways to measure angles. Consider an angle whose initial and terminal side overlap after one full rotation, creating (for all intents and purposes) a circle. It is fairly commonly known that a circle measures 360°, but less commonly that the circle measures 2π radians. Therefore, one radian equals $\frac{360}{2\pi}$ (approximately 57.296) degrees. Therefore, 15 radians equals slightly more than 859°, and m∠B is greater than m∠A.

7.2   Convert 270° into radians.

To convert a degree measurement into radians, multiply it by $\frac{\pi}{180}$.

> *It's fine to write $\frac{3\pi}{2}$ as your final answer—you don't have to actually write the word "radians."*

$$270 \cdot \frac{\pi}{180} = \frac{270\pi}{180}$$
$$= \frac{\cancel{90} \cdot 3 \cdot \pi}{\cancel{90} \cdot 2}$$
$$= \frac{3\pi}{2} \text{ radians}$$

7.3   Convert 144° into radians.

Multiply the degree measurement by $\frac{\pi}{180}$.

$$144 \cdot \frac{\pi}{180} = \frac{144\pi}{180}$$
$$= \frac{\cancel{36} \cdot 4 \cdot \pi}{\cancel{36} \cdot 5}$$
$$= \frac{4\pi}{5} \text{ radians}$$

7.4   Convert $\frac{\pi}{3}$ radians into degrees.

To convert a radian angle measurement into degrees, multiply it by $\frac{180}{\pi}$.

$$\frac{\pi}{3} \cdot \frac{180}{\pi} = \frac{180\pi}{3\pi}$$
$$= \frac{60 \cdot \cancel{3} \cdot \cancel{\pi}}{\cancel{3} \cdot \cancel{\pi}}$$
$$= 60°$$

7.5   Convert 4 radians into degrees, and express the answer accurate to three decimal places.

Although this radian angle measurement does not contain "π," you should still apply the method from Problem 7.4—multiply by $\frac{180}{\pi}$.

$$4 \cdot \frac{180}{\pi} = \left(\frac{720}{\pi}\right)^{\circ}$$

Use a calculator to determine the quotient: 4 radians $\approx 229.183°$.

7.6   Convert 900° into revolutions.

A revolution is one circular loop, so one revolution equals 360°.

To convert an angle measured in degrees into revolutions, divide by 360°.

$$\frac{900}{360} = \frac{180 \cdot 5}{180 \cdot 2} = \frac{5}{2} \text{ revolutions}$$

7.7   Convert $\frac{27\pi}{4}$ into revolutions.

To convert a radian angle into revolutions, divide it by 2π, or (to facilitate simplification) multiply it by $\frac{1}{2\pi}$.

$$\frac{27\pi}{4} \div 2\pi = \frac{27\pi}{4} \cdot \frac{1}{2\pi} = \frac{27 \cdot \pi}{8 \cdot \pi} = \frac{27}{8} \text{ revolutions}$$

Some textbooks discuss revolutions, but this method of measuring angles is sort of outdated, which is why it's not mentioned anywhere else in the book.

Angle Relationships
Coterminal, complementary, and supplementary angles

7.8   Calculate the complement of $\theta = 34°$.

Complementary angles expressed in degrees have a sum of 90°, so if α is the complement of θ, then $\alpha + \theta = 90°$. Substitute $\theta = 34°$ into the equation and solve for α.

$$\alpha + 34° = 90°$$
$$\alpha = 90° - 34°$$
$$\alpha = 56°$$

The complement of $\theta = 34°$ is 56°.

7.9   Calculate the compliment of $\theta = \frac{4\pi}{9}$.

Complementary angles expressed in radians have a sum of $\frac{\pi}{2}$. Express this as an equation such that α is the complement of θ.

$$\frac{4\pi}{9} + \alpha = \frac{\pi}{2}$$

$$\alpha = \frac{\pi}{2} - \frac{4\pi}{9}$$

$$\alpha = \frac{9\pi}{18} - \frac{8\pi}{18}$$

$$\alpha = \frac{\pi}{18}$$

$\frac{\pi}{2}$ radians and 90° mean the same thing, so this process is the same as the one from Problem 7.8— you're just using radians instead of degrees.

Therefore, $\frac{\pi}{18}$ and $\frac{4\pi}{9}$ are complementary angles.

7.10 Calculate the supplement of $\theta = 125°$.

When expressed in degrees, an angle θ and its supplement α have a sum of 180°.

$$125° + \alpha = 180°$$
$$\alpha = 180° - 125°$$
$$\alpha = 55°$$

The supplement of 125° is 55°.

7.11 Calculate the supplement of $\theta = \frac{11\pi}{17}$.

An angle θ and its supplement α, when expressed in radians, have a sum of π.

$$\frac{11\pi}{17} + \alpha = \pi$$

$$\alpha = \pi - \frac{11\pi}{17}$$

$$\alpha = \frac{17\pi}{17} - \frac{11\pi}{17}$$

$$\alpha = \frac{6\pi}{17}$$

Usually, you only talk about coterminal angles when angles are in standard position— for more info, see Problem 7.15.

Therefore, $\frac{6\pi}{17}$ and $\frac{11\pi}{17}$ are supplementary angles.

7.12 Identify the three smallest positive angles coterminal with $\theta = -70°$.

Coterminal angles share the same terminal ray, so consecutive coterminal angles differ in magnitude by a single revolution, 360°. Add 360° to (or subtract 360° from) an angle to generate a coterminal angle.

$$-70° + 360° = 290° \qquad 290° + 360° = 650° \qquad 650° + 360° = 1,010°$$

7.13 Identify the angle coterminal to $\theta = 1,265°$ that belongs to the interval $[0°, 360°)$.

Subtract 360° from the angle to identify a coterminal angle: $1,265° - 360° = 905°$. That coterminal angle is still too large for the interval $[0°, 360°)$, so continue subtracting 360° until the result is between the endpoints of the interval.

$$905° - 360° = 545° \qquad 545° - 360° = 185°$$

The angle measuring 185° is coterminal with 1,265° and belongs to the interval [0°, 360°).

7.14 Identify the three greatest negative angles coterminal with $\theta = \dfrac{7\pi}{2}$.

Subtract one revolution (2π) from the angle at a time until you get a negative coterminal angle.

$$\frac{7\pi}{2} - 2\pi = \frac{7\pi}{2} - \frac{4\pi}{2} = \frac{3\pi}{2} \qquad \frac{3\pi}{2} - 2\pi = \frac{3\pi}{2} - \frac{4\pi}{2} = -\frac{\pi}{2}$$

Subtract 2π twice more to generate the second and third greatest negative coterminal angles.

$$-\frac{\pi}{2} - 2\pi = -\frac{\pi}{2} - \frac{4\pi}{2} = -\frac{5\pi}{2} \qquad -\frac{5\pi}{2} - 2\pi = -\frac{5\pi}{2} - \frac{4\pi}{2} = -\frac{9\pi}{2}$$

The three greatest negative coterminal angles to $\theta = \dfrac{7\pi}{2}$ are $-\dfrac{\pi}{2}, -\dfrac{5\pi}{2}$, and $-\dfrac{9\pi}{2}$.

Evaluating Trigonometric Functions
Right triangle trig and reference angles

7.15 Evaluate $\sin\dfrac{\pi}{2}$ based on the unit circle.

Graph $\theta = \dfrac{\pi}{2}$ in standard position on the same axes as a unit circle with center (0,0); its terminal ray intersects the circle at (0,1). The y-coordinate of the point of intersection represents the sine value of $\theta = \dfrac{\pi}{2}$, so $\sin\dfrac{\pi}{2} = 1$.

If an angle is in standard position, one of its rays overlaps the positive (right-hand) part of the x-axis and its vertex is located at the origin, (0,0).

7.16 Evaluate $\cos\dfrac{7\pi}{6}$ based on the unit circle.

The terminal ray of $\theta = \dfrac{7\pi}{6}$ intersects the unit circle centered at the origin at the point $\left(-\dfrac{\sqrt{3}}{2}, -\dfrac{1}{2}\right)$. The x-coordinate of the intersection point represents $\cos\dfrac{7\pi}{6}$, so $\cos\dfrac{7\pi}{6} = -\dfrac{\sqrt{3}}{2}$.

If you haven't memorized the cosine and sine values of the unit circle, check out Appendix B in the back of the book.

7.17 Evaluate $\cos\left(-\dfrac{\pi}{3}\right)$ based on the unit circle.

The unit circle provides only cosine and sine values for angles on the interval $[0,2\pi]$. Add 2π to the angle in order to calculate a positive coterminal angle for $\theta = -\dfrac{\pi}{3}$ that belongs to the interval.

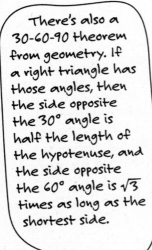

RULE OF THUMB: Positive angles are graphed counter-clockwise from the x-axis, and negative angles are graphed clockwise.

$$-\frac{\pi}{3}+2\pi=-\frac{\pi}{3}+\frac{6\pi}{3}=\frac{5\pi}{3}$$

Because $-\frac{\pi}{3}$ and $\frac{5\pi}{3}$ are coterminal angles, they have the same trigonometric values: $\cos\left(-\frac{\pi}{3}\right)=\cos\frac{5\pi}{3}$. According to the unit circle, $\cos\frac{5\pi}{3}=\frac{1}{2}$.

7.18 Verify the cosine and sine values of $\theta=\frac{\pi}{4}$ reported by the unit circle.

Draw the angle $\theta=\frac{\pi}{4}$ in standard position and a unit circle centered at $(0,0)$ on the same coordinate plane. Construct a right triangle, as illustrated by Figure 7-1. Because two angles of the triangle are known (45° and 90°), subtract their sum from 180° to calculate the remaining angle: $180°-(45°+90°)=45°$.

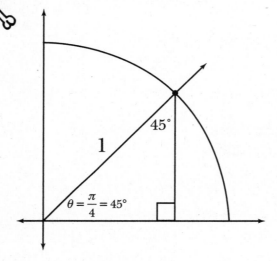

There's also a 30-60-90 theorem from geometry. If a right triangle has those angles, then the side opposite the 30° angle is half the length of the hypotenuse, and the side opposite the 60° angle is $\sqrt{3}$ times as long as the shortest side.

Figure 7-1
The radian measurement $\theta=\frac{\pi}{4}$ is equivalent to 45°, so this right triangle must be a 45-45-90 right triangle, whose side lengths are governed by specific geometric principles.

The hypotenuse of the right triangle is also a radius of the unit circle, so its length is 1. According to the 45-45-90 right triangle theorem, the legs of the triangle are $\frac{1}{\sqrt{2}}$ times as long as the hypotenuse, so multiply the hypotenuse length (1) by that fraction to calculate the lengths of the legs of the right triangle in Figure 7-1.

$$1\cdot\frac{1}{\sqrt{2}}=\frac{1}{\sqrt{2}}$$

Multiply the numerator and denominator of the fraction by $\sqrt{2}$ to rationalize it.

$$\frac{1}{\sqrt{2}}\left(\frac{\sqrt{2}}{\sqrt{2}}\right)=\frac{\sqrt{2}}{\sqrt{4}}=\frac{\sqrt{2}}{2}$$

This means you must travel $\frac{\sqrt{2}}{2}$ units right and $\frac{\sqrt{2}}{2}$ units up from the origin to reach the point of intersection between the angle and the unit circle, and the coordinates of the intersection point are $\left(\frac{\sqrt{2}}{2},\frac{\sqrt{2}}{2}\right)$—the same values reported by the unit circle for $\cos\frac{\pi}{4}$ and $\sin\frac{\pi}{4}$.

7.19 If an airborne kite 80 feet high is staked to the ground so that the fixed length of string forms a 50° angle of elevation, how long is the string that tethers the kite? Provide an answer accurate to three decimal places.

Consider Figure 7-2, which illustrates the situation described.

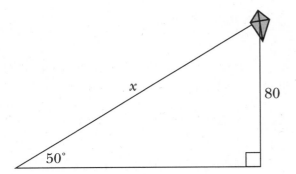

Figure 7-2

The kite is 80 feet high with a 50° angle of elevation; x represents the length of the string.

The three most common trig ratios are

$$\text{cosine} = \frac{\text{adjacent}}{\text{hypotenuse}},$$

$$\text{sine} = \frac{\text{opposite}}{\text{hypotenuse}}, \text{ and}$$

$$\text{tangent} = \frac{\text{opposite}}{\text{adjacent}}.$$

Note that the known side length, 80, is *opposite* the known angle and the side to be calculated is the *hypotenuse* of the right triangle. Of cosine, sine, and tangent, the only trigonometric ratio that describes the relationship between a right triangle's *opposite* side and *hypotenuse* is the sine. Substitute the known values and solve for *x*.

$$\sin 50° = \frac{80}{x}$$

$$x(\sin 50°) = 80$$

$$x = \frac{80}{\sin 50°}$$

Use a calculator to determine the quotient accurate to three decimal places: $x \approx 104.433$ feet.

Make sure your calculator or computer is set for degrees mode, because the angle in this problem is measured in degrees.

7.20 A stationary submarine (located at point *S* in Figure 7-3) has received intelligence indicating that a hostile submerged mine (point *M*) is located directly below a buoy (point *B*), which is exactly 2,500 feet from their current position. If the crew is instructed to fire a torpedo at a 9° angle of declension to detonate it from a safe distance, how far will the torpedo travel before it impacts the mine? Provide an answer accurate to three decimal places.

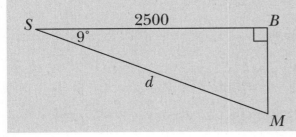

Figure 7-3

The distance d from the submarine S to the mine M can be calculated using a trigonometric ratio.

You're given a side length that is *adjacent* to the given angle and asked to calculate the length of the hypotenuse of the right triangle. You should apply the cosine ratio, since $\text{cosine} = \frac{\text{adjacent}}{\text{hypotenuse}}$, and then solve for *d*.

$$\cos 9° = \frac{2,500}{d}$$

$$d\left(\cos 9°\right) = 2,500$$

$$d = \frac{2,500}{\cos 9°}$$

$$d \approx 2,531.163 \text{ feet}$$

7.21 Evaluate $\cot \dfrac{11\pi}{6}$ based on the unit circle.

The tangent of an angle is equal to the quotient of its sine and cosine.

$$\tan \frac{11\pi}{6} = \frac{\sin\left(11\pi/6\right)}{\cos\left(11\pi/6\right)}$$

Substitute the sine and cosine values from the unit circle.

$$\tan \frac{11\pi}{6} = \frac{-1/2}{\sqrt{3}/2}$$

Simplify the complex fraction by multiplying its numerator and denominator by the reciprocal of the denominator.

$$\tan \frac{11\pi}{6} = \frac{-\dfrac{1}{2}\left(\dfrac{2}{\sqrt{3}}\right)}{\dfrac{\sqrt{3}}{2}\left(\dfrac{2}{\sqrt{3}}\right)} = \frac{\left(-\dfrac{1}{\sqrt{3}}\right)}{\left(\dfrac{1}{1}\right)} = -\frac{1}{\sqrt{3}}$$

The cotangent of an angle is equal to the reciprocal of its tangent. Bvecause $\tan \dfrac{11\pi}{6} = -\dfrac{1}{\sqrt{3}}$, $\cot \dfrac{11\pi}{6} = -\dfrac{\sqrt{3}}{1} = -\sqrt{3}$.

> And the secant function is the reciprocal of the cosine function.

7.22 Evaluate $\csc \dfrac{2\pi}{3}$ based on the unit circle.

The cosecant function is the reciprocal of the sine function.

$$\csc \frac{2\pi}{3} = \frac{1}{\sin\left(2\pi/3\right)}$$

According to the unit circle, $\sin \dfrac{2\pi}{3} = \dfrac{\sqrt{3}}{2}$.

$$\csc \frac{2\pi}{3} = \frac{1}{\sin\left(2\pi/3\right)} = \frac{1}{\sqrt{3}/2}$$

Multiply the numerator and denominator by the reciprocal of the denominator.

$$\left(\frac{\dfrac{1}{1}}{\dfrac{\sqrt{3}}{2}}\right)\left(\dfrac{\dfrac{2}{\sqrt{3}}}{\dfrac{2}{\sqrt{3}}}\right) = \frac{\dfrac{2}{\sqrt{3}}}{\dfrac{1}{1}} = \frac{2}{\sqrt{3}}$$

Therefore, $\csc \dfrac{2\pi}{3} = \dfrac{2}{\sqrt{3}}$. Multiply the numerator and denominator by $\sqrt{3}$ to rationalize the expression: $\csc \dfrac{2\pi}{3} = \dfrac{2}{\sqrt{3}}\left(\dfrac{\sqrt{3}}{\sqrt{3}}\right) = \dfrac{2\sqrt{3}}{3}$.

7.23 Identify the reference angle for $\theta = 140°$.

Every angle whose measure is greater than $90°$ has a corresponding reference angle—an acute angle used to calculate its trigonometric values. Each reference angle is formed by the terminal side of the angle and the x-axis, so the reference angle of $\theta = 140°$ is $40°$, as illustrated by Figure 7-4.

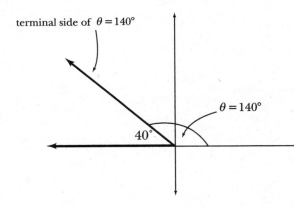

terminal side of $\theta = 140°$

$\theta = 140°$

$40°$

Figure 7-4

The reference angle for $\theta = 140°$ measures $40°$.

If the second-quadrant angle θ is measured in radians, its reference angle is $\alpha = \pi - \theta$.

As Figure 7-4 demonstrates, an obtuse angle θ whose terminal side lies in the second quadrant will have a supplementary reference angle α: $\alpha = 180° - \theta$. In this problem, $\alpha = 180° - 140° = 40°$.

7.24 Identify the reference angle for $\theta = \dfrac{5\pi}{4}$.

The terminal side of angle $\theta = \dfrac{5\pi}{4}$ lies in the third quadrant, so the reference angle is the acute angle formed by its terminal side and the negative x-axis, as illustrated by Figure 7-5.

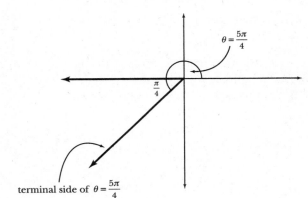

$\theta = \dfrac{5\pi}{4}$

$\dfrac{\pi}{4}$

terminal side of $\theta = \dfrac{5\pi}{4}$

Figure 7-5

The reference angle for $\theta = \dfrac{5\pi}{4}$ measures $\dfrac{\pi}{4}$ radians.

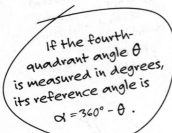

If the third-quadrant angle θ is measured in degrees, its reference $\alpha = \theta - 180°$.

As Figure 7-5 demonstrates, an angle θ whose terminal side lies in the third quadrant will have a reference angle α defined by the formula: $\alpha = \theta - \pi$.

$$\alpha = \frac{5\pi}{4} - \pi = \frac{5\pi}{4} - \frac{4\pi}{4} = \frac{\pi}{4}$$

7.25 Identify the reference angle for $\frac{11\pi}{6}$.

The terminal side of angle $\theta = \frac{11\pi}{6}$ lies in the fourth quadrant, so the reference angle is the acute angle formed by its terminal side and the positive x-axis, as illustrated by Figure 7-6.

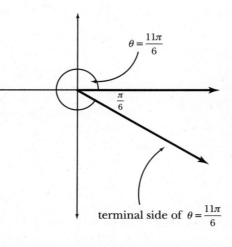

$\theta = \frac{11\pi}{6}$

$\frac{\pi}{6}$

terminal side of $\theta = \frac{11\pi}{6}$

Figure 7-6
The reference angle for $\theta = \frac{11\pi}{6}$ measures $\frac{\pi}{6}$ radians.

If the fourth-quadrant angle θ is measured in degrees, its reference angle is $\alpha = 360° - \theta$.

As Figure 7-6 demonstrates, an angle θ whose terminal side lies in the fourth quadrant has the reference angle $\alpha = 2\pi - \theta$.

$$\alpha = 2\pi - \theta = 2\pi - \frac{11\pi}{6} = \frac{12\pi}{6} - \frac{11\pi}{6} = \frac{\pi}{6}$$

7.26 If $\tan\theta = -\frac{5}{12}$ and $\frac{3\pi}{2} < \theta < 2\pi$, evaluate $\sin\theta$.

In the first quadrant cosine, sine, and tangent are all positive. However, sine's also positive in the second quadrant, tangent's also positive in the third, and cosine is positive in the fourth.

Angles whose tangents are negative have terminal sides that fall either in the second or fourth quadrant, but this problem specifically identifies the fourth quadrant's boundaries: $\frac{3\pi}{2} < \theta < 2\pi$.

Draw a right triangle based on the fourth quadrant reference angle (like the reference angle in Problem 7.25). Although angle θ in Figure 7-7 is a reference angle (and not the original θ from the problem), it has the same trigonometric values.

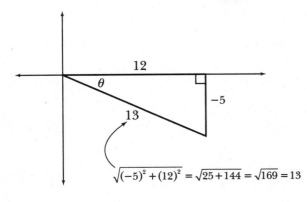

Figure 7-7

Apply the Pythagorean theorem to determine the length of the hypotenuse.

$$\sqrt{(-5)^2 + (12)^2} = \sqrt{25 + 144} = \sqrt{169} = 13$$

Because $\tan \theta = -\frac{5}{12}$, and tangent $= \frac{\text{opposite}}{\text{adjacent}}$, the side opposite θ has length 5 and the side adjacent has length 12. It is important to label the vertical side –5, because you must travel *down* from the y-axis to reach the end of the segment.

Once you've identified the lengths of all three sides of the right triangle, you can evaluate $\sin \theta$. According to Figure 7-7, the side opposite θ equals –5, the side adjacent to θ equals 12, and the hypotenuse equals 13.

$$\sin \theta = \frac{\text{side opposite } \theta}{\text{hypotenuse}} = \frac{-5}{13}$$

Make vertical segments negative in the third and fourth quadrants, and make horizontal segments negative in the second and third quadrants. The hypotenuse is never negative.

7.27 If $\sin \theta = -\dfrac{2}{9}$ and $\tan \theta > 0$, evaluate $\sec \theta$.

Note that $\sin \theta$ is negative in the third and fourth quadrants, and the tangent function is positive ($\tan \theta > 0$) in the first and third quadrants. Therefore, the terminal side of θ must fall in the third quadrant in order to meet both conditions. Draw a right triangle using a third-quadrant reference angle for θ (like the reference angle in Problem 7.24). Calculate the remaining side of the right triangle using the Pythagorean theorem, as illustrated in Figure 7-8.

$$\sqrt{9^2 - (-2)^2} = \sqrt{81 - 4} = \sqrt{77}$$

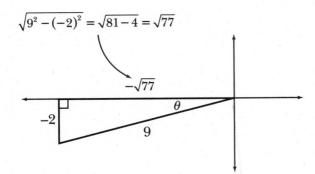

Figure 7-8

Both legs in the right triangle are labeled with negative values because all points in the third quadrant have negative x- and y-values.

The side opposite θ equals –2, the hypotenuse equals 9, and the side adjacent to θ equals $-\sqrt{77}$. In order to calculate $\sec \theta$, you should first calculate its reciprocal: $\cos \theta$.

$$\cos \theta = \frac{\text{side adjacent to } \theta}{\text{hypotenuse}} = \frac{-\sqrt{77}}{9}$$

Take the reciprocal of $\cos \theta$ to calculate $\sec \theta$.

$$\sec \theta = -\frac{9}{\sqrt{77}}$$

Rationalize the denominator.

$$\sec \theta = -\frac{9}{\sqrt{77}}\left(\frac{\sqrt{77}}{\sqrt{77}}\right) = -\frac{9\sqrt{77}}{77}$$

Inverse Trigonometric Functions

Input a number and output an angle for a change

See Problem 4.24.

7.28 The function $f(\theta) = \sin \theta$ is not one-to-one. As evidence, consider $\theta_1 = \frac{\pi}{4}$ and $\theta_2 = \frac{3\pi}{4}$; because $f(\theta_1) = f(\theta_2) = \frac{\sqrt{2}}{2}$, each input does not correspond to a unique output. How, then, can an inverse function exist?

Only one-to-one functions may possess inverses but functions can be defined creatively to ensure they pass the horizontal line test and thus have an inverse. In the case of $y = \sin x$, restricting the domain to $-\frac{\pi}{2} \le \theta \le \frac{\pi}{2}$ produces a graph upon which any horizontal line drawn across the graph intersects it only once (see Figure 7-9).

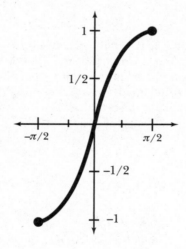

Figure 7-9

Restricting the domain of $y = \sin x$ to $-\frac{\pi}{2} \le \theta \le \frac{\pi}{2}$ ensures that it passes the horizontal line test and is thus a one-to-one function with an inverse.

Arccsc and arctan have the same restricted range as arcsin: $-\frac{\pi}{2} \le \theta \le \frac{\pi}{2}$. The other three inverse trig functions (arccos, arccot, and arcsec) have a restricted range of $0 \le \theta \le \pi$.

Therefore, the inverse function of $f(x)$, usually labeled $f^{-1}(x) = \arcsin x$ or $f^{-1}(x) = \sin^{-1} x$, has a restricted *range* of $-\frac{\pi}{2} \le \arcsin x \le \frac{\pi}{2}$.

7.29 A vacation resort in a mountain town has installed a zip line (a sturdy wire, down which customers in harnesses can quickly descend from high altitudes) to attract patrons. One zip line is 1,750 feet long and allows its rider to descend from a ski slope to the ground, a vertical drop of 450 feet. Calculate the angle of declension of the wire in radians, accurate to three decimal places.

Consider Figure 7-10, which illustrates the given information.

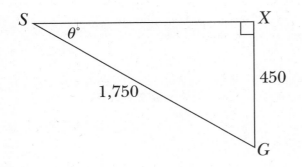

Figure 7-10

The zip line begins on the ski slope at point S and ends at point G on the ground. SG = 1,750 feet (the length of the zip line) and XG = 450 feet, the vertical distance from the ground to the top of the ski slope. The angle of declension of the zip line is θ°.

You're given information about the side opposite θ and the hypotenuse of the right triangle, so apply the sine ratio to calculate θ.

$$\sin\theta = \frac{\text{side opposite } \theta}{\text{hypotenuse}}$$

$$= \frac{450}{1,750}$$

$$= \frac{9}{35}$$

To solve for θ, apply the inverse sine function to both sides of the equation.

$$\arcsin(\sin\theta) = \arcsin\frac{9}{35}$$

$$\theta = \arcsin\frac{9}{35}$$

Use a calculator (in radians mode) to calculate θ: $\theta \approx 0.260$ radians.

That's about 15°.

7.30 Evaluate $\arccos\left(-\frac{1}{2}\right)$.

When you read "arc," think "What angles have this value?" In other words, $\arccos\left(-\frac{1}{2}\right)$ is asking "What angles have a cosine value of $-\frac{1}{2}$?"

Two angles on the unit circle have a cosine value of $-\frac{1}{2}$: $\theta = \frac{2\pi}{3}$ and $\theta = \frac{4\pi}{3}$. However, only $\theta = \frac{2\pi}{3}$ belongs to the restricted range of arccosine ($0 \le \theta \le \pi$), so discard the solution $\frac{4\pi}{3}$.

7.31 Evaluate $\operatorname{arccsc}\left(-\sqrt{2}\right)$.

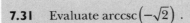

If, $\operatorname{arccsc}\left(-\sqrt{2}\right) = \theta$, then $\csc\theta = -\sqrt{2}$. Recall that $\sin\theta$ is the reciprocal of $\csc\theta$: $\sin\theta = -\frac{1}{\sqrt{2}} = -\frac{\sqrt{2}}{2}$. Therefore, $\operatorname{arccsc}\left(-\sqrt{2}\right) = \arcsin\left(-\frac{\sqrt{2}}{2}\right)$. Two angles on the

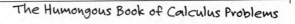

unit circle have a sine value of $-\frac{\sqrt{2}}{2}$: $\theta = \frac{5\pi}{4}$ and $\theta = \frac{7\pi}{4}$. Neither of those angles falls in the restricted range of arcsine $\left(-\frac{\pi}{2} \le \theta \le \frac{\pi}{2}\right)$, but a coterminal angle of $\theta = \frac{7\pi}{4}$ will. Subtract 2π from $\frac{7\pi}{4}$ to identify it.

$$\frac{7\pi}{4} - 2\pi = \frac{7\pi}{4} - \frac{8\pi}{4} = -\frac{\pi}{4}$$

Therefore, $\operatorname{arccsc}\left(-\sqrt{2}\right) = -\frac{\pi}{4}$.

$\theta = \frac{5\pi}{4}$ doesn't have any coterminal angles that fit in the interval: $\frac{5\pi}{4} - 2\pi = -\frac{3\pi}{4}$. Because $\theta = -\frac{3\pi}{4}$ is less than $-\frac{\pi}{2}$, it's too small for the interval, and you have to throw it out as a possible answer.

7.32 Evaluate arctan 1.

Because tangent is defined as the quotient of sine and cosine, the only way angle θ can have a tangent of 1 is if $\sin \theta = \cos \theta$. Consider an angle θ such that $\cos \theta = \sin \theta = c$, where c is a real number. Evaluate $\tan \theta$.

$$\tan \theta = \frac{\sin \theta}{\cos \theta} = \frac{c}{c} = 1$$

There are only two angles on the unit circle whose cosine and sine values are equal: $\theta = \frac{\pi}{4}$ (where $\cos \theta = \sin \theta = \frac{\sqrt{2}}{2}$) and $\theta = \frac{5\pi}{4}$ (where $\cos \theta = \sin \theta = -\frac{\sqrt{2}}{2}$). Of those two solution candidates, only the first falls within the restricted range of arctangent $\left(-\frac{\pi}{2} \le \theta \le \frac{\pi}{2}\right)$, so $\arctan 1 = \frac{\pi}{4}$.

7.33 Evaluate $\arcsin\left(\sin \frac{11\pi}{6}\right)$.

Because $y = \arcsin \theta$ and $y = \sin \theta$ are inverse functions, you may be tempted to report that they cancel one another out, leaving $\frac{11\pi}{6}$ as the answer. Though they are inverse functions, the restrictions placed upon them to ensure they are one-to-one invalidate that approach.

Begin by evaluating the expression inside parentheses: $\sin \frac{11\pi}{6} = -\frac{1}{2}$.

$$\arcsin\left(\sin \frac{11\pi}{6}\right) = \arcsin\left(-\frac{1}{2}\right)$$

Sometimes inverse trig functions WILL behave well and cancel each other out. If you reverse these functions, everything cancels nicely:

$\sin\left(\arcsin \frac{1}{2}\right)$

$= \sin\left(\frac{\pi}{6}\right)$

$= \frac{1}{2}$

Two angles on the unit circle have a sine value of $-\frac{1}{2}$: $\theta = \frac{7\pi}{6}$ and $\theta = \frac{11\pi}{6}$, but neither of those angles fall within the restricted range of arcsine $\left(-\frac{\pi}{2} \le \theta \le \frac{\pi}{2}\right)$. Calculate a coterminal angle for $\theta = \frac{11\pi}{6}$ to find a suitable angle, as demonstrated in Problem 7.31.

$$\frac{11\pi}{6} - 2\pi = \frac{11\pi}{6} - \frac{12\pi}{6} = -\frac{\pi}{6}$$

Therefore, $\arcsin\left(\sin \frac{11\pi}{6}\right) = -\frac{\pi}{6}$.

Chapter 8
TRIGONOMETRIC GRAPHS, IDENTITIES, AND EQUATIONS

Tricky graphs, trig equations, and identity proofs

After you have mastered the rudimentary trigonometric concepts of Chapter 7, you are sufficiently prepared to consider more rigorous, and significantly more useful, trigonometric principles. This chapter begins by extending the process of graphing by transformations (discussed in Chapter 4) to the realm of periodic functions. Deeper in the chapter, you'll manipulate trigonometric identities, in order to simplify trigonometric expressions and to verify identities.

Though many students report that proving trigonometric identities is one of the most memorable topics of a calculus preparation course (due to its foundation in logical proof as opposed to arithmetic fluency), far more useful is the ability to solve trigonometric equations, so the chapter will culminate accordingly.

Even though trig functions are slightly more complicated than the functions from Chapter 4, they'll still follow the same basic transformation rules for shifting up and down, left and right, reflecting about axes, blah blah blah, the whole bit. Next you'll look at trig identities (really short equations that are always true, no matter what you plug into the variable). You'll have to memorize a bunch of them before you get started, so I listed the major ones in Appendix C at the end of the book. Finally, you'll use the inverse trig function skills from Chapter 7 to solve some trig equations.

Graphing Trigonometric Transformations

Stretch and shift wavy graphs

8.1   Sketch the graph of $f(\theta) = 3\sin\theta - 1$.

> You can find all the major trig graphs (sine, cosine, tangent, cotangent, secant, and cosecant) in Appendix A, at the back of the book.

Transforming the standard sine graph ($y = \sin\theta$) into $f(\theta) = 3\sin\theta - 1$ requires two steps: (1) multiply $\sin\theta$ by 3 (which stretches the graph to heights three times as high and as low as the original graph); and (2) subtract 1 from $3\sin\theta$ (which shifts the entire graph down one unit). Both $y = \sin\theta$ and $f(\theta)$ are graphed in Figure 8-1.

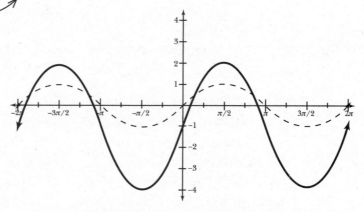

Figure 8-1   *The dotted graph of $y = \sin\theta$, and the solid graph of $f(\theta) = 3\sin\theta - 1$.*

8.2   The graph of the function $g(\theta) = a\sin(b \cdot \theta + c) + d$ has an amplitude of $|a|$. Verify this is true for the graph of $f(\theta) = 3\sin\theta - 1$ in Figure 8-1.

> If you drew a horizontal line through the middle of a periodic graph, the amplitude of the graph would be the distance from that line to the highest and lowest points on the curve.

To determine the amplitude of a periodic graph, calculate the difference between its highest and lowest y-values, divide the result by 2, and then take the absolute value. The graph of $f(\theta)$ in Figure 8-1 reaches a maximum height of 2 and a minimum height of -4.

$$\text{amplitude} = \left| \frac{\text{max height} - \text{min height}}{2} \right|$$
$$= \left| \frac{2 - (-4)}{2} \right|$$
$$= \left| \frac{6}{2} \right|$$
$$= 3$$

Therefore, the amplitude of the graph is, indeed, equal to the coefficient of the trigonometric function: $|3| = 3$.

8.3 Calculate the period of $f(\theta) = -5 \sin(2\theta + 1) - 6$.

To calculate the period of a periodic function, divide the period of the untransformed function by the coefficient of the variable in the transformed version. In this example, the untransformed function $y = \sin\theta$ has a period of 2π. To calculate the period of $f(\theta) = -5 \sin(2\theta + 1) - 6$, divide 2π by the coefficient of θ.

$$\text{new period} = \frac{\text{untransformed period}}{\text{coefficient of } \theta}$$

$$= \frac{2\pi}{2}$$

$$= \pi$$

A periodic graph repeats itself over and over. Each repetition of the graph measures the same horizontal distance from beginning to end, which is called the period of the graph.

8.4 Sketch the graph of $f(\theta) = \left|\tan\left(\theta - \dfrac{\pi}{2}\right)\right|$.

In order to transform the graph of $y = \tan\theta$ into $f(\theta)$, you must subtract $\dfrac{\pi}{2}$ from the argument of the function (which results in a phase shift of $\dfrac{\pi}{2}$ units to the right). Then, take the absolute value of the result (which reflects any portion of the graph for which $y < 0$ across the x-axis). Figure 8-2 contains both the untransformed graph of $y = \tan\theta$ and the graph of $f(\theta) = \left|\tan\left(\theta - \dfrac{\pi}{2}\right)\right|$.

"Argument" is a fancy way of saying "whatever's plugged into the function."

Fancy way to say "horizontal shift."

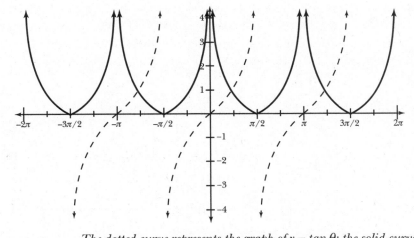

Figure 8-2 *The dotted curve represents the graph of $y = \tan\theta$; the solid curve is the graph of $f(\theta) = \left|\tan\left(\theta - \dfrac{\pi}{2}\right)\right|$.*

8.5 Sketch the graph of $g(\theta) = -\sec(-\theta)$.

In order to transform the graph of $y = \sec\theta$ into $g(\theta)$, you first multiply the argument of the function by -1, which reflects the graph across the y-axis. Notice that this has no effect on the graph, because it is symmetric about the y-axis. However, the second transformation will affect the graph—multiplying a function by -1 reflects its graph across the x-axis. Figure 8-3 contains the untransformed graph of $y = \sec\theta$ and the graph of $g(\theta) = -\sec(-\theta)$.

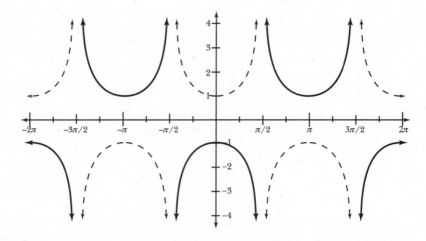

Here are the periods of all six trig functions: sine = 2π, cosine = 2π, tangent = π, cotangent = π, secant = 2π, and cosecant = 2π.

Figure 8-3   *The dotted curve represents the graph of y = sec θ; the solid curve is the graph of g(θ) = –sec (–θ).*

8.6   Calculate the period and amplitude of $h(\theta) = -4\cos(3\theta)$ and sketch the graph over the interval $[-2\pi, 2\pi]$.

The amplitude equals the absolute value of the function's coefficient: $|-4| = 4$; this changes the range of the graph from $[-1,1]$ to $[-4,4]$, and the negative sign reflects the graph about the *x*-axis. The period of $h(\theta)$ equals the original period of cosine (2π) divided by the coefficient of θ: $\frac{2\pi}{3}$; because of this transformation, three full periods of $h(\theta)$ fit into the same interval as a single period of $y = \cos\theta$. The graph of $h(\theta)$ appears in Figure 8-4.

Just like Problem 8.3, the coefficient of θ tells you how many periods of the new graph fit into one period of the old graph.

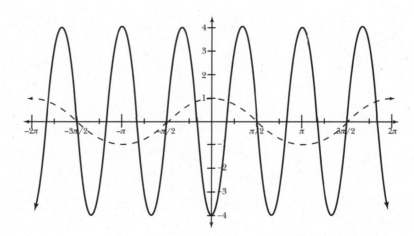

Figure 8-4   *The dotted curve represents the graph of y = cos θ; the solid curve is the graph of h(θ) = –4 cos (3θ).*

8.7 The function $f(\theta) = a \sin (b \cdot \theta) + c$ is graphed in Figure 8-5. Determine the values of a, b, and c.

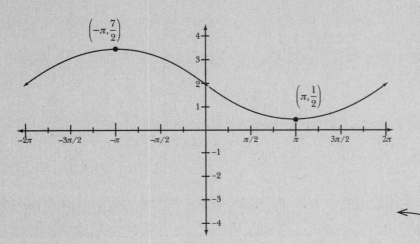

Figure 8-5 *The graph of $f(\theta) = a \sin (b \cdot \theta) + c$.*

Use the formula from Problem 8.2 to calculate the amplitude.

$$\text{amplitude} = \left| \frac{\text{max height} - \text{min height}}{2} \right| = \left| \frac{7/2 - 1/2}{2} \right| = \frac{3}{2}$$

This formula has one limitation—it cannot determine whether or not a should be positive or negative. Note that $y = \sin \theta$ increases at $\theta = 0$ but $f(\theta)$ decreases there, indicating that the graph has been reflected across the x-axis. Therefore, $a < 0$: $a = -\frac{3}{2}$.

The period of $f(\theta)$ is larger than 2π (the period of $y = \sin \theta$). Specifically, one period of the graph stretches from -2π to 2π; calculate the difference of those θ-values and take the absolute value to determine the period of $f(\theta)$.

$$\text{period} = |2\pi - (-2\pi)| = |4\pi| = 4\pi$$

Use this result and the formula from Problem 8.3 to determine the value of b.

$$\text{new period} = \frac{\text{untransformed period}}{\text{coefficient of } \theta}$$

$$4\pi = \frac{2\pi}{b}$$

$$4\pi b = 2\pi$$

$$b = \frac{2\cancel{\pi}}{4\cancel{\pi}}$$

$$b = \frac{1}{2}$$

To determine the value of c, take the average of the maximum and minimum values of the graph.

$$c = \frac{7/2 + 1/2}{2} = \frac{8/2}{2} = \frac{4}{2} = 2$$

Therefore, $a = -\frac{3}{2}$, $b = \frac{1}{2}$, and $c = 2$.

> Figure 8-5 is also the graph of $y = \frac{3}{2} \sin \left(\frac{1}{2} \theta \pm \pi \right) + 2$. You can move the sine graph π units left (or right) instead of flipping it upside down with a negative a-value.

> The max and the min of this graph are $\frac{3}{2}$ units (the amplitude) above and below $c = 2$.

Applying Trigonometric Identities
Simplify expressions and prove identities

8.8   Simplify the expression: $\cos(-\theta) \cdot \csc(-\theta)$.

Cosine is an even function, so $\cos(-\theta) = \cos\theta$, but cosecant is odd, so $\csc(-\theta) = -\csc\theta$.

$$\cos(-\theta) \cdot \csc(-\theta) = \cos\theta \cdot (-\csc\theta)$$

The cosecant function is defined as the reciprocal of the sine function.

$$= \frac{\cos\theta}{1} \cdot \left(-\frac{1}{\sin\theta}\right)$$

$$= -\frac{\cos\theta}{\sin\theta}$$

$$= -\cot\theta$$

Therefore, $\cos(-\theta) \cdot \csc(-\theta) = -\cot\theta$.

> These are even/odd identities. Check out the full list of the identities you need to know in Appendix C.

8.9   Simplify the expression: $\dfrac{\sin^2(\pi/2 - \theta)}{\sec(-\theta)}$.

Apply the cofunction identity $\sin\left(\dfrac{\pi}{2} - \theta\right) = \cos\theta$ to the numerator; the denominator is equal to $\sec\theta$, as secant is an even function.

$$\frac{\sin^2\left(\dfrac{\pi}{2} - \theta\right)}{\sec(-\theta)} = \frac{\cos^2\theta}{\sec\theta}$$

$$= \frac{\cos^2\theta}{1/\cos\theta}$$

Reduce the complex fraction by multiplying its numerator and denominator by $\cos\theta$, the reciprocal of the denominator.

$$= \frac{\dfrac{\cos^2\theta}{1}}{\dfrac{1}{\cos\theta}} \cdot \left(\frac{\dfrac{\cos\theta}{1}}{\dfrac{1}{\cos\theta}}\right)$$

$$= \cos^3\theta$$

Therefore, $\dfrac{\sin^2(\pi/2 - \theta)}{\sec(-\theta)} = \cos^3\theta$.

> If you multiply $\left(\frac{\pi}{2} - \theta\right)$ by -1, you get $\left(\theta - \frac{\pi}{2}\right)$.

8.10   Simplify the expression: $\cot\left(\theta - \dfrac{\pi}{2}\right) \cdot \tan\left(\dfrac{\pi}{2} - \theta\right)$.

The argument of the cotangent function is the opposite of the argument in the cofunction identities. To remedy this, factor -1 out of the argument.

$$\cot\left(\theta - \frac{\pi}{2}\right) \cdot \tan\left(\frac{\pi}{2} - \theta\right) = \cot\left[-\left(-\theta + \frac{\pi}{2}\right)\right] \cdot \tan\left(\frac{\pi}{2} - \theta\right)$$

$$= \cot\left[-\left(\frac{\pi}{2} - \theta\right)\right] \cdot \tan\left(\frac{\pi}{2} - \theta\right)$$

Since cotangent is an odd function $\cot(-\alpha) = -\cot\alpha$. In this case, $\alpha = \frac{\pi}{2} - \theta$.

$$= -\cot\left(\frac{\pi}{2} - \theta\right) \cdot \tan\left(\frac{\pi}{2} - \theta\right)$$

The tangent and cotangent functions (of the same angle) are reciprocals.

$$\cot\left(\theta - \frac{\pi}{2}\right) \cdot \tan\left(\frac{\pi}{2} - \theta\right) = -1 \ .$$

Rewrite $\cot\theta$ as the reciprocal of $\tan\theta$.

$$\cot\theta + \tan\theta = \frac{1}{\tan\theta} + \tan\theta$$

8.11 Simplify the expression: $\cot\theta + \tan\theta$.

Add the terms together using common denominators.

$$= \frac{1}{\tan\theta} + \frac{\tan\theta}{1} \cdot \frac{\tan\theta}{\tan\theta}$$

$$= \frac{1}{\tan\theta} + \frac{\tan^2\theta}{\tan\theta}$$

$$= \frac{1 + \tan^2\theta}{\tan\theta}$$

According to a Pythagorean identity, $1 + \tan^2\theta = \sec^2\theta$.

$$= \frac{\sec^2\theta}{\tan\theta}$$

Dividing by a quantity is the equivalent of multiplying by its reciprocal; the reciprocal of $\tan\theta$ is $\cot\theta$.

$$= \sec^2\theta \cdot \cot\theta$$

Rewrite the expression in terms of sine and cosine and simplify.

$$= \frac{1}{\cos^2\theta} \cdot \frac{\cos\theta}{\sin\theta}$$

$$= \frac{\cancel{\cos\theta}}{\cancel{\cos\theta}\cos\theta\sin\theta}$$

$$= \frac{1}{\cos\theta\sin\theta}$$

> When reducing a fraction leaves nothing behind in the numerator (or in the denominator, for that matter), make sure you write 1 (not 0) in the empty spot.

Therefore, $\cot\theta + \tan\theta = \dfrac{1}{\cos\theta\sin\theta}$.

8.12 Simplify the expression: $1 - 3\sin^2\theta + 2\sin^4\theta$.

Factor the expression, just as you would factor $1 - 3x + 2x^2$ (in this case, $x = \sin^2\theta$).

$$1 - 3\sin^2\theta + 2\sin^4\theta = (1 - \sin^2\theta)(1 - 2\sin^2\theta)$$

> Because $1 - 3x + 2x^2$ factors into $(1 - x)(1 - 2x)$, just plug in $x = \sin^2\theta$ to factor the trig expression.

According to a Pythagorean identity, $\cos^2 \theta + \sin^2 \theta = 1$. If you subtract $\sin^2 \theta$ from both sides of that identity, you get $\cos^2 \theta = 1 - \sin^2 \theta$; use that identify to replace the left factor. According to a double angle identity, $\cos 2\theta = 1 - 2\sin^2 \theta$; use this to replace the right factor with $\cos 2\theta$.

$$= (\cos^2 \theta)(\cos 2\theta)$$

Therefore, $1 - 3\sin^2 \theta + 2\sin^4 \theta = (\cos^2 \theta)(\cos 2\theta)$.

8.13 Simplify the expression: $\cot \theta \div \dfrac{1}{2}\csc^2 \theta$.

You can replace cos 2x with one of three things: $\cos^2 x - \sin^2 x$, $2\cos^2 x - 1$, or $1 - 2\sin^2 x$, as indicated in Appendix C. You'll use different versions of cos 2x for different problems.

Rewrite the quotient as a complex fraction in terms of $\sin \theta$ and $\cos \theta$.

$$\frac{\cot \theta}{\dfrac{\csc^2 \theta}{2}} = \frac{\dfrac{\cos \theta}{\sin \theta}}{\dfrac{1}{2\sin^2 \theta}}$$

Reduce the complex fraction using the method described in Problem 8.9.

$$\frac{\dfrac{\cos \theta}{\sin \theta}}{\dfrac{1}{2\sin^2 \theta}} \left(\frac{\dfrac{2\sin^2 \theta}{1}}{\dfrac{2\sin^2 \theta}{1}} \right) = \frac{\dfrac{2\cos \theta \sin^2 \theta}{\sin \theta}}{\dfrac{2\sin^2 \theta}{2\sin^2 \theta}} = \frac{2 \cdot \cos \theta \cdot \sin \theta \cdot \sin \theta}{\sin \theta} = 2\cos \theta \sin \theta$$

According to a double-angle identity, $2\sin \theta \cos \theta = \sin 2\theta$. Therefore, $\cot \theta \div \dfrac{1}{2}\csc^2 \theta = \sin 2\theta$.

8.14 Factor and simplify the expression: $\cos^4 \theta - \sin^4 \theta$.

Factor the difference of perfect squares.

$$\cos^4 \theta - \sin^4 \theta = (\cos^2 \theta + \sin^2 \theta)(\cos^2 \theta - \sin^2 \theta)$$

According to a Pythagorean identity, $\cos^2 \theta + \sin^2 \theta = 1$, and a double angle identity states that $\cos 2\theta = \cos^2 \theta - \sin^2 \theta$. Substitute those values into the expression.

$$= (1)(\cos 2\theta)$$

Therefore, $\cos^4 \theta - \sin^4 \theta = \cos 2\theta$.

8.15 Generate the identity $\cos^2 \theta + \sin^2 \theta = 1$ by examining an acute, positive angle θ graphed in standard position and its terminal side's intersection point with the unit circle.

That's because you have to go right $\cos \theta$ units and up $\sin \theta$ units from the origin to reach the point $(\cos \theta, \sin \theta)$.

Draw θ and the unit circle described by the Problem, as illustrated by Figure 8-6. An angle in standard position intersects the unit circle at the point $(\cos \theta, \sin \theta)$. Thus, the horizontal leg of the right triangle in Figure 8-6 has length $\cos \theta$ and the vertical leg has length $\sin \theta$.

According to the Pythagorean theorem, the sum of the squares of a right triangle's legs is equal to the square of its hypotenuse, so $\cos^2 \theta + \sin^2 \theta = 1$. It is no coincidence that the identity is classified as a Pythagorean identity, as its proof depends on the Pythagorean theorem.

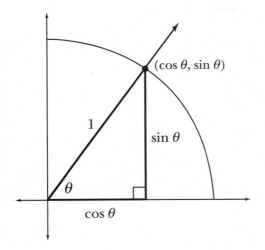

Figure 8-6

A right triangle created by the intersection point of θ's terminal side and the unit circle.

8.16 Verify the sine double angle identity $\sin 2\theta = 2 \sin \theta \cos \theta$ using a sum or difference identity.

The sum-to-product formula for sine is $\sin (a \pm b) = \sin a \cos b \pm \cos a \sin b$. Rewrite $\sin 2\theta$ as $\sin (\theta + \theta)$ and apply the sum formula for sine. (Note that $a = b = \theta$ in this example, but a and b need not be equal for the formula to apply.)

$$\sin(\theta + \theta) = \sin\theta\cos\theta + \cos\theta\sin\theta$$
$$= \sin\theta\cos\theta + \sin\theta\cos\theta$$
$$= 2\sin\theta\cos\theta$$

8.17 Evaluate $\cos \dfrac{5\pi}{12}$ without a calculator, given $\dfrac{5\pi}{12} = \dfrac{2\pi}{3} - \dfrac{\pi}{4}$.

The only trigonometric values you are normally expected to memorize are those on the unit circle, so $\dfrac{5\pi}{12}$ is a troublesome angle until you note that it is equal to the difference of two unit circle angles: $\dfrac{5\pi}{12} = \dfrac{2\pi}{3} - \dfrac{\pi}{4}$. Apply the cosine difference formula for cosine: $\cos(a \pm b) = \cos a \cos b \mp \sin a \sin b$.

> The upside-down "∓" sign means you should write the opposite of the "∓" sign earlier in the formula.

$$\cos\frac{5\pi}{12} = \cos\left(\frac{2\pi}{3} - \frac{\pi}{4}\right)$$
$$= \cos\frac{2\pi}{3}\cos\frac{\pi}{4} + \sin\frac{2\pi}{3}\sin\frac{\pi}{4}$$
$$= \left(-\frac{1}{2}\right)\left(\frac{\sqrt{2}}{2}\right) + \left(\frac{\sqrt{3}}{2}\right)\left(\frac{\sqrt{2}}{2}\right)$$
$$= \frac{\sqrt{6} - \sqrt{2}}{4}$$

8.18 Verify the identity: $2 - \cos^2 \theta - (1 - \cos^2 \theta) = 1$.

Problems 8.18–8.21 are different than Problems 8.8–8.17, because they contain equations, not expressions.

Expand and simplify the expression.

$$2 - \cancel{\cos^2 \theta} - 1 + \cancel{\cos^2 \theta} = 1$$
$$2 - 1 = 1$$
$$1 = 1$$

By applying valid operations to the equation, you have demonstrated that the equation $2 - \cos^2 \theta - (1 - \cos^2 \theta) = 1$ is equivalent to the equation $1 = 1$. Because the latter equation is always true, so is the former, regardless of the θ-value substituted into the equation.

8.19 Verify the identity: $\tan \theta + \tan^3 \theta = \dfrac{\sec^2 \theta}{\cot \theta}$.

Factor $\tan \theta$ out of the left side of the equation.

$$\tan \theta \left(1 + \tan^2 \theta \right) = \frac{\sec^2 \theta}{\cot \theta}$$

According to a Pythagorean identity, $1 + \tan^2 \theta = \sec^2 \theta$.

$$\tan \theta \left(\sec^2 \theta \right) = \frac{\sec^2 \theta}{\cot \theta}$$

Cross multiply to eliminate fractions.

$$(\tan \theta)(\sec^2 \theta)(\cot \theta) = \sec^2 \theta$$

Note that $\cot \theta$ is the reciprocal of $\tan \theta$.

$$\cancel{\tan \theta} \cdot \sec^2 \theta \cdot \frac{1}{\cancel{\tan \theta}} = \sec^2 \theta$$
$$1 \cdot \sec^2 \theta = \sec^2 \theta$$

8.20 Verify the identity: $\sin^2 \theta \cos \theta = \cos \theta - \cos^3 \theta$.

The greatest common factor on the right side of the equation is $\cos \theta$.

$$\sin^2 \theta \cos \theta = \cos \theta (1 - \cos^2 \theta)$$

Subtract $\cos^2 \theta$ from both sides of the Pythagorean identity $\cos^2 \theta + \sin^2 \theta = 1$ to generate an equally valid identity: $\sin^2 \theta = 1 - \cos^2 \theta$. Use this identity to replace the parenthetical quantity on the right side of the equation.

$$\sin^2 \theta \cos \theta = \cos \theta (\sin^2 \theta)$$

Multiplication is commutative, so the identity is verified.

It doesn't matter if you write $\sin^2 \theta$ before or after $\cos \theta$—you get the same product.

8.21 Verify the identify: $\dfrac{\sin 2\theta - \cos \theta (\sin \theta - 1)}{\cos \theta} = \sin \theta + \sec^2 \theta - \tan^2 \theta$.

According to a Pythagorean identity, $1 + \tan^2 \theta = \sec^2 \theta$. Subtract $\tan^2 \theta$ from both sides of that identity to generate an equally valid identity: $1 = \sec^2 \theta - \tan^2 \theta$; use it to rewrite the right side of the identity.

$$\frac{\sin 2\theta - \cos\theta(\sin\theta - 1)}{\cos\theta} = \sin\theta + 1$$

Cross multiply to eliminate the fraction.

$$\sin 2\theta - \cos\theta(\sin\theta - 1) = \cos\theta(\sin\theta + 1)$$

$$\sin 2\theta - \cos\theta\sin\theta + \cos\theta = \cos\theta\sin\theta + \cos\theta$$

Isolate $\sin 2\theta$ by adding $\cos\theta\sin\theta$ to, and subtracting $\cos\theta$ from, both sides of the equation.

$$\sin 2\theta = \cos\theta\sin\theta + \cos\theta\sin\theta + \cos\theta - \cos\theta$$

$$\sin 2\theta = 2\cos\theta\sin\theta$$

> You can stop working as soon as you get something that's definitely true, like a basic equation (0 = 0) or an identity (like sin 2θ = 2 sin θ cos θ in this problem).

Solving Trigonometric Equations

Solve for THETA instead of x

8.22 Find all solutions to the equation $3(\sin\theta + 1) = \sin\theta + 4$ on the interval $[0, 2\pi)$.

Distribute the constant and isolate $\sin\theta$ on the left side of the equation.

$$3\sin\theta + 3 = \sin\theta + 4$$

$$3\sin\theta - \sin\theta = 4 - 3$$

$$2\sin\theta = 1$$

$$\sin\theta = \frac{1}{2}$$

> This problem is asking for all the angles on the unit circle that have a sine value of $\frac{1}{2}$, no matter how many there are or what quadrant they're in.

You are instructed to identify all solutions on the interval $[0, 2\pi)$; there are two such angles whose sine value equals $\frac{1}{2}$: $\theta = \frac{\pi}{6}$ and $\theta = \frac{5\pi}{6}$.

8.23 Calculate the exact solution to the equation: $\sqrt{3}\cot\theta - 1 = 0$.

> The exact solution is the single answer you get from an inverse trig function that's on the restricted ranges discussed in Problem 7.28.

Isolate $\cot\theta$ on the left side of the equation.

$$\sqrt{3}\cot\theta = 1$$

$$\cot\theta = \frac{1}{\sqrt{3}}$$

Solve for θ by applying the inverse cotangent function.

$$\text{arccot}(\cot\theta) = \text{arccot}\frac{1}{\sqrt{3}}$$

$$\theta = \text{arccot}\frac{1}{\sqrt{3}}$$

In order to better understand the solution, multiply the numerator and denominator by $\frac{1}{2}$. Recall that the cotangent is defined as the quotient of the cosine and sine functions.

$$\frac{\cos\theta}{\sin\theta} = \frac{1/2}{\sqrt{3}/2}$$

The only angle θ with a cosine value of $\frac{1}{2}$ and a sine value of $\frac{\sqrt{3}}{2}$ on the restricted cotangent range of $[0,\pi]$ is $\theta = \frac{\pi}{3}$.

8.24 Find the general solution to the equation: $4\cos^2\theta - 2 = 0$.

Isolate $\cos^2\theta$ on the left side of the equation.

$$4\cos^2\theta = 2$$
$$\cos^2\theta = \frac{1}{2}$$

Take the square root of both sides of the equation.

$$\sqrt{\cos^2\theta} = \pm\sqrt{\frac{1}{2}}$$
$$\cos\theta = \pm\frac{1}{\sqrt{2}}$$

> Don't forget the "±" sign when you take an even root of both sides of an equation.

Rationalize the fraction and the constant will mirror unit circle values.

$$\cos\theta = \pm\frac{1}{\sqrt{2}}\cdot\frac{\sqrt{2}}{\sqrt{2}} = \pm\frac{\sqrt{2}}{2}$$

List all the angles on one period of cosine, $[0,2\pi)$, whose cosine is either $\frac{\sqrt{2}}{2}$ or $-\frac{\sqrt{2}}{2}$.

$$\theta = \frac{\pi}{4}, \frac{3\pi}{4}, \frac{5\pi}{4}, \frac{7\pi}{4}$$

> To get coterminal angles, you just add or subtract 2π over and over again (see Problem 7.14). That's what "+ $2k\pi$" means— add or subtract 2π to your heart's content to get an infinite number of answers.

Indicate that these angles and all of their coterminal angles are valid solutions.

$$\theta = \frac{\pi}{4} + 2k\pi, \frac{3\pi}{4} + 2k\pi, \frac{5\pi}{4} + 2k\pi, \frac{7\pi}{4} + 2k\pi, \text{ where } k \text{ is an integer}$$

8.25 Identify all the points at which the graphs of $f(\theta) = \sin\theta$ and $g(\theta) = \csc\theta$ intersect.

Set the functions equal and use a reciprocal identity to rewrite $\csc\theta$.

$$\sin\theta = \csc\theta$$
$$\sin\theta = \frac{1}{\sin\theta}$$

Cross multiply and solve for θ.

$$\sin^2\theta = 1$$
$$\sqrt{\sin^2\theta} = \pm\sqrt{1}$$
$$\sin\theta = \pm 1$$
$$\theta = \frac{\pi}{2} + 2k\pi, \frac{3\pi}{2} + 2k\pi$$

> Since you're supposed to find ALL intersection points, you should include ALL the coterminal angles of the solutions, like in Problem 8.24.

The functions intersect at $\theta = \frac{\pi}{2}$ (where $f(\theta) = g(\theta) = 1$), $\theta = \frac{3\pi}{2}$ (where $f(\theta) = g(\theta) = -1$), and the infinitely many coterminal angles:

$$\left\{\cdots, \left(-\frac{7\pi}{2}, 1\right), \left(-\frac{5\pi}{2}, -1\right), \left(-\frac{3\pi}{2}, 1\right), \left(-\frac{\pi}{2}, -1\right), \left(\frac{\pi}{2}, 1\right), \left(\frac{3\pi}{2}, -1\right), \left(\frac{5\pi}{2}, 1\right), \left(\frac{7\pi}{2}, -1\right), \cdots\right\}$$

8.26 Calculate the exact solution(s) to the equation: $\tan^2 \theta = \tan \theta$.

Set the equation equal to 0 and factor.

$$\tan^2 \theta - \tan \theta = 0$$
$$\tan \theta (\tan \theta - 1) = 0$$

Set each factor equal to 0 and solve the individual equations; ensure that you only include solutions from the restricted range of arctangent, as the problem specifies *exact* solutions.

$$\tan \theta = 0 \qquad\qquad \tan \theta - 1 = 0$$
$$\theta = \arctan 0 \qquad\qquad \tan \theta = 1$$
$$\theta = 0 \qquad \text{or} \qquad \theta = \arctan 1$$
$$\theta = \frac{\pi}{4}$$

The solutions to the equation are $\theta = 0$ or $\theta = \dfrac{\pi}{4}$.

8.27 Calculate the exact solution(s) to the equation: $3 \cos^2 \theta - 10 \cos \theta + 3 = 0$.

Factor the expression and set each factor equal to 0.

$$(\cos \theta - 3)(3 \cos \theta - 1) = 0$$
$$\cos \theta - 3 = 0 \qquad\qquad 3 \cos \theta - 1 = 0$$
$$\cos \theta = 3 \qquad \text{or} \qquad \cos \theta = \frac{1}{3}$$
$$\theta = \arccos 3 \qquad\qquad \theta = \arccos \frac{1}{3}$$

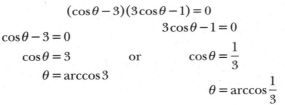

Think about the graph $y = \cos \theta$. It stretches from $y = -1$ to $y = 1$. Not only doesn't it reach 3, it never gets higher than 1!

The domain of arccosine (which is also the range of cosine, its inverse function) is $[-1,1]$; note that 3 does not belong to that interval, so $\theta = \mathrm{arcccos}\ 3$ is an invalid solution. However, $-1 \le \frac{1}{3} \le 1$, so $\theta = \arccos \frac{1}{3} \approx 1.231$ is a valid solution.

8.28 Calculate the exact solution(s) to the equation $2 \tan^2 \theta - \tan \theta - 5 = 0$, accurate to three decimal places.

This equation cannot be factored, so apply the quadratic formula instead.

$$\tan \theta = \frac{-(-1) \pm \sqrt{(-1)^2 - 4(2)(-5)}}{2(2)}$$
$$\tan \theta = \frac{1 \pm \sqrt{41}}{4}$$
$$\theta = \arctan \left(\frac{1 - \sqrt{41}}{4} \right), \ \arctan \left(\frac{1 + \sqrt{41}}{4} \right)$$

Don't round here— use all of the decimals your calculator spits out. The more you round in this step, the more inaccurate your final answer will be.

Use a calculator to evaluate the inverse tangent functions.

$$\theta \approx \arctan(-1.35078105935821) \qquad \theta \approx \arctan(1.85078105935821)$$
$$\theta \approx -0.933 \qquad \text{or} \qquad \theta \approx 1.075$$

8.29 Calculate the exact solution(s) to the equation $3 \sin^2 \theta = -3 \sin \theta + 1$, accurate to three decimal places.

Set the equation equal to 0: $3 \sin^2 \theta + 3 \sin \theta - 1 = 0$. Because the expression is not factorable, solve using the quadratic formula.

$$\sin \theta = \frac{-3 \pm \sqrt{9 - 4(3)(-1)}}{2(3)}$$

$$\sin \theta = \frac{-3 \pm \sqrt{21}}{6}$$

$$\theta = \arcsin\left(\frac{-3 - \sqrt{21}}{6}\right) \qquad \theta = \arcsin\left(\frac{-3 + \sqrt{21}}{6}\right)$$

$$\theta \approx \arcsin(-1.26376261583) \quad \text{or} \quad \theta \approx \arcsin(0.263762615826)$$

$$\text{no solution} \qquad\qquad\qquad \theta \approx 0.267$$

Note that arcsin (-1.26376261583) does not exist because the domain of $y = \arcsin \theta$ is $[-1,1]$ and $-1.26376261583 < -1$.

8.30 Identify all solution(s) to the equation $\tan 2\theta + \sqrt{3} = 0$ on the interval $\left(-\frac{\pi}{2}, \frac{\pi}{2}\right)$.

Notice that this equation contains the double angle 2θ within the trigonometric function. Begin by isolating the trigonometric expression.

$$\tan 2\theta = -\sqrt{3}$$

It is sometimes helpful to divide the numerator and denominator of the constant by $\frac{1}{2}$ when solving equations involving tangent and cotangent.

$$\tan 2\theta = -\frac{\sqrt{3}}{1} = -\frac{\sqrt{3}/2}{1/2}$$

As $\tan \theta = \dfrac{\sin \theta}{\cos \theta}$, identify the angle θ such that $\cos \theta = \dfrac{1}{2}$ and $\sin \theta = -\dfrac{\sqrt{3}}{2}$.

$$2\theta = -\frac{\pi}{3}$$

> *In other words, add the period of tangent (π) to the answer.*

Because the coefficient of θ is 2, you should list *twice* as many solutions. Whereas this equation normally has only one solution on $\left(-\frac{\pi}{2}, \frac{\pi}{2}\right)$, you should include one more—the smallest coterminal angle that's greater than the solution $-\frac{\pi}{3}$.

$$-\frac{\pi}{3} + \pi = \frac{2\pi}{3}$$

List the original solution and its coterminal angle.

$$2\theta = -\frac{\pi}{3}, \frac{2\pi}{3}$$

Multiply the entire equation by $\frac{1}{2}$ to isolate θ and thereby solve the equation.

$$\frac{1}{2}(2\theta) = \left(-\frac{\pi}{3}\right)\left(\frac{1}{2}\right), \left(\frac{2\pi}{3}\right)\left(\frac{1}{2}\right)$$

$$\theta = -\frac{\pi}{6}, \frac{\pi}{3}$$

8.31 Identify all solution(s) to the equation $3(\sin 3\theta + 1) - 5 = -2$ on the interval $[0,2\pi)$.

Isolate the trigonometric function on the left side of the equation.

$$3\sin 3\theta + 3 - 5 = -2$$
$$3\sin 3\theta - 2 = -2$$
$$3\sin 3\theta = 0$$
$$\sin 3\theta = 0$$

Identify all of the angles on the unit circle that have a sine value of 0.

$$3\theta = 0, \pi \quad \leftarrow$$

> Even though $\sin 2\pi$ also equals 0, the problem says you've got to stick to the interval $[0,2\pi)$. 2π is not included because of the parenthesis next to it.

Rather than two answers, list three times as many (because the coefficient of θ is 3). Calculate two coterminal angles for $\theta = 0$ and two coterminal angles for $\theta = \pi$.

$$3\theta = 0, \pi, 2\pi, 3\pi, 4\pi, 5\pi$$

Do not be concerned that these answers are out of the interval dictated by the problem; when you divide each by 3 to solve for θ, they are bounded correctly.

$$\theta = 0, \frac{\pi}{3}, \frac{2\pi}{3}, \pi, \frac{4\pi}{3}, \frac{5\pi}{3}$$

> This time add 2π to each angle since that's the period of sine.

8.32 Determine the general solution to the equation: $\tan^2 \theta - 4\sec \theta = -5$.

Rewrite the equation in terms of a single trigonometric function. To accomplish this, apply a Pythagorean identity: if $1 + \tan^2 \theta = \sec^2 \theta$, then $\tan^2 \theta = \sec^2 \theta - 1$.

$$\left(\sec^2 \theta - 1\right) - 4\sec \theta = -5$$
$$\sec^2 \theta - 4\sec \theta + 4 = 0$$

Factor and solve for θ.

$$(\sec \theta - 2)(\sec \theta - 2) = 0$$
$$\sec \theta = 2$$

Recall that secant and cosine are reciprocal functions.

$$\cos \theta = \frac{1}{2}$$

Find all angles on $[0,2\pi)$, one period of cosine, and express the solution in general form, as directed by the problem.

$$\theta = \frac{\pi}{3} + 2k\pi, \frac{5\pi}{3} + 2k\pi$$

> This procedure can introduce false answers, so make sure you plug your results back into the original equation and toss out anything that doesn't work.

8.33 Determine the general solution to the equation: $1 - \sin \theta = \cos \theta$.

It is easier to rewrite an entire equation in terms of a single trigonometric function when at least one of the functions is squared, because it allows you to apply Pythagorean identities. To introduce squared functions, square both sides of the equation.

$$(1 - \sin\theta)^2 = (\cos\theta)^2$$
$$1 - 2\sin\theta + \sin^2\theta = \cos^2\theta$$

According to a Pythagorean identity, $\cos^2\theta + \sin^2\theta = 1$; therefore, $\cos^2\theta = 1 - \sin^2\theta$. Use this identity to rewrite the entire equation in terms of sine.

$$1 - 2\sin\theta + \sin^2\theta = 1 - \sin^2\theta$$

Set the equation equal to 0, and solve by factoring.

$$2\sin^2\theta - 2\sin\theta = 0$$
$$2\sin\theta(\sin\theta - 1) = 0$$
$$2\sin\theta = 0 \qquad \sin\theta - 1 = 0$$
$$\sin\theta = 0 \quad \text{or} \quad \sin\theta = 1$$
$$\theta = 0 \qquad\qquad \theta = \frac{\pi}{2}$$

The general solution is $\theta = 0 + 2k\pi$, $\theta = \pi + 2k\pi$, or $\theta = \frac{\pi}{2} + 2k\pi$. Condense the notation by combining the first two solutions: $\theta = 0 + k\pi$ or $\theta = \frac{\pi}{2} + 2k\pi$.

$0 + 2k\pi$ are all the angles that end on the positive x-axis, and $\pi + 2k\pi$ are all the angles that end on the negative x-axis. $0 + k\pi$ lists all of those angles at once.

8.34 Identify all solution(s) to the equation $\sin\left(\theta - \dfrac{\pi}{3}\right) = \dfrac{1}{2}$ on the interval $[0, 2\pi)$.

Apply the difference formula for sine described in Problem 8.16.

$$\sin\left(\theta - \frac{\pi}{3}\right) = \frac{1}{2}$$
$$\sin\theta\cos\frac{\pi}{3} - \cos\theta\sin\frac{\pi}{3} = \frac{1}{2}$$
$$\sin\theta\left(\frac{1}{2}\right) - \cos\theta\left(\frac{\sqrt{3}}{2}\right) = \frac{1}{2}$$

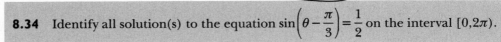

$\sin(a \pm b) = \sin a \cos b \pm \cos a \sin b$

Multiply the entire equation by 2 to eliminate fractions.

$$\sin\theta - \sqrt{3}\cos\theta = 1$$

Separate the trigonometric expressions and square both sides of the equation.

$$(\sin\theta)^2 = \left(\sqrt{3}\cos\theta + 1\right)^2$$
$$\sin^2\theta = 3\cos^2\theta + 2\sqrt{3}\cos\theta + 1$$

Because a Pythagorean identity states that $\cos^2\theta + \sin^2\theta = 1$, you can conclude that $\sin^2\theta = 1 - \cos^2\theta$. Use this identity to rewrite the equation in terms of cosine.

$$\left(1 - \cos^2\theta\right) = 3\cos^2\theta + 2\sqrt{3}\cos\theta + 1$$

Set the equation equal to 0 and solve by factoring.

$$4\cos^2\theta + 2\sqrt{3}\cos\theta = 0$$

$$2\cos\theta\left(2\cos\theta + \sqrt{3}\right) = 0$$

$$\begin{array}{cc}
2\cos\theta = 0 & 2\cos\theta + \sqrt{3} = 0 \\
\cos\theta = 0 & 2\cos\theta = -\sqrt{3} \\
\theta = \dfrac{\pi}{2} \quad \text{or} & \cos\theta = -\dfrac{\sqrt{3}}{2} \\
 & \theta = \dfrac{7\pi}{6}
\end{array}$$

The solutions to the equation are $\theta = \dfrac{\pi}{2}$ or $\dfrac{7\pi}{6}$.

Chapter 9
INVESTIGATING LIMITS
What height does a function INTEND to reach?

The concept of limits, though typically presented to students prior to the study of differentiation and integration, was the final component of calculus theory to fall into place. Though the theory allowing the calculation of instantaneous rates of change and the determination of area based upon infinite series existed, they could gain no credibility without a rigorous and systematic set of theorems concerning the existence and behavior of infinitely small or infinitely large quantities. The modern (epsilon-delta) definition of limits bears an undeniable, though less obfuscated, resemblance to the breakthrough characterization of limits whose discovery led to the establishment of calculus, and is a foundational concept for much of theoretical mathematics—even courses of study to which calculus is merely peripherally related.

Limits are important—really important. They're the spine that holds the rest of the calculus skeleton upright. Basically, a limit is just a statement that tells you what height a function is headed for, as you get close to a specific x-value. It doesn't matter if the function actually gets there (maybe there's a hole there instead of a function value, like in Problem 9.4). All that matters is that you can tell where the function INTENDS to go. This chapter starts really easy (you're just looking at graphs of curves and dots) and then gets tricky when it comes time for epsilon-delta problems. Push through those, because it gets easier again when Chapter 10 starts.

Evaluating One-Sided and General Limits Graphically

Find limits on a function graph

Note: Problems 9.1–9.9 refer to the graph of f(x) in Figure 9-1.

9.1 Evaluate $\lim\limits_{x \to 1^+} f(x)$.

> Trace your finger along the graph of f(x) FROM the right, not TO the right, and stop at the vertical line x = 1. The y-value you end up with is the limit.

The positive sign in the limit notation indicates a right-hand limit. As you approach $x = 1$ from the right, the function approaches (and in fact reaches) $y = 4$. Therefore, $\lim\limits_{x \to 1^+} f(x) = 4$.

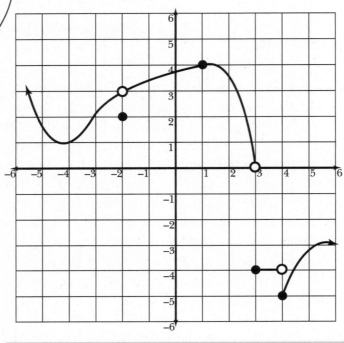

Figure 9-1
The graph of a function f(x).

Note: Problems 9.1–9.9 refer to the graph of f(x) in Figure 9-1.

9.2 Does $\lim\limits_{x \to 1^+} f(x) = \lim\limits_{x \to 1^-} f(x)$? Justify your answer.

> If the left- and right-hand limits at an x-value are equal, then a "general limit" (with no teeny "+" or "−" in it) exists: $\lim\limits_{x \to 1} f(x) = 4$.

Problem 9.1 states that $\lim\limits_{x \to 1^+} f(x) = 4$. The left hand limit, $\lim\limits_{x \to 1^-} f(x)$, also equals 4; the direction from which you approach $x = 1$ along the graph of $f(x)$ is irrelevant as either leads to the point $(1,4)$ on the graph of $f(x)$.

Note: Problems 9.1–9.9 refer to the graph of f(x) in Figure 9-1.

9.3 Evaluate $f(-2)$.

The graph of $f(x)$ contains the point $(-2,2)$, so $f(-2) = 2$.

Note: *Problems 9.1–9.9 refer to the graph of f(x) in Figure 9-1.*

9.4   Evaluate $\lim\limits_{x \to -2} f(x)$.

As x approaches -2 from the left and from the right, $f(x)$ approaches 3. Therefore, $\lim\limits_{x \to -2} f(x) = 3$.

Think of the function as a road. Even though there's a major pothole in the graph at $(-2, 3)$, that's still where the road is leading, so the limit exists at $x = -2$.

Note: *Problems 9.1–9.9 refer to the graph of f(x) in Figure 9-1.*

9.5   Determine the value of c that makes the statement true: $\lim\limits_{x \to c^-} f(x) = 0$.

The statement requires you to identify the x-value at which the graph of $f(x)$ approaches 0 (i.e. the x-intercept) as you approach c from the left. The correct answer is $c = 3$; the function approaches the point $(3,0)$ as you approach $x = 3$ from the left.

Note: *Problems 9.1–9.9 refer to the graph of f(x) in Figure 9-1.*

9.6   Evaluate $f(3)$.

Two coordinates of interest fall on the vertical line $x = 3$: $(3,0)$ and $(3,-4)$. Clearly, $f(3)$ cannot equal both 0 and -4, as that violates the fundamental definition of a function (each input must correspond to exactly one output). Note that $(3,-4)$ is actually on the graph of $f(x)$, whereas $(3,0)$ is essentially a "hole" in the curve. Therefore, $f(3) = -4$.

Note: *Problems 9.1–9.9 refer to the graph of f(x) in Figure 9-1.*

9.7   Evaluate $\lim\limits_{x \to 4^-} f(x)$.

As you approach $x = 4$ from the left, the graph of $f(x)$ approaches the point $(4,-4)$, so $\lim\limits_{x \to 4^-} f(x) = -4$.

Note: *Problems 9.1–9.9 refer to the graph of f(x) in Figure 9-1.*

9.8   Evaluate $\lim\limits_{x \to 4^+} f(x)$.

As you approach $x = 4$ from the right, the graph of the function approaches the point $(4,-5)$, so $\lim\limits_{x \to 4^+} f(x) = -5$.

So, $\lim\limits_{x \to 4} f(x)$ does not exist, because the right- and left-hand limits are different as x approaches 4.

Note: *Problems 9.1–9.9 refer to the graph of f(x) in Figure 9-1.*

9.9   Find two values of k such that the following statement is true: $\lim\limits_{x \to k^-} f(x) \neq \lim\limits_{x \to k^+} f(x)$.

As demonstrated by Problems 9.7 and 9.8, the left- and right-hand limits of $f(x)$ as x approaches 4 are unequal; thus, $k = 4$. As for the other correct value of k, notice that one-sided limits are unequal at $x = 3$: $\lim\limits_{x \to 3^-} f(x) = 0$ but $\lim\limits_{x \to 3^+} f(x) = -4$. Therefore, $k = 3$ or $k = 4$.

Note: Problems 9.10–9.11 reference the function g(x) defined below.

$$g(x) = \frac{2x^2 - x - 6}{x - 2}$$

9.10 Graph $g(x)$ without using a graphing calculator.

> Plugging 2 into the denominator means you'd end up dividing by zero, which is not allowed.

Factor the numerator of the function.

$$g(x) = \frac{(2x+3)(x-2)}{x-2}$$

Note that the numerator and denominator contain the same factor, but before you eliminate it to reduce the fraction, take note of the domain restriction dictated by the denominator: $x \neq 2$.

$$g(x) = \frac{(2x+3)\,\cancel{(x-2)}}{\cancel{x-2}}$$

$$g(x) = 2x + 3, \text{ if } x \neq 2$$

The graph of $y = \dfrac{2x^2 - x - 6}{x - 2}$ (pictured in Figure 9-2) is simply the graph of $y = 2x + 3$ with one difference—there is a hole in the graph when $x = 2$, due to the restriction $x \neq 2$.

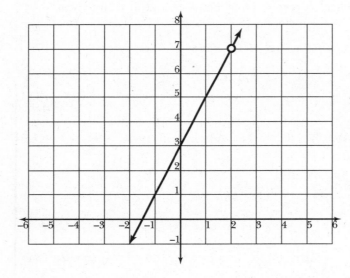

Figure 9-2

The graph of $g(x) = \dfrac{2x^2 - x - 6}{x - 2}$;

notice the hole at (2,7).

Note: Problems 9.10–9.11 reference the function g(x) defined in Problem 9.10.

9.11 Evaluate $\lim\limits_{x \to 2} g(x)$.

> If you plug $x = 2$ into $2x + 3$, the simplified version of g(x), you get $g(2) = 2(2) + 3 = 7$, which is the correct limit. You'll use this technique in Problems 10.11–10.15.

Substituting $x = 2$ into the function produced an indeterminite result: $g(2) = \dfrac{0}{0}$. (Note that zero divided by itself does *not* equal one.) However, you can determine $\lim\limits_{x \to 2} g(x)$ based upon its graph in Figure 9-2. Although the function is not defined when $x = 2$ (due to a hole in the graph), $g(x)$ clearly approaches the same y-value from the left and from the right: $y = 7$.

9.12 Given the piecewise-defined function $h(x)$ defined below, evaluate $\lim\limits_{x\to 0^-} h(x)$ and $\lim\limits_{x\to 0^+} h(x)$.

$$h(x) = \begin{cases} x^3, & x < 0 \\ \sqrt{x}, & x \geq 0 \end{cases}$$

Consider the graph of $h(x)$ in Figure 9-3. To generate this graph, first plot $y = x^3$, but only draw the portion of the graph for which $x < 0$—the portion of the graph that is left of the y-axis. Technically, this graph segment should end with an open dot on the y-axis, because x is less than *but not equal to* 0. Next, graph the function $y = \sqrt{x}$ on the same coordinate plane. Because its domain exactly matches the restricted domain assigned to it by $h(x)$ $(x \geq 0)$, the graph appears in its entirety, including a solid dot on the y-axis because the restriction $x \geq 0$ includes 0.

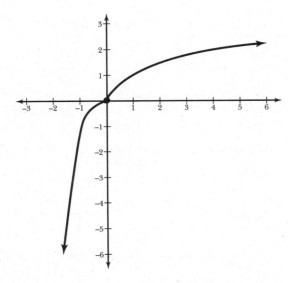

Figure 9-3

The graph of $h(x) = \begin{cases} x^3, & x < 0 \\ \sqrt{x}, & x \geq 0 \end{cases}$.

As x approaches 0 from the left *and* the right, $h(x)$ approaches $y = 0$. Therefore, $\lim\limits_{x\to 0^-} h(x) = \lim\limits_{x\to 0^+} h(x) = 0$.

9.13 Given the piecewise-defined function $j(x)$ defined below, evaluate $\lim\limits_{x\to -2^-} j(x)$ and $\lim\limits_{x\to -2^+} j(x)$.

$$j(x) = \begin{cases} (x+3)^2 - 4, & x \leq -2 \\ -x + 5, & x > -2 \end{cases}$$

Calculate the one-sided limits as x approaches -2 by substituting $x = -2$ into both of the rules that define $j(x)$.

$$\lim_{x\to -2^-} j(x) = (-2+3)^2 - 4 \quad \Big| \quad \lim_{x\to -2^+} j(x) = -(-2) + 5$$
$$= 1 - 4 \qquad\qquad\qquad = 2 + 5$$
$$= -3 \qquad\qquad\qquad\quad = 7$$

> Anything left of $x = -2$ is defined by the top rule, so use it to find the left-hand limit. Anything right of $x = -2$ is defined by the bottom rule, so that's where the right-hand limit comes from.

9.14   Evaluate $\lim\limits_{x \to 0} \dfrac{|x|}{x}$.

If $x > 0$, the graph of $f(x) = \dfrac{|x|}{x}$ will look like Figure 9-4.

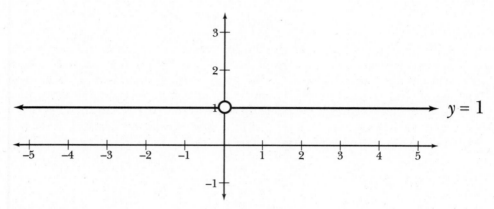

$y = 1$

Figure 9-4 *The graph of $f(x) = \dfrac{|x|}{x}$ matches the graph of $y = 1$ when x is a positive number.*

Any positive number divided by itself is 1, so when $x > 0$, $f(x) = 1$, and it is clear from the graph that $\lim\limits_{x \to 0^+} f(x) = 1$. However, if $x < 0$, the graph of $f(x)$ looks like Figure 9-5.

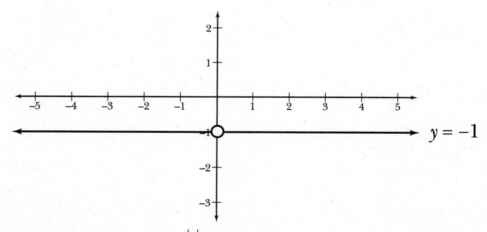

$y = -1$

When x is negtive, |x| is the opposite of x. For example, if x = –5, then |x| = |–5| = 5.

Figure 9-5 *The graph of $f(x) = \dfrac{|x|}{x}$ matches the graph of $y = -1$ when x is a negative number.*

A number divided by its opposite equals –1, so when $x < 0$, $f(x) = -1$ and $\lim\limits_{x \to 0^-} f(x) = 1$. Because $\lim\limits_{x \to 0^-} f(x) \neq \lim\limits_{x \to 0^+} f(x)$, $\lim\limits_{x \to 0} \dfrac{|x|}{x}$ does not exist.

Limits and Infinity

What happens when x or f(x) gets huge?

9.15 Evaluate $\lim\limits_{\theta \to (3\pi/2)^-} \tan\theta$ and $\lim\limits_{\theta \to (3\pi/2)^+} \tan\theta$.

Consider the graph of $y = \tan\theta$ in Figure 9-6.

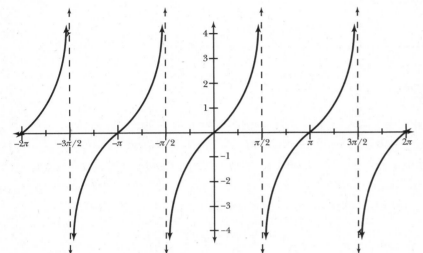

Figure 9-6

The graph of $y = \tan\theta$, with asymptotes at $\theta = \dfrac{k\pi}{2}$, if k is an odd integer.

As θ approaches $\dfrac{3\pi}{2}$ from the left, the tangent graph increases without bound, so $\lim\limits_{\theta \to 3\pi/2^-} \tan\theta = \infty$. However, as θ approaches $\dfrac{3\pi}{2}$ from the right, the function values decrease without bound, so $\lim\limits_{\theta \to 3\pi/2^+} \tan\theta = -\infty$. Note that a "limit" of ∞ or $-\infty$ is actually an indication that a real number limit does not exist and a justification for the nonexistence of a limit. ←

> Limits are real numbers, and ∞ is not a real number. It doesn't make sense to say "the limit is that it's infinitely unlimited."

9.16 Based on Problem 9.15, describe the relationship between infinite limits and vertical asymptotes.

If $\lim\limits_{x \to c} f(x) = \infty$ or $\lim\limits_{x \to c} f(x) = -\infty$ (and c is a real number), then $x = c$ is a vertical asymptote of $f(x)$.

9.17 Evaluate $\lim\limits_{x \to -\infty} e^{x-2} + 3$.

Graph $y = e^{x-2} + 3$ by applying two transformations to the graph of $y = e^x$: a horizontal shift of 2 units to the right and a vertical shift of 3 units up, as illustrated in Figure 9-7.

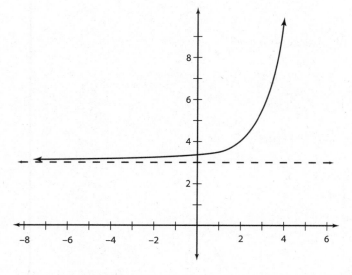

Figure 9-7

The graph of $y = e^{x-2} + 3$.

The untransformed graph of $y = e^x$ has a horizontal asymptote of $y = 0$ (the x-axis), so shifting the graph up 3 units results in a new horizontal asymptote of $y = 3$. As θ approaches $-\infty$, the graph gets infinitely close to, but never intersects, that asymptote. Therefore, $\lim\limits_{x \to -\infty} e^{x-2} + 3 = 3$. Note that $\lim\limits_{x \to \infty} e^{x-2} + 3 = \infty$, because the function increases without bound as x gets infinitely large.

9.18 Based on Problem 9.17, describe the relationship between limits at infinity and horizontal asymptotes.

> You're evaluating "limits at infinity" when x gets really positive or really negative (x → ∞ or x → −∞).

If $\lim\limits_{x \to \infty} f(x) = c$ or $\lim\limits_{x \to -\infty} f(x) = c$ (and c is a real number), then $y = c$ is a horizontal asymptote of $f(x)$.

Note: Problems 9.19–9.21 refer to the graph of g(x) in Figure 9-8.

9.19 Evaluate $\lim\limits_{x \to \infty} g(x)$.

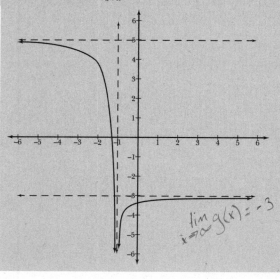

$$\lim_{x \to \infty} g(x) = -3$$

Figure 9-8

The graph of a function g(x) and its three asymptotes.

As x increases infinitely, the function approaches the horizontal asymptote $y = -3$. Therefore, $\lim_{x \to \infty} g(x) = -3$.

Note: Problems 9.19–9.21 refer to the graph of g(x) in Figure 9-8.

9.20 Evaluate $\lim_{x \to -\infty} g(x)$.

> Look at the right edge of the graph—you can tell that g(x) is getting closer and closer to y = -3.

As x becomes more and more negative, the function approaches the horizontal asymptote $y = 5$. Therefore, $\lim_{x \to \infty} g(x) = 5$.

Note: Problems 9.19–9.21 refer to the graph of g(x) in Figure 9-8.

9.21 Does $\lim_{x \to -1} g(x)$ exist? Justify your answer.

In order for a limit to exist, the corresponding one-sided limits must exist and be equal. According to the graph of $g(x)$, as x approaches -1 from both the left and the right, the function values decrease without bound; therefore $\lim_{x \to -1^-} g(x) = \lim_{x \to -1^+} g(x) = -\infty$. Although the limits are the same, they do not represent a finite limit (see Problem 9.15 for further explanation). Because $g(x)$ does not possess a finite limit at $x = -1$, $\lim_{x \to -1} g(x)$ does not exist.

9.22 Given a function $f(x)$ such that $\lim_{x \to k} f(x) = \infty$ or $\lim_{x \to k} f(x) = -\infty$, how many unique values of k are possible? If $\lim_{x \to \infty} f(x) = c$ or $\lim_{x \to -\infty} f(x) = c$, how many unique values of c are possible?

According to Problem 9.16, if $\lim_{x \to k} f(x) = \infty$ or $\lim_{x \to k} f(x) = -\infty$, then a vertical asymptote of $f(x)$ exists at $x = k$. The first part of the question, then, asks you to determine how many unique vertical asymptotes a function can possess. There are no limitations on this number—a graph could have an infinite number of vertical asymptotes. In fact, four of the trigonometric graphs have infinitely many vertical asymptotes; the graphs of $y = \tan \theta$ and $y = \sec \theta$ have vertical asymptotes at $x = \dfrac{n \cdot \pi}{2}$ (where n is an odd integer), and the graphs of $y = \cot \theta$ and $y = \csc \theta$ have vertical asymptotes at $y = n\pi$, where n is an integer. Therefore, k may have an infinite number of unique values.

> In other words, at $x = \cdots, -\dfrac{3\pi}{2}, -\dfrac{\pi}{2}, \dfrac{\pi}{2}, \dfrac{3\pi}{2}, \cdots$.

As for the second part of the question, if $\lim_{x \to \infty} f(x) = c$ or $\lim_{x \to -\infty} f(x) = c$, then $y = c$ is a horizontal asymptote of the function (according to Problem 9.18). A function may have, at most, two horizontal asymptotes, one which the function approaches as x gets infinitely positive, and one which it approaches in the negative direction. Therefore, c has a maximum of two unique values.

> Most functions only have one horizontal asymptote, if any. For instance, rational functions have the same limit at ∞ and -∞ (see Problem 9.25).

9.23 Evaluate $\lim\limits_{x \to -3^+} \dfrac{x^2 + 6x - 16}{x + 3}$.

Factor the numerator of the fraction.

$$\lim_{x \to -3^+} \frac{(x+8)(x-2)}{x+3}$$

Any value that causes the denominator to equal 0 *but not the numerator* represents a vertical asymptote of the function (see problem 4.29); in this problem such a value exists: $x = -3$. According to Problem 9.16, the limit of a function, as x approaches a vertical asymptote, either equals ∞ or $-\infty$. The answer depends upon the function and the direction from which x approaches the asymptote.

To determine whether the values of a rational function increases or decreases without bound, substitute a value *slightly* larger than -3, such as $x = -2.999$, into the function.

$$\lim_{x \to -3^+} \frac{(x+8)(x-2)}{x+3} \approx \frac{(-2.999+8)(-2.999-2)}{-2.999+3}$$
$$\approx \frac{(5.001)(-4.999)}{0.001}$$
$$\approx -24{,}999.998$$

As x approaches -3 from the right, the function values are becoming infinitely negative. Therefore, $\lim\limits_{x \to -3^+} \dfrac{(x+8)(x-2)}{x+3} = -\infty$.

> In Chapter 13, you'll figure out the direction of a function in a more satisfying way—using signs of derivatives instead of plugging in numbers like -2.99999 and -3.00001.

9.24 Evaluate $\lim\limits_{x \to \infty} \dfrac{2x^2 - 5x + 6}{-x^3 - 6x^2 - x + 2}$.

According to Problem 4.31, if the degree of a rational function's denominator is greater than the degree of its numerator, the function has a horizontal asymptote of $y = 0$. Therefore, $\lim\limits_{x \to \infty} \dfrac{2x^2 - 5x + 6}{-x^3 - 6x^2 - x + 2} = 0$.

9.25 Evaluate $\lim\limits_{x \to -\infty} \dfrac{5x^2 - 9x + 1}{5 - 3x - 6x^2}$.

According to Problem 4.32, if a rational function contains a numerator and denominator of equal degree, the function has a horizontal asymptote equal to the quotient of their leading coefficients. The leading coefficient of the numerator is 5 and the leading coefficient of the denominator is -6, so

$$\lim_{x \to \infty} \frac{5x^2 - 9x + 1}{5 - 3x - 6x^2} = \lim_{x \to -\infty} \frac{5x^2 - 9x + 1}{5 - 3x - 6x^2} = -\frac{5}{6}.$$ (Rational functions have the same limit as $x \to \infty$ and $x \to -\infty$.)

> If the highest powers in the numerator and denominator match, divide the coefficients attached to those powers to get the limit as x approaches $+\infty$ and $-\infty$.

Note: Problems 9.26–9.27 reference the function f(x) defined below. Note that a, b, c, d, and k are real numbers.

$$f(x) = \frac{(2x + a)(x - b)}{(cx + d)(3x - k)}$$

9.26 Evaluate $\lim_{x \to \infty} f(x)$.

Expand the products in the numerator and the denominator.

$$f(x) = \frac{2x^2 - 2bx + ax - ab}{3cx^2 - ckx + 3dx - dk}$$

The degrees of the numerator and denominator are equal, so the limit at infinity equals the quotient of the leading coefficients, as explained in Problem 9.25. (Only compare the powers of x when determining the degree, because a, b, c, d, and k are constants.)

$$\lim_{x \to \infty} \frac{2x^2 - 2bx + ax - ab}{3cx^2 - ckx + 3dx - dk} = \frac{2}{3c}$$

Note: Problems 9.26–9.27 reference the function f(x) defined in Problem 9.26.

9.27 Identify all values of n such that $\lim_{x \to n} f(x) = \infty$ or $\lim_{x \to n} f(x) = -\infty$.

An infinite limit indicates the presence of a vertical asymptote (see Problem 9.16). To find values of x that make the denominator equal 0, set both factors of the denominator equal to 0 and solve.

$$cx + d = 0 \qquad 3x - k = 0$$
$$cx = -d \quad \text{or} \quad 3x = k$$
$$x = -\frac{d}{c} \qquad x = \frac{k}{3}$$

It is not sufficient merely to identify these x-values—you must also ensure they do not, in turn, make the numerator 0 as well. Set the factors of the numerator equal to 0 and solve.

$$2x + a = 0 \qquad x - b = 0$$
$$2x = -a \quad \text{or} \quad x = b$$
$$x = -\frac{a}{2}$$

> If an x-value makes the numerator AND the denominator equal 0, it usually means there's a hole in the graph, not a vertical asymptote, so a limit still exists there.

Therefore, $\lim_{x \to -d/c} f(x) = -\infty$ or ∞ (it's not possible to determine which because you don't know the signs of a, b, c, d, and k) and $\lim_{x \to k/3} f(x) = -\infty$ or ∞, as long as neither $-\frac{d}{c}$ nor $\frac{k}{3}$ equals $-\frac{a}{2}$ or b.

Formal Definition of the Limit

Epsilon-delta problems are no fun at all

9.28. The formal definition of limit contains the expression $\left|f(x)-L\right|<\varepsilon$. Interpret this statement geometrically.

> One other important thing—this is the VERTICAL distance between a function and a limit L.

The absolute value of a difference is often used to describe the distance of the quantities that are subtracted. Accordingly, the expression $\left|f(x)-L\right|<\varepsilon$ states that the distance between a function $f(x)$ and a constant L is less than the constant ε. You can also conclude that $\varepsilon > 0$, because ε is greater that a quantity within absolute values (which must be nonnegative).

9.29. The formal definition of a limit contains the expression "$0<\left|x-a\right|<\delta$." Interpret this expression geometrically.

> This time you're measuring a horizontal distance.

Like the expression in Problem 9.28, this expression also describes a distance. It states that the distance between x and the constant a is less than a constant called δ, which must be a positive number. Unlike Problem 9.25, the expression explicitly states that the distance between x and a cannot equal 0, so $x \neq a$.

9.30. The formal definition of a limit (also called the epsilon-delta definition due to the variables customarily assigned to it) states that $\lim_{x\to a} f(x) = L$ if and only if for each real number $\varepsilon > 0$ there exists a corresponding real number $\delta > 0$ such that $0<\left|x-a\right|<\delta$ implies that $\left|f(x)-L\right|<\varepsilon$. Interpret this theorem geometrically.

> Here's an interpretation: "If you trace your finger along the graph of f(x) and your finger is getting close to x = a along the x-axis, it better be getting closer to a function height of y = L as well, or there's no limit as x approaches a."

The limit of $f(x)$, as x approaches a, is equal to L if the following requirement is met: assuming the function $f(x)$ and the limit L it approaches are within a fixed distance ε of one another, x and the value a it approaches must be within a corresponding fixed distance δ.

9.31. Calculate the value of δ that corresponds to ε given $\lim_{x\to 6}(2x+1)=13$, according to the definition of limits.

> You don't really need the full version of the expression $(0<\left|x-6\right|<\delta)$ as long as you're careful not to let x = 6—the distance just needs to be a positive number.

According to the definition of limits (stated in Problem 9.30), you must find δ that corresponds to ε given $\left|f(x)-L\right|<\varepsilon$. Substitute $f(x)$ and L into the expression.

$$\left|f(x)-L\right|<\varepsilon$$
$$\left|(2x+1)-13\right|<\varepsilon$$
$$\left|2x-12\right|<\varepsilon$$

Your goal is to generate an expression for δ that mimics the expression $\left|x-6\right|<\delta$ (since $a=6$). Factor 2 out of the left side of the inequality and isolate the absolute value expression.

$$2|x-6| < \varepsilon$$

$$|x-6| < \frac{\varepsilon}{2}$$

Therefore, $|x-6| < \frac{\varepsilon}{2}$. Compare that result to the expression $|x-6| < \delta$ to conclude that $\delta = \frac{\varepsilon}{2}$.

9.32 Calculate the value of δ that corresponds to ε given $\lim\limits_{x\to 2}\dfrac{x^2-7x+10}{x-2}=-3$, according to the definition of limits.

Substitute $f(x)$ and L into the ε expression in the limit definition.

$$|f(x)-L| < \varepsilon$$

$$\left|\frac{x^2-7x+10}{x-2}-(-3)\right| < \varepsilon$$

$$\left|\frac{x^2-7x+10}{x-2}+3\right| < \varepsilon$$

Simplify the expression on the left side of the inequality using the common denominator $x-2$.

$$\left|\frac{x^2-7x+10}{x-2}+\frac{3}{1}\left(\frac{x-2}{x-2}\right)\right| < \varepsilon$$

$$\left|\frac{x^2-7x+10+3x-6}{x-2}\right| < \varepsilon$$

$$\left|\frac{x^2-4x+4}{x-2}\right| < \varepsilon$$

Factor and simplify.

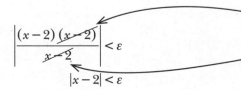

$$\left|\frac{(x-2)\,(x-2)}{x-2}\right| < \varepsilon$$

$$|x-2| < \varepsilon$$

It's okay to cancel the terms as long as you don't let $x = 2$.

By comparing the statements $|x-2| < \varepsilon$ and $|x-a| < \delta$, you can conclude that $\delta = \varepsilon$.

Remember, a is the number x approaches in the limit, so in this problem a = 2.

9.33 Identify a value of δ that corresponds to $\varepsilon = 0.0001$, such that $\lim\limits_{x\to 6}(15-4x)=-9$ according to the definition of limits.

Although the value of ε is specifically stated, begin this problem using the same method described in Problems 9.31 and 9.32—substitute into the ε statement of the limit definition.

$$\left|f(x) - L\right| < \varepsilon$$
$$\left|(15 - 4x) - (-9)\right| < 0.0001$$
$$\left|24 - 4x\right| < 0.0001$$

Factor −4 out of the expression within the absolute value symbols; note that factoring a negative number out of a difference reverses the order in which the terms are subtracted.

$$\left|-4(-6 + x)\right| < 0.0001$$
$$\left|-4(x - 6)\right| < 0.0001$$
$$\left|-4\right| \cdot \left|x - 6\right| < 0.0001$$
$$4\left|x - 6\right| < 0.0001$$
$$\left|x - 6\right| < \frac{0.0001}{4}$$
$$\left|x - 6\right| < 0.000025$$

By comparing the statements $|x - 6| < 0.000025$ and $|x - a| < \delta$, you can conclude that $\delta = 0.000025$.

9.34 Identify a value of δ (accurate to two decimal places) that corresponds to $\varepsilon = 0.01$ given $\lim\limits_{x \to -1} \left(x^2 + 3\right) = 4$, according to the definition of limits.

Substitute known values into the ε expression of the limit definition.

$$\left|f(x) - L\right| < \varepsilon$$
$$\left|(x^2 + 3) - 4\right| < 0.01$$
$$\left|x^2 - 1\right| < 0.01$$
$$\left|x + 1\right|\left|x - 1\right| < 0.01$$

Unlike Problems 9.31–9.33, the expression on the left side of the inequality is not immediately in the form $|x - a| < \delta$. In order to reach this form, begin by assuming that $\delta < 1$, which is reasonable because δ should represent an extremely small distance. If $\delta < 1$, then $|x + 1| < 1$, which means that $-1 < x + 1 < 1$. Subtract 2 from each of those expressions to get $-3 < x - 1 < -1$. Therefore, $|x - 1| < 3$.

> The smallest x can be is −1.9999..., and that number minus 1 is −2.999999, which is less than 3.

Your goal is to produce the expression $|x - a|$ (which equals $|x + 1|$ because $a = -1$) in the middle of the compound expression. Recall that $|x - 1| < 3$ and substitute 3 into the inequality.

$$\left|x - 1\right|\left|x + 1\right| < 0.01$$
$$3\left|x + 1\right| < 0.01$$
$$\left|x + 1\right| < \frac{0.01}{3}$$

> If 3|x+1| is less than 0.01 and 3 is bigger than |x−1|, then |x−1| |x+1| is also less than 0.01.

Therefore, $\delta = \dfrac{0.01}{3} \approx 0.00\overline{3}$.

Chapter 10
EVALUATING LIMITS

Calculating limits without a graph of the function

Though the formal definition of a limit provides the fundamental foundation, it is highly impractical to apply epsilon-delta proofs in order to verify that limits exist. In fact, one of the assumptions of such proofs is that the limit is already known. In order to calculate the vast majority of basic limits, you need only apply a few basic techniques or identify an elementary limit theorem.

If you had to draw a graph for every single limit you were asked to calculate, it would get old pretty fast. Graphing calculators make things a lot easier, not only because they can graph faster, but because they can do ridiculously complicated calculations instantly. However, you won't always be allowed to use them on quizzes and tests for this very reason. Besides, unless the limit is a really obvious number (like 2 or 4.5), calculator answers can be lousy. Sure you could find out that a function's limit is approximately 13.228757, but isn't it better to know that the limit is EXACTLY $5\sqrt{7}$ instead of that meaningless string of decimals? Decimals are good for checking an answer, but not so good for the answer itself. This chapter shows you ways to find limits that are technology-free.

Substitution Method

As easy as plugging in for x

10.1 Evaluate $\lim\limits_{x \to 4}\left(x^2 - 3x + 2\right)$.

Notice that substituting $x = 4$ into the expression results in a real, finite value, which indicates that the substitution method is applicable.

$$\lim\limits_{x \to 4}\left(x^2 - 3x + 2\right) = 4^2 - 3(4) + 2 = 16 - 12 + 2 = 6$$

> Real as in "real number," not real as in "not fake."

10.2 Evaluate $\lim\limits_{x \to -1} 3^x$.

Substituting $x = -1$ into the expression results in a real, finite value.

$$\lim\limits_{x \to -1} 3^x = 3^{-1} = \frac{1}{3}$$

10.3 Evaluate $\lim\limits_{x \to e} \arctan(\ln x)$.

> Remember, $\ln x$ and e^x are inverse functions, so $\ln e^x = x$ (see Problem 5.30). In this problem, $x = 1$ ($\ln e = \ln e^1$), so $\ln e = 1$.

Substitute $x = e$ into the expression $\ln x$. Note that $\ln e = 1$.

$$\lim\limits_{x \to e} \arctan(\ln x) = \arctan(\ln e)$$
$$= \arctan(1)$$

According to Problem 7.32, $\arctan 1 = \dfrac{\pi}{4}$, so $\lim\limits_{x \to e} \arctan(\ln x) = \dfrac{\pi}{4}$.

Note: Problems 10.4–10.6 refer to the functions f(x) and g(x) as defined by the table:

x	-1	0	1	2	3
$f(x)$	5	-3	-1	0	6
$g(x)$	-7	1	3	-4	2

Table 10-1: Although only selected values of functions f(x) and g(x) are given, f(x) and g(x) are continuous for all real numbers.

10.4 Evaluate $\lim\limits_{x \to -1}\left(f(x) - 3g(x)\right)$.

The limit of a sum (or difference) is equal to the sum of the individual limits.

$$\lim\limits_{x \to -1}\left(f(x) - 3g(x)\right) = \lim\limits_{x \to -1} f(x) - \lim\limits_{x \to -1} 3g(x)$$

The limit of a product is equal to the product of the limits.

$$= \lim_{x \to -1} f(x) - \left(\lim_{x \to -1} 3\right)\left(\lim_{x \to -1} g(x)\right)$$

$$= \lim_{x \to -1} f(x) - 3 \lim_{x \to -1} g(x)$$

Substituting –1 into the expression results in a real, finite value.

$$= f(-1) - 3g(-1)$$

$$= 5 - 3(-7)$$

$$= 26$$

> *lim 3 = 3 because the function (y = 3) is a horizontal line. No matter what x you approach, the function always has a value of exactly 3. It doesn't even APPROACH that height—it EQUALS that height.*

Note: Problems 10.4–10.6 refer to the functions f(x) and g(x) defined in Problem 10.4.

10.5 Find the value of k for which $\lim_{x \to k} \dfrac{g(x)}{f(x)}$ does not exist.

If $k = 2$, calculating the limit requires you to divide by 0, which is not a valid operation.

$$\lim_{x \to 2} \frac{g(x)}{f(x)} = \frac{g(2)}{f(2)} = \frac{-4}{0}$$

The value $\dfrac{-4}{0}$ is not a real, finite value, so $\lim_{x \to k} \dfrac{g(x)}{f(x)}$ does not exist when $x = k$.

> *You know that f(–1) = 5 and g(–1) = –7 because those are the numbers in the x = –1 column for f(x) and g(x).*

Note: Problems 10.4–10.6 refer to the functions f(x) and g(x) defined in Problem 10.4.

10.6 Evaluate $\lim_{x \to 3} f\big(g(x)\big)$.

Note that $g(3) = 2$; substitute this value into the expression.

$$\lim_{x \to 3} f\big(g(x)\big) = f\big(g(3)\big) = f(2)$$

Therefore, $\lim_{x \to 3} f\big(g(x)\big) = f(2) = 0$.

Note: Problems 10.7–10.10 refer to the piecewise-defined function h(x) defined below:

$$h(x) = \begin{cases} 4 - x^3, & x < -1 \\ -10, & x = -1 \\ 4 - x^3, & -1 < x < 2 \\ 6 - 5x, & x \geq 2 \end{cases}$$

10.7 Evaluate $\lim_{x \to -3} h(x)$.

According to the piecewise-defined function, the expression $4 - x^3$ generates the values of $h(x)$ when $x < -1$.

$$\lim_{x \to -3} h(x) = 4 - (-3)^3 = 4 - (-27) = 31$$

> *Think of h(x) as four rules that tell you what h(x) is, based on what you PLUG IN for x. The input x = –3 falls in the top category (x < –1).*

Note: Problems 10.7–10.10 refer to the piecewise-defined function h(x) defined in Problem 10.7.

10.8 Evaluate $\lim\limits_{x \to 11} h(x)$.

According to the piecewise-defined function, the expression $6 - 5x$ generates the values of $h(x)$ when $x \geq 2$.

$$\lim_{x \to 11} h(x) = 6 - 5(11) = 6 - 55 = -49$$

Note: Problems 10.7–10.10 refer to the piecewise-defined function h(x) defined in Problem 10.7.

10.9 Evaluate $\lim\limits_{x \to -1} h(x)$.

If you plug x = –1.0001 into h(x) to estimate the left-hand limit, it falls in the first input category: x < –1. If you plug in x = –0.999 to estimate the right-hand limit, it falls in the third category: –1 < x < 2.

Even though $h(-1) = 10$, that is not the limit as x *approaches* –1. If a limit exists at $x = -1$, the left- and right-hand limits as x approaches –1 must be equal. Note that $h(x)$ is defined by the expression $4 - x^3$ as x approaches –1 from the left and the right, so substitute $x = -1$ into that expression to evaluate the limit.

$$\lim_{x \to -1^-} h(x) = \lim_{x \to -1^+} h(x) = 4 - (-1)^3 = 4 + 1 = 5$$

Note: Problems 10.7–10.10 refer to the piecewise-defined function h(x) defined in Problem 10.7.

10.10 Evaluate $\lim\limits_{x \to 2} h(x)$.

The rule defining $h(x)$ changes when $x = 2$. For inputs slightly less that $x = 2$, $h(x)$ is defined as $4 - x^3$. Substitute $x = 2$ into that expression; the result is the left-hand limit as x approaches 2.

$$\lim_{x \to 2^-} h(x) = 4 - (2)^3 = 4 - 8 = -4$$

For inputs slightly greater than $x = 2$, $h(x) = 6 - 5x$. Substitute $x = 2$ into that expression to determine the limit as x approaches 2 from the right.

$$\lim_{x \to 2^+} h(x) = 6 - 5(2) = 6 - 10 = -4$$

Because the left- and right-hand limits are equal as x approaches 2, the general limit exists: $\lim\limits_{x \to 2} h(x) = -4$.

Factoring Method

The first thing to try if substitution doesn't work

Note: Problems 10.11–10.12 refer to the function f(x) defined below:

$$f(x) = \frac{(x-4)(x+6)}{x-4}$$

10.11 Evaluate $\lim\limits_{x \to 4} f(x)$.

Simplify the expression by canceling out the matching factors in the numerator and the denominator.

$$\lim_{x \to 4} \frac{\cancel{(x-4)}(x+6)}{\cancel{x-4}} = \lim_{x \to 4} \frac{x+6}{1}$$

Now substituting $x = 4$ into the expression will not return an indeterminate result: $\lim\limits_{x \to 4}(x+6) = 10$. Therefore, $\lim\limits_{x \to 4} \frac{(x-4)(x+6)}{x-4} = 10$.

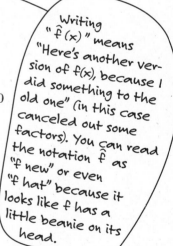

You can't plug x = 4 into the fraction because you'll get 0/0, which is called an "indeterminate" value. Basically, that's math's way of saying "Right now, I have no idea what 0/0 is equal to."

10.12 Graph $f(x)$.

The functions $f(x) = \frac{(x-4)(x+6)}{x-4}$ and $\hat{f}(x) = x+6$ are equivalent except at $x = 4$, as $f(x)$ is undefined at that value, but $\hat{f}(x)$ is not. (Recall from Problem 10.11 that $\hat{f}(x)$ is the reduced version of the rational function $f(x)$.) Therefore, their graphs have equal values except at $x = 4$; both are a line with slope 1 and y-intercept 6.

Although $f(x)$ is undefined at $x = 4$ (which means $f(4)$ doesn't exist), $\lim\limits_{x \to 4} f(x) = 10$ according to Problem 10.11, so the function approaches a height of 10 as $x \to 4$ from the left and the right. Therefore, the graph of $f(x)$ contains a "hole," as illustrated by Figure 10-1.

Writing "$\hat{f}(x)$" means "Here's another version of f(x), because I did something to the old one" (in this case canceled out some factors). You can read the notation $\hat{f}$ as "f new" or even "f hat" because it looks like f has a little beanie on its head.

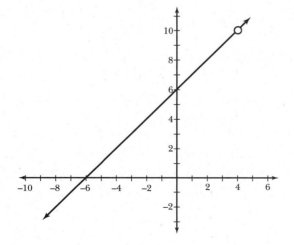

Figure 10-1

The graph of $f(x) = \frac{(x-4)(x+6)}{x-4}$

contains a hole at (4,10).

10.13 Evaluate $\lim\limits_{x \to 0} \dfrac{4x^5 - x^2}{x^2}$.

Substituting $x = 0$ into the expression returns an indeterminate result. Factor the numerator and reduce the fraction, thereby creating a new rational function for which the substitution method is a valid approach.

$$\lim_{x \to 0} \frac{4x^5 - x^2}{x^2} = \lim_{x \to 0} \frac{x^2\left(4x^3 - 1\right)}{x^2} = \lim_{x \to 0}\left(4x^3 - 1\right) = 4(0)^3 - 1 = -1$$

Therefore, $\lim\limits_{x \to 0} \dfrac{4x^5 - x^2}{x^2} = -1$.

10.14 Evaluate $\lim\limits_{x \to -9} \dfrac{x^2 + 5x - 36}{x + 9}$.

Factor the numerator, reduce the fraction, and evaluate the resulting limit using substitution.

$$\lim_{x \to -9} \frac{(x - 4)(x + 9)}{x + 9} = \lim_{x \to -9} \frac{(x - 4)\,\cancel{(x + 9)}}{\cancel{x + 9}} = \lim_{x \to -9}(x - 4) = -9 - 4 = -13$$

If you can't factor $8x^2 - 26x + 15$ by experimenting, you can factor it by decomposition—flip back to Problem 2.27 for more details.

10.15 Evaluate $\lim\limits_{x \to 3/4} \dfrac{8x^2 - 26x + 15}{4x - 3}$.

Factor the numerator, simplify, and apply the substitution method.

$$\lim_{x \to 3/4} \frac{(4x - 3)(2x - 5)}{4x - 3} = \lim_{x \to 3/4} \frac{\cancel{(4x - 3)}(2x - 5)}{\cancel{4x - 3}} = \lim_{x \to 3/4}(2x - 5) = \left(2\left(\frac{3}{4}\right) - 5\right) = \frac{6}{4} - 5 = -\frac{7}{2}$$

Therefore, $\lim\limits_{x \to 3/4} \dfrac{8x^2 - 26x + 15}{4x - 3} = -\dfrac{7}{2}$.

10.16 Evaluate $\lim\limits_{x \to 1}\left(\dfrac{7x^2 - 10x + 3}{x - 1} + \dfrac{2 - 7x}{x + 1}\right)$.

Notice that the substitution method fails only for the left expression. Recall that the limit of a sum is equal to the sum of the individual limits.

$$\lim_{x \to 1}\left(\frac{7x^2 - 10x + 3}{x - 1} + \frac{2 - 7x}{x + 1}\right) = \lim_{x \to 1}\frac{7x^2 - 10x + 3}{x - 1} + \lim_{x \to 1}\frac{2 - 7x}{x + 1}$$

Factor the left expression and substitute $x = 1$ into the right expression.

$$=\lim_{x\to 1}\frac{(7x-3)(x-1)}{x-1}+\frac{2-7(1)}{(1)+1}$$

$$=\lim_{x\to 1}(7x-3)+\left(-\frac{5}{2}\right)$$

$$=7(1)-3-\frac{5}{2}$$

$$=4-\frac{5}{2}$$

$$=\frac{3}{2}$$

Therefore, $\lim_{x\to 1}\left(\frac{7x^2-10x+3}{x-1}+\frac{2-7x}{x+1}\right)=\frac{3}{2}$.

10.17 Evaluate $\lim_{x\to -3/a}\dfrac{a^2x^2+3ax-4a^2bx-12ab}{ax+3}$ given $a\neq 0$.

> If a were equal to 0, x would be approaching $-\frac{3}{a}=-\frac{3}{0}$, which is not possible.

Factor the numerator by grouping.

$$\lim_{x\to -3/a}\frac{a^2x^2+3ax-4a^2bx-12ab}{ax+3}=\lim_{x\to -3/a}\frac{ax(ax+3)-4ab(ax+3)}{ax+3}$$

$$=\lim_{x\to -3/a}\frac{(ax-4ab)\,(ax+3)}{ax+3}$$

$$=\lim_{x\to -3/a}(ax-4ab)$$

> See Problem 2.26 if you don't know how to do this.

Substituting $x=-\dfrac{3}{a}$ for x now results in a real, finite value.

$$=\frac{a}{1}\left(-\frac{3}{a}\right)-4ab$$

$$=-\frac{3a}{a}-4ab$$

$$=-3-4ab$$

Therefore, $\lim_{x\to -3/a}\dfrac{a^2x^2+3ax-4a^2bx-12ab}{ax+3}=-3-4ab$.

10.18 Evaluate $\lim_{x\to -2}\dfrac{x^3+8}{x+2}$.

> In Chapter 14, you'll use L'Hôpital's rule to figure out limits like these without having to factor.

Factor the sum of perfect cubes in the numerator.

$$\lim_{x\to -2}\frac{x^3+8}{x+2}=\lim_{x\to -2}\frac{(x+2)(x^2-2x+4)}{x+2}$$

$$=\lim_{x\to -2}\frac{(x+2)\,(x^2-2x+4)}{x+2}$$

$$=\lim_{x\to -2}(x^2-2x+4)$$

$$=(-2)^2-2(-2)+4$$

$$=12$$

> This is a formula you should memorize: $x^3+y^3=(x+y)(x^2-xy+y^2)$. Look at Problem 2.25 for more info.

Therefore, $\displaystyle\lim_{x \to -2} \frac{x^3 + 8}{x + 2} = 12$.

10.19 Evaluate $\displaystyle\lim_{x \to -3} \frac{2x^3 - 7x^2 - 33x + 18}{x^2 - 9}$.

See Problems 2.18 and 2.19 if you need a refresher on synthetic division.

Substituting $x = -3$ into the numerator and denominator results in $\frac{0}{0}$. Therefore, -3 is a root of both functions ($y = 2x^3 - 7x^2 - 33x + 18$ and $y = x^2 - 9$). All functions with a root of $x = a$ must contain the factor $(x - a)$. Therefore, $(x - (-3)) = (x + 3)$ is a factor of $2x^3 - 7x^2 - 33x + 18$. Use synthetic division to identify the remaining two factors; factor the difference of perfect squares within the denominator.

$$\begin{array}{r|rrrr} -3 & 2 & -7 & -33 & 18 \\ & & -6 & 39 & -18 \\ \hline & 2 & -13 & 6 & 0 \end{array}$$

Factor the quotient: $2x^2 - 13x + 6 = (2x - 1)(x - 6)$.

$$\lim_{x \to -3} \frac{2x^3 - 7x^2 - 33x + 18}{x^2 - 9} = \lim_{x \to -3} \frac{(x + 3)(2x - 1)(x - 6)}{(x + 3)(x - 3)}$$

Note that $x + 3$ is a factor of the numerator and the denominator, so it can be eliminated.

$$= \lim_{x \to -3} \frac{\cancel{(x + 3)}(2x - 1)(x - 6)}{\cancel{(x + 3)}(x - 3)}$$

$$= \lim_{x \to -3} \frac{(2x - 1)(x - 6)}{x - 3}$$

Substituting $x = -3$ into the expression results in a real, finite value.

$$= \frac{(2(-3) - 1)(-3 - 6)}{-3 - 3} = \frac{(-7)(-9)}{-6} = -\frac{63}{6} = -\frac{21}{2}$$

Therefore, $\displaystyle\lim_{x \to -3} \frac{2x^3 - 7x^2 - 33x + 18}{x^2 - 9} = -\frac{21}{2}$.

10.20 Calculate the exact value of $\displaystyle\lim_{x \to 3/8} \frac{64x^3 - 8x^2 - 366x + 135}{48x^3 + 86x^2 - 23x - 6}$.

This is a nice way of saying "Don't even think about using your calculator to figure out this problem, because you'll get a decimal instead of a fraction and I'll know you couldn't do it by hand."

Substituting $x = \frac{3}{8}$ into the numerator and denominator produces the indeterminate result $\frac{0}{0}$. Apply synthetic division.

$$\begin{array}{r|rrrr} \frac{3}{8} & 64 & -8 & -366 & 135 \\ & & 24 & 6 & -135 \\ \hline & 64 & 16 & -360 & 0 \end{array} \qquad \begin{array}{r|rrrr} \frac{3}{8} & 48 & 86 & -23 & -6 \\ & & 18 & 39 & 6 \\ \hline & 48 & 104 & 16 & 0 \end{array}$$

Rewrite the original limit statement in factored form.

$$\lim_{x \to 3/8} \frac{64x^3 - 8x^2 + 366x + 135}{48x^3 + 86x^2 - 23x - 6} = \lim_{x \to 3/8} \frac{(x - 3/8)(64x^2 + 16x - 360)}{(x - 3/8)(48x^2 + 104x + 16)}$$

The greatest common factor of both quadratics is 8.

$$= \lim_{x \to 3/8} \frac{(8x - 3)(8x^2 + 2x - 45)}{(8x - 3)(6x^2 + 13x + 2)}$$

Multiply the linear factor $(x - 3/8)$ by 8.

$$= \lim_{x \to 3/8} \frac{(x - 3/8)(8)(8x^2 + 2x - 45)}{(x - 3/8)(8)(6x^2 + 13x + 2)}$$

Reduce the rational expression and substitute $x = \dfrac{3}{8}$ into the result.

$$= \lim_{x \to 3/8} \frac{\cancel{(8x - 3)}(8x^2 + 2x - 45)}{\cancel{(8x - 3)}(6x^2 + 13x + 2)}$$

$$= \lim_{x \to 3/8} \frac{8x^2 + 2x - 45}{6x^2 + 13x + 2}$$

$$= \frac{8(3/8)^2 + 2(3/8) - 45}{6(3/8)^2 + 13(3/8) + 2}$$

$$= \frac{\dfrac{9}{8} + \dfrac{6}{8} - \dfrac{45}{1}\left(\dfrac{8}{8}\right)}{\dfrac{54}{64} + \dfrac{39}{8}\left(\dfrac{8}{8}\right) + \dfrac{2}{1}\left(\dfrac{64}{64}\right)}$$

> $\dfrac{8}{8}$ and $\dfrac{64}{64}$ are in there to help out with common denominators.

$$= \frac{\dfrac{9}{8} + \dfrac{6}{8} - \dfrac{360}{8}}{\dfrac{54}{64} + \dfrac{312}{64} + \dfrac{128}{64}}$$

$$= \frac{\left(-\dfrac{345}{8}\right)}{\left(\dfrac{494}{64}\right)}$$

> To simplify, divide the top and bottom by 16.

Dividing by a fraction is equivalent to multiplying by its reciprocal.

$$= -\frac{345}{8} \cdot \frac{64}{494}$$

$$= -\frac{22,080}{3,952}$$

$$= -\frac{1,380}{247}$$

Therefore, $\displaystyle\lim_{x \to 3/8} \frac{64x^3 - 8x^2 - 366x + 135}{48x^3 + 86x^2 - 23x - 6} = -\frac{1,380}{247}$.

10.21 Under what conditions does $\lim\limits_{x \to c} \dfrac{(x+a)(x+b)}{x-c} = L$, if a, b, c, and L are real numbers and $c < 0 < a < b$?

> A number divided by 0 usually means an asymptote, but 0/0 usually means a hole on the graph.

From the statement $c < 0 < a < b$, you can conclude that a, b, and c are distinct (unequal) numbers, a and b are positive numbers, and c is a negative number. Substituting $x = c$ into the expression causes the denominator to equal 0 (since $c - c = 0$). If the numerator does not also equal 0 when $x = c$, then $x = c$ is a vertical asymptote to the graph of $y = \dfrac{(x+a)(x+b)}{x-c}$, and either $L = \infty$ or $L = -\infty$, neither of which is a real, finite number.

Therefore, one of the two factors in the numerator must equal 0 when $x = c$ in order for a limit to exist. Set both factors equal to 0, substitute $x = c$ in both equations, and solve.

$$\begin{array}{ccc} x + a = 0 & & x + b = 0 \\ c + a = 0 & \text{or} & c + b = 0 \\ c = -a & & c = -b \end{array}$$

Thus, $\lim\limits_{x \to c} \dfrac{(x+a)(x+b)}{x-c} = L$ if c is either the opposite of a or the opposite of b.

Conjugate Method
Break this out to deal with troublesome radicals

10.22 Simplify the expression: $\left(\sqrt{x} + 3\right)\left(\sqrt{x} - 3\right)$.

Apply the FOIL method to multiply the factors. (See Problem 2.13 for an explanation of the FOIL method.)

$$\left(\sqrt{x} + 3\right)\left(\sqrt{x} - 3\right) = \sqrt{x} \cdot \sqrt{x} - 3\sqrt{x} + 3\sqrt{x} + 3(-3)$$
$$= \sqrt{x^2} - 3\sqrt{x} + 3\sqrt{x} - 9$$
$$= x - 9$$

> You'd be taking the square root of a negative number.

It is not necessary to write $|x| - 9$ even though $\sqrt{x^2}$ is typically simplified as $|x|$. Notice that the original expression is invalid when $x < 0$, so it is appropriate to assume that $x > 0$ and omit the explicit absolute value indicators in the final solution.

10.23 Evaluate $\lim\limits_{x \to 16} \dfrac{x - 16}{\sqrt{x} - 4}$.

> The conjugate of $\sqrt{x} - 4$ is $\sqrt{x} + 4$. Just change the sign between the radical $\left(\sqrt{x}\right)$ and the number (4).

Substituting $x = 16$ results in $\dfrac{16 - 16}{4 - 4} = \dfrac{0}{0}$, an indeterminate value. To apply the conjugate method, multiply the numerator and denominator by the conjugate of the denominator. This method enables you to apply the substitution method without altering the value of the function or its limit.

$$\lim_{x \to 16} \frac{x-16}{\sqrt{x}-4} = \lim_{x \to 16} \frac{x-16}{\sqrt{x}-4} \cdot \frac{\sqrt{x}+4}{\sqrt{x}+4}$$

Multiply the denominators using the method of Problem 10.22. To speed up the simplification process, do not expand the product in the numerator.

$$= \lim_{x \to 16} \frac{(x-16)\left(\sqrt{x}+4\right)}{x-16}$$

Notice that the term $x-16$ appears in both the numerator and denominator; reduce the fraction by eliminating it.

$$= \lim_{x \to 16} \frac{\cancel{(x-16)}\left(\sqrt{x}+4\right)}{\cancel{x-16}}$$

$$= \lim_{x \to 16} \left(\sqrt{x}+4\right)$$

$$= \sqrt{16}+4$$

$$= 8$$

Therefore, $\displaystyle\lim_{x \to 16} \frac{x-16}{\sqrt{x}-4} = 8$.

See why you shouldn't multiply the numerator? If you did everything right, the factor without a square root in it gets cancelled out.

10.24 Evaluate $\displaystyle\lim_{x \to -4} \frac{-2+\sqrt{-x}}{8+2x}$.

Substitution produces an indeterminate result, so apply the conjugate method as described in Problem 10.23. Multiply the numerator and denominator by the conjugate of the numerator and simplify.

This time the conjugate is $-2 - \sqrt{-x}$.

$$\lim_{x \to -4} \frac{-2+\sqrt{-x}}{8+2x} = \lim_{x \to -4} \frac{-2+\sqrt{-x}}{8+2x} \cdot \frac{\left(-2-\sqrt{-x}\right)}{\left(-2-\sqrt{-x}\right)}$$

$$= \lim_{x \to -4} \frac{(-2)(-2)+2\sqrt{-x}-2\sqrt{-x}-\sqrt{(-x)^2}}{(8+2x)\left(-2-\sqrt{-x}\right)}$$

To facilitate simplification, do not expand the product in the denominator. Multiply only the part of the fraction that contains the conjugate pair.

$$= \lim_{x \to -4} \frac{4+\cancel{2\sqrt{-x}}-\cancel{2\sqrt{-x}}-(-x)}{(8+2x)\left(-2-\sqrt{-x}\right)}$$

$$= \lim_{x \to -4} \frac{4+x}{(8+2x)\left(-2-\sqrt{-x}\right)}$$

Factor 2 out of the expression $8+2x$ in the denominator.

$$= \lim_{x \to -4} \frac{\cancel{4+x}}{2\cancel{(4+x)}\left(-2-\sqrt{-x}\right)}$$

$$= \lim_{x \to -4} \frac{1}{2\left(-2-\sqrt{-x}\right)}$$

Substitute $x = -4$ into the expression.

$$= \frac{1}{2\left(-2 - \sqrt{-(-4)}\right)}$$

$$= \frac{1}{2\left(-2 - \sqrt{4}\right)}$$

$$= -\frac{1}{8}$$

Therefore, $\displaystyle\lim_{x \to -4} \frac{-2 + \sqrt{-x}}{8 + 2x} = -\frac{1}{8}$.

10.25 Evaluate $\displaystyle\lim_{x \to 5} \frac{x - 5}{\sqrt{x - 5}}$.

The square root of x equals $x^{1/2}$.

The conjugate method is not necessary to evaluate this limit. Instead, rewrite the radical expression using a rational exponent.

$$\lim_{x \to 5} \frac{x - 5}{\sqrt{x - 5}} = \lim_{x \to 5} \frac{(x - 5)^1}{(x - 5)^{1/2}}$$

To calculate the quotient of exponential expressions with the same base, subtract the powers.

$$= \lim_{x \to 5} (x - 5)^{1 - 1/2} = \lim_{x \to 5} (x - 5)^{1/2} = \lim_{x \to 5} \sqrt{x - 5}$$

Substitute $x = 5$ into the expression to calculate the limit.

$$\lim_{x \to 5} \sqrt{x - 5} = \sqrt{5 - 5} = \sqrt{0} = 0$$

10.26 Evaluate the limit: $\displaystyle\lim_{x \to 7} \frac{\sqrt{x - 6} - 1}{7 - x}$.

Multiply the numerator and denominator by $\sqrt{x - 6} + 1$, the conjugate of $\sqrt{x - 6} - 1$.

$$\lim_{x \to 7} \frac{\sqrt{x - 6} - 1}{7 - x} = \lim_{x \to 7} \frac{\sqrt{x - 6} - 1}{7 - x} \cdot \frac{\sqrt{x - 6} + 1}{\sqrt{x - 6} + 1}$$

$$= \lim_{x \to 7} \frac{(x - 6) - 1}{(7 - x)\left(\sqrt{x - 6} + 1\right)}$$

$$= \lim_{x \to 7} \frac{x - 7}{(7 - x)\left(\sqrt{x - 6} + 1\right)}$$

You could also factor –1 out of 7 – x in the denominator to get –(x – 7). You'll end up with the same final answer.

Factor –1 out of the terms in the numerator to reverse their order. This allows you to reduce the fraction.

$$= \lim_{x \to 7} \frac{-(7 - x)}{(7 - x)\left(\sqrt{x - 6} + 1\right)}$$

$$= \lim_{x \to 7} \frac{-1}{\sqrt{x - 6} + 1}$$

Apply the substitution method.

$$= \frac{-1}{\sqrt{7-6}+1} = \frac{-1}{\sqrt{1}+1} = -\frac{1}{2}$$

10.27 Evaluate the limit: $\lim\limits_{x \to 19} \dfrac{x-19}{5-\sqrt{x+6}}$.

Apply the conjugate method using the expression $5 + \sqrt{x+6}$.

$$\lim_{x \to 19} \frac{x-19}{5-\sqrt{x+6}} = \lim_{x \to 19} \frac{x-19}{5-\sqrt{x+6}} \cdot \frac{5+\sqrt{x+6}}{5+\sqrt{x+6}}$$

$$= \lim_{x \to 19} \frac{(x-19)\left(5+\sqrt{x+6}\right)}{25-(x+6)}$$

$$= \lim_{x \to 19} \frac{(x-19)\left(5+\sqrt{x+6}\right)}{19-x}$$

Factor –1 out of the denominator in order to reduce the fraction, as previously demonstrated by Problem 10.26.

$$= \lim_{x \to 19} \frac{\cancel{(x-19)}\left(5+\sqrt{x+6}\right)}{-\cancel{(x-19)}} = \lim_{x \to 19} -\left(5+\sqrt{x+6}\right) = -\left(5+\sqrt{25}\right) = -10$$

Special Limit Theorems

Limit formulas you should memorize

10.28 Evaluate $\lim\limits_{x \to \infty} \dfrac{c}{x^2}$, if c is a positive real number.

The degree of the denominator is 2, and the degree of the numerator is 0—technically, $c = c \cdot x^0 = c \cdot 1$. According to Problem 4.31, when the degree of the numerator is greater than the degree of the numerator, the limit at infinity equals 0: $\lim\limits_{x \to \infty} \dfrac{c}{x^2} = 0.$ ◄

Basically, any finite number divided by a humongous infinite number equals 0.

10.29 Evaluate $\lim\limits_{\theta \to 0} \dfrac{\sin 7\theta}{\theta}$.

A common limit formula states that $\lim\limits_{x \to 0} \dfrac{\sin x}{x} = 1$. Multiply the numerator and denominator by 7 in order to force the denominator to match the argument of $\sin 7\theta$.

$$\lim_{\theta \to 0} \frac{\sin 7\theta}{\theta} = \lim_{\theta \to 0} \frac{\sin 7\theta}{\theta} \cdot \frac{7}{7}$$

$$= \lim_{\theta \to 0} \frac{7 \sin 7\theta}{7\theta}$$

That formula works only if x approaches 0 and the value inside the sine function matches the denominator.

The limit of a product is equal to the product of its limits.

$$= \left(\lim_{\theta \to 0} 7\right)\left(\lim_{\theta \to 0} \frac{\sin 7\theta}{7\theta}\right)$$

$$= 7 \cdot 1 = 7$$

Therefore, $\lim_{\theta \to 0} \dfrac{\sin 7\theta}{\theta} = 7$.

10.30 Evaluate $\lim_{\theta \to 0} \dfrac{\cos 8\theta - 1}{2\theta}$.

Like Problem 10.29, this problem requires the knowledge of a common limit formula: $\lim_{x \to 0} \dfrac{\cos x - 1}{x} = 0$. Again, your goal is to match the argument of the trigonometric formula and the denominator of the expression. As the denominator of the expression is far easier to manipulate than the argument of cosine, force 2θ to become 8θ by multiplying the numerator and denominator by 4.

> **Multiply both parts of the fraction by 4, not just the bottom, or you'll change the function and its limit.**

$$\lim_{\theta \to 0} \frac{\cos 8\theta - 1}{2\theta} \cdot \frac{4}{4} = \lim_{\theta \to 0} \frac{4(\cos 8\theta - 1)}{8\theta} = \left(\lim_{\theta \to 0} 4\right)\left(\lim_{\theta \to 0} \frac{\cos 8\theta - 1}{8\theta}\right) = 4 \cdot 0 = 0$$

10.31 Evaluate $\lim_{x \to \infty} \left(1 + \dfrac{1}{x}\right)^{x(\ln 8)}$.

> **In other words, you should either recognize it or (if you don't recognize it) memorize it.**

Rewrite the function so that it more closely resembles a common limit: $\lim_{x \to \infty} \left(1 + \dfrac{1}{x}\right)^x = e$. According to exponential properties, an exponential expression raised to a power is equal to the original base raised to the product of the powers: $\left[f(x)\right]^{a \cdot b} = \left[\left(f(x)\right)^a\right]^b$. In this problem, $f(x) = \left(1 + \dfrac{1}{x}\right)$, $a = x$, and $b = \ln 8$, so

$$\lim_{x \to \infty} \left(1 + \frac{1}{x}\right)^{x(\ln 8)} = \lim_{x \to \infty} \left[\left(1 + \frac{1}{x}\right)^x\right]^{\ln 8}.$$

The power rule for limits states that $\lim_{x \to c} f(x)^a = \left(\lim_{x \to c} f(x)\right)^a$.

> **So the exponent ln 8 is not inside the limit anymore—the limit is raised to that power of ln 8.**

$$\lim_{x \to \infty} \left[\left(1 + \frac{1}{x}\right)^x\right]^{\ln 8} = \left[\lim_{x \to \infty} \left(1 + \frac{1}{x}\right)^x\right]^{\ln 8}$$

As stated earlier, $\lim_{x \to \infty} \left(1 + \dfrac{1}{x}\right)^x = e$.

$$\left[\lim_{x \to \infty} \left(1 + \frac{1}{x}\right)^x\right]^{\ln 8} = (e)^{\ln 8}$$

Because $y = \ln x$ and $y = e^x$ are inverse functions, $\ln\left(e^x\right) = e^{\ln x} = x$. Therefore, $e^{\ln 8} = 8$.

Chapter II
CONTINUITY AND THE DIFFERENCE QUOTIENT

Unbreakable graphs and a preview of derivatives

Though limits are exceedingly useful, there are few calculus applications that make *explicit* use of them. That does not imply that limits are unimportant, but that they are typically one feature of a far more substantial concept—continuity. The graph of an everywhere continuous function is best characterized by this property: it can be drawn without lifting your pencil from the graph. The predictable behavior of continuous graphs guarantees the functionality of key calculus theorems, such as the intermediate value theorem, and substantiates meaningful conclusions, such as the difference quotient.

Continuous functions don't have any undefined points, asymptotes, or breaks in the graph—they're just run-of-the-mill curves that, like roads, should be unbroken, un-potholed, and certainly un-suddenly-ending-in-a-cliff. If a function is continuous at every x in its domain, then you call the function "everywhere continuous." That doesn't mean the function's domain has to be all real numbers, only that the function is continuous everywhere it is defined. Once you understand continuity, it's time to start finding derivatives using the difference quotient, even if it is extraordinarily painful to do so. (Not because it's hard, but because you'll find out there are much shorter and easier ways to take derivatives discussed in Chapter 12.)

Continuity

Limit exists + function defined = continuous

A limit tells you what height a function INTENDS to reach. If the function is continuous, that means not only did the function intend to get there, it actually got there.

11.1 If $f(-2) = 7$, what other conditions must be met to ensure $f(x)$ is continuous at $x = -2$?

If $f(x)$ is continuous at $x = c$, three conditions must be met:

(1) $\lim\limits_{x \to c} f(x) = L_1$, such that L_1 is a real number

(2) $f(c) = L_2$, such that L_2 is a real number

(3) $L_1 = L_2$

Because $f(-2) = 7$, condition (2) is met. In order for $f(x)$ to be continuous at $x = -2$, the function must approach 7 as x approaches -2: $\lim\limits_{x \to -2} f(x) = 7$.

11.2 A function $r(x)$ is defined by the set of ordered pair listed below. At which value(s) in the domain of $r(x)$ is the function continuous?

$$r(x): \{(-2,6),(1,7),(4,6)\}$$

Even if you defined $r(3.999) = 5.999$, $r(3.9999) = 5.9999$, etc., $r(x)$ would still be discontinuous at $x = 4$ because you wouldn't be able to define an infinite number of points.

A function defined as a finite set of discrete points is not continuous at any of those points. In order to possess a limit at $x = c$ (one of the qualifying conditions for continuity at $x = c$), a function must approach the same value as x approaches c from the left and from the right. It is impossible to approach c from either direction, as the function is undefined immediately to the left and right of $x = c$, and no finite number of points could remedy that.

11.3 Which of the following trigonometric functions are continuous?

$y = \cos \theta, y = \sin \theta, y = \tan \theta, y = \cot \theta, y = \sec \theta, y = \csc \theta$

This is kind of a trick question, unless you remembered that functions aren't considered discontinuous at x-values you're not allowed to plug in.

This question is posed in a purposefully vague fashion. You may be tempted to classify $y = \tan \theta$, $y = \cot \theta$, $y = \sec \theta$, and $y = \csc \theta$ as discontinuous because their graphs contain vertical asymptotes. Not only do graphs lack function values at vertical asymptotes, they also have infinite limits, so all the conditions of continuity are violated.

However, it is only appropriate to judge a function's continuity over the interval for which it is defined. The functions $y = \tan \theta$ and $y = \sec \theta$ are undefined at $x = \dfrac{k\pi}{2}$ (when k is an odd integer), and $y = \cot \theta$ and $y = \csc \theta$ are undefined at $x = k\pi$ (for any integer k). These, not coincidentally, are the x-values for which the functions don't meet the requirements of continuity.

Therefore, the most appropriate answer is that each trigonometric function is continuous over its entire domain.

11.4   At what value(s) of x is the greatest integer function, $y = [\![x]\!]$, discontinuous?

The greatest integer function, whose graph appears in Figure 11-1, outputs the greatest integer that is less than or equal to the input.

So $[\![6.8]\!] = 6$ and $\left[\!\!\left[2\frac{1}{3}\right]\!\!\right] = 2$, but $[\![-1.3]\!] = -2$, not -1. The output is the biggest integer LESS than the input, and $-1 > -1.3$.

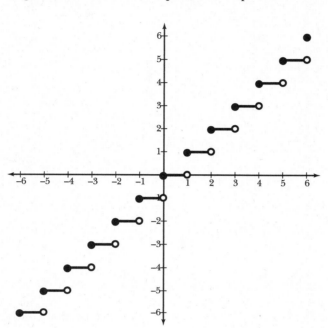

Figure 11-1   *The graph of $y = [\![x]\!]$, the greatest integer function.*

Though $y = [\![x]\!]$ is defined for all real numbers, at each integer j, $\lim\limits_{x \to j^-} [\![x]\!] \neq \lim\limits_{x \to j^+} [\![x]\!]$. Specifically, $\left(\lim\limits_{x \to j^-} [\![x]\!]\right) + 1 = \lim\limits_{x \to j^+} [\![x]\!]$. Because a general limit does not exist at each integer in its domain, $y = [\![x]\!]$ is discontinuous at those values.

Types of Discontinuity
Holes vs. breaks, removable vs. nonremovable

Note: Problems 11.5–11.13 refer to the graph of f(x) in Figure 11-2.

11.5   Is $f(x)$ continuous at $x = -3$? Explain your answer using the definition of continuity.

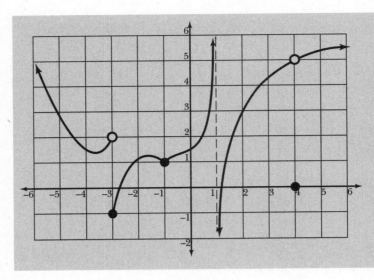

Figure 11-2

The graph of a function f(x).

Because $\lim\limits_{x \to -3^-} f(x) \neq \lim\limits_{x \to -3^+} f(x)$, the general limit $\lim\limits_{x \to -3} f(x)$ does not exist. Even though the function is defined at $x = -3$, the absence of a limit means that $f(x)$ is discontinuous there.

Note: Problems 11.5–11.13 refer to the graph of f(x) in Figure 11-2.

11.6   Is $f(x)$ continuous at $x = -1$? Explain your answer using the definition of continuity.

Yes, $f(x)$ is continuous at $x = -1$, as $\lim\limits_{x \to -1} f(x) = f(-1) = 1$.

Note: Problems 11.5–11.13 refer to the graph of f(x) in Figure 11-2.

11.7   Is $f(x)$ continuous at $x = 1$? Explain your answer using the definition of continuity.

No, $f(x)$ is discontinuous at $x = 1$, because $\lim\limits_{x \to 1} f(x)$ is not a finite number: $\lim\limits_{x \to 1^-} f(x) = \infty$ and $\lim\limits_{x \to 1^+} f(x) = -\infty$. Furthermore, $f(1)$ is not in the domain of $f(x)$.

> *f(x) increases without bound as x approaches 1 from the left and decreases without bound as x approaches 1 from the right.*

> *The function never intersects the vertical line x = 1, so as far as f(x) is concerned, x = 1 doesn't exist.*

Note: Problems 11.5–11.13 refer to the graph of f(x) in Figure 11-2.

11.8   Is $f(x)$ continuous at $x = 4$? Explain your answer using the definition of continuity.

No, $f(x)$ is not continuous at $x = 4$ because $\lim\limits_{x \to 4} f(x) = 5$ but $f(4) = 0$. The limit and function value must be equal in order to classify $f(x)$ continuous at $x = 4$.

Note: Problems 11.5–11.13 refer to the graph of f(x) in Figure 11-2.

11.9   At what value of x does the graph of $f(x)$ exhibit one-sided continuity (i.e., the graph is continuous only as you approach x from the left *or* from the right)?

The graph of $f(x)$ is continuous from the right at $x = -3$, because $\lim_{x \to -3^+} f(x) = f(-3) = -1$. One-sided continuity differs from standard continuity in that the general limit need not exist; it is sufficient that a one-sided limit at that x-value exists and is equal to the function value.

Note: Problems 11.5–11.13 refer to the graph of f(x) in Figure 11-2.

11.10   At what value of x does the graph of $f(x)$ exhibit point discontinuity?

If $\lim_{x \to c} f(x)$ exists but does not equal $f(c)$, $f(x)$ is said to demonstrate point discontinuity at $x = c$. Whether or not $f(c)$ actually exists is irrelevant. In Figure 11-2, $\lim_{x \to 4} f(x) = 5$ but $f(4) = 0$, so $f(x)$ exhibits point discontinuity at $x = 4$ (and still would even if $f(4)$ were undefined).

> Point discontinuity happens when a curve has a hole in it but doesn't actually break into two different pieces.

Note: Problems 11.5–11.13 refer to the graph of f(x) in Figure 11-2.

11.11   At what value of x does the graph of $f(x)$ exhibit jump discontinuity?

If $\lim_{x \to c^-} f(x) = L_1$ and $\lim_{x \to c^+} f(x) = L_2$ (where L_1 and L_2 are real numbers) but $L_1 \neq L_2$, then $f(x)$ exhibits jump discontinuity at $x = c$. In Figure 11-2, $\lim_{x \to -3^-} f(x) = 2$ but $\lim_{x \to -3^+} f(x) = -1$, so $f(x)$ exhibits jump discontinuity at $x = -3$.

> Think of the function as a sidewalk. It's called "jump" discontinuity because you'd have to jump at x = -3 if you wanted to stay on the sidewalk.

11.12   At what value of x does the graph of $f(x)$ exhibit infinite discontinuity?

If $\lim_{x \to c} f(x) = \infty$ or $\lim_{x \to c} f(x) = -\infty$ (i.e., when a function increases or decreases without bound near a vertical asymptote), $f(x)$ is said to demonstrate infinite discontinuity at $x = c$. In Figure 11-2, $f(x)$ exhibits infinite discontinuity at $x = 1$.

> Some books call it "essential discontinuity" instead of "infinite discontinuity," but they're the same thing.

Note: Problems 11.5–11.13 refer to the graph of f(x) in Figure 11-2.

11.13   Classify each instance of discontinuity in the graph of $f(x)$ as either removable or nonremovable.

Functions that are discontinuous despite the existence of a limit (i.e., functions with point discontinuity) are said to be "removably discontinuous," because redefining the function to correspond with the existing limit would, effectively, "remove" the discontinuity from the function. The graph of $f(x)$ possesses removable discontinuity at $x = 4$, because replacing $f(4) = 0$ with $f(4) = 5$ would make $f(x)$ continuous at $x = 4$.

> It would "fill in the hole" at the point (4,5) on the graph.

On the other hand, discontinuity caused by the nonexistence of a limit (i.e. jump and infinite discontinuity) is classified as "nonremovable," as redefining a finite number of points will not "remove" the discontinuity. Because no general limits exist on the graph of $f(x)$ at $x = -3$ and $x = 1$, $f(x)$ is nonremovably discontinuous at those values.

Note: Problems 11.14–11.16 refer to the graph of g(x) in Figure 11-3.

11.14 Identify the x-values at which $g(x)$ is discontinuous, and classify each instance of discontinuity as point, jump, or infinite.

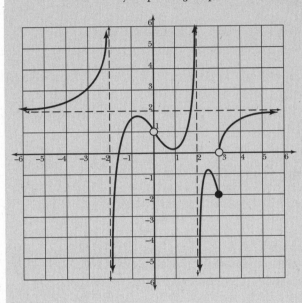

Figure 11-3

The graph of g(x) has horizontal asymptote y = 2 and vertical asymptotes x = –2 and x = 2.

The graph of $g(x)$ exhibits infinite discontinuities at $x = -2$ and $x = 2$, point discontinuity at $x = 0$, and jump discontinuity at $x = 3$.

Note: Problems 11.14–11.16 refer to the graph of g(x) in Figure 11-3.

11.15 At what value(s) of x is $g(x)$ nonremovably discontinuous?

> $g(x)$ is removably discontinuous at $x = 0$ because $\lim_{x \to 0} g(x) = 1$ (even though $g(0)$ is undefined).

Because $g(x)$ has no general limit as x approaches -2, 2, and 3, $g(x)$ is nonremovably discontinuous at those x-values.

Note: Problems 11.14–11.16 refer to the graph of g(x) in Figure 11-3.

11.16 At what value(s) of x does $g(x)$ exhibit one-sided continuity?

> $g(x)$ is continuous from the LEFT because the LEFT-hand limit equals g(3).

The function $g(x)$ is continuous from the left at $x = 3$, because $\lim_{x \to 3^-} g(x) = g(3) = -2$.

11.17 Given the piecewise-defined function $h(x)$ defined below, identify any value(s) of x at which $h(x)$ is discontinuous and describe the discontinuity exhibited.

$$h(x) = \begin{cases} 3x - x^2, & x < -1 \\ 6x + 2, & x > -1 \end{cases}$$

The function $h(x)$ is comprised of two polynomial functions, $y = 3x - x^2$ and $y = 6x + 2$, both of which are continuous over their entire domains. Therefore, the only x-value at which $h(x)$ could be discontinuous is $x = -1$, where the rule defining $h(x)$ changes from the quadratic to the linear equation.

> Polynomial, rational, trigonometric, exponential, and logarithmic functions are always continuous over their entire domains.

In order for $h(x)$ to be continuous at $x = -1$, the limit as x approaches -1 must exist and it must be equal to $h(-1)$. To determine if the limit exists, substitute -1 into both formulas; if the results are equal, $\lim\limits_{x \to -1} h(x)$ exists.

Substitute $x = -1$ into $3x - x^2$:   Substitute $x = -1$ into $6x + 2$:

$$3(-1) - (-1)^2 = -3 - (1) = -4 \qquad 6(-1) + 2 = -6 + 2 = -4$$

Therefore, $\lim\limits_{x \to -1^-} h(x) = \lim\limits_{x \to -1^+} h(x) = -4$, so $\lim\limits_{x \to -1} h(x) = -4$. Notice, however, that $h(-1)$ is undefined. According to the definition of $h(x)$, the quadratic rule $y = 3x - x^2$ applies when $x < -1$ and the linear rule $y = 6x + 2$ applies when $x > -1$. No rule addresses the case of $x = -1$, so $h(-1)$ does not exist, and $h(x)$ is discontinuous at $x = -1$.

Because a limit exists for $h(x)$ as x approaches -1, this is an example of point discontinuity, which is removable.

> Changing one of the inequality signs in $h(x)$ would make it continuous. Replacing $<$ with $\leq$ or $>$ with $\geq$ would mean $h(-1) = -4$, which would make $h(x)$ continuous.

11.18 Given the piecewise-defined function $f(x)$ defined below, identify any value(s) of x at which $f(x)$ is discontinuous and describe the discontinuity exhibited.

$$f(x) = \begin{cases} x^2 - 5x + 3, & x \leq 2 \\ x^3 - 12, & x > 2 \end{cases}$$

Notice that $f(x)$ is comprised of two polynomial functions, both of which are continuous over their entire domains; therefore, the only location at which $f(x)$ could be discontinuous is $x = 2$. Begin by evaluating $f(2)$.

$$f(2) = (2)^2 - 5(2) + 3 = 4 - 10 + 3 = -3$$

In order for $f(x)$ to be continuous at $x = 2$, $\lim\limits_{x \to 2} f(x)$ must also equal -3. Use the method described in Problem 11.17 (substituting $x = 2$ into both rules of the function) to determine whether or not the limit exists as x approaches 2. Notice that, by calculating $f(2)$, you've already substituted x into one of the functions. Simply evaluate the remaining function at $x = 2$.

$$y = x^3 - 12 = (2)^3 - 12 = 8 - 12 = -4$$

Therefore, $\lim\limits_{x \to 2^-} f(x) = -3$ but $\lim\limits_{x \to 2^+} f(x) = -4$. Because the left- and right-hand limits are unequal, the general limit $\lim\limits_{x \to 2} f(x)$ does not exist and $h(x)$ has a nonremovable jump discontinuity at $x = 2$.

> Even though the limit doesn't exist as x approaches 2, the right- and left-hand limits are FINITE, so there's a jump discontinuity. INFINITE one-sided limits are a sign of infinite discontinuity.

11.19 Determine the value of c that makes the piecewise-defined function $g(x)$ everywhere continuous.

$$g(x) = \begin{cases} \sqrt{2x - c}, & x < 0 \\ 3x^2 + 1, & x \geq 0 \end{cases}$$

The function $g(x)$ is comprised of a radical expression and a quadratic expression, both of which are continuous over their entire domains. Therefore, the only location at which $g(x)$ may be discontinuous is at $x = 0$. Begin by evaluating $g(0)$.

> Continuous over its entire domain.

$$g(0) = 3(0)^2 + 1 = 0 + 1 = 1$$

This result represents both $g(0)$ and $\lim\limits_{x \to 0^+} g(x)$, since $y = 3x^2 + 1$ generates the function values for all x's to the *right* of $x = 0$). If $g(x)$ is to be made continuous at $x = 0$, the left-hand limit as x approaches 0 must equal the right–hand limit. Calculate the left-hand limit by substituting $x = 0$ into $\sqrt{2x - c}$.

> The equation equals 1 because you just figured out that $\lim\limits_{x \to 0^+} g(x) = 1$ a few lines up, so you should substitute that in.

$$\lim_{x \to 0^-} g(x) = 1$$
$$\sqrt{2(0) - c} = 1$$
$$\sqrt{-c} = 1$$

Solve the equation for c.

$$\left(\sqrt{-c}\right)^2 = (1)^2$$
$$-c = 1$$
$$c = -1$$

When $c = -1$, $g(x)$ is continuous at every real number.

11.20 Calculate the value of k that makes $h(\theta)$ continuous over the interval $\left(-\dfrac{\pi}{2}, \dfrac{\pi}{2}\right)$.

$$h(\theta) = \begin{cases} 2(\tan \theta + 1), & \theta \leq -\dfrac{\pi}{4} \\ \tan(k\theta - 1), & \theta > -\dfrac{\pi}{4} \end{cases}$$

In order for $h(\theta)$ to be continuous at $x = -\dfrac{\pi}{4}$, substituting $x = -\dfrac{\pi}{4}$ into both pieces of the piecewise-defined function should produce the same result.

> That would mean the left- and right-hand limits are the same and they both equal the function value because the top rule includes $\theta = \dfrac{\pi}{4}$.

$$2\left(\tan\left(-\frac{\pi}{4}\right) + 1\right) = \tan\left(k \cdot \left(-\frac{\pi}{4}\right) - 1\right)$$

Solve for k.

$$2\tan\left(-\frac{\pi}{4}\right)+2=\tan\left(\frac{-k\pi}{4}-1\right)$$

$$2(-1)+2=\tan\left(\frac{-k\pi}{4}-1\right)$$

$$0=\tan\left(\frac{-k\pi}{4}-1\right)$$

$$\arctan 0=\arctan\left[\tan\left(\frac{-k\pi}{4}-1\right)\right]$$

$$0=\frac{-k\pi}{4}-1$$

$$4=-k\pi$$

$$-\frac{4}{\pi}=k$$

$$\tan\left(-\frac{\pi}{4}\right)$$
$$=\tan\left(-\frac{\pi}{4}+\pi\right)$$
$$=\tan\frac{3\pi}{4}$$
$$=\frac{\sqrt{2}/2}{-\sqrt{2}/2}$$
$$=-1$$

When $k=-\dfrac{4}{\pi}$, $h(\theta)$ is continuous over $\left(-\dfrac{\pi}{2},\dfrac{\pi}{2}\right)$.

11.21 Calculate the value of c that makes $f(x)$ everywhere continuous.

$$f(x)=\begin{cases}\ln\left(3c-2x^2\right), & x<7 \\ \ln\left(x+2c\right), & x\geq 7\end{cases}$$

Substitute $x=7$ into the expressions and set them equal. This effectively forces the left- and right-hand limits to be equal at $x=7$, which in turn forces $f(x)$ to be continuous there.

$$\ln\left(3c-2\cdot 7^2\right)=\ln\left(7+2c\right)$$

$$\ln\left(3c-98\right)=\ln\left(7+2c\right)$$

Solve the equation for c. In order to eliminate the natural logarithmic functions, exponentiate the equation using e. ←

$$e^{\ln(3c-98)}=e^{\ln(7+2c)}$$

$$3c-98=7+2c$$

$$c=105$$

When $c=105$, $f(x)$ is continuous for all real numbers.

Look at Problem 5.34 if you're not sure what exponentiate means, or why you use e.

11.22 Calculate the values of a and b that make $g(x)$, as defined below, everywhere continuous.

$$g(x)=\begin{cases}x^2, & x<-4 \\ ax+b, & -4\leq x<5 \\ \sqrt{x+31}, & x\geq 5\end{cases}$$

In order to ensure that the functions $y = x^2$ and $y = ax + b$ have the same limit as x approaches -4, substitute $x = -4$ into both and set them equal.

$$x^2 = ax + b$$
$$(-4)^2 = a(-4) + b$$
$$16 = -4a + b$$

> Because $g(x)$ changes at $x = -4$, you plug that into the two function rules that contain -4 in their boundaries and set them equal. Do the same thing for the second rule change, at $x = 5$.

The functions $y = ax + b$ and $y = \sqrt{x + 31}$ must also have matching limits, as x approaches 5.

$$ax + b = \sqrt{x + 31}$$
$$a(5) + b = \sqrt{5 + 31}$$
$$5a + b = 6$$

You now know that $-4a + b = 16$ and $5a + b = 6$. Solve the system of two equations in two variables to calculate a and b. To solve the system using elimination, subtract the equation $-4a + b = 16$ from $5a + b = 6$.

$$
\begin{array}{rrrr}
5a & + \ b & = & 6 \\
4a & - \ b & = & -16 \\
\hline
9a & & = & -10 \\
& a & = & -\dfrac{10}{9}
\end{array}
$$

Substitute a into either equation in the system to determine the corresponding value of b.

$$5a + b = 6$$
$$5\left(-\frac{10}{9}\right) + b = 6$$
$$-\frac{50}{9} + b = 6$$
$$b = \frac{54}{9} + \frac{50}{9}$$
$$b = \frac{104}{9}$$

When $a = -\dfrac{10}{9}$ and $b = \dfrac{104}{9}$, $g(x)$ is continuous over its entire domain.

11.23 Given a function $f(x)$ that is continuous over the closed interval $[a,b]$, what conclusions can be drawn about d, if d is a real number in the closed interval $[f(a), f(b)]$?

According to the intermediate value theorem, there must exist a value c in the closed interval $[a,b]$ such that $f(c) = d$.

> The bar above 6704 means those four numbers repeat infinitely. In other words, $f(c) = 5.9243186704046704...$

11.24 Apply the intermediate value theorem to verify the following statement:

Given the function $f(x) = x^2$, there exists some number c on the interval $[-1,3]$, such that $f(c) = 5.9243186\overline{704}$.

You can only apply the intermediate value theorem to functions if they are continuous over a specifically identified closed interval. In this case, $f(x) = x^2$ is a polynomial function whose domain is all real numbers, so $f(x)$ is continuous on any closed interval, including the interval specified: $[-1,3]$. Once continuity is assured, evaluate $f(x)$ at the endpoints of the closed interval.

$$f(-1) = (-1)^2 = 1 \qquad\qquad f(3) = (3)^2 = 9$$

For any real number d in the interval $[1,9]$, the intermediate value theorem guarantees that there exists a value c on $[-1,3]$ for which $f(c) = d$. Notice that the relatively absurd number 5.9243186704 falls on the interval $[1,9]$, so some value c on the interval $[-1,3]$ exists such that $f(c) = 5.9243186704$.

> In other words, you can plug any real number into x² and you're sure to get a real number back out.

11.25 A calculus student is adamant that the intermediate value theorem is fundamentally flawed. Locate and explain the logical flaw in his "counterexample" below.

The function $g(x) = \sec x$ is a trigonometric function, and like all trigonometric functions, is continuous over its entire domain or any piece of that domain, including $[0,\pi]$. If you evaluate the endpoints of the interval, you see that $\sec 0 = 1$ and $\sec \pi = -1$. Therefore, according to the intermediate value theorem, I can choose any number d from the interval $[-1,1]$, such as the number 0, and some other number c must exist in the interval $[0,\pi]$ such that $\sec c = 0$. However, $g(x) = \sec x$ never equals 0! Therefore, the intermediate value theorem is not necessarily true, especially in the case of trig functions. ←

> It would take a lot of guts to suddenly announce that you'd disproved a calculus theorem that's been around for eons, don't you think?

Although each trigonometric function is, indeed, continuous over its entire *domain*, $g(x) = \sec x$ is not continuous over the entire *interval* $[0,\pi]$—it is undefined at $x = \dfrac{\pi}{2}$. Therefore, the intermediate value theorem cannot be applied.

11.26 Use the intermediate value theorem to prove that the number $\sqrt[3]{20}$ exists and has a value greater than 2 but less than 3.

> The problem asks you to show that $\sqrt[3]{20}$ is greater than 2 but less than 3, so you have to prove that the endpoints don't give you 20 when you plug them into f(x).

Given a function $f(x)$ that's continuous on the closed interval $[a,b]$, any number d between $f(a)$ and $f(b)$ has a corresponding value c between a and b such that $f(c) = d$ (according to the intermediate value theorem). In this problem, $f(x) = x^3$, $a = 2$, $b = 3$, $f(a) = f(2) = 8$, and $f(b) = f(3) = 27$. Thus, for any value d you choose in the interval $[8,27]$ (such as $d = 20$), there exists a corresponding c in the interval $[2,3]$ such that $f(c) = 20$ ($c = \sqrt[3]{20}$). Because $f(2) = 8$ and $f(3) = 27$ (and neither of them equal 20), $\sqrt[3]{20}$ actually belongs to the interval $(2,3)$.

11.27 Use the intermediate value theorem to verify that $h(x) = (x-3)^2 - 7$ has a root between $x = 5$ and $x = 6$.

If you were wet at 5 p.m. and dry at 6 p.m., then at some point between 5 and 6 p.m., you got out of the pool. At some point between $x = 5$ and $x = 6$, $f(x)$ got out of the negative.

Because $h(5) = -3$ and $h(6) = 2$, the intermediate value theorem guarantees that $d = 0$ (which belongs to the interval $[-3,2]$) will correspond to some c in the interval $[5,6]$ such that $h(c) = 0$. In other words, there's some value c between $x = 5$ and $x = 6$ that makes the function equal 0, and is therefore a root of the function. Note that this problem does not ask you to find c—the intermediate value theorem is an *existence theorem* that merely guarantees the *existence* of c but falls short of actually identifying it.

11.28 If $f(x)$ is continuous over the interval $[0,1]$ such that $f(0)$ and $f(1)$ also belong to the interval $[0,1]$, prove that there exists some value c in $[0,1]$ such that $f(c) = c$.

Your goal is to demonstrate that $f(x) = x$ for some x in the interval $[0,1]$, which is the equivalent of demonstrating that $f(x) - x = 0$ for some x in $[0,1]$. To simplify later calculations, write the difference as a new function, $g(x) = f(x) - x$, and attempt to verify that $g(x)$ has a root in $[0,1]$.

This fact has a formal name: the fixed point theorem.

If $f(x)$ is a continuous function over $[0,1]$, then $g(x)$ is continuous as well—the difference of continuous functions is, itself, continuous—so you can apply the intermediate value theorem. Begin by evaluating $g(x)$ at the specified endpoints, much like you evaluated $h(5)$ and $h(6)$ in Problem 11.27.

$$g(0) = f(0) - 0 \qquad g(1) = f(1) - 1$$

This warrants closer inspection. Remember that both $f(0)$ and $f(1)$ are between 0 and 1, so they're positive numbers. If $f(0)$ is a positive number, then so is $g(0)$, because $g(0)$ instructs you to subtract 0 from a positive number, which won't change its value. On the other hand, $g(1)$ must be a negative number. Remember, $f(1)$, like $f(0)$, is a positive number less than 1, so $f(1) - 1$ must be a negative number.

Actually, they're nonnegative numbers, so they're either positive or 0, but if either one equals 0, you've already proven that $g(x)$ has a 0, and the proof is already over!

If $g(0) > 0$ and $g(1) < 0$, then you can choose $d = 0$, and according to the intermediate value theorem, some c must exist in $[0,1]$ such that $g(c) = 0$. Essentially, once you find function values that have different signs, you can conclude that a continuous function must cross the x-axis somewhere between those function values, thereby possessing a root between them as well. Therefore, $g(x) = f(x) - x$ has a root in $[0,1]$, and thus $c = f(c)$.

The Difference Quotient
The "long way" to find the derivative

11.29 Explain the relationship between differentiation and calculating limits.

Derivatives are indelibly tied to limits at the most foundational level—derivatives are defined as limits of specific fractions called "difference quotients." The most common difference quotient is a formula containing "Δx" (read "delta x") used to calculate the general derivative $f'(x)$ of a function $f(x)$.

$$f'(x) = \lim_{\Delta x \to 0} \frac{f(x + \Delta x) - f(x)}{\Delta x}$$

In order to quickly evaluate a derivative for a specific x-value, a second version of the difference quotient is sometimes employed, one which contains a constant (like c in the formula below) representing the value at which you are evaluating the derivative.

$$f'(c) = \lim_{x \to c} \frac{f(x) - f(c)}{x - c}$$

Note: Problems 11.30–11.33 refer to the function $f(x) = 7x^2$.

11.30 Use the definition of the derivative to find $f'(x)$. ←

> The three most common ways to indicate a derivative are f'(x), y', and dy/dx.

Apply the general difference quotient, as described in Problem 11.29. To do so, first substitute $x + \Delta x$ into $f(x) = 7x^2$, then subtract $f(x) = 7x^2$ from the result, and finally divide by Δx.

$$f'(x) = \lim_{\Delta x \to 0} \frac{f(x + \Delta x) - f(x)}{\Delta x}$$

$$= \lim_{\Delta x \to 0} \frac{7(x + \Delta x)^2 - 7x^2}{\Delta x}$$

Expand the expression $(x + \Delta x)^2$. Note that Δx should be considered a single value, not the product $\Delta \cdot x$. Therefore, $(\Delta x)(\Delta x) = (\Delta x)^2$, not $\Delta^2 x^2$.

> Write $(x + \Delta x)^2$ as $(x + \Delta x)(x + \Delta x)$ and use the FOIL method.

$$= \lim_{\Delta x \to 0} \frac{7\left(x^2 + 2x\Delta x + (\Delta x)^2\right) - 7x^2}{\Delta x}$$

$$= \lim_{\Delta x \to 0} \frac{7x^2 + 14x\Delta x + 7(\Delta x)^2 - 7x^2}{\Delta x}$$

$$= \lim_{\Delta x \to 0} \frac{14x\Delta x + 7(\Delta x)^2}{\Delta x}$$

Evaluate the limit by factoring. ←

> Pull the greatest common factor out of the numerator, just like we did back in Problem 10.13 when there weren't any triangles in the problems yet.

$$= \lim_{\Delta x \to 0} \frac{\Delta x(14x + 7\Delta x)}{\Delta x}$$

$$= \lim_{\Delta x \to 0} (14x + 7\Delta x)$$

Substitute $\Delta x = 0$ into the expression to evaluate the limit.

$$= 14x + 7(0)$$
$$= 14x$$

Therefore, if $f(x) = 7x^2$, then $f'(x) = 14x$.

Note: Problems 11.30–11.33 refer to the function $f(x) = 7x^2$.

11.31 Calculate $f'(3)$.

According to Problem 11.30, $f'(x) = 14x$. To evaluate $f'(3)$, substitute 3 into the derivative function.

$$f'(3) = 14(3) = 42$$

If you're a fan of the book Hitchhiker's Guide to the Galaxy, *you know that the answer to "life, the universe, and everything" is 42. Now you know the question: "If $f(x) = 7x^2$, what is $f'(3)$?" Try not to look so disappointed.*

Note: Problems 11.30–11.33 refer to the function $f(x) = 7x^2$.

11.32 Given a is a real number, calculate $f'(a-4)$.

Replace the x in the derivative formula with the quantity $(a-4)$.

$$f'(x) = 14x$$
$$f'(a-4) = 14(a-4)$$
$$f'(a-4) = 14a - 64$$

To find the derivative at a specific number (like $x = 3$), either use the $\Delta x \to 0$ formula (like Problem 11.30) and then plug 3 into the answer (like Problem 11.31), or just use the $x \to c$ formula (like Problem 11.32).

Note: Problems 11.30–11.33 refer to the function $f(x) = 7x^2$.

11.33 Use the specific value difference quotient to calculate $f'(3)$ and verify that the solution matches the solution to Problem 11.31.

Substitute $f(x) = 7x^2$ and $c = 3$ into the difference quotient.

$$f'(c) = \lim_{x \to c} \frac{f(x) - f(c)}{x - c}$$
$$f'(3) = \lim_{x \to 3} \frac{7x^2 - f(3)}{x - 3}$$
$$= \lim_{x \to 3} \frac{7x^2 - 7(3)^2}{x - 3}$$
$$= \lim_{x \to 3} \frac{7x^2 - 63}{x - 3}$$

$a^2 - b^2 = (a+b)(a-b)$

Factor the numerator completely, including the difference of perfect squares that arises once 7 is factored out of the quadratic expression.

$$= \lim_{x \to 3} \frac{7(x^2 - 9)}{x - 3}$$

$$= \lim_{x \to 3} \frac{7(x + 3)(x - 3)}{x - 3}$$

$$= \lim_{x \to 3} 7(x + 3)$$

$$= 7(3 + 3)$$

$$= 42$$

Therefore, $f'(3) = 42$, a result that matches Problem 11.31.

Note: Problems 11.34–11.36 refer to the function $g(x) = x^2 - 5x + 9$.

11.34 Use the definition of the derivative to find $g'(x)$.

Apply the general difference quotient formula, as demonstrated by Problem 11.30.

$$g'(x) = \lim_{\Delta x \to 0} \frac{g(x + \Delta x) - g(x)}{\Delta x}$$

$$= \lim_{\Delta x \to 0} \frac{\left[(x + \Delta x)^2 - 5(x + \Delta x) + 9\right] - (x^2 - 5x + 9)}{\Delta x}$$

Make sure you plug $x + \Delta x$ into both of the x's in $g(x)$.

Simplify the expression, expanding $(x + \Delta x)^2$ and distributing –5 and –1.

$$= \lim_{\Delta x \to 0} \frac{\left[x^2 + 2x\Delta x + (\Delta x)^2 - 5x - 5\Delta x + 9\right] - x^2 + 5x - 9}{\Delta x}$$

$$= \lim_{\Delta x \to 0} \frac{x^2 - x^2 + 2x\Delta x + (\Delta x)^2 - 5x + 5x - 5\Delta x + 9 - 9}{\Delta x}$$

$$= \lim_{\Delta x \to 0} \frac{2x\Delta x + (\Delta x)^2 - 5\Delta x}{\Delta x}$$

At this point in a difference quotient problem, everything inside these parentheses needs to have Δx in it. If one of your terms is missing a Δx, go back and fix the mistake before you go on.

Apply the factoring method to evaluate the limit.

$$= \lim_{\Delta x \to 0} \frac{\Delta x (2x + \Delta x - 5)}{\Delta x}$$

$$= \lim_{\Delta x \to 0} (2x + \Delta x - 5)$$

$$= 2x + 0 - 5$$

$$= 2x - 5$$

If $g(x) = x^2 - 5x + 9$, then $g'(x) = 2x - 5$.

Note: Problems 11.34–11.36 refer to the function $g(x) = x^2 - 5x + 9$.

11.35 Evaluate $g'(-2)$.

Substitute $x = -2$ into the derivative function from Problem 11.34.

$$g'(x) = 2x - 5$$

$$g'(-2) = 2(-2) - 5$$

$$g'(-2) = -9$$

Note: Problems 11.34–11.36 refer to the function $g(x) = x^2 - 5x + 9$.

11.36 Use the specific difference quotient to calculate $g'(-2)$ and thereby verify the solution to Problem 11.35.

Use the method described in Problem 11.33, this time using the function $g(x) = x^2 - 5x + 9$ and the constant $c = -2$.

$$g'(c) = \lim_{x \to c} \frac{g(x) - g(c)}{x - c}$$

$$= \lim_{x \to -2} \frac{(x^2 - 5x + 9) - g(-2)}{x - (-2)}$$

$$= \lim_{x \to -2} \frac{(x^2 - 5x + 9) - \left[(-2)^2 - 5(-2) + 9\right]}{x + 2}$$

$$= \lim_{x \to -2} \frac{(x^2 - 5x + 9) - [4 + 10 + 9]}{x + 2}$$

$$= \lim_{x \to -2} \frac{x^2 - 5x - 14}{x + 2}$$

> Make sure to plug –2 into x^2 AND $-5x$.

Use the factoring method to evaluate the limit.

$$= \lim_{x \to -2} \frac{(x - 7)(x + 2)}{x + 2}$$

$$= \lim_{x \to -2} \frac{(x - 7)\,\cancel{(x + 2)}}{\cancel{x + 2}}$$

$$= \lim_{x \to -2} (x - 7)$$

$$= -2 - 7$$

$$= -9$$

Therefore, if $g(x) = x^2 - 5x + 9$, then $g'(-2) = -9$, which matches the solution to Problem 11.35.

Differentiability

When does a derivative exist?

11.37 The derivative describes what geometric characteristic of a function's graph?

The derivative of $f(x)$ at $x = c$, written $f'(c)$, is equal to the slope of the tangent line to $f(x)$ at the point $(c, f(c))$.

11.38 Describe the relationship between the continuity of a function and its differentiability.

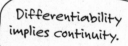

> Differentiability implies continuity.

If a function is differentiable at $x = c$, it must also be continuous at $x = c$. The converse is not true: a function continuous at $x = c$ is not necessarily differentiable at $x = c$. See Problem 11.41 for an example of a continuous but nondifferentiable function.

Note: Problems 11.39–11.42 refer to the graph of f(x) in Figure 11-4.

11.39 Given $\lim\limits_{x \to 6^+} \dfrac{f(x) - f(6)}{x - 6} = \lim\limits_{x \to 6^-} \dfrac{f(x) - f(6)}{x - 6} = \dfrac{1}{2}$, does $f'(6) = \dfrac{1}{2}$? Why or why not?

> Do you recognize the difference quotient from Problems 11.33 and 11.36? These are the derivatives of the last two pieces of the function, the ones that have 6 in their boundaries.

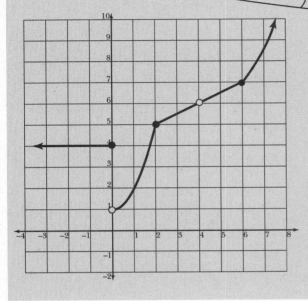

Figure 11-4

The graph of the piecewise-defined function f(x).

$$f(x) = \begin{cases} 4, & x \leq 0 \\ x^2 + 1, & 0 < x \leq 2 \\ (1/2)x + 4, & 2 < x < 4 \\ (1/2)x + 4, & 4 < x < 6 \\ x^2 - (23/2)x + 40, & x \geq 6 \end{cases}$$

The given left- and right-hand limits represent the left- and right-hand derivatives of $f(x)$ as x approaches 6. According to the information given,

$\lim\limits_{x \to 6^+} \dfrac{f(x) - f(6)}{x - 6} = \dfrac{1}{2}$, so the derivative of $y = x^2 - \dfrac{23}{2}x + 40$ equals $\dfrac{1}{2}$ when $x = 6$.

Additionally, $\lim\limits_{x \to 6^-} \dfrac{f(x) - f(6)}{x - 6} = \dfrac{1}{2}$, so the derivative of $y = \dfrac{1}{2}x + 4$ equals $\dfrac{1}{2}$ at

$x = 6$.

Because $f(x)$ is continuous at $x = 6$ (which is another matter entirely, but the function is continuous because $\lim\limits_{x \to 6^-} f(x) = \lim\limits_{x \to 6^+} f(x) = f(6) = 7$) and the left- and right-hand derivatives are equal at $x = 6$, $f(x)$ is differentiable at $x = 6$; specifically, $f'(6) = \dfrac{1}{2}$.

> For a piecewise-defined function to be differentiable at an x-value "break point" (like x = 0, 2, 4, or 6 in Figure 11-4), the left- and right-hand limits as x approaches c must equal f(c), and the left- and right-hand derivatives as x approaches c must be equal.

Note: Problems 11.39–11.42 refer to the graph of f(x) in Figure 11-4.

11.40 Given $\lim\limits_{x \to 4^+} \dfrac{f(x) - f(4)}{x - 4} = \lim\limits_{x \to 4^-} \dfrac{f(x) - f(4)}{x - 4} = \dfrac{1}{2}$, does $f'(4) = \dfrac{1}{2}$? Why or why not?

> The derivative of a linear function is the slope of the line.

While the left- and right-hand derivatives of $f(x)$ are equal as x approaches 4, $f(x)$ is discontinuous at $x = 4$. Though a limit exists $\left(\lim\limits_{x \to 4} f(x) = 6\right)$, $f(4)$ is undefined, which results in point discontinuity. Therefore, $f'(4)$ does not exist—in order for a function to be differentiable at $x = c$, it must also be continuous at $x = c$ (as stated in Problem 11.38).

Note: Problems 11.39–11.42 refer to the graph of f(x) in Figure 11-4.

11.41 Given $\lim\limits_{x \to 2^+} \dfrac{f(x) - f(2)}{x - 2} = \dfrac{1}{2}$ and $\lim\limits_{x \to 2^-} \dfrac{f(x) - f(2)}{x - 2} = 4$, evaluate $f'(2)$.

Although $f(x)$ is continuous at $x = 2$, the left- and right-hand derivatives are not equal as x approaches 2. Therefore, $f(x)$ is not differentiable at $x = 2$ and $f'(2)$ does not exist.

Note: Problems 11.39–11.42 refer to the graph of f(x) in Figure 11-4.

11.42 Given $\lim\limits_{x \to 0^+} \dfrac{f(x) - f(0)}{x - 0} = 0$ and $\lim\limits_{x \to 0^-} \dfrac{f(x) - f(0)}{x - 0} = 0$, evaluate $f'(0)$.

Notice that $f(x)$ is discontinuous as x approaches 0 because the right hand limit $\lim\limits_{x \to 0^+} f(x) = 1$ does not equal the left-hand limit $\lim\limits_{x \to 0^-} f(x) = f(0) = 4$. Although the right- and left-hand derivatives of $f(x)$ are equal as x approaches 0, the right- and left-hand limits of $f(x)$ are unequal as x approaches 0, so $f'(0)$ does not exist.

Ending a chapter with two trick questions in a row is just rude.

Chapter 12
BASIC DIFFERENTIATION METHODS

The four heavy hitters for finding derivatives

Though conceptually gratifying and a fitting culmination of prior limit evaluation techniques (most specifically the factoring method), calculating derivatives by means of the difference quotient is, at best, prohibitively time-consuming and, at worst, nontrivial. It behooves a student of calculus to embrace procedural algorithms once an underlying concept is understood, as the determination of derivatives is but one of the foundational skills to be learned in this course. Undoubtedly, the investment of time required to truly master archaic limit expressions is not proportional to the worth of such an undertaking.

In math talk, "nontrivial" means "pretty hard."

The difference quotient is only good for calculating easy derivatives. And I mean really, really easy derivatives. If you're trying to find the derivative of x^2, the difference quotient is no big deal, but what about x^5? Do you really feel like expanding $(x + \Delta x)^5$? It doesn't sound like a good time to me. In this chapter, you'll learn to find derivatives without having to resort to the difference quotient.

Different techniques apply in different circumstances, so you'll learn how to differentiate functions that are raised to powers, multiplied by other functions, divided by other functions, and composed with (plugged into) other functions. But before all that, you need to learn how to take derivatives of functions that aren't polynomials. If you have trouble with the problems in the first section of the chapter, flip back to Appendix D—it contains all the important derivative formulas.

Trigonometric, Logarithmic, and Exponential Derivatives
Memorize specific formulas for these functions

12.1 Differentiate with respect to x: $f(x) = \sin x$.

The derivative of the sine function, as well as the derivatives of the other five basic trigonometric functions, are used throughout calculus and should be memorized: $f'(x) = \cos x$.

12.2 Differentiate with respect to x: $y = 5 \tan \theta$.

This means you can ignore the coefficient of a function when you take a derivative. Once you figure out what the derivative is, just multiply it by the coefficient.

The derivative of a function multiplied by a constant is equal to the constant multiplied by the derivative of the function.

$$\frac{d}{dx}\big[k \cdot f(x)\big] = k \cdot \frac{d}{dx}\big[f(x)\big]$$

Multiply the derivative of $\tan \theta$ by 5.

$$\frac{dy}{d\theta} = 5 \cdot \frac{d}{d\theta}(\tan\theta)$$
$$= 5\big(\sec^2\theta\big)$$
$$= 5\sec^2\theta$$

12.3 Find the second derivative of $y = \cos x$, with respect to x.

Differentiate both sides of the equation with respect to x. (Note that the derivative of y with respect to y is written $\dfrac{dy}{dx}$.)

$$\frac{dy}{dx} = -\sin x$$

The derivative of $\dfrac{dy}{dx}$ is written $\dfrac{d^2y}{dx^2}$. As explained in Problem 12.2, the derivative of $(-1)\cdot \sin x$ is equal to $(-1)\cdot \dfrac{d}{dx}(\sin x)$.

$$\frac{d^2y}{dx^2} = (-1)\left[\frac{dy}{dx}(\sin x)\right]$$
$$= (-1)[\cos x]$$

If $y = \cos x$, then $\dfrac{d^2y}{dx^2} = -\cos x$.

12.4 Differentiate with respect to θ: $y = \sec\theta - \csc\theta$.

The derivative of a difference is equal to the difference of the derivatives. (The same is true for the sum of derivatives, but is *not true* for the product or the quotient of derivatives.)

$$\frac{dy}{d\theta} = \frac{d}{d\theta}(\sec\theta) - \frac{d}{d\theta}(\csc\theta)$$

$$= (\sec\theta\tan\theta) - (-\csc\theta\cot\theta)$$

$$= \sec\theta\tan\theta + \csc\theta\cot\theta$$

Find the derivative of $\sec\theta$ and subtract the derivative of $\csc\theta$.

12.5 Given $g(x) = 4^x$, find $g''(x)$.

The derivative of the exponential function a^x is $a^x(\ln a)$, the function itself times the natural logarithm of its base.

$$g'(x) = \left(4^x\right)\ln 4$$

$$g''(x) = (\ln 4)\cdot\frac{d}{dx}\left(4^x\right)$$

$$= (\ln 4)\cdot\left(4^x\cdot\ln 4\right)$$

$$= (\ln 4)^2\cdot 4^x$$

Take the derivative of 4^x again while ignoring ln 4 (because it's a constant). You'll get the same thing you got the first time: $4^x(\ln 4)$.

12.6 Differentiate with respect to x: $y = 4e^x$.

Recall that the derivative of an exponential function is the function itself times the natural logarithm of its base.

$$\frac{dy}{dx} = 4\cdot\frac{d}{dx}\left(e^x\right)$$

$$= 4\left(e^x\cdot\ln e\right)$$

$$= 4\left(e^x\cdot 1\right)$$

$$= 4e^x$$

Because ln x and e^x are inverse functions, $\ln e^x = x$. That means $\ln e^1 = 1$.

This problem demonstrates an important differentiation formula: $\frac{d}{dx}\left(e^x\right) = e^x$; e^x is its own derivative.

12.7 Differentiate with respect to x: $y = \log_7 x$.

Note that $\frac{d}{dx}\left(\log_a x\right) = \frac{1}{x\cdot\ln a}$; the derivative of a logarithmic function is the reciprocal of the logarithmic argument $\left(\frac{1}{x}\right)$ divided by the natural logarithm of the base ($\ln a$). Therefore, $\frac{d}{dx}\left(\log_7 x\right) = \frac{1}{x(\ln 7)}$.

12.8   Differentiate with respect to x: $h(x) = \ln x$.

> This is another important derivative formula to memorize: if $f(x) = \ln x$, then $f'(x) = \frac{1}{x}$.

Apply the formula discussed in Problem 12.7.

$$h'(x) = \frac{1}{x \cdot \ln e}$$
$$= \frac{1}{x \cdot 1}$$
$$= \frac{1}{x}$$

The Power Rule
A shortcut for differentiating x^n

12.9   Differentiate with respect to x: $y = 5x^3$.

> Pull the exponent down in front of the x, and multiply it by the coefficient (if there is one). The new power will be one less than the old one.

The power rule for differentiation provides a simple method of differentiating a single variable (with or without a coefficient) that is raised to a power:
$\frac{d}{dx}\left(ax^n\right) = (n \cdot a)x^{n-1}$.

$$\frac{dy}{dx} = (3 \cdot 5)x^{3-1}$$
$$= 15x^2$$

12.10   Differentiate with respect to x: $f(x) = x^9$.

The expression x^9 has an implied coefficient of 1, so the coefficient of $f'(x)$ is $9 \cdot 1 = 9$.

$$f'(x) = (9 \cdot 1)x^{9-1}$$
$$= 9x^8$$

12.11   Differentiate with respect to x: $g(x) = -6x^{-4}$.

Remember to *subtract* one from the exponent when applying the power rule, even when the exponent is negative. In this problem, x will be raised to the $-4 - 1 = -5$ power.

$$g'(x) = (-4)(-6)x^{-4-1}$$
$$= 24x^{-5}$$
$$= \frac{24}{x^5}$$

12.12 Differentiate with respect to w: $y = \dfrac{6}{w}$.

Rewrite the function using a negative exponent. ←

$$y = 6 \cdot w^{-1}$$

Apply the power rule. Note that this function is written in terms of w, not x, so the derivative of y is $\dfrac{dy}{dw}$ instead of $\dfrac{dy}{dx}$.

$$\frac{dy}{dw} = (-1 \cdot 6) w^{-1-1}$$
$$= -6w^{-2}$$
$$= -\frac{6}{w^2}$$

> But don't leave negative exponents in a final answer.

12.13 Differentiate $f(x) = 2$ with respect to x.

Even though $f(x)$ contains only a constant, it can be written with an x-term. Because $x^0 = 1$, multiplying 2 by x^0 does not change its value: $f(x) = 2x^0$. Apply the power rule.

$$f'(x) = (0 \cdot 2) x^{0-1}$$
$$= 0x^{-1}$$
$$= \frac{0}{x}$$
$$= 0, \quad \text{if } x \neq 0$$

The domain of $f(x)$ is all real numbers, but the power rule cannot determine the derivative when $x = 0$. Consider the graph of $f(x)$, the horizontal line $y = 2$. Horizontal lines have a slope of 0 for all real numbers, so $f'(x)$—the slope of the tangent line to $f(x)$—equals 0 for all real numbers, including $x = 0$.

> In these two problems, the power rule helps you figure out the derivative, but it stops just short of giving you the whole answer. In both cases, the power rule can't calculate the derivative when $x = 0$, so you have to resort to looking at the actual graphs of the functions and figuring out what the slope of the tangent line is when $x = 0$.

12.14 Use the power rule to differentiate $f(x) = -6x$ with respect to x. Generalize the solution to construct a corollary of the power rule concerning the derivatives of linear terms.

Include the implied exponent of 1 when writing the function: $f(x) = -6x^1$. Apply the power rule.

$$f'(x) = 1 \cdot (-6) x^{1-1}$$
$$= -6x^0$$
$$= -6 \quad \text{if } x \neq 0 \ \leftarrow$$

Therefore, the derivative of a linear function is the slope of the line. Although the power rule leaves $f'(x)$ undefined at $x = 0$ (because $0^0 \neq 1$), the slope of the line $f(x) = -6x$ is -6 for all xs, including $x = 0$.

12.15 Differentiate $h(x) = ax^{a+1}$ with respect to x, assuming a is a real number.

Apply the power rule.

$$h'(x) = (a+1)(a)x^{(a+1)-1} = (a^2 + a)x^a$$

12.16 Differentiate with respect to x: $y = \sqrt[3]{x^2}$.

Rewrite the function using a rational exponent.

$$y = x^{2/3}$$

In case you forgot, $x^{a/b} = \sqrt[b]{x^a} = \left(\sqrt[b]{x}\right)^a$. See Problem 2.10 for more information.

Apply the power rule.

$$\frac{dy}{dx} = \left(\frac{2}{3} \cdot 1\right)x^{(2/3)-1}$$

Subtract 1 from the rational exponent: $\dfrac{2}{3} - 1 = \dfrac{2}{3} - \dfrac{3}{3} = -\dfrac{1}{3}$.

$$\frac{dy}{dx} = \left(\frac{2}{3}\right)x^{-1/3}$$

$$= \frac{2}{3x^{1/3}}$$

It is equally correct to write your solution as a radical: $\dfrac{dy}{dx} = \dfrac{2}{3\sqrt[3]{x}}$. However, rational solutions are usually presented in rationalized form.

That means you should get rid of the radical in the denominator.

$$\frac{2}{3\sqrt[3]{x}} = \frac{2}{3\sqrt[3]{x}} \cdot \frac{\sqrt[3]{x^2}}{\sqrt[3]{x^2}} = \frac{2\sqrt[3]{x^2}}{3\sqrt[3]{x^3}} = \frac{2\sqrt[3]{x^2}}{3x}$$

12.17 Differentiate with respect to t: $y = \dfrac{1}{\sqrt{t}}$.

Recall that 2 is the implied radical index when no other index is specified.

$$y = \frac{1}{\sqrt[2]{t^1}}$$

$$= t^{-1/2}$$

Apply the power rule.

$$\frac{dy}{dt} = -\frac{1}{2}t^{(-1/2)-1}$$

$$= -\frac{1}{2}t^{-3/2}$$

$$= -\frac{1}{2t^{3/2}}$$

12.18 Differentiate with respect to x: $f(x) = \sqrt{x}\left(\sqrt[5]{x^4} - \sqrt[7]{x^3}\right)$.

Rewrite the radical expressions using rational exponents.

$$f(x) = x^{1/2}\left(x^{4/5} - x^{3/7}\right)$$

Distribute $x^{1/2}$ through the parentheses.

$$= \left(x^{1/2}\right)\left(x^{4/5}\right) - \left(x^{1/2}\right)\left(x^{3/7}\right)$$
$$= x^{13/10} - x^{13/14}$$

The derivative of a difference is equal to the difference of the individual derivatives. Calculate those derivatives using the power rule.

$$f'(x) = \frac{d}{dx}\left(x^{13/10}\right) - \frac{d}{dx}\left(x^{13/14}\right)$$
$$= \frac{13}{10}x^{(13/10)-1} - \frac{13}{14}x^{(13/14)-1}$$
$$= \frac{13}{10}x^{(13/10)-(10/10)} - \frac{13}{14}x^{(13/14)-(14/14)}$$
$$= \frac{13}{10}x^{3/10} - \frac{13}{14}x^{-1/14}$$
$$= \frac{13}{10}x^{3/10} - \frac{13}{14x^{1/14}}$$

> Add the exponents when you're multiplying two things with the same base:
> $$\frac{1}{2} + \frac{4}{5} = \frac{5}{10} + \frac{8}{10} = \frac{13}{10}$$
> $$\frac{1}{2} + \frac{3}{7} = \frac{7}{14} + \frac{6}{14} = \frac{13}{14}$$

The Product and Quotient Rules
Differentiate functions that are multiplied or divided

12.19 Differentiate with respect to θ: $y = \sin\theta\cos\theta$.

The derivative of a product must be calculated using the product rule, stated below.

$$\frac{d}{dx}\left(f(x)\cdot g(x)\right) = f(x)\cdot g'(x) + g(x)\cdot f'(x)$$

> In other words, multiply the first function by the derivative of the second one, then add the second function multiplied by the derivative of the first one.

To apply the product rule formula, set $f(\theta) = \sin\theta$, $g(\theta) = \cos\theta$, $f'(\theta) = \cos\theta$, and $g'(\theta) = -\sin\theta$.

$$\frac{dy}{d\theta} = f(\theta)\cdot g'(\theta) + g(\theta)\cdot f'(\theta)$$
$$= \sin\theta(-\sin\theta) + (\cos\theta)(\cos\theta)$$
$$= \cos^2\theta - \sin^2\theta$$
$$= \cos 2\theta$$

> This comes from the double angle identity $\cos^2\theta - \sin^2\theta = \cos 2\theta$. Look it up in Appendix C or flip back to Problems 8.12 and 8.14 for examples that use this identity.

12.20 Differentiate with respect to x: $y = 2^x \cot x$.

Apply the product rule, as described in Problem 12.19. To apply the product rule formula, set $f(x) = 2^x$ and $g(x) = \cot x$.

$$\frac{dy}{dx} = f(x)g'(x) + g(x)f'(x)$$

$$= 2^x\left(-\csc^2 x\right) + \cot x\left(2^x \cdot \ln 2\right)$$

$$= -2^x \csc^2 x + (\ln 2)2^x \cot x$$

12.21 Differentiate with respect to x: $f(x) = 2x^3 e^x - 7x^2 e^x + 10xe^x - e^x$.

Rather than differentiating each term separately, first factor e^x out of the expression.

$$f(x) = e^x\left(2x^3 - 7x^2 + 10x - 1\right)$$

Now $f(x)$ is written explicitly as the product of two functions: e^x and $2x^3 - 7x^2 + 10x - 1$. Apply the product rule.

$$f'(x) = e^x \cdot \frac{d}{dx}\left(2x^3 - 7x^2 + 10x - 1\right) + \left(2x^3 - 7x^2 + 10x - 1\right) \cdot \frac{d}{dx}\left(e^x\right)$$

$$= e^x\left(6x^2 - 14x + 10\right) + \left(2x^3 - 7x^2 + 10x - 1\right)\left(e^x\right)$$

Distribute e^x through both quantities and simplify the expression by combining like terms.

$$= 6x^2 e^x - 14xe^x + 10e^x + 2x^3 e^x - 7x^2 e^x + 10xe^x - e^x$$

$$= 2x^3 e^x + \left(6x^2 e^x - 7x^2 e^x\right) + \left(-14xe^x + 10xe^x\right) + \left(10e^x - e^x\right)$$

$$= 2x^3 e^x - x^2 e^x - 4xe^x + 9e^x$$

This way, you only have to do the product rule once. If you take the derivatives of the terms separately, you'll have to apply the product rule three times.

12.22 Problem 12.9 used the power rule to determine that $\frac{d}{dx}\left(5x^3\right) = 15x^2$. Use the product rule to verify this result.

Interpret $5x^3$ as the product $f(x)g(x)$, where $f(x) = 5$ and $g(x) = x^3$.

$$\frac{d}{dx}\left(f(x)g(x)\right) = f(x)g'(x) + g(x)f'(x)$$

$$\frac{d}{dx}\left(5x^3\right) = (5) \cdot \frac{d}{dx}\left(x^3\right) + \left(x^3\right) \cdot \frac{d}{dx}(5)$$

$$= 5\left(3x^2\right) + x^3(0)$$

$$= 15x^2 + 0$$

$$= 15x^2$$

According to Problem 12.13, the derivative of 5 (or any constant) is 0.

12.23 Differentiate with respect to x: $y = \dfrac{\cos x}{3x}$.

The derivative of the quotient $h(x) = \dfrac{f(x)}{g(x)}$ is calculated according to the quotient rule, defined below.

$$h'(x) = \frac{g(x)\,f'(x) - f(x)\,g'(x)}{\left(g(x)\right)^2}$$

That's the bottom of the fraction times the derivative of the top minus the top times the derivative of the bottom, all divided by the bottom squared.

To apply the quotient rule, set $f(x) = \cos x$, $g(x) = 3x$, $f'(x) = -\sin x$, and $g'(x) = 3$.

$$\frac{d}{dx}\left(\frac{\cos x}{3x}\right) = \frac{3x \cdot \dfrac{d}{dx}(\cos x) - (\cos x)\dfrac{d}{dx}(3x)}{(3x)^2} = \frac{3x(-\sin x) - \cos x(3)}{9x^2} = \frac{-3x\sin x - 3\cos x}{9x^2}$$

Reduce the fraction by factoring -3 out of the numerator.

$$= \frac{-\cancel{3}\,(x\sin x + \cos x)}{\cancel{3} \cdot 3x^2} = -\frac{x\sin x + \cos x}{3x^2}$$

12.24 Differentiate with respect to x: $y = \dfrac{\ln x}{x}$.

Use the formula from Problem 12.23, setting $f(x) = \ln x$ and $g(x) = x$.

Apply the quotient rule.

$$\frac{dy}{dx} = \frac{x \cdot \dfrac{d}{dx}(\ln x) - (\ln x)\dfrac{d}{dx}(x)}{(x)^2}$$

$$= \frac{x\left(\dfrac{1}{x}\right) - (\ln x)(1)}{x^2}$$

$$= \frac{\dfrac{x}{x} - \ln x}{x^2}$$

$$= \frac{1 - \ln x}{x^2}$$

According to Problem 12.8, the derivative of $\ln x$ is $\dfrac{1}{x}$.

12.25 Differentiate with respect to x: $f(x) = \dfrac{x^3 - 2x^2 - 5x - 12}{4x^2 - 9}$.

Apply the quotient rule to set up the derivative and use the power rule to differentiate the individual polynomials.

$$f'(x) = \frac{\left(4x^2 - 9\right)\left(3x^2 - 4x - 5\right) - \left(x^3 - 2x^2 - 5x - 12\right)(8x)}{\left(4x^2 - 9\right)^2}$$

$$= \frac{12x^4 - 16x^3 - 47x^2 + 36x + 45 - \left(8x^4 - 16x^3 - 40x^2 - 96x\right)}{16x^4 - 72x^2 + 81}$$

$$= \frac{4x^4 - 7x^2 + 132x + 45}{16x^4 - 72x^2 + 81}$$

12.26 Verify the trigonometric derivative using the quotient rule: $\frac{d}{d\theta}(\tan\theta) = \sec^2\theta$.

As the problem implies, you must first rewrite $\tan\theta$ as a quotient.

$$\frac{d}{d\theta}(\tan\theta) = \frac{d}{d\theta}\left(\frac{\sin\theta}{\cos\theta}\right)$$

Apply the quotient rule.

> There aren't any derivatives in the denominator of the quotient rule—just the original denominator squared.

$$= \frac{\cos\theta \cdot \frac{d}{d\theta}(\sin\theta) - \sin\theta \cdot \frac{d}{d\theta}(\cos\theta)}{(\cos\theta)^2}$$

$$= \frac{\cos\theta(\cos\theta) - \sin\theta(-\sin\theta)}{\cos^2\theta}$$

$$= \frac{\cos^2\theta + \sin^2\theta}{\cos^2\theta}$$

According to a trigonometric identity, $\cos^2\theta + \sin^2\theta = 1$.

$$= \frac{1}{\cos^2\theta}$$

$$= \sec^2\theta$$

Note: Problems 12.27–12.29 refer to the functions f(x) and g(x) and their derivatives, f'(x) and g'(x). All four functions are differentiable on (−∞,∞) and selected values are listed in the table below.

x	-2	-1	0	1	2	3
$f(x)$	5	9	6	2	-1	-4
$f'(x)$	8	$\frac{1}{2}$	-2	-1	-3	-9
$g(x)$	1	3	-2	5	6	-1
$g'(x)$	2	0	-3	12	1	-6

12.27 Evaluate $\frac{d}{dx}\big(f(x) - 5g(x)\big)$ at $x = 1$.

> Look in the third column from the right. It gives you all the function values when x = 1.

The derivative of $f(x)$ is $f'(x)$ and the derivative of $-5g(x)$ is $-5g'(x)$.

$$\frac{d}{dx}\big(f(x) - 5g(x)\big) = f'(x) - 5g'(x)$$

Substitute $x = 1$ into the expression and use the above table to determine that $f'(1) = -1$ and $g'(1) = 12$.

$$f'(1) - 5g'(1) = (-1) - 5(12)$$

$$= -1 - 60$$

$$= -61$$

Note: Problems 12.27–12.29 reference the table of values provided in Problem 12.27.

12.28 Evaluate $(fg)'(2)$.

This problem asks you to differentiate $f(x) \cdot g(x)$ and then substitute $x = 2$ into your answer. You cannot simply multiply $f'(2)$ by $g'(2)$—you must apply the product rule.

$$(fg)'(2) = f(2) \cdot g'(2) + g(2) \cdot f'(2)$$
$$= (-1)(1) + (6)(-3)$$
$$= -19$$

Note: Problems 12.27–12.29 reference the table of values provided in Problem 12.27.

12.29 Evaluate $\left(\dfrac{g}{f}\right)'(-1)$.

As this derivative is the quotient of two functions, it requires the quotient rule.

$$\left(\frac{g}{f}\right)'(-1) = \frac{f(-1) \cdot g'(-1) - g(-1) \cdot f'(-1)}{[f(-1)]^2}$$
$$= \frac{(9)(0) - 3(1/2)}{(9)^2}$$
$$= \frac{-3/2}{81}$$
$$= -\frac{3}{162}$$
$$= -\frac{1}{54}$$

When a numerator has its own denominator (-3/2 has a denominator of 2), move that denominator down and multiply it by the denominator of the whole fraction: 2(81) = 162.

The Chain Rule
Differentiate functions that are plugged into functions

12.30 Assuming $f(x)$ and $g(x)$ are differentiable functions, differentiate $f(g(x))$ with respect to x.

According to the chain rule, the derivative of the composition of functions $f(g(x))$ equals $f'(g(x)) \cdot g'(x)$. In other words, the derivative of the "outer function" $f(x)$ evaluated at the "inner" function $g(x)$, which is then multiplied by the derivative of the inner function $g'(x)$.

You can assume that "differentiable" means "differentiable over their entire domains."

12.31 Differentiate $y = (2x-3)^2$ using the power rule and then verify the result using the chain rule.

Expand the polynomial using the FOIL method and differentiate.

$$y = 4x^2 - 12x + 9$$

$$\frac{dy}{dx} = 8x - 12$$

It is helpful to rewrite $y = (2x-3)^2$ as a composition of functions before differentiating it. The linear expression $2x - 3$ is squared, so in essence, you're substituting $2x - 3$ into the function x^2. The function being substituted is the "inner" function, $g(x) = 2x - 3$, and the function it's substituted into is the "outer" function, $f(x) = x^2$. Note that $f(g(x)) = f(2x - 3) = (2x - 3)^2$. Use the chain rule to calculate the derivative of $f(g(x))$.

$$\frac{d}{dx}\left[f(g(x)) \right] = f'(g(x)) \cdot g'(x)$$

The derivative of $f(x) = (x)^2$ is $f'(x) = 2(x)$; therefore, the derivative of $f(g(x)) = (g(x))^2$ equals $2(g(x))$.

$$\frac{d}{dx}\left[(2x-3)^2 \right] = 2(g(x)) \cdot g'(x)$$

Substitute $g(x) = 2x - 3$ and $g'(x) = 2$ into the expression.

$$\frac{d}{dx}\left[(2x-3)^2 \right] = 2(2x-3) \cdot 2 = (4x-6) \cdot 2 = 8x - 12$$

The power and chain rules produce identical derivatives: $\frac{d}{dx}\left[f(g(x)) \right] = 8x - 12$.

Rule of thumb: If the function you're differentiating has something besides x plugged into it, use the chain rule. For example, x^2 doesn't require the chain rule, but anything else squared will, like $(3x)^2$, $(x + 7)^2$, and $(\tan x)^2$.

12.32 Differentiate with respect to θ and simplify the result: $y = \sqrt{\sec \theta}$.

Rewrite the function using a rational exponent.

$$y = (\sec \theta)^{1/2}$$

Because the function $\sec \theta$, not just the single variable θ, is raised to a power, you must apply the chain rule. Rewrite the function, indicating the composition within explicitly. Note that the inner function is $g(\theta) = \sec \theta$ and the outer function is $f(\theta) = \theta^{1/2}$.

$$y = f(g(\theta))$$

$$\frac{dy}{d\theta} = f'(g(\theta)) \cdot g'(\theta)$$

Differentiate $f(\theta)$ using the power rule: $f'(\theta) = \frac{1}{2}\theta^{-1/2}$.

$$\frac{dy}{d\theta} = \frac{1}{2}(g(\theta))^{-1/2} \cdot g'(\theta)$$

This just means rewrite it as $f(g(\theta))$. You don't have to do this if you understand what's going on, but it helps if the chain rule confuses you.

Substitute $g(\theta) = \sec\theta$ and $g'(\theta) = \sec\theta\tan\theta$ into the chain rule formula.

$$\frac{dy}{d\theta} = \frac{1}{2}(\sec\theta)^{-1/2} \cdot (\sec\theta\tan\theta)$$

$$= \frac{\sec\theta\tan\theta}{2(\sec\theta)^{1/2}}$$

You can reduce the $\sec\theta$ factors by applying exponential properties:

$$\frac{(\sec\theta)^1}{(\sec\theta)^{1/2}} = (\sec\theta)^{1-(1/2)} = (\sec\theta)^{1/2} = \sqrt{\sec\theta}.$$

$$\frac{dy}{d\theta} = \frac{\sqrt{\sec\theta}\cdot\tan\theta}{2}$$

12.33 Differentiate with respect to x: $f(x) = \ln(\sin x)$.

Because a *function* is substituted into the natural logarithm function instead of a single *variable*, you must apply the chain rule. Differentiate the outer function $\ln x$, leaving the inner function $\sin x$ inside that derivative. Then multiply by the derivative of the inner function.

$$f'(x) = \frac{1}{\sin x} \cdot \frac{d}{dx}(\sin x)$$

$$= \frac{1}{\sin x}(\cos x)$$

$$= \frac{\cos x}{\sin x}$$

$$= \cot x$$

The derivative of the natural log of something equals 1 divided by that something. (In this case, the "something" is sin x.)

12.34 Differentiate with respect to x: $g(x) = e^{4x+1}$.

Because the exponent of e is a function $(4x + 1)$, not a single variable like x, you must apply the chain rule.

$$g'(x) = e^{4x+1} \cdot \frac{d}{dx}(4x+1)$$

$$= e^{4x+1} \cdot 4$$

$$= 4e^{4x+1}$$

Remember e^x is its own derivative, so the derivative of e to some power (in this case 4x + 1) equals e to that power times the derivative of that power.

12.35 Differentiate with respect to x: $y = \csc\left(e^{4x+1}\right)$.

Apply the chain rule formula such that $f(x) = \csc x$ and $g(x) = e^{4x+1}$.

$$\frac{d}{dx}\left[f(g(x))\right] = f'(g(x)) \cdot g'(x)$$

$$\frac{d}{dx}\left[\csc\left(e^{4x+1}\right)\right] = -\csc\left(e^{4x+1}\right)\cot\left(e^{4x+1}\right) \cdot \frac{d}{dx}\left(e^{4x+1}\right)$$

The derivative of csc x is –csc x cot x. Make sure to substitute e^{4x+1} for x in both csc x and cot x.

According to Problem 12.34, $\dfrac{d}{dx}\left(e^{4x+1}\right) = 4e^{4x+1}$.

$$= -\csc\left(e^{4x+1}\right)\cot\left(e^{4x+1}\right) \cdot 4e^{4x+1}$$
$$= -4e^{4x+1}\csc\left(e^{4x+1}\right)\cot\left(e^{4x+1}\right)$$

12.36 Differentiate with respect to x: $f\left(g\left(h(x)\right)\right)$.

Begin by differentiating the outermost function, leaving everything "inside it" alone, and then multiply by the derivative of the quantity "inside."

$$\frac{d}{dx}\left[f\left(g\left(h(x)\right)\right)\right] = f'\left[g\left(h(x)\right)\right] \cdot \frac{d}{dx}\left[g\left(h(x)\right)\right]$$

Notice that $\dfrac{d}{dx}\left[g\left(h(x)\right)\right]$ also requires the chain rule.

$$\frac{d}{dx}\left[g\left(h(x)\right)\right] = g'\left(h(x)\right) \cdot h'(x)$$

Therefore, $\dfrac{d}{dx}\left[f\left(g\left(h(x)\right)\right)\right] = f'\left[g\left(h(x)\right)\right] \cdot g'\left(h(x)\right) \cdot h'(x)$.

12.37 Differentiate with respect to θ: $j(\theta) = -3\sin\left(\cos\dfrac{\theta}{2}\right)$.

Apply the method outlined in Problem 12.36; begin by differentiating the outermost function.

$$j'(\theta) = -3\cos\left(\cos\frac{\theta}{2}\right) \cdot \frac{d}{d\theta}\left[\cos\frac{\theta}{2}\right]$$

Differentiating $\cos\dfrac{\theta}{2}$ also requires the chain rule.

$$j'(\theta) = -3\cos\left(\cos\frac{\theta}{2}\right) \cdot \left[-\sin\frac{\theta}{2} \cdot \frac{d}{d\theta}\left(\frac{\theta}{2}\right)\right]$$

Note that $\dfrac{\theta}{2} = \dfrac{1}{2}\theta$, so $\dfrac{d}{d\theta}\left(\dfrac{1}{2}\theta\right) = \dfrac{1}{2}$.

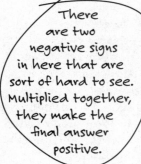

There are two negative signs in here that are sort of hard to see. Multiplied together, they make the final answer positive.

$$j'(\theta) = -3\cos\left(\cos\frac{\theta}{2}\right) \cdot \left[\left(-\sin\frac{\theta}{2}\right) \cdot \frac{1}{2}\right]$$
$$= \left(\frac{3}{2} \cdot \sin\frac{\theta}{2}\right)\cos\left(\cos\frac{\theta}{2}\right)$$

12.38 According to the quotient rule, $\dfrac{d}{dx}\left[\dfrac{f(x)}{g(x)}\right] = \dfrac{g(x)f'(x) - f(x)g'(x)}{\left(g(x)\right)^2}$. Because dividing and multiplying by a reciprocal are equivalent operations, generate the quotient rule formula by differentiating $f(x) \cdot \dfrac{1}{g(x)}$ with respect to x.

Rewrite $g(x)$ using a negative exponent.

$$f(x) \cdot \frac{1}{g(x)} = f(x) \cdot \left[g(x) \right]^{-1}$$

Apply the product rule.

$$\frac{d}{dx}\left(f(x) \cdot \frac{1}{g(x)} \right) = f(x) \cdot \frac{d}{dx}\left[(g(x))^{-1} \right] + \left[g(x) \right]^{-1} \cdot \frac{d}{dx}(f(x))$$

In order to differentiate $[g(x)]^{-1}$, you must apply the chain rule.

$$= f(x)\left[-\frac{g'(x)}{(g(x))^2} \right] + \left[g(x) \right]^{-1} \cdot f'(x)$$

> Here are the steps for finding that derivative:
> $$\frac{d}{dx}[(g(x))^{-1}]$$
> $$= (-1)[g(x)]^{-2} \cdot g'(x)$$
> $$= -\frac{g'(x)}{(g(x))^2}$$

Rewrite the derivative, eliminating the negative exponent.

$$= f(x)\left[-\frac{g'(x)}{(g(x))^2} \right] + \frac{1}{g(x)} \cdot f'(x)$$

$$= -\frac{f(x)g'(x)}{(g(x))^2} + \frac{f'(x)}{g(x)}$$

The least common denominator is $(g(x))^2$.

$$= -\frac{f(x)g'(x)}{(g(x))^2} + \frac{f'(x)}{g(x)} \cdot \frac{g(x)}{g(x)}$$

$$= -\frac{f(x)g'(x)}{(g(x))^2} + \frac{f'(x)g(x)}{(g(x))^2}$$

According to the commutative property, you can reorder the terms.

$$= \frac{f'(x)g(x)}{(g(x))^2} - \frac{f(x)g'(x)}{(g(x))^2}$$

$$= \frac{g(x)f'(x) - f(x)g'(x)}{(g(x))^2}$$

12.39 Differentiate with respect to x: $f(x) = x^3 \cos 2x$.

Because $f(x)$ is the product of two functions, you must apply the product rule.

$$f'(x) = x^3 \cdot \frac{d}{dx}[\cos 2x] + \cos 2x \cdot \frac{d}{dx}(x^3)$$

Use the power rule to differentiate x^3 and the chain rule to differentiate $\cos 2x$.

$$f'(x) = x^3 \cdot -\sin 2x \cdot 2 + (\cos 2x)(3x^2)$$
$$= -2x^3 \sin 2x + 3x^2 \cos 2x$$

> The derivative of $\cos 2x$ is $-2\sin 2x$. If you didn't get the 2 out front, you forgot to take the derivative of the inner function.

12.40 Verify the solution to Problem 12.39 by differentiating $f(x) = x^3 \cos 2x$ using the quotient rule.

You must first rewrite $f(x)$ as a quotient—otherwise, the quotient rule won't apply. To do so, utilize a reciprocal identity, rewriting $\cos 2x$ as $\dfrac{1}{\sec 2x}$.

$$f(x) = x^3 \cdot \frac{1}{\sec 2x}$$
$$= \frac{x^3}{\sec 2x}$$

Apply the quotient rule.

$$f'(x) = \frac{\sec 2x \cdot \frac{d}{dx}(x^3) - x^3 \cdot \frac{d}{dx}(\sec 2x)}{(\sec 2x)^2}$$
$$= \frac{\sec 2x \cdot 3x^2 - x^3 \cdot \sec 2x \tan 2x \cdot 2}{\sec^2 2x}$$
$$= \frac{3x^2 \sec 2x - 2x^3 \sec 2x \tan 2x}{\sec^2 2x}$$

Factor $\sec 2x$ out of the numerator and reduce the fraction.

$$f'(x) = \frac{\sec 2x \left[3x^2 - 2x^3 \tan 2x\right]}{\sec 2x \cdot \sec 2x}$$
$$= \frac{3x^2 - 2x^3 \tan 2x}{\sec 2x}$$

If $\dfrac{a}{b} + \dfrac{z}{b} = \dfrac{a+z}{b}$ because they have common denominators, you can work backward and say that $\dfrac{a+z}{b} = \dfrac{a}{b} + \dfrac{z}{b}$.

Write the derivative as two separate fractions, each with denominator $\sec x$.

$$f'(x) = \frac{3x^2}{\sec 2x} - \frac{2x^3 \tan 2x}{\sec 2x}$$

Rewrite the function in terms of sine and cosine: $\dfrac{1}{\sec 2x} = \cos 2x$ and $\tan 2x = \dfrac{\sin 2x}{\cos 2x}$.

$$f'(x) = 3x^2 \left(\frac{1}{\sec 2x}\right) - 2x^3 \tan 2x \left(\frac{1}{\sec 2x}\right)$$
$$= (3x^2 \cos 2x) - \left[2x^3 \cdot \frac{\sin 2x}{\cos 2x} \cdot (\cos 2x)\right]$$
$$= \left[3x^2 \cos 2x\right] - \left[\frac{2x^3 \cdot \sin 2x \cos 2x}{\cos 2x}\right]$$
$$= 3x^2 \cos 2x - 2x^3 \sin 2x$$

12.41 Differentiate with respect to x: $j(x) = \left(\dfrac{x}{e^x}\right) \ln x$.

Because $j(x)$ is a product of functions, you must apply the product rule to differentiate it.

$$j'(x) = \left(\frac{x}{e^x}\right) \cdot \frac{d}{dx}(\ln x) + (\ln x) \cdot \frac{d}{dx}\left(\frac{x}{e^x}\right)$$

The derivative of $\dfrac{x}{e^x}$ requires the quotient rule: $\dfrac{d}{dx}\left(\dfrac{x}{e^x}\right) = \dfrac{1-x}{e^x}$. $\longleftarrow$

Here's how you get that derivative:

$$\frac{d}{dx}\left(\frac{x}{e^x}\right) = \frac{e^x \cdot 1 - x \cdot e^x}{(e^x)^2}$$

$$= \frac{e^x - xe^x}{(e^x)(e^x)}$$

$$= \frac{\cancel{e^x}(1-x)}{(\cancel{e^x})(e^x)}$$

$$= \frac{1-x}{e^x}$$

$$j'(x) = \left[\left(\frac{\cancel{x}}{e^x}\right)\left(\frac{1}{\cancel{x}}\right)\right] + \left[(\ln x)\left(\frac{1-x}{e^x}\right)\right]$$

$$= \frac{1}{e^x} + \frac{(\ln x)(1-x)}{e^x}$$

$$= \frac{1 + \ln x - x\ln x}{e^x}$$

12.42 Differentiate with respect to θ: $y = (\sin 3\theta \tan 3\theta)^{12}$.

Apply the chain rule such that θ^{12} is the "outer" function and $\sin 3\theta \tan 3\theta$ is the "inner" function.

$$\frac{dy}{d\theta} = 12(\sin 3\theta \tan 3\theta)^{11} \cdot \frac{d}{d\theta}(\sin 3\theta \tan 3\theta)$$

Apply the product rule to calculate $\dfrac{d}{d\theta}(\sin 3\theta \tan 3\theta)$.

$$= 12(\sin 3\theta \tan 3\theta)^{11} \cdot \left[\sin 3\theta \cdot \frac{d}{d\theta}(\tan 3\theta) + \tan 3\theta \cdot \frac{d}{d\theta}(\sin 3\theta)\right]$$

Differentiating $\tan 3\theta$ and $\sin 3\theta$ requires the chain rule.

$$= 12(\sin 3\theta \tan 3\theta)^{11} \cdot \left[\sin 3\theta \cdot \sec^2 3\theta \cdot \frac{d}{d\theta}(3\theta) + \tan 3\theta \cdot \cos 3\theta \cdot \frac{d}{d\theta}(3\theta)\right]$$

$$= 12(\sin 3\theta \tan 3\theta)^{11} \cdot \left[\sin 3\theta \cdot (\sec^2 3\theta) \cdot 3 + \tan 3\theta \cdot (\cos 3\theta) \cdot 3\right]$$

$$= 12(\sin 3\theta \tan 3\theta)^{11} \cdot \left(3\sin 3\theta \sec^2 3\theta + 3\tan 3\theta \cos 3\theta\right)$$

Note: Problems 12.43–12.44 reference the functions f(x) and g(x) and their derivatives, f'(x) and g'(x). All four of the functions are continuous and differentiable on (−∞,∞), and a few of their values are listed in the table below.

x	-2	-1	0	1	2	3
$f(x)$	5	9	6	2	-1	-4
$f'(x)$	8	$\frac{1}{2}$	-2	-1	-3	-9
$g(x)$	1	3	-2	5	6	-1
$g'(x)$	2	0	-3	12	1	-6

12.43 Evaluate $\left(f \circ g\right)'(-2)$.

This is just the chain rule. You're differentiating the outer function f, leaving g alone, and then multiplying by the derivative of g.

Note that $(f \circ g)(x) = f(g(x))$.

$$(f \circ g)'(-2) = f'(g(-2)) \cdot g'(-2)$$

According to the table, $g(-2) = 1$, so $f'(g(-2)) = f'(1)$.

$$= f'(1) \cdot g'(-2)$$
$$= (-1)(2)$$
$$= -2$$

Note: Problems 12.43–12.44 reference the table of values provided in Problem 12.43.

12.44 If $h(x) = \dfrac{e^{f(x)}}{\ln(g(x))}$, evaluate $h'(2)$ accurate to five decimal places.

Because $h(x)$ is a quotient, you must apply the quotient rule.

$$h'(x) = \frac{\ln(g(x)) \cdot \dfrac{d}{dx}(e^{f(x)}) - e^{f(x)} \cdot \dfrac{d}{dx}[\ln(g(x))]}{[\ln(g(x))]^2}$$

The derivative of e to the f(x) power is e^{f(x)} times the derivative of the power, f'(x). The derivative of ln(g(x)) is 1 over g(x) times the derivative of the function you're logging: g'(x).

Differentiating $e^{f(x)}$ and $\ln(g(x))$ requires the chain rule.

$$h'(x) = \frac{\ln(g(x)) \cdot e^{f(x)} \cdot f'(x) - e^{f(x)} \cdot \dfrac{1}{g(x)} \cdot g'(x)}{[\ln(g(x))]^2}$$

Evaluate the derivative at $x = 2$.

$$h'(2) = \frac{\ln(g(2)) \cdot e^{f(2)} \cdot f'(2) - e^{f(2)} \cdot \dfrac{g'(2)}{g(2)}}{[\ln(g(2))]^2}$$

Evaluate the functions using the table of values from Problem 12.43.

$$= \frac{(\ln 6)(e^{-1})(-3) - (e^{-1})\left(\dfrac{1}{6}\right)}{(\ln 6)^2}$$

$$= \frac{\dfrac{-3\ln 6}{e} - \dfrac{1}{6e}}{(\ln 6)^2}$$

Use a calculator to determine that $h'(2) \approx -0.63505$.

Chapter 13
DERIVATIVES AND FUNCTION GRAPHS

What signs of derivatives tell you about graphs

Once you've mastered the procedural skills required to differentiate, your next objective is to explore uses and applications of derivatives. The most rudimentary and immediately gratifying application is the correspondence between a function's behavior and the signs of its first and second derivatives. Specifically, this chapter investigates how the signs of $f'(x)$ describe the direction of $f(x)$ and how the signs of $f''(x)$ describe the concavity of $f(x)$.

In a nutshell, you'll be finding first and second derivatives of a function, figuring out whether those derivatives are positive or negative for various values of x, and then learning why the heck that even matters. However, before you can delve into any of that, you need to remember how to find critical numbers—a skill that (believe it or not) first popped up in the review section of the book, in Chapter 3.

Critical Numbers

Numbers that break up wiggle graphs

13.1 If a is a critical number of $f(x)$, evaluate $f(a)$.

If a is a critical number of $f(x)$, then either a is a zero of the function (meaning $f(a) = 0$) or the function is discontinuous at a (meaning $f(a)$ does not exist).

13.2 Identify the critical numbers of $y = 2x^2 - 21x + 27$.

> There's nothing you can plug in for x that will make the function undefined, like negatives inside square roots or zeroes in denominators.

All polynomial functions, including this quadratic, are continuous over their entire domains; the domain of this function is $(-\infty, \infty)$. Therefore, the only critical numbers will be its roots. Identify those roots by setting $y = 0$ and solving the equation.

$$2x^2 - 21x + 27 = 0$$
$$(2x - 3)(x - 9) = 0$$
$$2x - 3 = 0 \qquad x - 9 = 0$$
$$x = \frac{3}{2} \quad \text{or} \quad x = 9$$

The critical numbers of $y = 2x^2 - 21x + 27$ are $x = \frac{3}{2}$ and $x = 9$.

13.3 Identify the critical numbers of $f(x) = x^2 \cos x$.

Note that $f(x)$ is the product of two functions (one polynomial and one trigonometric), both of which are continuous over their entire domains. Furthermore, x^2 and $\cos x$ are both defined on $(-\infty, \infty)$, so $f(x)$ is everywhere continuous. Therefore, the only critical numbers of $f(x)$ are its roots.

$$x^2 \cos x = 0$$
$$x^2 = 0 \qquad \cos x = 0$$
$$\sqrt{x^2} = \pm\sqrt{0} \quad \text{or} \quad x = \frac{k\pi}{2}, \ k \text{ an odd integer}$$
$$x = 0$$

Therefore, $f(x)$ has infinitely many critical numbers: $x = 0$ and
$$x = \cdots - \frac{5\pi}{2}, -\frac{3\pi}{2}, -\frac{\pi}{2}, \frac{\pi}{2}, -\frac{3\pi}{2}, \frac{5\pi}{2}, \cdots.$$

13.4 Identify the critical numbers of $h(x) = \dfrac{x^2 - 9}{9x^2 + 30x + 25}$.

> A root of the numerator is also a root of the whole fraction.

Zeroes of the numerator are roots of the function, and zeroes of the denominator represent undefined values of $h(x)$. In both cases, those x-values are critical numbers, so set the numerator and denominator equal to 0 and solve.

$$\frac{x^2 - 9}{9x^2 + 30x + 25} = 0$$

The numerator is a difference of perfect squares, and the denominator is, itself, a perfect square. Factor both and set all three unique factors equal to 0. (No need to set the repeated factor in the denominator equal to zero twice—you'll get the same double root both times.)

> If x is a root of the denominator, $h(x)$ is undefined at x because you're dividing by 0.

$$\frac{(x+3)(x-3)}{(3x+5)(3x+5)} = 0$$

$$
\begin{array}{ccccc}
x + 3 = 0 & & x - 3 = 0 & & 3x + 5 = 0 \\
x = -3 & \text{or} & x = 3 & \text{or} & 3x = -5 \\
& & & & x = -\dfrac{5}{3}
\end{array}
$$

The critical numbers of $h(x)$ are $x = -3$, $x = -\dfrac{5}{3}$, and $x = 3$.

13.5 Identify the critical numbers of $g(x) = \dfrac{x-1}{\ln x}$.

Use the method described in Problem 13.4: Set the numerator and denominator of a rational function equal to 0 and solve both equations to determine the critical numbers.

$$
\begin{array}{ccc}
x - 1 = 0 & & \ln x = 0 \\
x = 1 & \text{or} & e^{\ln x} = e^0 \\
& & x = e^0 = 1
\end{array}
$$

> Like Problems 10.11 and 10.12.

The only critical number of $g(x)$ is $x = 1$. Certainly, $x = 1$ cannot be both a root *and* an undefined value. (If $x = 1$ is a root then $g(1) = 0$, and 0 is a real number, not an undefined value.) This problem represents an indeterminate case: $g(1) = \dfrac{0}{0}$. Previous examples suggest that indeterminate values represent "holes" in the function—point discontinuity at which a limit exists. Regardless, the function is undefined (and therefore discontinuous) at this x-value, so $x = 1$ is a critical number of $g(x)$.

13.6 Given the function $f(x) = (x+4)(2x-9)$, create a new function $\hat{f}(x)$ such that $\hat{f}(x)$ has three critical numbers: $x = -4$, $x = \dfrac{9}{2}$, and $x = 8$, and is equivalent to $f(x)$ at all $x \neq 8$.

Consider the function $\hat{f}(x) = \dfrac{(x+4)(2x-9)(x-8)}{x-8}$, the rational function defined such that $\hat{f}(x) = f(x)\left(\dfrac{x-8}{x-8}\right)$. Multiplying $f(x)$ by $\dfrac{x-8}{x-8}$ will not change its function values, since $x-8$ divided by itself, like any nonzero quantity divided by itself, equals 1, and $f(x) \cdot 1 = f(x)$. However, the expression $\dfrac{x-8}{x-8}$ does carry with it one restriction: x can no longer equal 8. Therefore, $\hat{f}(x)$ is equivalent to $f(x)$ for every real number $x \neq 8$, including matching critical numbers $x = -4$ and $x = \dfrac{9}{2}$. However, since $\hat{f}(x)$ is undefined at $x = 8$, that x-value is a critical number unique to $\hat{f}(x)$.

Since $\hat{f}(8) = \dfrac{0}{0}$, $\hat{f}(x)$ has a hole at $x = 8$. Other than that, it looks exactly like the graph of $f(x)$.

13.7 Identify the critical numbers of the function $j(x)$, graphed in Figure 13-1.

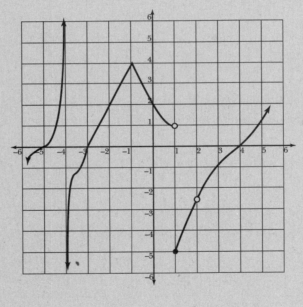

Figure 13-1

The graph of a function $j(x)$.

The critical numbers of a function are the values at which the function equals 0 or is discontinuous. A function equals 0 at its roots, which are also the x-intercepts of its graph. Because $j(x)$ intersects the x-axis at $x = -5$, -3, and 4, those are critical numbers of $j(x)$. A function is discontinuous at any holes, breaks, or vertical asymptotes on the graph, so $j(x)$ also has critical numbers $x = -4$, 1, and $x = 2$. Therefore, $j(x)$ has a total of six critical numbers: $x = -5$, -4, -3, 1, 2, and 4.

Signs of the First Derivative
Use wiggle graphs to determine function direction

> *Note: Problems 13.8–13.11 refer to the graph of f(x) in Figure 13-2.*

13.8 Sketch the tangent lines to $f(x)$ at every integer value of x on the interval $[-8,8]$. ←

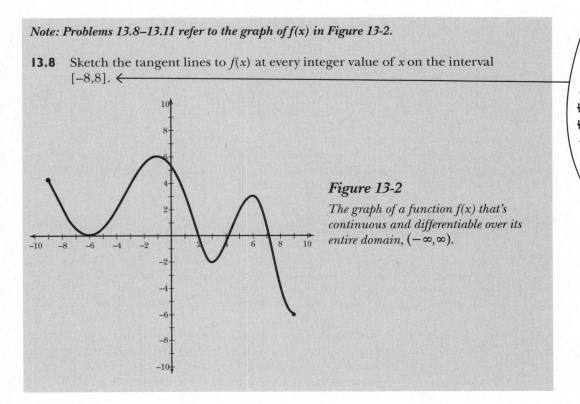

Figure 13-2

The graph of a function f(x) that's continuous and differentiable over its entire domain, $(-\infty, \infty)$.

> This is not "busywork"—there's a point to this, so do it carefully. Pay close attention to the signs of the tangent slopes. Do the lines slope up to the right or down to the left?

This problem requires you to draw 17 tangent lines, all of which are illustrated in Figure 13-3.

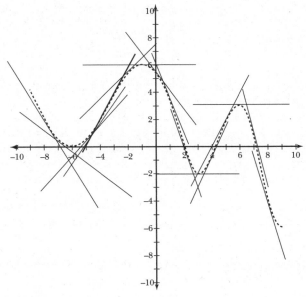

Figure 13-3 *The tangent lines to f(x) at x = −8, −7, −6, ⋯, 6, 7, 8.*

Note: Problems 13.8–13.11 refer to the graph of f(x) in Figure 13-2.

13.9   On what intervals do the tangent lines drawn in Problem 13.8 have positive slopes? What tangent lines have negative slopes? 0 slopes?

A line with a positive slope increases in height as *x* increases. In other words, as you travel from left to right, a line with a positive slope heads in the positive *y* direction. Notice that the tangent lines at *x* = –5, –4, –3, –2, 4, and 5 have positive slopes.

Lines with negative slopes slant downward as you travel along the line from left to right. In Problem 13.8, the tangent lines at *x* = –9, –8, –7, 0, 1, 2, 7, and 8 have negative slopes. Horizontal lines have a slope of 0. Therefore, the tangent lines at *x* = –6, –1, 3, and 6 have slopes equal to 0.

Note: Problems 13.8–13.11 refer to the graph of f(x) in Figure 13-2.

13.10   Based on your answers to Problems 13.8 and 13.9, describe the relationship between the direction of *f(x)* and the sign of its derivative as you travel along the *x*-axis from *x* = –9 to *x* = 9.

> Trace your finger along the graph from left to right. Whenever your finger is going up, the tangent line drawn there has a positive slope. When your finger goes down, the slopes are negative.

Recall that the derivative is defined as the slope of the tangent line to a curve. Travel along the graph of *f(x)* from left to right. Whenever *f(x)* is increasing at an *x*-value, the tangent line at that *x*-value has a positive slope. On the other hand, whenever *f(x)* is decreasing at an *x*-value, the slope of the tangent line (i.e., derivative) is negative there. The slope of the tangent line to *f(x)* equals 0 at a relative maximum or a relative minimum, when the graph changes from increasing to decreasing or from decreasing to increasing.

Note: Problems 13.8–13.11 refer to the graph of f(x) in Figure 13-2.

13.11.   Draw a sign graph for $f'(x)$.

> Also called a "wiggle graph," because the sign of f'(x) helps you figure out which way f(x) is "wiggling," up or down.

A sign graph is a number line that identifies the intervals upon which a function is positive or negative. To construct one, begin by identifying the critical numbers for the function. Remember, you are constructing the sign graph of $f'(x)$, so the critical numbers are the *x*-values at which the slope of the tangent line to the graph, $f'(x)$, equals 0: *x* = –6, –1, 3, and 6. (There are no values for which $f'(x)$ is undefined, as Figure 13-2 describes *f(x)* as everywhere differentiable.) Plot the critical numbers as points on the sign graph, as illustrated by Figure 13-4.

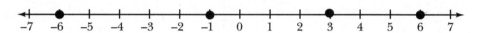

Figure 13-4   *The four critical numbers of f'(x) split the number line into five distinct intervals.*

The critical numbers act as boundaries that define these intervals: (–9,–6), (–6,–1), (–1,3), (3,6), and (6,9). The sign of a function can only change at one of its critical numbers (though, at some critical numbers, the sign will not change).

According to Problem 13.9, the slopes of the tangent lines to $f(x)$ (i.e., the values of $f'(x)$) are positive on $(-6,-1)$ and $(3,6)$. Similarly, $f'(x)$ is negative on the intervals $(-9,-6)$, $(-1,3)$, and $(6,9)$. Write the sign that describes $f'(x)$, either "+" or "–," above each interval to complete the sign graph, as illustrated by Figure 13-5.

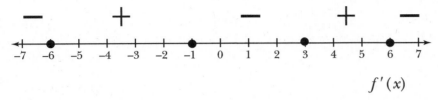

Figure 13-5 *The sign graph of $f'(x)$.*

Note: Problems 13.12–13.13 refer to the function $f(x) = 3x^2 - 4x + 2$.

13.12 Construct the sign graph of $f'(x)$.

To create the sign graph for $f'(x)$, you must first find the derivative: $f'(x) = 6x - 4$. Now set the derivative equal to 0 and solve to determine its critical number.

$$6x - 4 = 0$$
$$6x = 4$$
$$x = \frac{4}{6} = \frac{2}{3}$$

As Figure 13-6 illustrates, this critical number splits the number line into two intervals: $\left(-\infty, \frac{2}{3}\right)$ and $\left(\frac{2}{3}, \infty\right)$. To determine the sign of each interval, choose one "test value" from each interval and substitute them into $f'(x)$. Note that $x = 0$ belongs to the left interval and $x = 1$ belongs to the right interval.

$$f'(0) = 6(0) - 4 = -4 \qquad f'(1) = 6(1) - 4 = 2$$

Because $f'(x) < 0$ for $x = 0$, the left interval should be negative; similarly, the right interval $\left(\frac{2}{3}, \infty\right)$ is positive because $f'(x) > 0$ for $x = 1$. The completed sign graph appears in Figure 13-6.

Even though you plug only one test value from each interval into f' (x), all the other values in each interval will give you the same sign, + or –, that the test value did.

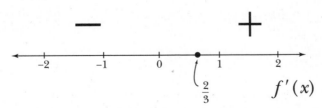

Figure 13-6 *The sign graph of $f'(x) = 6x - 4$.*

Note: Problems 13.12–13.13 refer to the function f(x) = 3x² – 4x + 2.

13.13 On what interval(s) is $f(x)$ increasing?

According to Problem 13.11, $f(x)$ is increasing whenever its derivative is positive. Problem 13.12 indicates that $f'(x) > 0$ on the interval $\left(\dfrac{2}{3}, \infty\right)$. Therefore, $f(x)$ is increasing on the interval $\left(\dfrac{2}{3}, \infty\right)$.

Note: Problems 13.14–13.15 reference the graph of g′(x) in Figure 13-7.

13.14 On what interval(s) is $g(x)$ increasing?

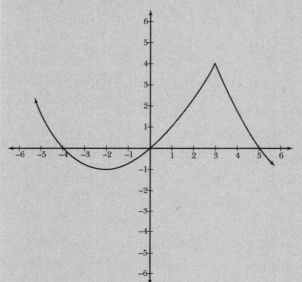

Figure 13-7

The graph of g′(x), the derivative of some function g(x).

> In other words, the intervals where g'(x) is above the x-axis.

This question does not ask where the *graph* is increasing, as the graph represents the derivative of $g(x)$, not $g(x)$ itself. Instead, note where the graph of $g'(x)$ is positive, as those intervals correspond to the intervals upon which $g(x)$ increases.

Since $g'(x)$ is positive on $(-\infty, -4) \cup (0, 5)$, $g(x)$ is increasing on those intervals.

Note: Problems 13.14–13.15 reference the graph of g′(x) in Figure 13-7.

13.15 Identify the relative extrema of $g(x)$.

> Relative extrema are humps in a function graph. Relative max points are the highest peaks of mountains, and minimums are the lowest points of valleys.

A function has a relative extreme point (i.e., a relative maximum or a relative minimum) when a function changes direction. Consider this: if a function changes from increasing to decreasing at $x = a$, then its graph will reach a peak at $x = a$. While this may not represent the *absolute* maximum value of the function, it is the highest function value reached by any x close to a.

Relative extrema can only occur at a function's critical numbers, and are indicated by a sign change in the derivative of the function.

Notice that at $x = -4$ and again at $x = 5$, the graph of $g(x)$ changes from positive to negative—it crosses from above to below the x-axis. Therefore, $g(x)$ has relative maximums at $x = -4$ and $x = 5$, as $g(x)$ changes from increasing to decreasing at those x-values. At $x = 0$, $g(x)$ has a relative minimum, because its derivative changes from negative to positive at $x = 0$ (indicating that the direction of $g(x)$ changes from decreasing to increasing there).

> Check out the sign graph. If the signs around a critical number are different, it's a relative extreme point. If the change is from + to −, it's a max, and if the change is from − to +, it's a min.

13.16 If Figure 13-8 is the graph of $f'(x)$, the derivative of some continuous and differentiable function $f(x)$, at what x-value(s) will $f(x)$ reach a relative minimum? A relative maximum?

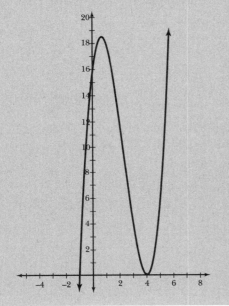

Figure 13-8

The graph of $f'(x)$, the derivative of some function $f(x)$.

The graph of $f'(x)$ intersects the x-axis at $x = -1$ and $x = 4$, so those are the critical numbers of $f'(x)$. Note that $f'(x)$ changes from negative to positive as its graph crosses the x-axis at $x = -1$. Therefore, $f(x)$ has a relative minimum at $x = -1$, as it will change from decreasing to increasing there. That is the only relative extreme point for $f(x)$. At $x = 4$, the only other critical number of $f'(x)$, the derivative does not change sign—it is positive both before and after $x = 4$. Therefore, $f(x)$ has no relative maximum points.

> Because it hits, but doesn't cross, the x-axis.

13.17 Find and classify the relative extrema of $g(x) = 4x^3 - 15x^2 + 12x - 8$.

Differentiate $g(x)$.

$$g'(x) = 12x^2 - 30x + 12$$

This derivative is defined for all real numbers, so its only critical numbers are its roots. Set $g'(x) = 0$ and solve.

$$\frac{12x^2 - 30x + 12}{6} = \frac{0}{6}$$
$$2x^2 - 5x + 2 = 0$$
$$(2x - 1)(x - 2) = 0$$
$$x = \frac{1}{2}, 2$$

Construct a sign graph for $g'(x)$, as illustrated by Figure 13-9.

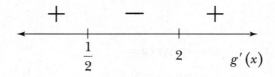

Figure 13-9 *The sign graph of $g'(x) = 12x^2 - 30x + 12$.*

According to the sign graph, $g(x)$ changes from increasing to decreasing at $x = \frac{1}{2}$, resulting in a relative maximum, and $g(x)$ changes from decreasing to increasing at $x = 2$, resulting in a relative minimum.

> Because the sign of $g'(x)$ changes from positive to negative. The opposite happens at $x = 2$.

13.18 Find and classify the relative extrema of $h(x) = -8x^3 + 11x^2 + 35x - 19$.

Differentiate $h(x)$.

$$h'(x) = -24x^2 + 22x + 35$$

This quadratic is defined for all real numbers, so its only critical points are its roots. Set $h'(x) = 0$ and solve using the quadratic formula.

$$-24x^2 + 22x + 35 = 0$$
$$x = \frac{-22 \pm \sqrt{22^2 - 4(-24)(35)}}{2(-24)}$$
$$x = \frac{-22 \pm \sqrt{3{,}844}}{-48}$$
$$x = \frac{-22 \pm 62}{-48}$$
$$x = \frac{-84}{-48} \text{ or } \frac{40}{-48}$$
$$x = \frac{7}{4} \text{ or } -\frac{5}{6}$$

Construct a sign graph for $h'(x)$, as illustrated by Figure 13-10.

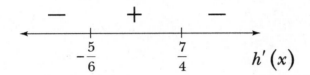

Figure 13-10 *The sign graph of $h'(x) = -24x^2 + 22x - 35$.*

According to Figure 13-10, $h(x)$ changes from decreasing to increasing at $x = -\dfrac{5}{6}$, resulting in a relative minimum, and changes from increasing to decreasing at $x = \dfrac{7}{4}$, resulting in a relative maximum.

Signs of the Second Derivative

Points of inflection and concavity

Note: Problems 13.19–13.20 refer to the function $g(x) = 18x^3 - 39x^2 + 8x + 16$.

13.19 Generate the sign graph of $g''(x)$.

Use the method described in Problem 13.12, although you should use the second derivative, $g''(x)$, instead of $g'(x)$.

$$g'(x) = 54x^2 - 78x + 8$$
$$g''(x) = 108x - 78$$

Set $g''(x)$ equal to 0 and solve for x.

$$108x - 78 = 0$$
$$108x = 78$$
$$x = \frac{78}{108}$$
$$x = \frac{13}{18}$$

> Divide the top and bottom by 6.

The critical number divides the number line into two intervals: $\left(-\infty, \dfrac{13}{18}\right)$ and $\left(\dfrac{13}{18}, \infty\right)$. Choose one test value from each interval (such as $x = 0$ and $x = 1$, respectively) to determine the sign of $g''(x)$ for each interval.

$$g''(0) = 108(0) - 78 = -78 \qquad g''(1) = 108(1) - 78 = 30$$

Therefore, $g''(x) < 0$ on $\left(-\infty, \dfrac{13}{18}\right)$ and $g''(x) > 0$ on $\left(\dfrac{13}{18}, \infty\right)$, as illustrated by the sign graph in Figure 13-11.

> Make sure you plug the test values into $g''(x)$, not $g'(x)$. Label the sign graph $g''(x)$ so it's clear what derivative you're talking about.

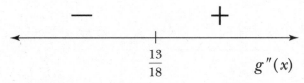

Figure 13-11   *The sign graph of g"(x) = 108x − 78.*

> The concavity of a function changes at an inflection point.

13.20 At what *x*-value(s) does $g(x)$ have an inflection point(s)?

If the sign of $g''(x)$—and therefore the concavity of $g''(x)$—changes at one of its critical numbers, that number represents an inflection point on the graph of $g(x)$. In Figure 13-11, the sign of $g''(x)$ changes from negative to positive at $x = \dfrac{13}{18}$, indicating an inflection point of $g(x)$ occurs at $x = \dfrac{13}{18}$.

> So, the function g(x) has an inflection point if the critical number of its SECOND DERIVATIVE changes sign.

Note: Problems 13.21–13.23 refer to the function f(x) = sin 3x.

13.21 Identify the *x*-values of the inflection points of $f(x)$ on the interval $\left[\dfrac{\pi}{6}, \dfrac{5\pi}{4}\right]$.

Find $f''(x)$.

$$f'(x) = 3\cos 3x$$
$$f''(x) = -9\sin 3x$$

Identify the critical numbers of $f''(x)$ by setting it equal to 0 and solving for *x*.

> Look at Problem 8.31 if you're not sure how to solve trig equations containing 3x instead of just x. That equation looks a lot like this one.

$$-9\sin 3x = 0$$
$$\sin 3x = 0$$
$$3x = 0, \pi, 2\pi, 3\pi, 4\pi, 5\pi$$
$$x = 0, \frac{\pi}{3}, \frac{2\pi}{3}, \pi, \frac{4\pi}{3}, \frac{5\pi}{3}$$

Eliminate *x*-values outside of $\left[\dfrac{\pi}{6}, \dfrac{5\pi}{4}\right]$.

$$x = \frac{\pi}{3}, \frac{2\pi}{3}, \pi$$

Choose test values from the intervals $\left(\dfrac{\pi}{6}, \dfrac{\pi}{3}\right)$, $\left(\dfrac{\pi}{3}, \dfrac{2\pi}{3}\right)$, $\left(\dfrac{2\pi}{3}, \pi\right)$, and $\left(\pi, \dfrac{5\pi}{4}\right)$ to generate the sign graph in Figure 13-12.

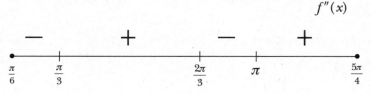

Figure 13-12   *The sign graph of f"(x) = −9 sin 3x over the interval $\left[\dfrac{\pi}{6}, \dfrac{5\pi}{4}\right]$.*

Because the sign of $f''(x)$ changes at all three critical numbers, each represents the location of an inflection point of $f(x)$: $x = \dfrac{\pi}{3}, \dfrac{2\pi}{3}, \pi$.

Note: Problems 13.21–13.23 refer to the function f(x) = sin 3x.

13.22 Based on the sign graph generated by Problem 13.21, indicate the intervals on which $f(x)$ is concave down.

A function is concave down on the same intervals its second derivative is negative. According to Figure 13-12, $f''(x)$ is negative on $\left(\dfrac{\pi}{6}, \dfrac{\pi}{3}\right)$ and $\left(\dfrac{2\pi}{3}, \pi\right)$, so $f(x)$ is concave down on those intervals.

Note: Problems 13.21–13.23 refer to the function f(x) = sin 3x.

13.23 Based on the sign graph generated by Problem 13.21, indicate the intervals on which $f'(x)$ is increasing.

Although $f''(x)$ is the second derivative of $f(x)$, note that it is also the *first* derivative of $f'(x)$. Therefore, the sign of $f''(x)$ also describes the direction of $f'(x)$. According to its sign graph, $f''(x)$ is positive on the intervals $\left(\dfrac{\pi}{3}, \dfrac{2\pi}{3}\right)$ and $\left(\pi, \dfrac{5\pi}{4}\right)$, so $f'(x)$ must be increasing there.

> Just like the signs of f'(x) describe the direction of f(x), the signs of f''(x) describe the direction of f'(x).

Note: Problems 13.24–13.25 reference the graph of h'(x) in Figure 13-13.

13.24 Identify the *x*-values at which the relative maximum(s) of $h(x)$ occur.

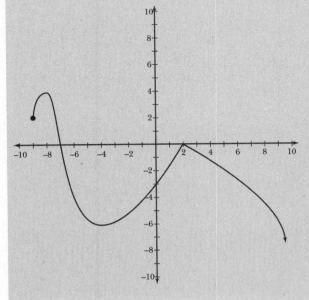

Figure 13-13

The graph of h'(x), the first derivative of some function h(x) defined on the interval [−9,∞).

Because a function has critical numbers wherever it equals 0 (or is discontinuous).

Relative extrema of $h(x)$ occur when $h'(x)$ changes sign at one or more critical numbers. The function $h'(x)$ has critical numbers at its x-intercepts ($x = -7$ and $x = 2$) and wherever it is undefined; however, $h'(x)$ is continuous and therefore defined over its entire domain $[-9,\infty)$.

Note that $h'(x)$ is positive (above the x-axis) on the interval $(-9,-7)$ only; it then crosses the x-axis and is nonpositive for $x > -7$. Therefore, $h(x)$ changes from increasing to decreasing at $x = -7$, indicating a relative maximum. (Although the graph of $h'(x)$ intersects the x-axis at $x = 2$ it does not cross the axis, so there is no sign change at $x = 2$ and no relative maximum or minimum point occurs there.)

Note: Problems 13.24–13.25 reference the graph of $h'(x)$ in Figure 13-13.

13.25 On what interval(s) is $h(x)$ concave up?

If you're confused, Problem 13.21 explains why that's true.

The function $h(x)$ is concave up wherever its second derivative, $h''(x)$, is positive. The sign of $h''(x)$ also describes the direction of $h'(x)$. Because $h'(x)$ is increasing on the intervals $(-9,-8)$ and $(-4,2)$, $h''(x)$ is positive on those intervals, and therefore $h(x)$ is concave up on those intervals as well.

Note: Problems 13.26–13.28 refer to the graph of $f''(x)$ in Figure 13-14.

13.26 On what interval(s) is the graph of $f(x)$ concave down?

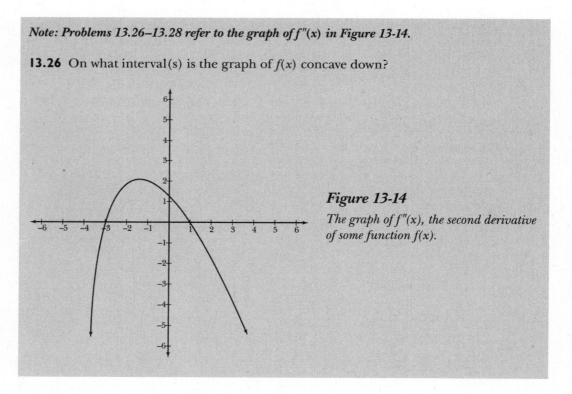

Figure 13-14

The graph of $f''(x)$, the second derivative of some function $f(x)$.

The graph of $f(x)$ is concave down on the same intervals that $f''(x)$ is negative (i.e., below the x-axis). Therefore, $f(x)$ is concave down on $(-\infty,-3)\cup(1,\infty)$.

Note: Problems 13.26–13.28 refer to the graph of f''(x) in Figure 13-14.

13.27 Which is greater: $f'(-2)$ or $f'(0)$?

Note that $f''(x)$ is positive on the interval $(-3,1)$, so $f'(x)$ must be increasing over the entire interval. If $f'(x)$ increases between $x = -2$ and $x = 0$, then $f'(-2) < f'(0)$.

> The sign of f''(x) describes the direction of f'(x). If f'(x) goes up as you travel from x = -2 to x = 0, then f'(0) is higher than f'(-2).

Note: Problems 13.26–13.28 refer to the graph of f''(x) in Figure 13-14.

13.28 Identify and classify the relative extrema of $f'(x)$.

The relative extrema of a function can only occur at the critical numbers of its derivative. The only critical numbers of $f''(x)$ are its x-intercepts: $x = -3$ and $x = 1$.

Because $f''(x)$ changes from negative to positive at $x = -3$, $f'(x)$ changes from decreasing to increasing, indicating a relative minimum. Additionally, $f''(x)$ changes from positive to negative at $x = 1$, so $f'(x)$ changes from increasing to decreasing there, indicating a relative maximum.

> You're looking for the max's and min's of f'(x), which has a derivative of f''(x).

13.29 Describe how the second derivative test classifies the relative extrema of a function $g(x)$.

Plug a critical number of $g'(x)$ into $g''(x)$. If the result is positive, then the critical number represents a relative minimum of $g(x)$. If the result is negative, it represents a relative maximum of $g(x)$. If, however, the result is 0, no conclusion can be drawn, and a sign graph must be used to determine the direction of $g(x)$ to the left and to the right of the critical number.

13.30 Identify the relative extrema of $f(x) = x^3 + x^2 - x + 9$ and classify each using the second derivative test.

Find the critical numbers of $f'(x)$.

$$f'(x) = 3x^2 + 2x - 1$$
$$0 = 3x^2 + 2x - 1$$
$$0 = (3x - 1)(x + 1)$$
$$3x - 1 = 0 \qquad x + 1 = 0$$
$$x = \frac{1}{3} \quad \text{or} \quad x = -1$$

> The actual numbers you'll get don't matter, only whether they are positive or negative.

Substitute those critical numbers into $f''(x) = 6x + 2$.

$$f''\left(\frac{1}{3}\right) = 6\left(\frac{1}{3}\right) + 2 = 4 \qquad f''(-1) = 6(-1) + 2 = -4$$

Because $f''\left(\frac{1}{3}\right) > 0$, $f(x)$ has a relative minimum at $x = \frac{1}{3}$, according to the second derivative test. Similarly, $f(x)$ has a relative maximum at $x = -1$ because $f''(-1) < 0$.

13.31 A parabola with a vertical axis of symmetry has standard form $f(x) = a(x-h)^2 + k$. Assuming $a < 0$, where does the relative maximum of $f(x)$ occur? Use the second derivative test to verify your answer.

Differentiate $f(x)$ and identify the critical numbers of $f'(x)$.

$$f'(x) = 2a(x-h)$$
$$0 = 2a(x-h)$$
$$\frac{0}{2a} + h = x$$
$$h = x$$

> To practice working with these kinds of parabolas, check out Problems 6.1 and 6.2.

> 2a is just a constant, and 0 divided by any nonzero number equals 0.

Substitute $x = h$ into $f''(x)$.

$$f''(x) = 2a \cdot \frac{d}{dx}(x-h)$$
$$f''(x) = 2a \cdot 1$$
$$f''(h) = 2a$$

> There aren't any x's in $f''(x) = 2a$, so no matter what x is, the second derivative at that x-value is $2a$.

No matter what x-value is plugged into the second derivative, $f''(x) < 0$—the problem states that $a < 0$, so $2a$ is negative as well. Therefore, according to the second derivative test, $f(x)$ has a relative maximum when $x = h$.

Function and Derivative Graphs
How are the graphs of f, f', and f" related?

13.32 A function and its derivative are graphed in Figure 13-15. Determine which graph represents $f(x)$ and which represents $f'(x)$.

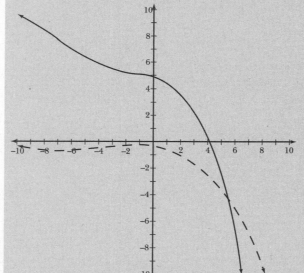

Figure 13-15

One of the functions graphed here is the derivative of the other.

> The solid curve can't be the derivative because the dotted curve is not always increasing when the solid curve is positive.

Notice that the solid graph decreases along its entire domain and the dotted graph is negative along its entire domain. Because $f'(x)$ is negative when $f(x)$ is decreasing, $f'(x)$ is the dotted graph and $f(x)$ is the solid graph in Figure 13-15.

13.33 A function and its derivative are graphed in Figure 13-16. Determine which graph represents $g(x)$ and which represents $g'(x)$.

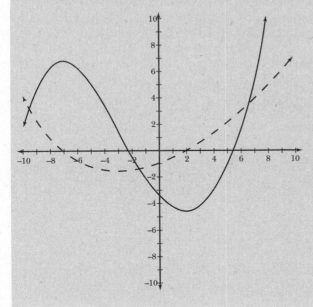

Figure 13-16

One of the functions graphed here is the derivative of the other.

The solid graph is $g(x)$ and the dotted graph is $g'(x)$. The solid graph reaches a relative maximum at $x = -7$. Meanwhile, the dotted graph intersects the x-axis as it changes from positive to negative at $x = -7$. Therefore, the relative maximum of the solid graph is reflected by a critical number sign change in the dotted graph. Similarly, the solid graph changes from decreasing to increasing at $x = 2$, and the dotted graph again intersects the x-axis at $x = 2$, this time changing from negative to positive to reflect the direction change of the solid graph.

This time you have the graph of a function and its SECOND derivative.

13.34 A function and its second derivative are graphed in Figure 13-17. Determine which graph represents $h(x)$ and which represents $h''(x)$.

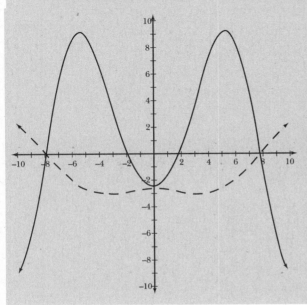

Figure 13-17

One of the functions graphed here is the second derivative of the other.

The solid graph appears to be concave up on the interval (−3,3), but the dotted graph is negative for those x-values. Therefore, the dotted graph *cannot* be the second derivative of the solid graph, as intervals on which $h(x)$ is concave up must correspond to intervals on which $h''(x)$ is positive. On the other hand, the dotted graph appears to be concave down from roughly $x = -2$ to $x = 2$, and the solid graph is negative for those x-values. Therefore, $h(x)$ is the dotted graph and $h''(x)$ is the solid graph.

13.35 Match the functions $j(x)$, $j'(x)$, and $j''(x)$ with their graphs in Figure 13-18.

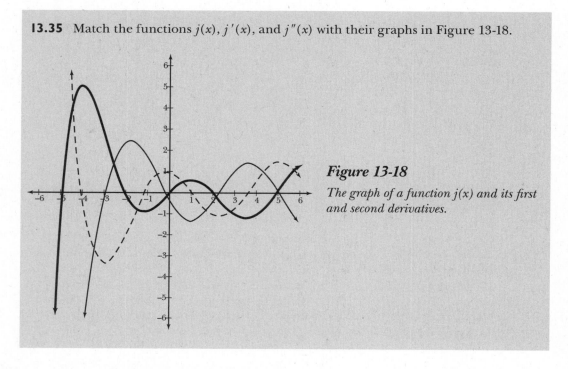

Figure 13-18

The graph of a function $j(x)$ and its first and second derivatives.

Each time the thick graph reaches a relative maximum or minimum, the dotted graph intersects the x-axis. Furthermore, each time the dotted graph reaches a relative maximum or minimum value, the thin graph intersects the x-axis. Therefore, the thick graph is $j(x)$, the dotted graph is $j'(x)$, and the thin graph is $j''(x)$.

So, the (dotted) derivative equals 0 whenever the (thick) function has a horizontal tangent line.

Chapter 14

BASIC APPLICATIONS OF DIFFERENTIATION

Put your derivatives skills to use

Though a conceptual understanding of how a function and the signs of its derivatives are illuminating and worthwhile, the concepts are very rudimentary. In this chapter, you will progress beyond direction and concavity, using derivatives to identify equations of tangent lines, approximate roots, and evaluate difficult limits. The skills of Chapter 13 are not to be shelved, however, as you will apply them to the extreme value theorem, and then again in Chapter 15 as you optimize functions.

Even though understanding what the derivatives of a function tell you about that function's direction and concavity is handy, it's honestly not all that interesting. However, derivatives can be used for more practical purposes, and in this chapter, you'll find out that derivatives make some tasks that used to be hard actually quite simple. For example, you can use them to estimate the roots of a function like $f(x) = x^7 - 4x^3 + 2x + 11$, which refuses to be factored and can't be solved by completing the square or the quadratic formula (because it's obviously not a quadratic expression). You'll even use derivatives to find limits that required specific tricks back in Chapter 10, but will now be significantly easier thanks to something called L'Hôpital's rule.

Equations of Tangent Lines

Point of tangency + derivative = equation of tangent

Problems 14.1–14.2 refer to the function f(x) = –2x² + 5x – 9.

14.1 If line *l* is tangent to $f(x)$ at $x = -1$, identify the point of tangency.

The point of tangency lies on the graph of $f(x)$ when $x = -1$, so evaluate $f(-1)$ to find the corresponding y-coordinate.

$$f(-1) = -2(-1)^2 + 5(-1) - 9$$
$$= -2(1) - 5 - 9$$
$$= -16$$

The point of tangency is $(-1, f(-1)) = (-1, -16)$.

Problems 14.1–14.2 refer to the function f(x) = –2x² + 5x – 9.

14.2 If line *l* is tangent to $f(x)$ at $x = -1$, identify the slope of *l*.

Calculate the derivative of $f(x)$ using the power rule and evaluate $f'(x)$ at $x = -1$ to determine the slope of the tangent line at the point identified in Problem 14.1: $(-1, -16)$.

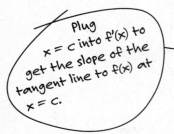

Plug $x = c$ into $f'(x)$ to get the slope of the tangent line to $f(x)$ at $x = c$.

$$f'(x) = -4x + 5$$
$$f'(-1) = -4(-1) + 5$$
$$f'(-1) = 9$$

The slope of tangent line *l* is 9.

14.3 Write the equation of the tangent line to some differentiable function $g(x)$ at $x = a$.

The point-slope formula is the most expedient way to determine the equation of the tangent line. It requires two components: a slope m and a point (x_1, y_1) on the line.

$$y - y_1 = m(x - x_1)$$

Begin by substituting the x-value a into $g(x)$ to get the corresponding y-value: $g(a)$. The point of tangency, $(a, g(a))$, is located on both the curve and the tangent line. Substitute that point and $g'(a)$, the derivative of $g(x)$ when $x = a$, into the point-slope formula. The result will be the equation of the tangent line to $g(x)$ at the point $g(a)$.

Substitute these values into the formula: $x_1 = a$, $y_1 = g(a)$, and $m = g'(a)$.

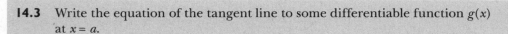

$$y - g(a) = g'(a)(x - a)$$

It is customary to solve for y when writing the equation of a tangent line, although it is not required.

$$y = g'(a)(x - a) + g(a)$$

14.4 Using the method outlined in Problem 14.3 (and the values generated by Problems 14.1 and 14.2), write the equation of the line tangent to $f(x) = -2x^2 + 5x - 9$ at $x = -1$.

Substitute the point of tangency $(x_1, y_1) = (-1, -16)$ and the slope of the tangent line $m = 9$ into the point-slope formula.

$$y - y_1 = m(x - x_1)$$
$$y - (-16) = 9(x - (-1))$$
$$y + 16 = 9(x + 1)$$

Solve the equation for y.

$$y + 16 = 9x + 9$$
$$y = 9x - 7$$

14.5 Determine the equation of the tangent line to $f(x) = 3\cos 2x$ at $x = \dfrac{5\pi}{6}$.

Evaluate $f\left(\dfrac{5\pi}{6}\right)$.

$$f\left(\frac{5\pi}{6}\right) = 3\cos\left(2 \cdot \frac{5\pi}{6}\right)$$
$$= 3\cos\left(\frac{5\pi}{3}\right)$$
$$= \frac{3}{2}$$

You need to know that $\cos\left(\dfrac{5\pi}{3}\right) = \dfrac{1}{2}$ from memory—it's a basic unit circle angle. All of the angles you need to memorize are in Appendix B.

Differentiate $f(x)$ using the chain rule.

$$f'(x) = 3 \cdot (-\sin 2x) \cdot 2$$
$$f'(x) = -6\sin 2x$$

Don't forget to multiply by the derivative of 2x, the inner function. That's where this 2 comes from.

Evaluate $f'\left(\dfrac{5\pi}{6}\right)$.

$$f'\left(\frac{5\pi}{6}\right) = -6\sin\left(2 \cdot \frac{5\pi}{6}\right)$$
$$= -6\sin\left(\frac{5\pi}{3}\right)$$
$$= -6\left(-\frac{\sqrt{3}}{2}\right)$$
$$= 3\sqrt{3}$$

Substitute $(x_1, y_1) = \left(\dfrac{5\pi}{6}, \dfrac{3}{2}\right)$ and $m = 3\sqrt{3}$ into the point-slope formula.

$$y - \frac{3}{2} = 3\sqrt{3}\left(x - \frac{5\pi}{6}\right)$$

$$y - \frac{3}{2} = 3\sqrt{3}x - \frac{15\pi\sqrt{3}}{6}$$

$$y = 3\sqrt{3}x + \frac{3}{2} - \frac{5\pi\sqrt{3}}{2}$$

$$y = 3\sqrt{3}x + \frac{3 - 5\pi\sqrt{3}}{2}$$

> The fractions have the same denominator, so you should combine them.

> The function is defined for ALL REAL NUMBERS, not just 3, 3.7, 4, 5, and 5.15. These are just five x-values and what you get when you plug each one into g(x).

Note: Problems 14.6 –14.7 refer to a differentiable function g(x) defined for all real numbers, and include the selected function values in the table below.

14.6 Estimate $g'(4)$.

x	3	3.7	4	5	5.15
$g(x)$	9	11.6	12.3	3	−0.4

Not much information is given about $g(x)$, so you must make the best use of the limited values given. Remember that the derivative represents the slope of the tangent line at a specific x-value. The best geometric approximation of the tangent line at $x = 4$ is the secant line connecting the points $(3.7, 11.6)$ and $(4, 12.3)$, because $x = 3.7$ is closer to $x = 4$ than any other known x-value. Calculate the slope of that secant line by dividing the difference of the y-values by the difference of the x-values.

> If you can't find the tangent slope, make due with a nearby secant slope.

$$g'(4) \approx \frac{12.3 - 11.6}{4 - 3.7}$$

$$\approx \frac{0.7}{0.3}$$

$$\approx 2.\overline{3}$$

Note: Problems 14.6 –14.7 refer to tables of values provided in Problem 14.6.

14.7 Estimate $g'(5)$.

Use the same technique demonstrated in Problem 14.6. The closest x-value to $x = 5$ is $x = 5.15$, so the best approximation of $g'(5)$ is the slope of the secant line passing through $(5, 3)$ and $(5.15, -0.4)$.

$$g'(5) \approx \frac{-0.4 - 3}{5.15 - 5}$$

$$\approx \frac{-3.4}{0.15}$$

$$\approx -22.\overline{6}$$

Note: Problems 14.8 and 14.9 refer to the graph of f(x) in Figure 14-1.

14.8 Estimate $f'(-6)$ and write the equation of the tangent line to $f(x)$ at $x = -6$.

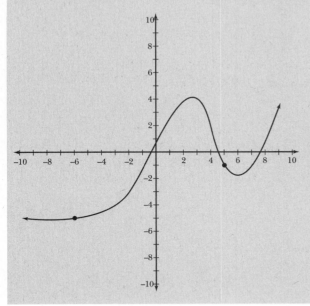

Figure 14-1

The graph of an everywhere differentiable function f(x).

There is no single correct answer to this problem, as very little information is known about $f(x)$. You could use the method of Problems 14.6 and 14.7—estimate $f(-5) \approx 4.8$ and then calculate a slope based on the points $(-6,-5)$ and $(-5,-4.8)$.

Students savvy with graphing calculators could use even more complex means to approximate the derivative. However, any method is entirely based on estimation and the accuracy of one valid technique is difficult to prove better than another. The most straightforward way to approximate the derivative is to sketch the tangent line at $x = -6$ (as demonstrated in Figure 14-2) and calculate its slope.

You could create a statistical plot, generate a regression curve, and then differentiate that curve at $x = -6$. That's a lot of work to do for an answer that's still just an estimate, though.

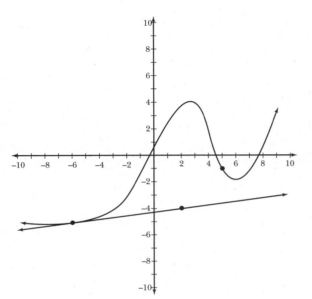

Figure 14-2

The graph of f(x) and a best guess at its tangent line at the point (−6,−5).

This is the same formula from Problems 14.6 and 14.7: $m = \dfrac{y_2 - y_1}{x_2 - x_1}$.

It appears that the tangent line drawn in Figure 14-2 passes through (2,–4), so calculate the slope of the line using that coordinate and the point of tangency (–6,–5).

$$m \approx \frac{-5 - (-4)}{-6 - 2}$$

$$\approx \frac{1}{8}$$

Note: Problems 14.8 and 14.9 refer to the graph of f(x) in Figure 14-1.

14.9 Estimate $f'(5)$ and write the equation of the tangent line to $f(x)$ at $x = 5$.

Draw a tangent line to $f(x)$ at $x = 5$, as illustrated by Figure 14-3.

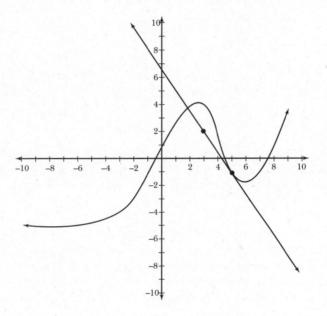

Figure 14-3

The graph of f(x) and a best guess at its tangent line at the point (5,–1).

The tangent line drawn appears to pass through (3,2); use that point and the point of tangency (5,–1) to calculate the approximate derivative.

$$m \approx \frac{-1 - 2}{5 - 3}$$

$$\approx -\frac{3}{2}$$

14.10 The equation of the tangent line to $f(x) = ax^2 + bx - 3$ at (–4,–31) is $y = 9x - 5$. Determine the values of a and b.

You are given $f(-4) = -31$, so substitute those values into the function.

$$f(-4) = a(-4)^2 + b(-4) - 3$$

$$-31 = 16a - 4b - 3$$

$$-28 = 16a - 4b$$

Differentiate $f(x)$. Because a is a real number, you should treat it like any other coefficient—take the derivative of x^2 and multiply it by a.

$$f'(x) = 2ax + b$$

Note that the tangent line $y = 9x - 5$ has slope 9, so $f'(-4) = 9$.

$$f'(-4) = 2a(-4) + b$$
$$9 = -8a + b$$

You now have two equations, $-28 = 16a - 4b$ and $9 = -8a + b$, each containing two unknowns (a and b). Solve the system of equations to determine the solution. One approach is to solve the second equation for b ($b = 9 + 8a$) and substitute it into the other equation.

$$-28 = 16a - 4b$$
$$-28 = 16a - 4(9 + 8a)$$
$$-28 = 16a - 36 - 32a$$
$$8 = -16a$$
$$-\frac{1}{2} = a$$

Substitute a into either equation of the system to determine the corresponding value of b.

$$9 = -8a + b$$
$$9 = -8\left(-\frac{1}{2}\right) + b$$
$$9 = 4 + b$$
$$5 = b$$

Therefore, the equation of $f(x)$ with the correct values of a and b is

$$f(x) = -\frac{1}{2}x^2 + 5x - 3.$$

> The line $y = 9x - 5$ is in slope-intercept form ($y = mx + b$), where $m = 9$ and $b = -5$. The slope is its x-coefficient.

> If you need to review solving systems of equations, look at Problems 1.28–1.30.

The Extreme Value Theorem
Every function has its highs and lows

14.11 If $f(x)$ is a continuous on the interval $[a,b]$, what is guaranteed by the extreme value theorem?

The extreme value theorem guarantees that $f(x)$ possesses both an absolute maximum and an absolute minimum on $[a,b]$. The absolute extrema are not guaranteed to be unique, however. In other words, $f(x)$ may reach its absolute maximum or minimum value more than once on the interval, but it will not surpass either.

> In other words, there's one function value that's higher and one function value that's lower than the other function values on the interval.

14.12 At what x-values can the absolute extrema guaranteed by the extreme value theorem occur, given a function $f(x)$ continuous on $[a,b]$?

> Remember, if c is a critical number of $f'(x)$, then either $f'(c) = 0$ or $f'(c)$ does not exist.

Absolute extrema occur either at an endpoint or a relative extreme point on the interval.

14.13 Identify the absolute maximum and the absolute minimum of $f(x)$, the function graphed in Figure 14-4.

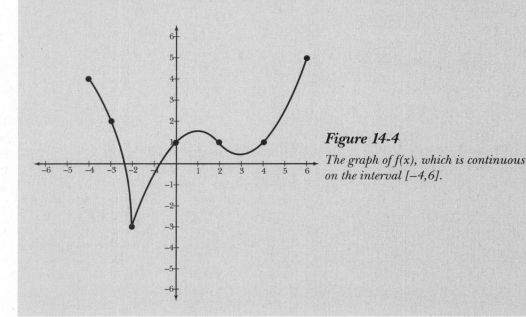

Figure 14-4

The graph of $f(x)$, which is continuous on the interval $[-4,6]$.

The highest y-value reached by $f(x)$, $y = 5$, occurs at $x = 6$, the right endpoint of the interval; the lowest y-value reached by $f(x)$, $y = -3$, occurs at the critical number $x = -2$.

14.14 Identify the absolute maximum and the absolute minimum of $f(x) = 3x^2 - 12x + 5$ on the interval $[1,4]$.

According to Problem 14.12, the absolute extrema occur at endpoints of the interval or relative extreme points of $f(x)$ (which must be located at critical numbers of $f'(x)$). Set $f'(x) = 0$ and solve for x to identify critical numbers of $f'(x)$.

$$f'(x) = 6x - 12$$
$$0 = 6x - 12$$
$$12 = 6x$$
$$2 = x$$

The absolute extrema of $f(x)$ could occur at $x = 1$ (an endpoint), $x = 2$ (a critical number of $f'(x)$) or $x = 4$ (an endpoint). Substitute each of those x-values into $f(x)$ and determine which generates the maximum and which generates the minimum function value.

$f(1) = 3(1)^2 - 12(1) + 5$	$f(2) = 3(2)^2 - 12(2) + 5$	$f(4) = 3(4)^2 - 12(4) + 5$
$= 3 - 12 + 5$	$= 12 - 24 + 5$	$= 3(16) - 48 + 5$
$= -4$	$= -7$	$= 5$

The absolute maximum of $f(x)$ on the interval $[1,4]$ is 5; it occurs at the right endpoint, $x = 4$. The absolute minimum value of $f(x)$ on the interval $[1,4]$ is –7; it occurs at $x = 2$.

14.15 Identify the absolute maximum and the absolute minimum of $g(x) = -3x^3 + 4x^2 - 1$ on the interval $[0,1]$.

Calculate the critical numbers of $g'(x)$.

$$-9x^2 + 8x = g'(x)$$
$$x(-9x + 8) = 0$$
$$-9x + 8 = 0$$
$$x = 0 \quad \text{or} \quad x = \frac{8}{9}$$

Evaluate $g(x)$ at those critical numbers and at the endpoints of the interval. Notice that $x = 0$ is both an endpoint and a critical point, so you need to calculate a total of three function values.

> Make sure you plug them into g(x) NOT g'(x). You're looking for the biggest and smallest FUNCTION values (not DERIVATIVE values), even though the critical number comes from f'(x).

$g(0) = -3(0)^3 + 4(0)^2 - 1$	$g\left(\frac{8}{9}\right) = -3\left(\frac{8}{9}\right)^3 + 4\left(\frac{8}{9}\right)^2 - 1$	$g(1) = -3(1)^3 + 4(1)^2 - 1$
$= 0 + 0 - 1$	$= -3\left(\frac{512}{729}\right) + 4\left(\frac{64}{81}\right) - 1$	$= -3 + 4 - 1$
$= -1$	$= -\frac{512}{243} + \frac{256}{81} - 1$	$= 0$
	$= -\frac{512}{243} + \frac{768}{243} - \frac{243}{243}$	
	$= \frac{13}{243}$	

The absolute maximum of $f(x)$ on $[0,1]$ is $\frac{13}{243}$; it occurs at $x = \frac{8}{9}$. The absolute minimum of $f(x)$ is –1, occurring at $x = 0$, the left-hand endpoint of the interval.

Newton's Method
Approximate the zeroes of a function

14.16 Identify the formula used in Newton's method and explain what is meant by an "iterative" calculation.

> Some textbooks call this the Newton-Raphson method.

An iterative calculation is used to produce a sequence of values, one at a time, so that each successive value is based on one or more of the values that precede it. It requires an initial value (often called a "seed value"), which is substituted into a formula to generate a new value, which is then substituted back into the formula, and the process repeats. Newton's method approximates the roots of functions. Based on some initial seed value x_1, it generates x_2 (a better approximation of the root) based on the formula below.

$$x_{n+1} = x_n - \frac{f(x_n)}{f'(x_n)}$$

You then substitute x_2 into the formula to generate x_3, substitute x_3 to generate x_4, and repeat the process until the desired degree of accuracy is achieved,

Note: Problems 14.17–14.19 refer to the function $f(x) = x^2 - 3$.

14.17 Perform one iteration of Newton's method to estimate a root of $f(x)$ using a seed value of $x_1 = 4$.

> The closer the seed value is to the actual root, the fewer the iterations required to reach an accurate answer.

Substitute $x_n = x_1 = 4$ into the formula for Newton's method to determine $x_{n+1} = x_2$. Note that $f(4) = 4^2 - 3 = 13$ and $f'(4) = 2(4) = 8$.

$$x_2 = x_1 - \frac{f(x_1)}{f'(x_1)}$$

$$x_2 = 4 - \frac{13}{8}$$

$$= \frac{32 - 13}{8}$$

$$= \frac{19}{8}$$

Therefore, $x_2 = \dfrac{19}{8}$ is closer to the positive root of $f(x) = x^2 - 3$ than the original seed value $x_1 = 4$.

Note: Problems 14.17–14.19 refer to the function $f(x) = x^2 - 3$.

14.18 Apply Newton's method using x_2 from Problem 14.17 to generate x_3. How accurately does x_3 estimate the root of $f(x) = x^2 - 3$? Use a calculator to represent the function and derivative values as decimals.

Substitute $n = 2$, $n + 1 = 3$, and $x_2 = \dfrac{19}{8} = 2.375$ into the Newton's method formula.

$$x_3 = 2.375 - \frac{f(2.375)}{f'(2.375)}$$

$$= 2.375 - \frac{(2.375)^2 - 3}{2(2.375)}$$

$$= 2.375 - \frac{2.640625}{4.75}$$

$$= 1.81907894737$$

In order to determine the accuracy of x_3, subtract it the actual root, $\sqrt{3}$.

$$\left| \sqrt{3} - 1.81907894737 \right| \approx 0.08703$$

Therefore, x_3 has an accuracy of 0.08703.

You're less than one-tenth away from the actual root after only two iterations.

> *Note: Problems 14.17–14.19 refer to the function $f(x) = x^2 - 3$.*

14.19 Calculate two iterations of Newton's method to estimate the root of $f(x)$, this time using a seed value of $x_1 = 2$. Determine the accuracy of x_3.

Substitute $x_1 = 2$ into Newton's method.

$$x_2 = x_1 - \frac{f(x_1)}{f'(x_1)}$$

$$= 2 - \frac{f(2)}{f'(2)}$$

$$= 2 - \frac{1}{4}$$

$$= 1.75$$

Now calculate x_3 given $x_2 = 1.75$.

$$x_3 = x_2 - \frac{f(x_2)}{f'(x_2)}$$

$$= 1.75 - \frac{f(1.75)}{f'(1.75)}$$

$$= 2 - \frac{0.0625}{3.5}$$

$$= 1.98214285714$$

Determine the accuracy of x_3 by subtracting it from the actual root, $\sqrt{3}$.

$$\left| \sqrt{3} - 1.98214285714 \right| \approx 0.25009$$

Accuracy describes the distance away from the actual answer, and distances are measured in positive numbers—that's why you need absolute values.

Typically, seed values closer to the actual roots produce more accurate approximations more quickly, but that is not always true. According to Problem 14.18, the seed value $x_1 = 4$ more accurately predicts the root than the seed value $x_1 = 3$ from this problem, if two iterations of Newton's method are performed for both.

14.20 How do you apply Newton's method to a function with multiple roots?

If a function contains more than one root, approximate each separately, being careful to choose a seed value that is as close as possible to the root you are tracking each time. Note that $f(x) = x^2 - 3$, the function in Problems 14.17–14.19, has two roots: $-\sqrt{3}$ and $\sqrt{3}$. The seed values of $x = 2$ and $x = 4$ will, after multiple iterations, tend toward the root $x = \sqrt{3}$. In order to approximate the negative root, you will need to choose a negative seed value, such as $x_1 = -2$.

14.21 What are the two most common reasons Newton's method fails to calculate a root?

If the seed value x_1 is too far away from the root, Newton's method may fail to locate it; each iteration will produce values that are farther and farther apart (rather than closer and closer together, like in Problems 14.17–14.19).

Newton's method cannot estimate non-real roots. If a function (such as $h(x) = x^2 + 3$) does not intersect the x-axis, it has no real roots; Newton's method cannot calculate imaginary roots.

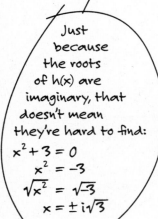

RULE OF THUMB: keep an eye on each iteration and make sure the difference between each x_n and the x_{n+1} it generates are closer together than x_n and x_{n-1}. If not, pick a better seed value and start over.

Just because the roots of $h(x)$ are imaginary, that doesn't mean they've hard to find:
$x^2 + 3 = 0$
$x^2 = -3$
$\sqrt{x^2} = \sqrt{-3}$
$x = \pm i\sqrt{3}$

14.22 Calculate $x = \sqrt[3]{21}$ accurate to five decimal places.

If you cube both sides of the equation and set it equal to 0, you create a function $g(x)$ whose root is $\sqrt[3]{21}$.

$$x^3 = \left(\sqrt[3]{21}\right)^3$$
$$x^3 = 21$$
$$x^3 - 21 = 0$$

Because the root of $g(x) = x^3 - 21$ is the value you are seeking, apply Newton's method to determine a decimal approximation. You're not given a seed value, so you must choose an appropriate value for x_1. Because $\sqrt[3]{8} = 2$ and $\sqrt[3]{27} = 3$, $\sqrt[3]{21}$ must be between 2 and 3. Furthermore, $\sqrt[3]{21}$ will be closer to 3 than 2 (because 21 is closer to 27 than 8). As such, $x_1 = 3$ is an appropriate seed value.

$$x_2 = 3 - \frac{g(3)}{g'(3)}$$
$$= 3 - \frac{6}{27}$$
$$= 3 - 0.\overline{2}$$
$$= 2.\overline{7}$$

Repeat the process until two consecutive iterations produce equal values for the first 5 decimal places.

$$x_3 = 2.75905185...$$
$$x_4 = 2.\underline{75892}418...$$
$$x_5 = 2.\underline{75892}417...$$

Because x_4 and x_5 have the required number of matching decimal places (in fact, their first *seven* decimal places match), $\sqrt[3]{21} \approx 2.75892$ according to Newton's method. ←

The actual value of $\sqrt[3]{21}$ is 2.75892417638, so in only four iterations, Newton's method already had it right to 8 decimal places!

14.23 The function $g(\theta) = \sin 3\theta - 2\cos 2\theta$ has two roots on the interval $[0,3]$, one approximately $\theta = 0.5$ radians and the other approximately $\theta = 2.5$ radians. Calculate both roots accurate to five decimal places.

You must apply Newton's method twice, once with a seed value of $\theta_1 = 0.5$, and once with a seed value of $\theta_1 = 2.5$. Differentiate $g(\theta)$ using the chain rule. ←

Make sure your calculator is set in radians mode. Leave it in radians unless a problem specifically mentions degrees.

$$g'(\theta) = \cos(3\theta) \cdot 3 - 2(-\sin 2\theta) \cdot 2$$
$$= 3\cos 3\theta + 4\sin 2\theta$$

Calculate a sufficient number of iterations of Newton's method for each seed value (i.e., until the first five decimal places of consecutive iterations match).

$$\theta_1 = 0.5$$
$$\theta_2 = 0.523227335$$
$$\theta_3 = 0.\underline{523598}676$$
$$\theta_4 = 0.\underline{523598}775$$

$$\theta_1 = 2.5$$
$$\theta_2 = 2.63258343$$
$$\theta_3 = 2.61814022$$
$$\theta_4 = 2.\underline{61799}389$$
$$\theta_5 = 2.\underline{61799}387$$

The roots of $g(\theta) = \sin 3\theta - 2\cos 2\theta$ are approximately $\theta = 0.52359$ and $\theta = 2.61799$.

Note: Problems 14.24–14.25 refer to the function $f(x) = -e^{\sqrt{x}} + 4$.

14.24 The graph of $f(x)$ intersects the x-axis only once, near $x = 2$. Calculate the root of $f(x)$ accurate to seven decimal places.

Differentiate $f(x)$. ←

The derivative of $-e^{\sqrt{x}}$ is itself times the derivative of the power.

$$f'(x) = -e^{\sqrt{x}} \cdot \frac{d}{dx}(x^{1/2}) = -e^{\sqrt{x}} \cdot \frac{1}{2}x^{(1/2)-1} = -e^{\sqrt{x}} \cdot \frac{1}{2x^{1/2}} = -\frac{e^{\sqrt{x}}}{2\sqrt{x}}$$

Apply Newton's method, given $f(x) = -e^{\sqrt{x}} + 4$, $f'(x) = \frac{-e^{\sqrt{x}}}{2\sqrt{x}}$, and $x_1 = 2$.

$$x_1 = 2$$
$$x_2 = 1.92212473$$
$$x_3 = 1.\underline{92181206}$$
$$x_4 = 1.\underline{92181205}$$

Therefore, the root of $f(x) = -e^{\sqrt{x}} + 4$ is approximately $x = 1.9218120$.

Note: *Problems 14.24–14.25 refer to the function* $f(x) = -e^{\sqrt{x}} + 4$.

14.25 Repeat Problem 14.24 using a seed value of $x_1 = 1$ to demonstrate that a seed value further away from the root requires additional iterations of Newton's method to reach the same degree of accuracy.

Use $f(x)$ and $f'(x)$ from Problem 14.24, but set $x_1 = 1$.

$$x_1 = 1$$
$$x_2 = 1.94303552$$
$$x_3 = 1.92183480$$
$$x_4 = 1.\underline{92181205}$$
$$x_5 = 1.\underline{92181205}$$

An additional iteration, x_5, is required to reach the same degree of accuracy as Problem 14.24.

The function only has complex roots because it doesn't intersect the x-axis anywhere.

14.26 Demonstrate that Newton's method fails to calculate the complex roots of $y = x^4 - 2x^2 + 2$ by calculating the first six iterations of the seed value $x_1 = -2$.

Plug $f(x) = x^4 - 2x^2 + 2$, $f'(x) = 4x^3 - 4x$, and $x_1 = -2$ into the Newton's method formula. Calculate x_2 through x_7, as they represent the first six iterations.

$$x_1 = -2$$
$$x_2 = -1.58\overline{3}$$
$$x_3 = -1.2406166$$
$$x_4 = -0.75820114$$
$$x_5 = -1.6739696$$
$$x_6 = -1.3219531$$
$$x_7 = -0.92760393$$

The terms of the sequence $\{x_1, x_2, x_3, x_4, x_5, x_6, x_7\}$ increase and decrease but never approach any real number limit. Furthermore, the iterations do not estimate any of the function's roots more and more accurately, because the difference of consecutive iterations is not decreasing.

That means you get one of these: $\frac{0}{0}, \frac{\infty}{\infty}, 0\cdot\infty, 0^0, 1^\infty,$ or ∞^0. In L'Hôpital's rule problems, you'll usually see either $\frac{0}{0}$ or $\frac{\infty}{\infty}$.

L'Hôpital's Rule
Find limits that used to be impossible

14.27 If $\lim\limits_{x\to\infty}\frac{f(x)}{g(x)}$, $\lim\limits_{x\to-\infty}\frac{f(x)}{g(x)}$, or $\lim\limits_{x\to c}\frac{f(x)}{g(x)}$ (where c is a real number) is indeterminate, how do you apply L'Hôpital's rule to calculate the limit?

According to L'Hôpital's rule, if a rational function has an indeterminate limit, you can differentiate the numerator and denominator of the function without altering the value of the limit. For example, if $\lim\limits_{x \to \infty} \dfrac{f(x)}{g(x)}$ is indeterminate, then $\lim\limits_{x \to \infty} \dfrac{f(x)}{g(x)} = \lim\limits_{x \to \infty} \dfrac{f'(x)}{g'(x)}$. ←

Don't use the quotient rule! Just take the derivative of the numerator then the derivative of the denominator—simple as that.

14.28 According to Problem 10.18, $\lim\limits_{x \to -2} \dfrac{x^3 + 8}{x + 2} = 12$. Use L'Hôpital's rule to verify the solution.

Substituting $x = -2$ into the rational function produces the indeterminate result $\dfrac{0}{0}$.

$$\lim_{x \to -2} \frac{x^3 + 8}{x + 2} = \frac{(-2)^3 + 8}{(-2) + 2} = \frac{-8 + 8}{-2 + 2} = \frac{0}{0}$$

Apply L'Hôpital's rule by differentiating the numerator and the denominator: $\dfrac{d}{dx}\left(x^3 + 8\right) = 3x^2$ and $\dfrac{d}{dx}\left(x + 2\right) = 1$. Replace each expression with its derivative.

$$\lim_{x \to -2} \frac{x^3 + 8}{x + 2} = \lim_{x \to -2} \frac{3x^2}{1} = \lim_{x \to -2} 3x^2$$

Substituting $x = -2$ no longer results in an indeterminate value.

$$\lim_{x \to -2} 3x^2 = 3(-2)^2 = 3(4) = 12$$

L'Hôpital's rule generates the same limit as Problem 10.18. ←

Plus you didn't have to have the "sum of perfect cubes" factoring formula memorized like you did in Problem 10.18.

14.29 According to Problem 10.13, $\lim\limits_{x \to 0} \dfrac{4x^5 - x^2}{x^2} = -1$. Use L'Hôpital's rule to verify the solution.

Substituting $x = 0$ into the rational expression produces an indeterminate result.

$$\lim_{x \to 0} \frac{4x^5 - x^2}{x^2} = \frac{4(0)^5 - (0)^2}{(0)^2} = \frac{0}{0}$$

Apply L'Hôpital's rule.

$$\lim_{x \to 0} \frac{4x^5 - x^2}{x^2} = \lim_{x \to 0} \frac{20x^4 - 2x}{2x}$$

Unfortunately, substituting $x = 0$ into the new limit expression also produces an indeterminate answer.

$$= \frac{20(0)^4 - 2(0)}{2(0)} = \frac{0}{0}$$

There is no restriction on the number of times you may apply L'Hôpital's rule, as long as the limit produces an indeterminate value each time.

$$\lim_{x \to 0} \frac{20x^4 - 2x}{2x} = \lim_{x \to 0} \frac{80x^3 - 2}{2} = \frac{80(0)^3 - 2}{2} = \frac{-2}{2} = -1$$

This solution matches Problem 10.13.

14.30 According to Problem 10.27, $\lim\limits_{x \to 19} \dfrac{x-19}{5-\sqrt{x+6}} = -10$. Use L'Hôpital's rule to verify the solution.

Substituting $x = 19$ into the rational expression produces an indeterminate result.

$$\lim_{x \to 19} \frac{x-19}{5-\sqrt{x+6}} = \frac{19-19}{5-\sqrt{19+6}} = \frac{0}{5-\sqrt{25}} = \frac{0}{0}$$

Apply L'Hôpital's rule by differentiating the numerator and denominator separately.

$$\lim_{x \to 19} \frac{x-19}{5-\sqrt{x+6}} = \lim_{x \to 19} \frac{1}{-\dfrac{1}{2}(x+6)^{-1/2} \cdot 1} = \lim_{x \to 19} \frac{1}{-\dfrac{1}{2\sqrt{x+6}}} = \lim_{x \to 19}\left(-2\sqrt{x+6}\right)$$

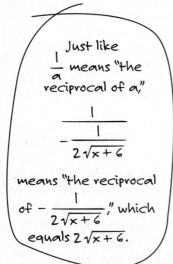

Just like $\dfrac{1}{a}$ means "the reciprocal of a,"

$$\dfrac{1}{-\dfrac{1}{2\sqrt{x+6}}}$$

means "the reciprocal of $-\dfrac{1}{2\sqrt{x+6}}$," which equals $2\sqrt{x+6}$.

Substitute $x = 19$ into the new limit expression.

$$= -2\sqrt{19+6} = -2\sqrt{25} = -2(5) = -10$$

The solution matches Problem 10.27.

14.31 According to Problem 10.29, $\lim\limits_{\theta \to 0} \dfrac{\sin 7\theta}{\theta} = 7$. Use L'Hôpital's rule to verify the solution.

Substituting $\theta = 0$ into the rational expression produces an indeterminate result.

$$\lim_{\theta \to 0} \frac{\sin 7\theta}{\theta} = \frac{\sin(7 \cdot 0)}{0} = \frac{\sin 0}{0} = \frac{0}{0}$$

Apply L'Hôpital's Rule; use the chain rule to differentiate $\sin 7\theta$.

$$\lim_{\theta \to 0} \frac{\sin 7\theta}{\theta} = \lim_{\theta \to 0} \frac{\cos(7\theta) \cdot 7}{1} = \lim_{\theta \to 0} 7\cos 7\theta$$

Substitute $\theta = 0$ into the limit expression.

According to the unit circle in Appendix B, $\cos 0 = \cos 2\pi = 1$ and $\sin 0 = \sin 2\pi = 0$.

$$= 7\cos(7 \cdot 0) = 7\cos 0 = 7(1) = 7$$

This matches the solution to Problem 10.29.

14.32 According to Problem 9.25, $\lim\limits_{x \to \infty} \dfrac{5x^2 - 9x + 1}{5 - 3x - 6x^2} = -\dfrac{5}{6}$. Use L'Hôpital's rule to verify the solution.

Substituting $x = \infty$ produces an indeterminate result.

$$\lim_{x \to \infty} \frac{5(\infty)^2 - 9 \cdot \infty + 1}{5 - 3\infty - 6(\infty)^2} = \frac{\infty}{-\infty}$$

Apply L'Hôpital's rule.

$$\lim_{x \to \infty} \frac{5x^2 - 9x + 1}{5 - 3x - 6x^2} = \lim_{x \to \infty} \frac{10x - 9}{-3 - 12x}$$

Substituting $x = \infty$ still produces the indeterminate result $-\dfrac{\infty}{\infty}$, so apply L'Hôpital's rule again.

$$\lim_{x \to \infty} \frac{10x - 9}{-3 - 12x} = \lim_{x \to \infty} \frac{10}{-12} = -\frac{5}{6}$$

This method is a much more mathematically satisfying technique than the process used to reach the same result in Problem 9.25, which simply presented the limit as a quotient of the leading coefficients of the numerator and denominator of the rational function.

14.33 Evaluate $\displaystyle\lim_{x \to -2} \frac{3x^2 + 8x + 4}{-2 + x}$.

Substituting $x = -2$ does *not* produce an indeterminate result, so you *cannot* apply L'Hôpital's rule. ←

$$\lim_{x \to -2} \frac{3x^2 + 8x + 4}{-2 + x} = \frac{3(-2)^2 + 8(-2) + 4}{-2 + (-2)} = \frac{12 - 16 + 4}{-4} = \frac{0}{-4} = 0$$

Therefore, $\displaystyle\lim_{x \to -2} \frac{3x^2 + 8x + 4}{-2 + x} = 0$.

> Trick question! You can only use L'Hôpital's rule if the limit is indeterminate, so if you got −4, you fell for the trick.

14.34 Evaluate $\displaystyle\lim_{x \to \infty}\left[\left(x^3 + 5x^2 - 3x - 9\right)\left(e^{-ax}\right)\right]$.

Substituting $x = \infty$ into the expression produces an indeterminate result.

$$\left(\infty^3 + 5\infty^2 - 3\infty - 9\right)\left(e^{-a \cdot \infty}\right) = (\infty)\left(\frac{1}{e^\infty}\right) = \infty \cdot 0 \;\longleftarrow$$

> Because e to a gigantic power is, as you would expect, really gigantic, 1 divided by that giant number is basically 0. (See Problem 10.28 for more explanation.)

Rewrite the expression as a quotient so you can apply L'Hôpital's rule.

$$\lim_{x \to \infty}\left[\left(x^3 + 5x^2 - 3x - 9\right)\left(e^{-a \cdot x}\right)\right] = \lim_{x \to \infty} \frac{x^3 + 5x^2 - 3x - 9}{e^{a \cdot x}}$$

You'll have to apply L'Hôpital's rule three times before the limit finally ceases to be indeterminate.

$$\lim_{x \to \infty} \frac{x^3 + 5x^2 - 3x - 9}{e^{a \cdot x}} = \lim_{x \to \infty} \frac{3x^2 + 10x - 3}{a \cdot e^{a \cdot x}}$$

$$= \lim_{x \to \infty} \frac{6x + 10}{a \cdot \left(a \cdot e^{a \cdot x}\right)}$$

$$= \lim_{x \to \infty} \frac{6}{a \cdot a \cdot \left(a \cdot e^{a \cdot x}\right)} = \lim_{x \to \infty} \frac{6}{a^3 \cdot e^{a \cdot x}}$$

Substitute $x = \infty$ into the expression.

$$\lim_{x \to \infty} \frac{6}{a^3 \cdot e^{a \cdot x}} = \frac{6}{a^3 \cdot e^{a \cdot \infty}} = 0$$

Any fixed value divided by a number that increases without bound has a limit of 0.

14.35 Evaluate $\lim_{x \to \infty} x^{1/x}$.

Substituting $x = \infty$ produces an indeterminate result.

$$\lim_{x \to \infty} x^{1/x} = \infty^{1/\infty} = \infty^0$$

If expressions in terms of x are raised to a powers in terms of x, it is often useful to employ natural logarithms to rewrite the exponential expression. Begin by setting the expression whose limit you are evaluating equal to y.

$$\lim_{x \to \infty} x^{1/x} = \lim_{x \to \infty} y \text{ when } y = x^{1/x}$$

For the moment, manipulate only the y equation. Begin by taking the natural logarithm of both sides.

$$\ln y = \ln x^{1/x}$$

> See Problem 5.23 for more details.

According to a property of logarithms that states $\log a^x = x \log a$, you can extricate the exponent from the logarithm and write it as the coefficient of the logarithmic expression.

$$\ln y = \frac{1}{x} \cdot \ln x = \frac{\ln x}{x}$$

Notice that $\lim_{x \to \infty} \ln y = \frac{\ln \infty}{\infty} = \frac{\infty}{\infty}$, so you should apply L'Hôpital's rule.

$$\lim_{x \to \infty} \ln y = \lim_{x \to \infty} \frac{\ln x}{x} = \lim_{x \to \infty} \frac{1/x}{1} = \lim_{x \to \infty} \frac{1}{x} = 0$$

> See Problem 5.34 for a definition of "exponentiating" and an example with natural logs and e^x.

Therefore, $\lim_{x \to \infty} \ln y = 0$. Exponentiate both sides of this equation using the natural exponential function.

$$\lim_{x \to \infty} \ln y = 0$$
$$\lim_{x \to \infty} e^{\ln y} = e^0$$
$$\lim_{x \to \infty} y = 1$$

Recall that $y = x^{1/x}$.

$$\lim_{x \to \infty} x^{1/x} = 1$$

Chapter 15
ADVANCED APPLICATIONS OF DIFFERENTIATION
Tricky but interesting uses for derivatives

In Chapter 14, you were presented with a variety of basic differentiation applications, each of which represented either an extension of a previously known concept (such as locating absolute versus relative maxima and minima) or a streamlined approach to solving problems (such as Newton's method to find real roots and L'Hôpital's rule to calculate indeterminate limits). In this chapter, you will explore uses of the derivative that extend beyond prior knowledge. Interpreting the derivative not only as the slope of a tangent line, but as an instantaneous rate of change, presents unique uses and applications concerning not only functions and graphs, but the actual mathematical modeling of physical phenomena.

In this chapter, you'll look at the derivative as a rate of change. The most basic rate of change is the slope of a line, which is constant over the entire line—it's always sloping at the same angle. However, curves have different rates of change, because their tangent lines are different for different values of x. You'll start with the mean value theorem, which has limited "real world" uses but is an important mathematical principle. After that, however, you'll use functions and derivatives to describe rectilinear motion (like the paths of projectiles that are thrown or fired), related rates (such as how quickly the water level of a pool is decreasing if you know how fast the water is leaking out), and optimization (calculating the biggest or smallest value of a function, like the lowest possible cost of manufacturing a product given specific restrictions). Basically, this chapter argues that derivatives are useful not only in the abstract world of math, but can shed light on the real world, too.

The Mean Value and Rolle's Theorems
Average slopes = instant slopes

15.1 If $f(x)$ is continuous and differentiable on the interval $[a,b]$ and $a < c < b$, calculate the average rate of change of $f(x)$ over $[a,b]$ and the instantaneous rate of change of $f(x)$ at $x = c$.

> From way back in Problem 1.5.

The average rate of change of a function is the slope of the secant line connecting the function values at the endpoints of the interval. In this problem, the endpoints of the function on $[a,b]$ are $(a,f(a))$ and $(b,f(b))$. Use the formula to calculate the slope of a line given two points.

$$m = \frac{y_2 - y_1}{x_2 - x_1} = \frac{f(b) - f(a)}{b - a}$$

The instantaneous rate of change at $x = c$ is the slope of the tangent line to $f(x)$ at $x = c$: $f'(c)$.

15.2 Explain the geometric implications of the mean value theorem.

> So the average (mean) rate of change equals the instantaneous rate of change there.

Given a function $f(x)$ that is continuous on $[a,b]$ and differentiable on (a,b), the mean value theorem states that there is at least one value $x = c$ between a and b such that $f'(c) = \frac{f(b) - f(a)}{b - a}$. According to Problem 15.1, that means there is at least one x-value on the interval at which the tangent line is parallel to the secant line that connects the endpoints of the interval.

15.3 How many times does the function $g(x)$ graphed in Figure 15-1 satisfy the mean value theorem?

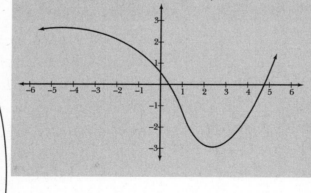

Figure 15-1

A function g(x) that is continuous and differentiable on [−5,5].

> Hold your pencil against the function so that it looks like a tangent line. Slide your pencil along the graph, twisting it as you go, so that it always looks tangent to the function, and count the number of times it is parallel to the secant.

Draw the secant line connecting the function values $g(-5)$ and $g(5)$, as illustrated by the dotted line in Figure 15-2. There are two values of x on the interval at which the tangent line to $g(x)$ is parallel to the secant line.

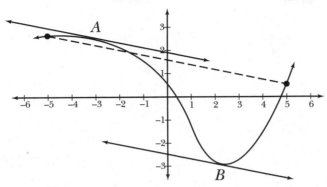

Figure 15-2 *The tangent lines to g(x) at points A and B are parallel to the secant line joining (−5,g(−5)) and (5,g(5)).*

15.4 How many times does the function $h(x)$ graphed in Figure 15-3 satisfy the mean value theorem on the interval $[-6,6]$?

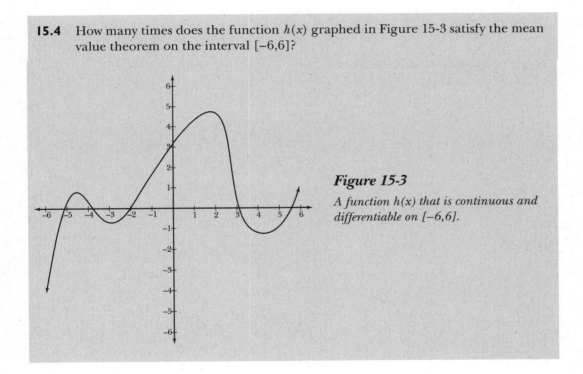

Figure 15-3

A function h(x) that is continuous and differentiable on [−6,6].

As illustrated by Figure 15-4, there are 4 values of x on the interval $[-6,6]$ at which the tangent line to the function is parallel to the secant line connecting the function values $h(-6)$ and $h(6)$.

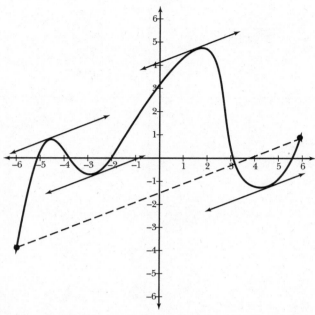

Figure 15-4 *At approximately x = −4.6, −2.75, 1.5, and 4.5, the tangent lines to h(x) are parallel to the dotted secant line connecting (−6,h(6)) and (6,h(6)).*

15.5 At what value(s) of x does $f(x) = x^3 - 2x^2 - 4x + 1$ satisfy the mean value theorem on the interval [0,1]?

According to the mean value theorem, there exists some value c between $x = 0$ and $x = 1$ such that $f'(c)$ equals the average rate of change of $f(x)$ over the interval [0,1].

$$f'(c) = \frac{f(b) - f(a)}{b - a} = \frac{f(1) - f(0)}{1 - 0}$$

Evaluate $f(1)$ and $f(0)$.

$$f'(c) = \frac{-4 - 1}{1 - 0} = \frac{-5}{1} = -5$$

Note that $f'(x) = 3x^2 - 4x - 4$.

$$f'(c) = -5$$
$$3c^2 - 4c - 4 = -5$$

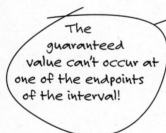

The guaranteed value can't occur at one of the endpoints of the interval!

Solve for c by setting the equation equal to 0 and factoring.

$$3c^2 - 4c + 1 = 0$$
$$(3c - 1)(c - 1) = 0$$
$$c = \frac{1}{3}, 1$$

Therefore, the mean value theorem is satisfied at $x = \frac{1}{3}$ but not at $x = 1$, because the c-value must belong to the *open* interval (0,1).

15.6 At what value(s) of x does the continuous and differentiable function $g(x) = ax^2 + bx + c$ satisfy the mean value theorem on the interval $[0,b]$? Assume that a, b, and c are nonzero real numbers.

Use the method outlined in Problem 15.5—differentiate $g(x)$ and set $g'(x)$ equal to the slope of the secant line connecting $(0,g(0))$ to $(b,g(b))$.

$$g'(x) = \frac{g(b) - g(0)}{b - 0}$$

$$2ax + b = \frac{(ab^2 + b^2 + c) - (a \cdot 0^2 + b \cdot 0 + c)}{b}$$

$$2ax + b = \frac{ab^2 + b^2 + \cancel{c} - \cancel{c}}{b}$$

$$2ax + b = \frac{\cancel{b}(ab + b)}{\cancel{b}}$$

$$2ax + b = ab + b$$

Solve for x.

$$2ax = ab + \cancel{b} - \cancel{b}$$

$$2ax = ab$$

$$x = \frac{\cancel{a} \cdot b}{2 \cdot \cancel{a}}$$

$$x = \frac{b}{2}$$

> This is a formula from algebra: distance traveled equals the rate traveled multiplied by the time traveled at that speed.

15.7 A policeman clocks a commuter's speed at 50 mph as he enters a tunnel whose length is exactly 0.75 miles. A second officer measures the commuter's speed at 45 mph as he exits the tunnel 43 seconds later and tickets the driver for exceeding the posted speed limit of 50 mph. Use the mean value theorem to justify the speeding charge levied by the officer, even though the driver was neither exceeding the posted speed limit while entering nor while exiting the tunnel.

Determine the average speed of the driver.

$$d = r \cdot t \leftarrow$$

The distance traveled is the length of the tunnel ($d = 0.75$ miles), and it took the driver 43 seconds to travel that distance. Convert $t = 43$ seconds into a consistent unit of measurement and substitute d and t into the distance formula. $\leftarrow$

$$0.75 = r\left(\frac{43}{3600}\right)$$

Solve the equation for r, the driver's average rate of speed, in miles per hour.

$$\left(\frac{3600}{43}\right)(0.75) = r$$

$$62.791 \approx r$$

> Because distance is measured in MILES and the speeds are measured in miles per HOUR, the time needs to be measured in HOURS, not seconds. There are 60 seconds in a minute and 60 minutes in an hour, so every second is $\frac{1}{60 \cdot 60} = \frac{1}{3600}$th of an hour.

His top speed in the tunnel was probably much faster than 62.7 mph. You know he was driving slower than that at the beginning and end of the tunnel, so he must have also driven pretty fast inside there to get such a high average.

If the driver's average rate of change is 62.791 mph, then (according to the mean value theorem) the driver's actual (instantaneous) speed was 62.791 mph at least once inside the tunnel, which violates the posted speed limit.

15.8 Describe the difference between the mean value theorem and Rolle's theorem.

Both theorems guarantee the existence of a value c on a function $f(x)$ that's continuous on $[a,b]$ and differentiable on (a,b) such that $a < c < b$. Furthermore, both guarantee that $f'(c) = \dfrac{f(b) - f(a)}{b - a}$. However, Rolle's theorem has one additional requirement: $f(a) = f(b)$. When $x = a$ and $x = b$ have equivalent function values, the slope of the secant line connecting $(a, f(a))$ and $(b, f(b))$ is $\dfrac{0}{b - a} = 0$, so $f'(c)$ must equal 0 somewhere on (a,b).

And if f'(c) = 0, then c is a critical number of f'(x), and therefore f(x) may have a relative max or a relative min at x = c.

15.9 Find the smallest positive value of b such that you can apply Rolle's theorem to $g(x) = x \sin x$ on the interval $[0,b]$?

In order to apply Rolle's theorem to $g(x)$, you must determine the value of $x = b$ such that $g(0) = g(b)$. Because $g(0) = (0)(\sin 0) = 0$, set $g(b) = 0$ and solve.

$$g(b) = 0$$
$$b \sin b = 0$$

The expression on the left side of the equation is a product, so set each factor equal to 0 and solve.

$$b = 0 \quad \text{or} \quad \sin b = 0$$
$$b = k\pi, \ k \text{ is an integer}$$

0 is not a positive or a negative number. It's neutral like Switzerland—only rounder.

You are asked to identify the smallest *positive* value of b, which corresponds to the smallest positive value of k, $k = 1$. Therefore, $b = 1 \cdot \pi = \pi$.

15.10 Verify that $f(x) = 40x^4 + 22x^3 - 91x^2 - 58x + 15$ has at least one horizontal tangent line on the interval $\left[-\dfrac{5}{4}, \dfrac{1}{5}\right]$.

These coefficients are ridiculous, but that's because they don't want you trying to graph this by hand or to do anything besides Rolle's theorem.

The endpoints of the given interval are also roots of $f(x)$, as $f\left(-\dfrac{5}{4}\right) = f\left(\dfrac{1}{5}\right) = 0$.

$$f\left(-\frac{5}{4}\right) = \frac{3125}{32} - \frac{1375}{32} - \frac{2275}{16} + \frac{145}{2} + 15$$
$$= \frac{3125 - 1375 - 2275(2) + 145(16) + 15(32)}{32}$$
$$= \frac{0}{32} = 0$$

$$f\left(\frac{1}{5}\right) = \frac{8}{125} + \frac{22}{125} - \frac{91}{25} - \frac{58}{5} + 15$$
$$= \frac{8 + 22 - 91(5) - 58(25) + 15(125)}{125}$$
$$= \frac{0}{125} = 0$$

Because $f\left(-\dfrac{5}{4}\right) = f\left(\dfrac{1}{5}\right) = 0$, then $f'(c) = 0$ for some value c between $-\dfrac{5}{4}$ and $\dfrac{1}{5}$ according to Rolle's theorem.

Rectilinear Motion

Position, velocity, and acceleration functions

Note: Problems 15.11–15.17 refer to the path of a baseball thrown from an initial height of 5 feet above the ground with an initial velocity of 100 ft/sec. For the sake of simplicity, discount the effect of wind resistance on the baseball.

15.11 Construct a function that models the height of the baseball, in feet, at t seconds.

The path of a projectile t seconds after its launch is modeled by the position equation $s(t) = -\dfrac{a}{2}t^2 + v_0 \cdot t + h_0$, where a is acceleration due to gravity, v_0 is the initial velocity, and h_0 is the initial height. Note that a is a constant value, 32 ft/sec², and the rest of the values are explicitly defined by the problem: $v_0 = 100$ and $h_0 = 5$.

$$s(t) = -\frac{a}{2}t^2 + v_0 \cdot t + h_0$$

$$s(t) = \frac{-32}{2}t^2 + 100 \cdot t + 5$$

$$s(t) = -16t^2 + 100t + 5$$

> If meters are used instead of feet, then $a = 9.8 \text{ m/sec}^2$.

Note: Problems 15.11–15.17 refer to the path of a baseball thrown from an initial height of 5 feet above the ground with an initial velocity of 100 ft/sec.

15.12 How high is the baseball exactly one second after it is thrown?

The moment the baseball is thrown is considered $t = 0$ seconds. Given $s(t)$ as defined in Problem 15.11, $s(1)$ is the height of the baseball one second after it is thrown.

$$s(1) = -16(1)^2 + 100(1) + 5$$
$$= -16 + 105$$
$$= 89 \text{ feet above the ground}$$

> That's why the initial height (h_0) and initial velocity (v_0) are usually written with the little 0's next to them—they're the height and velocity when the ball is thrown, when $t = 0$.

Note: Problems 15.11–15.17 refer to the path of a baseball thrown from an initial height of 5 feet above the ground with an initial velocity of 100 ft/sec.

15.13 What is the velocity of the baseball at $t = 1$?

The velocity is the rate at which the position of the projectile changes. Thus, the derivative of $s(t)$, $s'(t) = -32t + 100$ represents the baseball's velocity t seconds after it is thrown. Evaluate $s'(1)$ to determine the velocity of the baseball at $t = 1$.

$$s'(t) = -32(1) + 100 = 68 \text{ ft/sec}$$

Note: Problems 15.11–15.17 refer to the path of a baseball thrown from an initial height of 5 feet above the ground with an initial velocity of 100 ft/sec.

15.14 What is the acceleration of the baseball at $t = 1$?

*So,
$a(1) = -32$.
Actually, $a(t) = -32$
no matter what t
you plug in.*

The acceleration of an object is the rate at which its velocity changes. Therefore, the derivative of the velocity function is the acceleration function: $a(t) = v'(t)$ (and the second derivative of the position function: $a(t) = s''(t)$). Notice that $a(t) = -32$ ft/sec²; the only acceleration of the baseball is the acceleration due to gravity.

Note: Problems 15.11–15.17 refer to the path of a baseball thrown from an initial height of 5 feet above the ground with an initial velocity of 100 ft/sec.

15.15 When will the baseball hit the ground? (Provide an answer in seconds that is accurate to four decimal places.)

It doesn't make sense for the ball to hit the ground .0496 seconds before you throw it. If time starts at $t = 0$ in this problem, answers involving negative time should be ignored.

The baseball hits the ground when its height, $s(t)$, equals 0. Set $s(t) = 0$ and solve using the quadratic formula.

$$-16t^2 + 100t + 5 = 0$$

$$t = \frac{-100 \pm \sqrt{10000 - 4(-16)(5)}}{2(-16)}$$

Use a calculator to evaluate the expression: $t \approx -0.049606$ or 6.299606. Discard the negative answer and round to the correct number of decimal places. The baseball hits the ground approximately 6.2996 seconds after it is thrown.

Note: Problems 15.11–15.17 refer to the path of a baseball thrown from an initial height of 5 feet above the ground with an initial velocity of 100 ft/sec.

15.16 When does the baseball reach its maximum height?

The maximum height of the baseball is also the relative maximum of the function $s(t)$. To determine the location of the relative maximum, find the critical number of $s'(t)$.

$$s'(t) = -32t + 100$$
$$0 = -32t + 100$$
$$-100 = -32t$$
$$t = \frac{100}{32} = \frac{25}{8}$$

Note that the critical number $t = \dfrac{25}{8}$ produces a relative maximum according to the second derivative test. Therefore, the baseball reaches its maximum height $t = \dfrac{25}{8} = 3.125$ seconds after it is thrown.

> Because $s''(t) = -32$, the second derivative is negative no matter what t is. According to Problem 13.29, that makes the critical number a relative maximum.

Note: Problems 15.11–15.17 refer to the path of a baseball thrown from an initial height of 5 feet above the ground with an initial velocity of 100 ft/sec.

15.17 Assuming the baseball is thrown straight up, what is the maximum height it will reach?

According to Problem 15.16, the baseball reaches its maximum height at $t = 3.125$ seconds. To determine the height of the ball at that moment, evaluate $s(3.125)$.

$$s(3.125) = -16(3.125)^2 + 100(3.125) + 5 = 161.25 \text{ feet above ground}$$

Note: Problems 15.18–15.22 refer to the path of a particle traveling right and left along the x-axis. The position of the particle (in inches) after t seconds is given by the equation $s(t) = \cos(\ln(t + 0.5))$. Use a graphing calculator to solve equations and evaluate derivatives as necessary in these problems. Assume that all angles are measured in radians.

15.18 Construct the velocity function, $v(t)$, of the particle and use it to determine the particle's speed at $t = 3$ seconds. Provide an answer accurate to three decimal places.

The velocity function, $v(t)$, of the particle is the derivative of its position function, $s(t)$.

$$s'(t) = v(t) = -\sin(\ln(t + 0.5)) \cdot \frac{1}{t + 0.5} = -\frac{\sin(\ln(t + 0.5))}{t + 0.5}$$

> Use the chain rule to find the derivative. Differentiate cosine to get –sine, leaving the inner function ln (t + 0.5) alone, multiply by the derivative of ln (t + .5), which is 1/(t + .5), and then multiply by the derivative of (t + .5), which is just 1.

Evaluate $v(3)$ to determine the velocity at $t = 3$.

$$v'(3) = -\frac{\sin(\ln 3.5)}{3.5} \approx -0.271386$$

Therefore, the particle is traveling at a speed of approximately 0.271 inches/second. The negative sign is omitted when measuring speed; it merely indicates that the particle is traveling in a negative direction (in this case to the left).

> Speed is the absolute value of velocity—it's always positive.

Note: Problems 15.18–15.22 refer to the path of a particle traveling right and left along the x-axis whose position, in inches, at t seconds is given by the equation s(t) = cos (ln (t + 0.5)).

15.19 Calculate the particle's acceleration at $t = 6$. Provide an answer accurate to four decimal places.

The acceleration function of the particle is the derivative of the velocity function generated in Problem 15.18. Use the quotient rule to differentiate $v(t)$, and notice that the chain rule is required to differentiate the numerator.

$$a(t) = \frac{d}{dt}\left(-\frac{\sin\left(\ln\left(t+0.5\right)\right)}{t+0.5}\right)$$

$$a(t) = -\frac{(t+0.5)\left(\cos\left(\ln\left(t+0.5\right)\right)\cdot\left(\frac{1}{t+0.5}\right)\right) - \sin\left(\ln\left(t+0.5\right)\right)}{(t+0.5)^2}$$

The particle's acceleration is measured in inches per second per second, or in/sec². Just square the denominator of the units you used for velocity.

Evaluate $t = 6$.

$$a(6) = -\frac{(6.5)\left(\cos\left(\ln 6.5\right)\cdot\left(\frac{1}{6.5}\right)\right) - \sin\left(\ln 6.5\right)}{(6.5)^2} \approx 0.0296218$$

Therefore, the particle is accelerating at a rate of 0.0296 in/sec² when $t = 6$.

Note: Problems 15.18–15.22 refer to the path of a particle traveling right and left along the x-axis whose position, in inches, at t seconds is given by the equation s(t) = cos (ln (t + 0.5)).

15.20 How many times does the particle change direction (assuming $t > 0$)?

When s(t) is increasing, v(t) is positive and the particle is moving to the right; when s(t) is decreasing, v(t) is negative and the particle is heading left.

The particle changes direction whenever its velocity changes from positive to negative or vice versa. Notice that $\frac{1}{2}$ is a critical number of $v(t)$ because $-\frac{\sin\left(\ln\left(t+0.5\right)\right)}{t+0.5}$ equals 0 at that t-value.

The graph of $v(t)$ is above the x-axis, and therefore positive, when $t < \frac{1}{2}$; $v(t)$ then crosses the x-axis only once, at $t = \frac{1}{2}$, so $v'(t) < 0$ when $t > \frac{1}{2}$. Therefore, the particle changes direction once, at $t = \frac{1}{2}$.

Note: Problems 15.18–15.22 refer to the path of a particle traveling right and left along the x-axis whose position, in inches, at t seconds is given by the equation s(t) = cos (ln t + 0.5)).

15.21 When is the particle moving left?

The particle is moving left when $v(t) < 0$. According to Problem 15.20, $v(t) < 0$ on the interval $\left(\frac{1}{2}, \infty\right)$. Therefore, the particle travels left when $t > \frac{1}{2}$.

Note: Problems 15.18–15.22 refer to the path of a particle traveling right and left along the x-axis whose position, in inches, at t seconds is given by the equation s(t) = cos (ln (t + 0.5)).

15.22 What is the total distance traveled by the particle from $t = 0$ seconds to $t = 5$ seconds? Provide an answer accurate to three decimal places.

Determine the position of the particle when $t = 0$ and when $t = 5$.

$$s(0) = \cos(\ln 0.5) \approx 0.7692389 \qquad s(5) = \cos(\ln 5.5) \approx -0.1335515$$

You must also find the position of the particle wherever its direction changes. According to Problem 15.20, the particle changes direction at $t = 0.5$.

$$s(0.5) = \cos(\ln 1) = \cos 0 = 1$$

> You can't just subtract these numbers and say the particle traveled |-0.1335515 - 0.7692389| = 0.9027904 inches. That's the total DISPLACEMENT between its starting and ending points but not necessarily how far it actually traveled.

From $t = 0$ to $t = 0.5$ the particle travels right, from 0.7692389 units right of the origin to 1 unit right of the origin. Calculate the absolute value of the difference between those positions to find the distance the particle travels during that half second.

$$|1 - 0.7692389| = |-0.2307611| = 0.2307611$$

From $t = 0.5$ to $t = 5$ the particle travels left, from 1 one unit right of the origin to a position of −0.1335515, which is 0.1335515 units left of the origin. Find the absolute value of the difference between those positions to find the total distance the particle travels from $t = 0.5$ to $t = 5$ seconds.

$$|-0.1335515 - 1| = |-1.1335515| = 1.1335515$$

Sum the individual distances to calculate the total distance traveled.

$$0.2307611 + 1.1335515 = 1.3643126$$

The particle travels approximately 1.364 inches during the first 5 seconds.

Related Rates

Figure out how quickly the variables change in a function

15.23 A particle travels from left to right along the graph of $y = x^3 e^x$. Assuming its vertical rate of change is $\dfrac{dy}{dt} = 7$ ft/sec when $x = 1$, what is its horizontal rate of change at that moment?

Take the derivative of the equation with respect to t using the product rule.

$$\frac{dy}{dt} = \left(3x^2 \cdot \frac{dx}{dt}\right)e^x + x^3\left(e^x \cdot \frac{dx}{dt}\right)$$

$$\frac{dy}{dt} = \frac{dx}{dt}\left(3x^2 e^x + x^3 e^x\right)$$

> The derivative of x^2 with respect to x is $2x$, but the derivative of x^2 with respect to t is $2x \cdot \dfrac{dx}{dt}$. You have to use the chain rule, and when you take the derivative of x, the function "inside" x^2, it isn't 1 anymore—it's $\dfrac{dx}{dt}$.

According to the problem, $\frac{dy}{dt} = 7$ when $x = 1$. Substitute those values into the equation and solve for $\frac{dx}{dt}$, the horizontal rate of change.

$$7 = \frac{dx}{dt}\left(3(1)^2 e^1 + (1)^3 e^1\right)$$

$$7 = \frac{dx}{dt}(3e + e)$$

$$\frac{7}{4e}\,\text{ft/sec} = \frac{dx}{dt}$$

Your first job is to find an equation that contains all the variables you either know or want to find. Then take the derivative with respect to t. Make sure all of the rates you're working with (like dV/dt and ds/dt) appear in the derivative.

15.24 A bouillon cube with side length 0.8 cm is placed into boiling water. Assuming it roughly resembles a cube as it dissolves, at approximately what rate is its volume changing when its side length is 0.25 cm and is decreasing at a rate of 0.12 cm/sec?

The volume of a cube with side s is $V = s^3$. Differentiate this equation with respect to t.

$$V = s^3$$

$$\frac{dV}{dt} = 3s^2 \cdot \frac{ds}{dt}$$

When a value is DECREASING, its rate of change is NEGATIVE, so ds/dt = −0.12, not 0.12, and dV/dt will be negative as well.

You are given $s = 0.25$ and $\frac{ds}{dt} = -0.12$. Substitute these values into the equation and solve for $\frac{dV}{dt}$.

$$\frac{dV}{dt} = 3(0.25)^2 \cdot (-0.12) = -0.0225\,\text{cm}^3/\text{sec}$$

Thus, the volume is decreasing at a rate of 0.0225 cm³/sec.

15.25 Two hikers begin at the same location and travel in perpendicular directions. Hiker *A* travels due north at a rate of 5 miles per hour; Hiker *B* travels due west at a rate of 8 miles per hour. At what rate is the distance between the hikers changing 3 hours into the hike?

Figure 15-5 shows the positions of the hikers after three hours of traveling. Hiker *A* has traveled 5 mph for three hours, totaling 15 miles; Hiker *B* has traveled 8 mph for three hours, for a total of 24 miles. Apply the Pythagorean theorem to determine the distance between the hikers three hours into the hike (represented by the hypotenuse of the right triangle in Figure 15-5).

Side length a (in Figure 15-5) is the distance traveled by Hiker A, and b is the distance traveled by Hiker B.

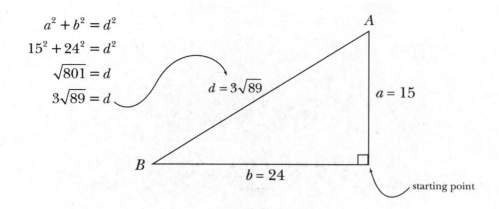

$a^2 + b^2 = d^2$

$15^2 + 24^2 = d^2$

$\sqrt{801} = d$

$3\sqrt{89} = d$

$d = 3\sqrt{89}$

A

$a = 15$

B

$b = 24$

starting point

Figure 15-5 *After three hours, the distance d between the hikers is $3\sqrt{89}$, according to the Pythagorean theorem.*

Because the hikers and the point from which they began walking create a right triangle, use the Pythagorean theorem to describe the relationship between the distances a, b, and d in Figure 15-5, and differentiate the equation with respect to t.

$$a^2 + b^2 = d^2$$

$$2a \cdot \frac{da}{dt} + 2b \cdot \frac{db}{dt} = 2d \cdot \frac{dd}{dt}$$

You are given $\dfrac{da}{dt} = 5$ and $\dfrac{db}{dt} = 8$. Substitute those values, and the values of a, b, and d from Figure 15-5, into the equation and solve for $\dfrac{dd}{dt}$.

> *Notice that you don't plug any given information into the formula until AFTER you take the derivative.*

$$2(15) \cdot 5 + 2(24) \cdot 8 = 2\left(3\sqrt{89}\right) \cdot \frac{dd}{dt}$$

$$534 = 6\sqrt{89} \cdot \frac{dd}{dt}$$

$$\frac{dd}{dt} = \frac{534}{6\sqrt{89}}$$

$$\frac{dd}{dt} = \frac{89}{\sqrt{89}} = \sqrt{89} \approx 9.434 \text{ miles/hour}$$

> *Nothing magical happened to get this number— 534 divided by 6 is 89.*

15.26 A 20-foot extension ladder propped up against the side of a house is not properly secured, causing the bottom of the ladder to slide away from the house at a constant rate of 2 ft/sec. How quickly is the top of the ladder falling at the exact moment the base of the ladder is 12 feet away from the house?

As illustrated in Figure 15-6, the ladder, ground, and house form a right triangle. While the length of the ladder remains fixed, you can calculate h when $g = 12$ using the Pythagorean theorem.

> *That means $dl/dt = 0$. The length of the ladder does not change, so it makes sense that its rate of change equals 0.*

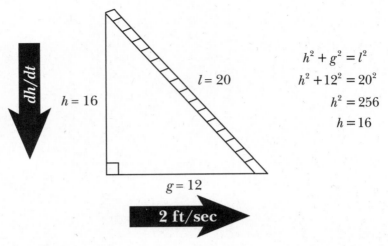

Figure 15-6 *The ladder is side l, h is the distance from the top of the ladder to the ground, and g is the distance from the house to the bottom of the ladder.*

Use the Pythagorean Theorem to express the relationship between h, g, and l, and differentiate the equation with respect to t.

$$h^2 + g^2 = l^2$$

$$2h \cdot \frac{dh}{dt} + 2g \cdot \frac{dg}{dt} = 2l \cdot \frac{dl}{dt}$$

Substitute the known rates of change. Note that $\frac{dg}{dt}$ is positive; as the ladder slides away from the house, the length g in Figure 15-6 is *increasing*, and increasing quantities have positive rates of change.

$$2(16) \cdot \frac{dh}{dt} + 2(12)(2) = 2(20)(0)$$

$$32\frac{dh}{dt} + 48 = 0$$

$$32\frac{dh}{dt} = -48$$

$$\frac{dh}{dt} = -\frac{48}{32} = -\frac{3}{2} \text{ ft/sec}$$

The ladder slides down the house at a rate of –1.5 ft/sec.

15.27 Water stored in an inverted right circular cone, as illustrated in Figure 15-7, leaks out at a constant rate of 2 gallons per day. Assuming the tank is 100 feet high and the radius of its base is 25 feet, at what rate $\frac{dd}{dt}$ is the depth of the water inside the tank decreasing at the moment it is 40 feet deep?

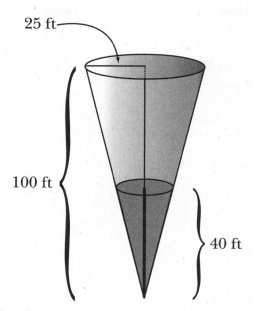

Figure 15-7 *The dimensions of a leaky water tank.*

The volume of a right circular cone is described by the formula $V = \frac{1}{3}\pi r^2 h$. Let h represent the height of the tank, d represent the depth of the water, R represent the radius of the tank's base, and r represent the radius of the water's surface (as illustrated by Figure 15-8). You are given $h = 100$, $d = 40$, $R = 25$, $\frac{dV}{dt} = -2$, and are asked to find $\frac{dd}{dt}$.

> dV/dt is negative because volume is decreasing as the water leaks out of the tank.

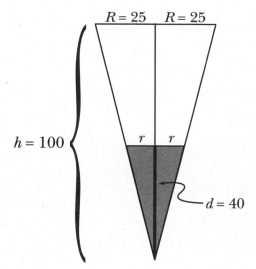

Figure 15-8 *A cross-section of the conical tank from Figure 15-7.*

No information is given about $\frac{dr}{dt}$, so it (and therefore r) must be eliminated from the problem—only one unknown can be left in the final equation or else you will be unable to solve it. Because the isosceles triangles in Figure 15-8 are similar,

> The large triangle (which includes the shaded one) and the shaded triangle are similar because they share the angle at the bottom and their bases are parallel (so their base angles are congruent).

you can set up a proportion relating the heights and the radii of the tank and the water it holds. Solve this proportion for r.

$$\frac{100}{25} = \frac{d}{r}$$

$$100r = 25d$$

$$r = \frac{d}{4}$$

Substitute this value of r into the volume formula to eliminate it from the equation. Note that the formula describes the volume of the *water*, not the tank; Figure 15-8 uses d to describe the depth of the water (not h).

$$V = \frac{1}{3}\pi r^2 d$$

$$V = \frac{1}{3}\pi\left(\frac{d}{4}\right)^2 d$$

$$V = \frac{\pi}{48}d^3$$

> Because $\pi/48$ is just a number (a coefficient), ignore it when you differentiate d^3 and multiply it in when you're done taking the derivative.

Differentiate with respect to t, substitute the known values into the formula, and solve for $\dfrac{dd}{dt}$.

$$\frac{dV}{dt} = \frac{\pi}{48}\left(3d^2 \cdot \frac{dd}{dt}\right)$$

$$-2 = \frac{\pi}{48}\left(3 \cdot 40^2 \cdot \frac{dd}{dt}\right)$$

$$-2 = \frac{4800\pi}{48} \cdot \frac{dd}{dt}$$

$$-2\left(\frac{1}{100\pi}\right) = \frac{dd}{dt}$$

$$-\frac{1}{50\pi} = \frac{dd}{dt}$$

> To keep the units consistent (velocity is in ft/sec), rewrite 1 mile in terms of feet (5,280 ft).

So the depth of the water in the tank is decreasing at a rate of $\dfrac{1}{50\pi} \approx 0.0064$ ft/day.

15.28 A camera exactly one mile away from the Space Shuttle's launch site tracks the ascent of the spacecraft for a network news program. At what rate is the camera's angle of elevation increasing in order to maintain its focus on the shuttle 30 seconds into the launch, once the shuttle has reached a height of 9,720 feet and is traveling 700 ft/sec? Report your answer in radians per second accurate to four decimal places.

Figure 15-9 summarizes the important information from the problem and verifies that a right triangle aptly illustrates the geometric relationship between the variables.

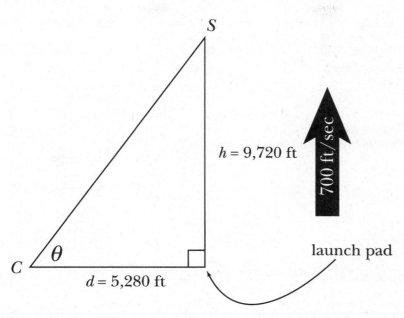

Figure 15-9 *The Space Shuttle S, the camera C, and the launch pad serve as the vertices of a right triangle. The height h of the shuttle varies over time, but the camera remains a fixed distance d from the launch pad. The angle of elevation of the camera is θ.*

Unlike Problems 15.25 and 15.26, the Pythagorean Theorem is not a good choice of equation, because it does not include θ, the variable whose rate of change you are directed to calculate. Therefore, you should apply the tangent trigonometric ratio, as you are given the lengths of the sides opposite and adjacent to θ.

$$\tan\theta = \frac{h}{5,280}$$

Because d is constant throughout the launch, its value is used in the equation, but h and θ vary at different times t throughout the launch, so they are expressed as variables. Differentiate this equation with respect to t.

$$\sec^2\theta \cdot \frac{d\theta}{dt} = \frac{1}{5,280} \cdot \frac{dh}{dt}$$

You are given $\dfrac{dh}{dt} = 700$. Calculate $\tan\theta$ exactly 30 seconds into the flight.

$$\tan\theta = \frac{9,720}{5,280}$$

$$\theta = \arctan\left(\frac{9,720}{5,280}\right)$$

$$\theta \approx 1.0731812$$

Substitute all known values into the equation you differentiated with respect to t and solve for $\dfrac{d\theta}{dt}$.

$$\sec^2(1.0731812) \cdot \frac{d\theta}{dt} = \frac{1}{5,280}(700)$$

$$4.38894628 \cdot \frac{d\theta}{dt} = \frac{700}{5,280}$$

$$\frac{d\theta}{dt} = \frac{700}{5,280} \cdot \frac{1}{4.38894628}$$

$$\frac{d\theta}{dt} = 0.0302 \text{ radians/sec}$$

Optimization

Find the biggest or smallest values of a function

15.29 Calculate the smallest possible product of two numbers, if one is exactly 9 greater than the other.

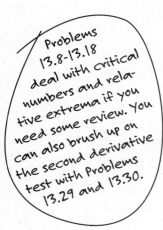

Problems 13.8–13.18 deal with critical numbers and relative extrema if you need some review. You can also brush up on the second derivative test with Problems 13.29 and 13.30.

Let x equal one of the numbers and $x + 9$ equal the other. You are asked to optimize the product, so you should optimize the function $p(x) = x(x+9) = x^2 + 9x$. To optimize $p(x)$, find its critical number(s) and determine whether each represents a relative extrema point.

$$p(x) = x^2 + 9x$$
$$p'(x) = 2x + 9$$
$$0 = 2x + 9$$
$$-\frac{9}{2} = x$$

Note that $p''(x) > 0$ for all x, so (according to the second derivative test) this critical number gives the relative minimum requested by the problem. One of the numbers in the minimum product is $-\dfrac{9}{2}$, and the other is exactly 9 greater.

$$-\frac{9}{2} + 9 = -\frac{9}{2} + \frac{18}{2} = \frac{9}{2}$$

The two numbers with the minimum product are $-\dfrac{9}{2}$ and $\dfrac{9}{2}$; the minimum product is $\left(-\dfrac{9}{2}\right)\left(\dfrac{9}{2}\right) = -\dfrac{81}{4}$.

15.30 The ACME company has begun selling an MP3 player called the FooPod at a price of $200 - 0.05x$, where x is the number of FooPods ACME produces each day. The parts and labor cost for each FooPod is \$140, and marketing and operational costs amount to an additional \$9,500 per day. Approximately how many FooPods should ACME produce and sell each day to maximize profit?

The total profit generated by an item is defined as the revenue minus the manufacturing cost: $p(x) = r(x) - c(x)$. The revenue function is equal to the number of units sold each day multiplied by the price: $r(x) = x(200 - 0.05x)$. Create the function $c(x)$ representing total cost per day: $c(x) = 140x + 9,500$.

Substitute $r(x)$ and $c(x)$ into the profit function.

$$p(x) = r(x) - c(x)$$
$$= x(200 - 0.05x) - (140x + 9,500)$$
$$= -0.05x^2 + 200x - 140x - 9,500$$
$$= -0.05x^2 + 60x - 9,500$$

> Each FooPod costs $140 to make, so multiply 140 by the total number x they make per day and add the fixed cost per day: $9,500.

Differentiate $p(x)$ and identify critical numbers.

$$p'(x) = -0.1x + 60$$
$$0 = -0.1x + 60$$
$$x = \frac{60}{0.1}$$

Apply the second derivative test. Because $p''(x) < 0$ for all real numbers, $p(x)$ has a relative maximum at $x = 600$. Therefore, ACME should attempt to manufacture and sell approximately 600 FooPods a day in order to maximize profit.

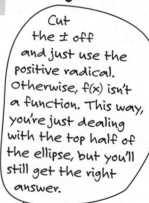

15.31 What are the dimensions of the largest rectangle that can be inscribed in the ellipse $x^2 + 4y^2 = 16$? Report each dimension accurate to three decimal places.

Solve the equation of the ellipse for y to express the conic section as a pair of functions.

$$x^2 + 4y^2 = 16$$
$$4y^2 = 16 - x^2$$
$$y^2 = \frac{16 - x^2}{4}$$
$$y = \pm\sqrt{\frac{16 - x^2}{4}}$$
$$y = \pm\frac{\sqrt{16 - x^2}}{2}$$

> Cut the ± off and just use the positive radical. Otherwise, f(x) isn't a function. This way, you're just dealing with the top half of the ellipse, but you'll still get the right answer.

For any x, $f(x) = \dfrac{\sqrt{16 - x^2}}{2}$ is the corresponding y-value on the of ellipse in either the first or second quadrant.

> To get to a corner of the rectangle from the origin, you have to go x units in either horizontal direction. That makes the length of the rectangle 2x. Same thing with the width, except you have to go up or down f(x) units.

The area of a rectangle is the product of its length and width, so optimize the area function $A(x) = l \times w$. To determine the values of l and w, consider Figure 15-10, where the ellipse $x^2 + 4y^2 = 16$ is graphed, and a sample rectangle is inscribed within. If you must travel x units right and $f(x)$ units up to reach the corner indicated, then $2x$ and $2f(x)$ represent the length and width of the rectangle.

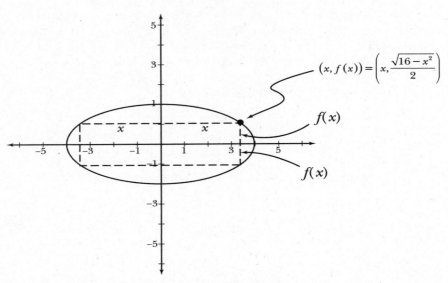

Figure 15-10   *A rectangle inscribed in the ellipse $x^2 + 4y^2 = 16$.*

Substitute l and w into $A(x)$.

$$A(x) = l \cdot w$$
$$= 2x \cdot 2f(x)$$
$$= 2x \cdot \cancel{2} \left(\frac{\sqrt{16 - x^2}}{\cancel{2}} \right)$$
$$= 2x\sqrt{16 - x^2}$$

> Two functions are multiplied together to get $A(x)$, so use the product rule.

Differentiate $A(x)$.

$$A'(x) = \cancel{2}x \cdot \frac{1}{\cancel{2}} \left(16 - x^2 \right)^{-1/2} \cdot (-2x) + \sqrt{16 - x^2} \cdot 2$$
$$= \frac{-2x^2}{\sqrt{16 - x^2}} + 2\sqrt{16 - x^2}$$
$$= \frac{-2x^2 + 2\left(16 - x^2\right)}{\sqrt{16 - x^2}}$$
$$= \frac{-4x^2 + 32}{\sqrt{16 - x^2}}$$

> The critical number 4 doesn't make sense, because if the rectangle had length 4, it wouldn't have any height! (Look at Figure 15-10.) You can also throw out the negative critical numbers—length is a positive number.

The critical numbers of $A'(x)$ are the x-values that cause either the numerator or the denominator to equal 0.

$$-4x^2 + 32 = 0 \qquad\qquad \sqrt{16 - x^2} = 0$$
$$-4x^2 = -32 \qquad\qquad 16 - x^2 = 0$$
$$x^2 = 8 \qquad\qquad\qquad x = \pm 4$$
$$x = \pm 2\sqrt{2}$$

Because $A'(x)$ changes from positive to negative at $x = 2\sqrt{2}$, the value represents a relative maximum. Calculate the dimensions of the rectangle when $x = 2\sqrt{2}$.

$$l = 2x = 2\left(2\sqrt{2}\right) = 4\sqrt{2}$$

$$w = 2f(x) = 2\left(\frac{\sqrt{16 - \left(2\sqrt{2}\right)^2}}{2}\right) = \sqrt{16 - 8} = 2\sqrt{2}$$

The dimensions of the largest inscribed rectangle are $2\sqrt{2}$ and $4\sqrt{2}$.

15.32 A farmer wishes to fence in a rectangular pasture on a 3,750 ft² piece of riverfront property. He also plans to separate the pasture into four regions, as illustrated by Figure 15-11. What is the least amount of fence (in feet) he will need to purchase, assuming that he will not erect a fence along the river?

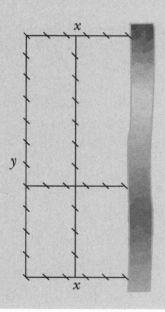

Figure 15-11

The pasture has width x and length y. The total amount of fence required is 3x + 2y.

Three sections of fence measuring x feet long are needed—one for the northern border, one for the southern, and one between those boundaries. Two lengths of fence measuring y feet are also required—one for the western border and one between that border and the river. In all, the farmer will need $f(x) = 3x + 2y$ feet of fencing material. However, you cannot optimize $f(x)$ without first replacing one of its variables. Because you know the area of the field, set up an equation representing that value and solve it for y.

$$\text{Area} = (\text{length})(\text{width})$$
$$A = yx$$
$$3,750 = yx$$
$$\frac{3,750}{x} = y$$

Before you take the derivative of a function to find its critical numbers, the function has to be in terms of one variable. It could have all x's or all y's but not both.

Substitute this value into $f(x)$ to eliminate y from the function.

$$f(x) = 3x + 2y$$

$$f(x) = 3x + 2\left(\frac{3,750}{x}\right)$$

$$f(x) = 3x + 7,500x^{-1}$$

Differentiate $f(x)$ and identify its critical number(s).

$$f'(x) = 3 - 7,500x^{-2}$$

$$0 = 3 - \frac{7,500}{x^2}$$

$$\frac{7,500}{x^2} = 3$$

$$3x^2 = 7,500$$

$$x^2 = 2,500$$

The length of the field is x feet, and you can't have a field with a negative length.

Because x must be a positive number, you don't have to indicate "±" when you take the square root of both sides of the equation.

$$x = \sqrt{2,500} = 50 \text{ feet}$$

Because $f''(x) = \frac{2(7,500)}{x^3} = \frac{15,000}{x^3}$, the second derivative is positive for all $x > 0$, which verifies that $x = 50$ corresponds to a relative minimum of $f(x)$ (according to the second derivative test). Evaluate $f(x)$ to determine the total amount of fence needed.

$$f(x) = 3x + 7,500x^{-1}$$

$$f(50) = 3(50) + \frac{7,500}{50}$$

$$f(50) = 150 + 150$$

$$f(50) = 300$$

The farmer needs a minimum of 300 feet of fence—50 feet for each horizontal section and 75 feet for each vertical section.

15.33 An open box with depth x can be created from a rectangular sheet of cardboard by cutting squares of side x from its corners, as illustrated by Figure 15-12. What is the largest volume of such a box given cardboard that measures 20×30 inches? Report your solution accurate to three decimal places.

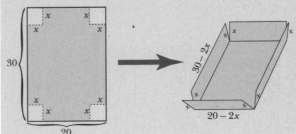

Figure 15-12

When the squares of side x are cut from the corners of the sheet of cardboard, they leave behind rectangles along the sides. Fold those rectangles up to construct an open box.

The volume function for a right rectangular prism is $v(x) = l \times w \times h$. Note that the length of the box is $30 - 2x$, because the cardboard originally measured 30 inches but you remove x inches from each end when the squares cut out. Similarly, the width of the box is $20 - 2x$ inches. Substitute the dimensions of the box into the volume function.

$$v(x) = l \cdot w \cdot h$$
$$= (30 - 2x)(20 - 2x)(x)$$
$$= 600x - 100x^2 + 4x^3$$

Find $v'(x)$ and its critical numbers.

$$v'(x) = 600 - 200x + 12x^2$$
$$0 = 600 - 200x + 12x^2$$

Solve the equation using a graphing calculator; the solutions are $x = 3.923747815$ or $x = 12.74291885$. Note that the latter is not a valid solution, but $v''(3.923747815) < 0$, so $x = 3.923747815$ corresponds to a relative maximum of $v(x)$. Evaluate $v(3.923747815)$.

$$v(3.923747815) \approx 600(3.923747815) - 100(3.923747815)^2 + 4(3.923747815)^3$$
$$v(3.923747815) \approx 1056.306 \text{ in}^3$$

> Really, really fancy way to say "box that's not necessarily a cube."

> The height of the box is x. The bigger the squares you cut out, the deeper the box will be.

> You can't cut more than 12 inches from the right and left corners of the cardboard (for a total of more than 24 inches) when the cardboard is only 20 inches wide!

Chapter 16
ADDITIONAL DIFFERENTIATION TECHNIQUES
Yet more ways to differentiate

Before concluding the comprehensive investigation of differentiation that began in Chapter 11, a few advanced differentiation techniques still merit mention. Although these methods will be required far less frequently than the differentiation methods outlined in Chapter 12, in specific situations, they represent the most fastidious and logical method for calculating derivatives. As such, knowing *when* to apply these techniques is as essential as knowing *how* to perform them.

When faced with a tricky derivative, sometimes the power, product, quotient, and chain rules are not enough. This chapter covers four specific techniques for calculating derivatives that address four specific situations. These methods don't replace the rules covered back in Chapter 12, which still work for about 90% of the derivative problems you'll deal with—they just give you options for the other 10%, problems that are nearly impossible without the methods covered in this chapter.

Implicit Differentiation
Essential when you can't solve a function for y

Technically, when you differentiate x^2 with respect to x, you are applying the chain rule. You first take the derivative of the "outer function" x^2 to get $2x$, leaving the "inner function" x alone. Then you take the derivative of the inner function with respect to x: $\dfrac{dx}{dx}$. For all practical purposes, $\dfrac{dx}{dx}$ (like any nonzero number divided by itself) equals 1.

$$\frac{d}{dx}\left(x^2\right) = 2x^{2-1} \cdot \frac{dx}{dx} = 2x^1 \cdot 1 = 2x$$

In other words, when an expression containing x is differentiated with respect to x, the answer is no different than the derivatives calculated in Chapter 12. Differentiate the other term in the expression containing x in a similar fashion.

When you differentiate an expression containing a variable other than x with respect to x, the derivative will contain differentials. For instance, the derivative of y with respect to x is simply $\dfrac{dy}{dx}$. This term cannot be simplified or omitted, because unlike $\dfrac{dx}{dx}$, this quotient does not necessarily equal 1.

To differentiate $-3y^2$, apply the chain rule.

$$\frac{d}{dx}\left(-3y^2\right) = -3 \cdot 2y^{2-1} \cdot \frac{dy}{dx} = -6y\frac{dy}{dx}$$

Combine the derivatives for each of the four terms in the expression, all of which are calculated above.

$$\frac{d}{dx}\left(x^2 - 3y^2 + 2x + y\right) = 2x - 6y\frac{dy}{dx} + 2 + \frac{dy}{dx}$$

> Differentials are things like dx and dy. Any time the variable in the expression doesn't match the variable you're "respecting," these pop up in a derivative.

> The derivative of y^2 is $2y$, but you should treat the y as an "inner function" and take the derivative of it (with respect to x) at the end.

Equations explicitly written in terms of x can be expressed in a single variable, x. For instance, $3x^2y - 7xy = 14$ can be explicitly written as a function of x by solving for y.

$$3x^2y - 7xy = 14$$
$$y\left(3x^2 - 7x\right) = 14$$
$$y = \frac{14}{3x^2 - 7x}$$

If an equation isn't written as a single equation in terms of x (i.e., the equation is not solved for the other variable, usually y, as illustrated in the above example), it is *implicitly* expressed in terms of *x*. For instance, circles, hyperbolas, and ellipses must be implicitly expressed in terms of *x*. Note that these conic sections are not functions, and neither are the vast majority of implicitly defined functions.

16.3 Differentiate the product *xy* with respect to *x*.

Neither *x* nor *y* is a constant, so the product must be differentiated using the product rule.

$$\frac{d}{dx}(xy) = x \cdot \frac{d}{dx}(y) + y \cdot \frac{d}{dx}(x)$$

$$= x \cdot \frac{dy}{dx} + y \cdot \frac{dx}{dx}$$

$$= x \frac{dy}{dx} + y$$

16.4 Given $25x^2 + 8x - 16y^2 - 4y - 9 = 0$, find $\frac{dy}{dx}$.

Differentiate each term with respect to *x*.

$$50x + 8 - 32y \cdot \frac{dy}{dx} - 4\frac{dy}{dx} = 0 \longleftarrow$$

The constant disappeared because the derivative of –9 (with respect to any variable) is 0.

Move all terms not containing $\frac{dy}{dx}$ to the right side of the equation.

$$-32y \cdot \frac{dy}{dx} - 4\frac{dy}{dx} = -50x - 8$$

Factor out $\frac{dy}{dx}$ and solve for it.

$$\frac{dy}{dx}(-32y - 4) = -50x - 8$$

$$\frac{dy}{dx} = \frac{-50x - 8}{-32y - 4}$$

$$\frac{dy}{dx} = \frac{-2(25x + 4)}{-2(16y + 2)}$$

$$\frac{dy}{dx} = \frac{25x + 4}{16y + 2}$$

All the terms are negative, which is kind of ugly. If you factor –1 out of the top and bottom, –1 divided by –1 equals 1, so they cancel out.

16.5 Given $y = 9\sqrt{x} - 2\sqrt[5]{y^3}$, find $\frac{dy}{dx}$.

Rewrite the radical expressions using rational exponents.

$$y = 9x^{1/2} - 2y^{3/5}$$

When you use the power rule, you subtract one from the power. Because the original power of y is 3/5, the new power is $\frac{3}{5} - 1 = \frac{3}{5} - \frac{5}{5} = -\frac{2}{5}$.

Differentiate each term with respect to x.

$$\frac{dy}{dx} = 9 \cdot \frac{1}{2} x^{-1/2} - 2 \cdot \frac{3}{5} y^{-2/5} \cdot \frac{dy}{dx}$$

$$\frac{dy}{dx} = \frac{9}{2}\sqrt{x} - \frac{6}{5\sqrt[5]{y^2}} \frac{dy}{dx}$$

Move all terms containing $\frac{dy}{dx}$ to the left side of the equation and factor out $\frac{dy}{dx}$.

$$\frac{dy}{dx} + \frac{6}{5\sqrt[5]{y^2}} \frac{dy}{dx} = \frac{9\sqrt{x}}{2}$$

$$\frac{dy}{dx}\left(1 + \frac{6}{5\sqrt[5]{y^2}}\right) = \frac{9\sqrt{x}}{2}$$

Rewrite 1 using a common denominator to get a single fraction here.

$$\frac{dy}{dx}\left(\frac{5\sqrt[5]{y^2}}{5\sqrt[5]{y^2}} + \frac{6}{5\sqrt[5]{y^2}}\right) = \frac{9\sqrt{x}}{2}$$

$$\frac{dy}{dx}\left(\frac{5\sqrt[5]{y^2} + 6}{5\sqrt[5]{y^2}}\right) = \frac{9\sqrt{x}}{2}$$

Solve for $\frac{dy}{dx}$.

$$\frac{dy}{dx} = \frac{9\sqrt{x}}{2}\left(\frac{5\sqrt[5]{y^2}}{5\sqrt[5]{y^2} + 6}\right)$$

$$\frac{dy}{dx} = \frac{45\sqrt{x}\sqrt[5]{y^2}}{10\sqrt[5]{y^2} + 12}$$

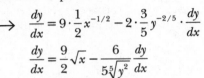

16.6 Given $\sin y - \cos xy = x - y$, find $\frac{dy}{dx}$.

The ARGUMENT of a trig function (or any function for that matter) is whatever's getting plugged into it. The argument of cos xy is xy.

The chain rule is required to differentiate both terms on the left side of the equation. Once you differentiate each of the trigonometric functions (momentarily leaving the argument unchanged), you must then multiply by the derivative of the argument.

$$\cos y \cdot \frac{d}{dx}(y) - \left[-\sin xy \cdot \frac{d}{dx}(xy)\right] = 1 - \frac{dy}{dx}$$

Note that $\frac{d}{dx}(xy)$ is calculated in Problem 16.3.

$$\cos y \frac{dy}{dx} + \sin xy\left(x\frac{dy}{dx} + y\right) = 1 - \frac{dy}{dx}$$

Distribute $\sin xy$.

$$\cos y \frac{dy}{dx} + x\sin xy \frac{dy}{dx} + y\sin xy = 1 - \frac{dy}{dx}$$

Move terms containing $\frac{dy}{dx}$ to the left side of the equation and all others to the right side.

$$\cos y \frac{dy}{dx} + x\sin xy \frac{dy}{dx} + \frac{dy}{dx} = 1 - y\sin xy$$

Factor out $\dfrac{dy}{dx}$ and then solve for it.

$$\frac{dy}{dx}\left[\cos y + x\sin xy + 1\right] = 1 - y\sin xy$$

$$\frac{dy}{dx} = \frac{1 - y\sin xy}{\cos y + x\sin xy + 1}$$

16.7 Given $e^x + \cos y = \ln y^6$, find $\dfrac{d^2 y}{dx^2}$.

Differentiate the equation with respect to x.

$$e^x - \sin y \cdot \frac{dy}{dx} = \frac{1}{y^6} \cdot 6y^5 \cdot \frac{dy}{dx}$$

$$e^x - \sin y \cdot \frac{dy}{dx} = \frac{6\,y^{5}}{y^{6} \cdot y} \cdot \frac{dy}{dx}$$

$$e^x - \sin y \cdot \frac{dy}{dx} = \frac{6}{y} \cdot \frac{dy}{dx}$$

> The second derivative

> The derivative of $\ln y^6$ is 1 divided by what's inside the log (y^6) times the derivative of what's inside ($6y^5$) times the derivative of what's inside THAT function—$6y^5$ contains y so the derivative is dy/dx.

Solve for $\dfrac{dy}{dx}$.

$$e^x = \sin y \frac{dy}{dx} + \frac{6}{y}\frac{dy}{dx}$$

$$e^x = \left(\sin y + \frac{6}{y}\right)\frac{dy}{dx}$$

$$e^x = \left(\frac{y \cdot \sin y}{y} + \frac{6}{y}\right)\frac{dy}{dx}$$

$$e^x = \left(\frac{y\sin y + 6}{y}\right)\frac{dy}{dx}$$

$$\frac{e^x y}{y\sin y + 6} = \frac{dy}{dx}$$

> Just like in Problem 16.5, it's a good idea to make this term one fraction by using common denominators. To isolate dy/dx, multiply both sides by the reciprocal of this big fraction.

To find the second derivative, apply the quotient rule. Note that the derivatives of the numerator and denominator each require the product rule—the numerator contains the product of e^x and y, and the denominator contains the product of y and $\sin y$.

$$\frac{d^2 y}{dx^2} = \frac{(y\sin y + 6)\cdot \dfrac{d}{dx}\left(e^x y\right) - e^x y \cdot \dfrac{d}{dx}\left(y\sin y + 6\right)}{\left(y\sin y + 6\right)^2}$$

$$\frac{d^2 y}{dx^2} = \frac{(y\sin y + 6)\left(e^x \dfrac{dy}{dx} + e^x y\right) - e^x y\left(y\cos y \dfrac{dy}{dx} + \sin y \dfrac{dy}{dx}\right)}{\left(y\sin y + 6\right)^2}$$

Recall that $\dfrac{dy}{dx} = \dfrac{e^x y}{y\sin y + 6}$.

$$\frac{d^2 y}{dx^2} = \frac{(y\sin y + 6)\left[\dfrac{e^{2x} y}{y\sin y + 6} + e^x y\right] - e^x y\left(\dfrac{e^x y^2 \cos y}{y\sin y + 6} + \dfrac{e^x y\sin y}{y\sin y + 6}\right)}{\left(y\sin y + 6\right)^2}$$

Note: Problems 16.8–16.10 refer to an ellipse centered at the origin with a horizontal major axis of length 16 and a major axis of length 12.

16.8 Write the equation of the ellipse, expressing each coefficient therein as an integer.

If you need to review ellipses and their standard form, look at Problems 6.17-6.20.

If the major axis of the ellipse is 16 units long, then $a = \dfrac{16}{2} = 8$; similarly, $b = 6$. Plug these values into the standard form equation of an ellipse with a horizontal major axis. (Note that $h = k = 0$ since the ellipse is centered at the origin.)

$$\frac{(x-h)^2}{a^2} + \frac{(y-k)^2}{b^2} = 1$$

$$\frac{x^2}{64} + \frac{y^2}{36} = 1$$

Multiply each term in the equation by the least common multiple: 576.

$$9x^2 + 16y^2 = 576$$

Note: Problems 16.8–16.10 refer to an ellipse centered at the origin with a horizontal major axis of length 16 and a major axis of length 12.

16.9 Differentiate the equation of the ellipse generated in Problem 16.8 with respect to x.

Don't forget to take the derivative of 576. Like all constants, its derivative is 0.

Differentiate $36x^2 + 64y^2 = 2{,}304$ with respect to x and solve for $\dfrac{dy}{dx}$.

$$18x + 32y\frac{dy}{dx} = 0$$

$$32y\frac{dy}{dx} = -18x$$

$$\frac{dy}{dx} = -\frac{18x}{32y}$$

$$\frac{dy}{dx} = -\frac{9x}{16y}$$

Note: Problems 16.8–16.10 refer to an ellipse centered at the origin with a horizontal major axis of length 16 and a major axis of length 12.

16.10 Calculate the slopes of the tangent lines to the ellipse when $x = 1$.

You first need to determine the points on the ellipse when $x = 1$. To do so, substitute $x = 1$ into its equation and solve for y.

$$9(1)^2 + 16y^2 = 576$$

$$16y^2 = 567$$

$$\sqrt{y^2} = \pm\sqrt{\frac{567}{16}}$$

$$y = \pm\frac{9\sqrt{7}}{4}$$

Therefore, the ellipse contains the points $\left(1, \frac{9\sqrt{7}}{4}\right)$ and $\left(1, -\frac{9\sqrt{7}}{4}\right)$. Evaluate $\frac{dy}{dx}$ at each point to determine the slope of the tangent line there.

> Plug the x- and y- values of each point into the derivative formula from Problem 16.9.

$$\boxed{\frac{dy}{dx} \ at \left(1, -\frac{9\sqrt{7}}{4}\right)}$$

$$\frac{dy}{dx} = -\frac{9(1)}{16\left(-\dfrac{9\sqrt{7}}{4}\right)}$$

$$\frac{dy}{dx} = -\frac{9}{4(-9)\sqrt{7}}$$

$$\frac{dy}{dx} = \frac{1}{4\sqrt{7}} \ or \ \frac{\sqrt{7}}{28}$$

$$\boxed{\frac{dy}{dx} \ at \left(1, \frac{9\sqrt{7}}{4}\right)}$$

$$\frac{dy}{dx} = -\frac{9(1)}{16\left(\dfrac{9\sqrt{7}}{4}\right)}$$

$$\frac{dy}{dx} = -\frac{9}{4(9)\sqrt{7}}$$

$$\frac{dy}{dx} = -\frac{1}{4\sqrt{7}} \ or \ -\frac{\sqrt{7}}{28}$$

Note: Problems 16.11–16.13 refer to a circle centered at the origin with radius r.

16.11 Identify the equation of the circle and differentiate it with respect to x.

> Treat r like a number, since the radius of a circle stays constant. That means the derivative of r^2 is 0.

The equation of a circle centered at $(0,0)$ with radius r is $x^2 + y^2 = r^2$. Differentiate implicitly with respect to x.

$$2x + 2y\frac{dy}{dx} = 0$$

$$2y\frac{dy}{dx} = -2x$$

$$\frac{dy}{dx} = \frac{-2x}{2y}$$

$$\frac{dy}{dx} = -\frac{x}{y}$$

Note: Problems 16.11–16.13 refer to a circle centered at the origin with radius r.

16.12 Calculate the slopes of the tangent lines to the circle when $x = c$.

Use the method outlined in Problem 16.10—begin by substituting $x = c$ into the equation to determine the points of tangency.

$$x^2 + y^2 = r^2$$
$$c^2 + y^2 = r^2$$
$$y^2 = r^2 - c^2$$
$$y = \pm\sqrt{r^2 - c^2}$$

Therefore, the circle passes through the points $\left(c, -\sqrt{r^2 - c^2}\right)$ and $\left(c, \sqrt{r^2 - c^2}\right)$. Substitute these points into the derivative calculated in Problem 16.11.

$$\boxed{\dfrac{dy}{dx} \text{ at } \left(c, -\sqrt{r^2 - c^2}\right)}$$

$$\frac{dy}{dx} = -\frac{x}{y}$$

$$\frac{dy}{dx} = -\frac{c}{-\sqrt{r^2 - c^2}}$$

$$\frac{dy}{dx} = \frac{c}{\sqrt{r^2 - c^2}}$$

$$\boxed{\dfrac{dy}{dx} \text{ at } \left(c, \sqrt{r^2 - c^2}\right)}$$

$$\frac{dy}{dx} = -\frac{x}{y}$$

$$\frac{dy}{dx} = -\frac{c}{\sqrt{r^2 - c^2}}$$

This means find the tangent line to the circle at the point $\left(c, \sqrt{r^2 - c^2}\right)$.

Note: Problems 16.11–16.13 refer to a circle centered at the origin with radius r.

16.13 Write the equation of the tangent line to the circle in the first quadrant given $x = c$.

According to Problem 16.12, the slope of the tangent line at point $\left(c, \sqrt{r^2 - c^2}\right)$ is $\dfrac{dy}{dx} = -\dfrac{c}{\sqrt{r^2 - c^2}}$. Use the point-slope formula to write the equation of the line.

$$y - y_1 = m\left(x - x_1\right)$$

$$y - \sqrt{r^2 - c^2} = -\frac{c}{\sqrt{r^2 - c^2}}\left(x - c\right)$$

Rationalize the expression on the right side of the equation and solve for y.

$$y - \sqrt{r^2 - c^2} = -\frac{c\sqrt{r^2 - c^2}}{r^2 - c^2}\left(x - c\right)$$

$$y = -\frac{c\sqrt{r^2 - c^2}}{r^2 - c^2}\left(x - c\right) + \sqrt{r^2 - c^2}$$

Logarithmic Differentiation

Use log properties to make complex derivatives easier

16.14 Given $y = x^x$, find $\dfrac{dy}{dx}$.

> Any time you need to find the derivative of an x raised to an x power, that's a dead giveaway you should use logarithmic differentiation.

Take the natural logarithm of both sides of the equation.

$$\ln y = \ln x^x$$

Using a property of logarithms (explained in Problem 5.23), $\ln a^b = b \ln a$.

$$\ln y = x \ln x$$

Differentiate both sides of the equation with respect to x, applying the product rule to differentiate $x \ln x$.

> If you're taking the log of something raised to a power, you can pull the power out in front of the log as a coefficient.

$$\frac{1}{y} \cdot \frac{dy}{dx} = x \cdot \frac{1}{x} + \ln x \cdot 1$$

$$\frac{1}{y} \cdot \frac{dy}{dx} = 1 + \ln x$$

Solve for $\dfrac{dy}{dx}$.

$$\frac{dy}{dx} = y(1 + \ln x)$$

The original problem states that $y = x^x$, so substitute that value into the equation.

$$\frac{dy}{dx} = x^x (1 + \ln x)$$

$$\frac{dy}{dx} = x^x + x^x \ln x$$

16.15 Given $y = 2x^{2x}$, find $\dfrac{dy}{dx}$.

Use the method described in Problem 16.14—take the natural logarithm of both sides of the equation and use a logarithmic property to change the exponent $2x$ into a coefficient.

$$\ln y = \ln 2x^{2x}$$
$$\ln y = 2x \ln 2x$$

Differentiate with respect to x, applying the product rule to differentiate the right side of the equation.

$$\frac{1}{y} \cdot \frac{dy}{dx} = 2x \left(\frac{1}{2x} \cdot 2 \right) + \ln 2x (2)$$

$$\frac{1}{y} \cdot \frac{dy}{dx} = 2 + 2 \ln 2x$$

Solve for $\dfrac{dy}{dx}$ and recall that $y = 2x^{2x}$.

$$\frac{dy}{dx} = y(2 + 2\ln 2x)$$

$$\frac{dy}{dx} = 2x^{2x}(2 + 2\ln 2x)$$

$$\frac{dy}{dx} = 4x^{2x} + 4x^{2x}\ln 2x$$

The problem is explicitly defined in terms of x. Sure it starts with "y =", but the function $2x^{2x}$ contains only x's, so the derivative should also contain only x's.

16.16 Given $y = x^{\ln x}$, find $\dfrac{dy}{dx}$.

Take the natural logarithm of both sides of the equation, apply a logarithmic property to relocate the exponent, differentiate with respect to x, solve for $\dfrac{dy}{dx}$, and replace y in the final answer with $x^{\ln x}$, as defined by the problem.

$$\ln y = \ln x^{\ln x}$$

$$\ln y = (\ln x)(\ln x)$$

$$\frac{1}{y} \cdot \frac{dy}{dx} = \ln x \cdot \frac{1}{x} + \ln x \cdot \frac{1}{x}$$

$$\frac{1}{y} \cdot \frac{dy}{dx} = 2\left(\frac{\ln x}{x}\right)$$

$$\frac{dy}{dx} = y\left(\frac{2\ln x}{x}\right)$$

$$\frac{dy}{dx} = \frac{2x^{\ln x}\ln x}{x}$$

You can also rewrite $(\ln x)(\ln x)$ as $(\ln x)^2$ and use the power rule (and chain rule) to find the derivative:

$$2(\ln x)^1 \cdot \frac{1}{x} = \frac{2\ln x}{x}.$$

16.17 Given $y = 5^x x^{5x}$, find $\dfrac{dy}{dx}$.

Take the natural logarithm of both sides of the equation.

$$\ln y = \ln\,(5^x x^{5x})$$

Expand the logarithm on the right side of the equation using the logarithmic property $\log ab = \log a + \log b$.

$$\ln y = \ln 5^x + \ln x^{5x}$$

Move the exponents using logarithmic properties.

$$\ln y = x\ln 5 + (5x)\ln x$$

The log of things that are multiplied equals the sum of the individual logs.

Differentiate with respect to x. Note that $\ln 5$ is a constant, so the product rule is not needed to differentiate $x \ln 5$; however, it is needed to differentiate $(5x)\ln x$.

$$\frac{1}{y} \cdot \frac{dy}{dx} = \ln 5 + 5x \cdot \frac{1}{x} + 5 \ln x$$

$$\frac{1}{y} \cdot \frac{dy}{dx} = \ln 5 + 5 + 5 \ln x$$

$$\frac{dy}{dx} = y(\ln 5 + 5 + 5 \ln x)$$

$$\frac{dy}{dx} = 5^x x^{5x}(\ln 5 + 5 + 5 \ln x)$$

> Since $\ln 5$ is a real number, the derivative of $(x)(\ln 5)$ equals $\ln 5$, just like the derivative of $(x)(12) = 12$.

16.18 Given $y = \ln x^{e^x}$, find $\dfrac{dy}{dx}$.

The right side of the equation already contains a logarithmic function, so apply a logarithmic property to move its exponent. There is no need to take the logarithm of both sides of the equation, and as such, this is not a true logarithmic differentiation problem (although you do apply logarithmic properties).

$$y = e^x \cdot \ln x$$

Apply the product rule to differentiate with respect to x

$$\frac{dy}{dx} = e^x \cdot \frac{1}{x} + e^x \ln x$$

$$\frac{dy}{dx} = \frac{e^x}{x} + e^x \ln x$$

> Trick problem alert! They threw this in to see if you were paying attention or just automatically logging both sides of the equation without thinking about it. Very sneaky.

16.19 Given $y = \sqrt[3]{x^9 \ln(x+2)}$, find $\dfrac{dy}{dx}$.

Take the natural logarithm of both sides of the equation and rewrite the radical expression using a rational exponent.

$$\ln y = \ln\left[\left(x^9 \ln(x+2)\right)^{1/3}\right]$$

Apply a logarithmic property to move the exponent.

$$\ln y = \frac{1}{3}\ln\left[x^9 \ln(x+2)\right]$$

Expand the natural logarithm of the product by expressing it as a sum of logarithms.

$$\ln y = \frac{1}{3}\left[\ln x^9 + \ln\left(\ln(x+2)\right)\right]$$

$$\ln y = \frac{1}{3}\ln x^9 + \frac{1}{3}\ln\left(\ln(x+2)\right)$$

Use a logarithmic property to move the remaining exponent.

$$\ln y = \frac{1}{3} \cdot 9 \cdot \ln x + \frac{1}{3}\ln\left(\ln(x+2)\right)$$

$$\ln y = 3\ln x + \frac{1}{3}\ln\left(\ln(x+2)\right)$$

> This time you use logarithmic differentiation not because an x is raised to an x power but because log properties can break this ridiculously ugly expression into bite-sized chunks.

The derivative of ln(ln (x + 2)) is 1 over what's inside the big log (ln (x + 2)) times the derivative of that inner quantity (1/(x + 2)) times the derivative of x + 2, the quantity inside the inside function!

Differentiate with respect to x.

$$\frac{1}{y} \cdot \frac{dy}{dx} = 3 \cdot \frac{1}{x} + \frac{1}{3} \cdot \frac{1}{\ln(x+2)} \cdot \frac{1}{x+2} \cdot 1$$

$$\frac{1}{y} \cdot \frac{dy}{dx} = \frac{3}{x} + \frac{1}{3(x+2)(\ln(x+2))}$$

$$\frac{dy}{dx} = y\left(\frac{3}{x} + \frac{1}{(3x+6)(\ln(x+2))}\right)$$

Substitute $y = \sqrt[3]{x^9 \ln(x+2)}$ into the derivative.

$$\frac{dy}{dx} = \sqrt[3]{x^9 \ln(x+2)}\left(\frac{3}{x} + \frac{1}{(3x+6)(\ln(x+2))}\right)$$

16.20 Given $y = 4^x \cdot e^{2x} \cdot \tan(5x-1)$, find $\frac{dy}{dx}$.

When you plug e^1 into ln x, all that's left is the exponent of e^1 because they are inverse functions. Check out Problem 5.30 for more info.

Take the natural logarithm of both sides of the equation and expand the logarithm of the product into a sum.

$$\ln y = \ln\left[4^x \cdot e^{2x} \cdot \tan(5x-1)\right]$$

$$\ln y = \ln 4^x + \ln e^{2x} + \ln\left(\tan(5x-1)\right)$$

Apply a logarithmic property to move the exponents.

$$\ln y = x \ln 4 + (2x)\ln e + \ln\left(\tan(5x-1)\right)$$

$$\ln y = x \ln 4 + 2x(1) + \ln\left(\tan(5x-1)\right)$$

Differentiate with respect to x.

$$\frac{1}{y} \cdot \frac{dy}{dx} = \ln 4 + 2 + \frac{1}{\tan(5x-1)} \cdot \sec^2(5x-1) \cdot 5$$

$$\frac{1}{y} \cdot \frac{dy}{dx} = \ln 4 + 2 + \frac{5\sec^2(5x-1)}{\tan(5x-1)}$$

$$\frac{dy}{dx} = y\left[\ln 4 + 2 + \frac{5\sec^2(5x-1)}{\tan(5x-1)}\right]$$

$$\frac{dy}{dx} = 4^x e^{2x}\tan(5x-1)\left[\ln 4 + 2 + \frac{5\sec^2(5x-1)}{\tan(5x-1)}\right]$$

16.21 Given $y = \dfrac{\left(\sec^2 x\right)\left(\log_4 x^6\right)}{(12x-7)^{19}}$, find $\dfrac{dy}{dx}$.

Take the natural logarithm of both sides of the equation.

$$\ln y = \ln \frac{\left(\sec^2 x\right)\left(\log_4 x^6\right)}{(12x-7)^{19}}$$

Expand the logarithm on the right side of the equation. ←

$$\ln y = \ln\left[\left(\sec^2 x\right)\left(\log_4 x^6\right)\right] - \ln(12x-7)^{19}$$

$$\ln y = \ln\left(\sec^2 x\right) + \ln\left(\log_4 x^6\right) - \ln(12x-7)^{19}$$

$$\ln y = 2\ln(\sec x) + \ln\left(6\log_4 x\right) - 19\ln(12x-7)$$

$$\ln y = 2\ln(\sec x) + \ln 6 + \ln\left(\log_4 x\right) - 19\ln(12x-7)$$

> This is a lot like Problem 5.27, which asks you to fully expand a logarithm. Review that problem if you're confused.

Differentiate each term with respect to x.

$$\frac{1}{y}\cdot\frac{dy}{dx} = 2\cdot\frac{1}{\sec x}\cdot\sec x\tan x + 0 + \frac{1}{\log_4 x}\cdot\frac{1}{x\cdot\ln 4} - 19\cdot\frac{1}{12x-7}\cdot 12$$

$$\frac{1}{y}\cdot\frac{dy}{dx} = \frac{2\sec x\tan x}{\sec x} + \frac{1}{x(\ln 4)(\log_4 x)} - \frac{19\cdot 12}{12x-7}$$

$$\frac{dy}{dx} = y\left[2\tan x + \frac{1}{x(\ln 4)(\log_4 x)} - \frac{228}{12x-7}\right]$$

> The 6 goes in front of $\log_4 x$ because x is raised to the 6th power. If the whole expression $\log_4 x$ was raised to the 6th power, only then would you pull the 6 out in front of $\ln(\log_4 x)$.

Recall that $y = \dfrac{\left(\sec^2 x\right)\left(\log_4 x^6\right)}{(12x-7)^{19}}$.

$$\frac{dy}{dx} = \frac{\left(\sec^2 x\right)\left(\log_4 x^6\right)}{(12x-7)^{19}}\left[2\tan x + \frac{1}{x(\ln 4)(\log_4 x)} - \frac{228}{12x-7}\right]$$

> The derivative of $\ln 6$ is 0.

16.22 Given $y = x^{x^2/(x-4)}$, find $\dfrac{dy}{dx}$.

Apply logarithmic differentiation.

$$\ln y = \ln x^{x^2/(x-4)}$$

$$\ln y = \frac{x^2}{x-4}\ln x$$

Differentiate with respect to x; apply the product rule on the right side of the equation.

$$\frac{1}{y}\cdot\frac{dy}{dx} = \frac{x^2}{x-4}\cdot\frac{d}{dx}(\ln x) + \ln x\cdot\frac{d}{dx}\left(\frac{x^2}{x-4}\right)$$

Note that differentiating $\dfrac{x^2}{x-4}$ requires the quotient rule.

$$\frac{1}{y}\cdot\frac{dy}{dx} = \frac{x^2}{x-4}\cdot\frac{1}{x} + \ln x\cdot\frac{(x-4)(2x) - x^2(1)}{(x-4)^2}$$

$$\frac{1}{y}\cdot\frac{dy}{dx} = \frac{\cancel{x}\cdot x}{\cancel{x}(x-4)} + \ln x\left(\frac{2x^2 - 8x - x^2}{(x-4)^2}\right)$$

$$\frac{dy}{dx} = y\left[\frac{x}{x-4} + \frac{(\ln x)\left(x^2 - 8x\right)}{(x-4)^2}\right]$$

$$\frac{dy}{dx} = x^{x^2/(x-4)}\left[\frac{x}{x-4} + \frac{(\ln x)\left(x^2 - 8x\right)}{(x-4)^2}\right]$$

16.23 Given $y = x^{x^3}$, find $\dfrac{dy}{dx}$.

Apply logarithmic differentiation.

$$\ln y = \ln x^{x^3}$$

$$\ln y = x^3 \ln x$$

Use the product rule.

$$\frac{1}{y} \cdot \frac{dy}{dx} = x^3 \cdot \frac{1}{x} + \ln x \left(3x^2\right)$$

$$\frac{1}{y} \cdot \frac{dy}{dx} = \frac{x^3}{x} + 3x^2 \ln x$$

$$\frac{dy}{dx} = y\left(x^2 + 3x^2 \ln x\right)$$

$$\frac{dy}{dx} = x^{x^3}\left(x^2 + 3x^2 \ln x\right)$$

Differentiating Inverse Trigonometric Functions
'Cause the derivative of $\tan^{-1} x$ ain't $\sec^{-2} x$

16.24 Differentiate $\arcsin x$ with respect to x.

Basically, the derivative is $\dfrac{1}{\sqrt{1-u^2}}$, where u is whatever's plugged into inverse sine. Then multiply that entire fraction by the derivative of u. In this problem, u = x, and the derivative of x is 1.

The derivative of $\arcsin u$, if u is a function of x, is $\dfrac{1}{\sqrt{1-u^2}} \cdot \dfrac{du}{dx}$; in this instance, $u = x$.

$$\frac{d}{dx}(\arcsin x) = \frac{1}{\sqrt{1-x^2}} \cdot \frac{d}{dx}(x)$$

$$= \frac{1}{\sqrt{1-x^2}} \cdot 1$$

$$= \frac{1}{\sqrt{1-x^2}}$$

16.25 Differentiate $\arccos 4x^3$ with respect to x.

All of the inverse trig function derivatives formulas are listed in Appendix D.

Apply the arccosine differentiation formula.

$$\frac{d}{dx}(\arccos u) = -\frac{1}{\sqrt{1-u^2}} \cdot \frac{du}{dx}$$

$$\frac{d}{dx}\left(\arccos 4x^3\right) = -\frac{1}{\sqrt{1-\left(4x^3\right)^2}} \cdot \frac{d}{dx}\left(4x^3\right)$$

$$\frac{d}{dx}\left(\arccos 4x^3\right) = -\frac{12x^2}{\sqrt{1-16x^6}}$$

16.26 Differentiate $\tan^{-1}\left(e^{\cos x}\right)$ with respect to x.

Apply the inverse tangent derivative formula.

$$\frac{d}{dx}(\arctan u) = \frac{1}{1+u^2}\cdot\frac{du}{dx}$$

$$\frac{d}{dx}\left(\tan^{-1}e^{\cos x}\right) = \frac{1}{1+\left(e^{\cos x}\right)^2}\cdot\frac{d}{dx}\left(e^{\cos x}\right)$$

Use the chain rule to differentiate $e^{\cos x}$.

$$\frac{d}{dx}\left(\tan^{-1}e^{\cos x}\right) = \frac{1}{1+e^{2\cos x}}\cdot\left(e^{\cos x}\right)(-\sin x)$$

$$\frac{d}{dx}\left(\tan^{-1}e^{\cos x}\right) = -\frac{\sin x\cdot e^{\cos x}}{1+e^{2\cos x}}$$

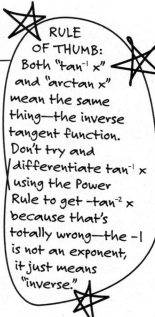

RULE OF THUMB: Both "$\tan^{-1}x$" and "$\arctan x$" mean the same thing—the inverse tangent function. Don't try and differentiate $\tan^{-1}x$ using the Power Rule to get $-\tan^{-2}x$ because that's totally wrong—the -1 is not an exponent, it just means "inverse."

16.27 Differentiate $\ln(\text{arccot } 2x)$ with respect to x.

Because this is a composite function, apply the chain rule.

$$\frac{d}{dx}\left[\ln(\text{arccot } 2x)\right] = \frac{1}{\text{arccot } 2x}\cdot\frac{d}{dx}(\text{arccot } 2x)$$

Apply the differentiation formula for the arccotangent function.

$$\frac{d}{dx}\left[\ln(\text{arccot } 2x)\right] = \left(\frac{1}{\text{arccot } 2x}\right)\left(-\frac{1}{1+(2x)^2}\cdot 2\right)$$

$$= -\frac{2}{\left(1+4x^2\right)(\text{arccot } 2x)}$$

The arccotangent derivative formula is the same as arctangent, only negative. All of the functions that start with "co" (cosine, cotangent, and cosecant) have negative derivatives, and the derivatives of their inverse functions (arccosine, arccotangent, and arccosecant) are negative as well!

16.28 Differentiate $\text{arcsec}\dfrac{\ln x}{x^2}$ with respect to x.

Apply the arcsecant differentiation formula.

$$\frac{d}{dx}(\text{arcsec } u) = \frac{1}{|u|\sqrt{u^2-1}}\cdot\frac{du}{dx}$$

$$\frac{d}{dx}\left(\text{arcsec}\left(\frac{\ln x}{x^2}\right)\right) = \frac{1}{\left|\dfrac{\ln x}{x^2}\right|\sqrt{\left(\dfrac{\ln x}{x^2}\right)^2-1}}\cdot\frac{d}{dx}\left(\frac{\ln x}{x^2}\right)$$

Simplify the complex fraction by combining the quantities within the square root into a single fraction.

$$= \frac{1}{\left|\dfrac{\ln x}{x^2}\right|\sqrt{\dfrac{(\ln x)^2}{x^4}-\dfrac{x^4}{x^4}}}\cdot\frac{d}{dx}\left(\frac{\ln x}{x^2}\right)$$

$$= \frac{1}{\left|\dfrac{\ln x}{x^2}\right|\dfrac{\sqrt{(\ln x)^2-x^4}}{\sqrt{x^4}}}\cdot\frac{d}{dx}\left(\frac{\ln x}{x^2}\right)$$

Note that neither x^2 nor $\sqrt{x^4} = |x^2|$ require absolute values, because x^2 must be a nonnegative number.

> You can't square a number and get a negative, so you don't need to keep dragging those absolute value bars with you. However, ln x can be negative (when x is between 0 and 1) so you need to leave its absolute values on.

$$= \frac{1}{\dfrac{|\ln x|}{x^2} \cdot \dfrac{\sqrt{(\ln x)^2 - x^4}}{x^2}} \cdot \frac{d}{dx}\left(\frac{\ln x}{x^2}\right)$$

$$= \frac{1}{\dfrac{|\ln x| \sqrt{(\ln x)^2 - x^4}}{x^4}} \cdot \frac{d}{dx}\left(\frac{\ln x}{x^2}\right)$$

Eliminate the complex fraction.

> I divided by a fraction equals the reciprocal of that fraction.

$$= \frac{x^4}{|\ln x| \sqrt{(\ln x)^2 - x^4}} \cdot \frac{d}{dx}\left(\frac{\ln x}{x^2}\right)$$

Note that the derivative has not yet fully been determined. Now that the rational expression is in an acceptable format (i.e., it doesn't contain a complex fraction), use the quotient rule to differentiate $\dfrac{\ln x}{x^2}$.

$$= \frac{x^4}{|\ln x| \sqrt{(\ln x)^2 - x^4}} \cdot \left(\frac{x^2 \cdot \dfrac{1}{x} - \ln x (2x)}{\left(x^2\right)^2}\right)$$

$$= \frac{x^4}{|\ln x| \sqrt{(\ln x)^2 - x^4}} \left(\frac{x - 2x \ln x}{x^4}\right)$$

$$= \frac{x^4 (x - 2x \ln x)}{x^4 \cdot |\ln x| \sqrt{(\ln x)^2 - x^4}}$$

$$= \frac{x - 2x \ln x}{|\ln x| \sqrt{(\ln x)^2 - x^4}}$$

Differentiating Inverse Functions
Without even knowing what they are!

> If you get an answer of 143, it's because you plugged x = 5 into f(x). You're supposed to plug it into the inverse function: $f^{-1}(x)$.

16.29 Given $f(x) = x^3 + 4x - 2$, use a graphing calculator to evaluate $f^{-1}(5)$ accurate to three decimal places.

> Functions and their inverses reverse each other's inputs and outputs. If inputting 5 into $f^{-1}(x)$ outputs some number c, then inputting c into the inverse function f(x) outputs 5.

Identifying an equation for $f^{-1}(x)$ is not a trivial matter. If you attempt to reverse x and y and solve for y (the technique used to determine inverse functions that is outlined in Problems 4.25 and 4.26), you will find that the equation cannot be solved for y easily. Therefore, you must use an alternate approach, one that will allow you to evaluate the inverse function for real number values without actually identifying the function $f^{-1}(x)$. Because $f(x)$ and $f^{-1}(x)$ are inverse functions, given $f^{-1}(5) = c$, it follows that $f(c) = 5$. Set $f(c) = 5$ and solve for c.

$$c^3 + 4c - 2 = 5$$

Set the equation equal to 0.

$$c^3 + 4c - 7 = 0$$

Graph the equation $y = x^3 + 4x - 7$ on a graphing calculator and calculate the x-intercept (i.e., the root or the zero); you should get $x \approx 1.255$. Therefore, $f(1.255) \approx 5$, so $f^{-1}(5) \approx 1.255$.

Note: Problems 16.30–16.31 refer to the function $g(x) = x^5 + 9x^3 - x^2 + 6x - 2$.

16.30 Prove that $g^{-1}(x)$ exists.

Only one-to-one functions have inverses. Graphically, this means any horizontal line drawn on the function may intersect it once, at most. Therefore, any function that changes direction will fail the horizontal line test, as demonstrated in Figure 16-1.

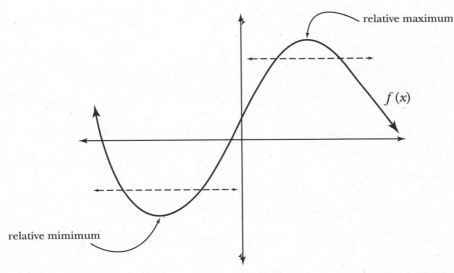

Figure 16-1 *The graph of a continuous function f(x) that fails the horizontal line test because of its relative extrema. The dotted horizontal lines represent only two of an infinite number of horizontal lines that intersect the graph of f(x) more than once.*

Because the graph of $g'(x) = 5x^4 + 27x^2 - 2x + 6$ falls entirely above the x-axis, $g'(x)$ is positive for all x and $g(x)$ is strictly increasing. All monotone increasing or decreasing functions possess inverses.

If a graph never stops increasing, the function is "strictly increasing" or "monotone increasing." If the function always decreases, it's classified "strictly decreasing" or "monotone decreasing."

Note: Problems 16.30–16.31 refer to the function $g(x) = x^5 + 9x^3 - x^2 + 6x - 2$.

16.31 Use a graphing calculator to evaluate $g^{-1}\left(\dfrac{7}{6}\right)$ accurate to three decimal places.

Use the method described in Problem 16.29, which states that given $g(c) = \dfrac{7}{6}$, it follows that $g^{-1}\left(\dfrac{7}{6}\right) = c$. Substitute c into $g(x)$, set the equation equal to $\dfrac{7}{6}$, subtract $\dfrac{7}{6}$ from both sides of the equation, and use technology to identify the x-intercept of the resulting function.

Rewrite 2 as 12/6 so that it has the same denominator as 7/6, allowing you to combine the fractions.

$$c^5 + 9c^3 - c^2 + 6c - 2 = \frac{7}{6}$$

$$c^5 + 9c^3 - c^2 + 6c - \frac{19}{6} = 0$$

$$c \approx 0.434$$

Because $g(0.434) \approx \dfrac{7}{6}$, it follows that $g^{-1}\left(\dfrac{7}{6}\right) \approx 0.434$.

16.32 If $h(x) = 7x^3 + 9x + 18$, evaluate $\left(h^{-1}\right)'(2)$, the derivative of $h^{-1}(x)$ with respect to x when $x = 2$.

This formula doesn't appear out of thin air—Problem 16.37 explains where it comes from.

Given a function $f(x)$, the derivative of its inverse, $\left(f^{-1}\right)'(x)$, is defined according to the formula below.

$$\left(f^{-1}\right)'(x) = \frac{1}{f'\left(f^{-1}(x)\right)}$$

Substitute $h(x)$ for $f(x)$ and plug $x = 2$ into the formula.

$$\left(h^{-1}\right)'(2) = \frac{1}{h'\left(h^{-1}(2)\right)}$$

Calculate $h^{-1}(2)$ using the technique described in Problems 16.29 and 16.31: if $h^{-1}(2) = c$, it follows that $h(c) = 2$.

$$7c^3 + 9c + 18 = 2$$

$$7c^3 + 9c + 16 = 0$$

$$c = -1$$

Substitute $h^{-1}(2) = -1$ into the inverse derivative formula.

$$\left(h^{-1}\right)'(2) = \frac{1}{h'\left(h^{-1}(2)\right)} = \frac{1}{h'(-1)}$$

Differentiate $h(x)$ and evaluate $h'(-1)$.

$$h'(x) = 21x^2 + 9$$

$$h'(-1) = 21(-1)^2 + 9$$

$$h'(-1) = 30$$

Substitute $h'(-1) = 30$ into the inverse derivative formula.

$$\left(h^{-1}\right)'(2) = \frac{1}{h'(-1)} = \frac{1}{30}$$

16.33 If $f(x) = 5x^9 - x^2 + x - 4$, evaluate $\left(f^{-1}\right)'(6)$ accurate to three decimal places.

Apply the method described in Problem 16.32.

$$\left(f^{-1}\right)'(6) = \frac{1}{f'\left(f^{-1}(6)\right)}$$

If $f^{-1}(6) = c$, then $f(c) = 6$.

$$5c^9 - c^2 + c - 4 = 6$$
$$5c^9 - c^2 + c - 10 = 0$$
$$c \approx 1.081107956$$

Substitute $f^{-1}(6) \approx 1.081107956$ into the inverse derivative formula.

$$\left(f^{-1}\right)'(6) = \frac{1}{f'(1.081107956)}$$

Even though the final answer should contain three decimal places, you shouldn't round any decimals until then—use as many decimal places as your calculator can spit at you, or you risk a less accurate answer.

Differentiate $f(x)$ and evaluate $f'(1.081107956)$.

$$f'(x) = 45x^8 - 2x + 1$$
$$f'(1.081107956) \approx 82.8156861426$$

Substitute $f'(1.081107956)$ into the inverse derivative formula.

$$\left(f^{-1}\right)'(6) \approx \frac{1}{82.8156861426}$$
$$\left(f^{-1}\right)'(6) \approx 0.012$$

Note: Problems 16.34–16.35 reference a one-to-one function $f(x)$ that is continuous and differentiable for all real numbers. Selected values of the function and its derivative are listed in the table below.

16.34 Evaluate $\left(f^{-1}\right)'(-2)$.

x	-2	-1	0	1	2	3
$f(x)$	6	2	1	-1	$-\dfrac{3}{2}$	-2
$f'(x)$	$-\dfrac{1}{2}$	-6	-1	$-\dfrac{5}{3}$	$-\dfrac{1}{8}$	$-\dfrac{3}{4}$

Apply the formula for the derivative of an inverse function.

$$\left(f^{-1}\right)'(-2) = \frac{1}{f'\left(f^{-1}(-2)\right)}$$

Since $f(3) = -2$, you can reverse the input and output to get values of the inverse function: $f^{-1}(-2) = 3$.

According to the table, $f^{-1}(-2) = 3$.

$$\left(f^{-1}\right)'(-2) = \frac{1}{f'(3)}$$

According to the table, $f'(3) = -\frac{3}{4}$.

$$\left(f^{-1}\right)'(-2) = \frac{1}{-3/4} = -\frac{4}{3}$$

Note: Problems 16.34–16.35 reference a one-to-one function f(x) that is continuous and differentiable for all real numbers. Selected values of the function and its derivative are listed in Problem 16.34.

16.35 Evaluate $\left(f^{-1}\right)'(-1)$.

Apply the formula for the derivative of an inverse function.

$$\left(f^{-1}\right)'(-1) = \frac{1}{f'\left(f^{-1}(-1)\right)}$$

According to the table, $f(1) = -1$; therefore, $f^{-1}(-1) = 1$.

$$\left(f^{-1}\right)'(-1) = \frac{1}{f'(1)}$$

According to the table, $f'(1) = -\frac{5}{3}$.

$$\left(f^{-1}\right)'(-1) = \frac{1}{-5/3} = -\frac{3}{5}$$

16.36 Given the function $k(x)$ graphed in Figure 16-2, estimate $\left(k^{-1}\right)'(0)$.

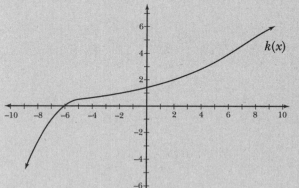

Figure 16-2

The graph of a continuous, one-to-one function k(x).

Apply the formula for the derivative of an inverse function.

$$\left(k^{-1}\right)'(0) = \frac{1}{k'\left(k^{-1}(0)\right)}$$

The function appears to have a root of $x = -6$; therefore, $k^{-1}(0) \approx -6$.

$$\left(k^{-1}\right)'(0) = \frac{1}{k'(-6)}$$

If $k(x)$ crosses the x-axis at $x = -6$, then $k(-6) = 0$. Reverse the numbers to get values for the inverse function: $k^{-1}(0) = -6$.

To approximate $k'(-6)$, estimate the slope of the tangent line to $k(x)$ at $x = -6$. One acceptable way to do so is to calculate the slope of the secant line connecting the points on $k(x)$ that are one unit left and one unit right of $x = -6$: $(-7, -1)$ and $\left(-5, \frac{1}{2}\right)$.

$$k'(-6) \approx \frac{k(-5) - k(-7)}{-5 - (-7)}$$
$$\approx \frac{(1/2) - (-1)}{-5 + 7}$$
$$\approx \frac{3/2}{2}$$
$$\approx \frac{3}{4}$$

The y-values -1 and $\frac{1}{2}$ are just guesses. There's no way of identifying the actual points on the graph, but guessing is okay because you're just doing an approximation.

Substitute this value into the inverse derivative formula.

$$\left(k^{-1}\right)'(0) = \frac{1}{k'(-6)} \approx \frac{1}{3/4} \approx \frac{4}{3}$$

16.37 Generate the formula for the derivative of an inverse function:

$$\left(f^{-1}\right)'(x) = \frac{1}{f'\left(f^{-1}(x)\right)}.$$

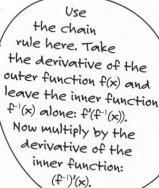

If $f^{-1}(x)$ is the inverse function of $f(x)$, then $f(f^{-1}(x)) = x$. Differentiate both sides of this equation with respect to x.

$$\frac{d}{dx}\left[f\left(f^{-1}(x)\right)\right] = \frac{d}{dx}(x)$$

$$f'\left(f^{-1}(x)\right) \cdot \left(f^{-1}\right)'(x) = 1$$

Use the chain rule here. Take the derivative of the outer function $f(x)$ and leave the inner function $f^{-1}(x)$ alone: $f'(f^{-1}(x))$. Now multiply by the derivative of the inner function: $(f^{-1})'(x)$.

Solve the equation for $\left(f^{-1}\right)'(x)$.

$$\frac{f'\left(f^{-1}(x)\right) \cdot \left(f^{-1}\right)'(x)}{f'\left(f^{-1}(x)\right)} = \frac{1}{f'\left(f^{-1}(x)\right)}$$

$$\left(f^{-1}\right)'(x) = \frac{1}{f'\left(f^{-1}(x)\right)}$$

Chapter 17
APPROXIMATING AREA

Estimate the area between a curve and the x-axis

The opposite of differentiation, a process appropriately titled "anti-differentiation" or (more commonly) "integration," is accompanied by its own robust set of theorems, rules, algorithms, and applications, all of which are discussed in Chapters 18 through 23. This chapter describes different ways to estimate the area between a function and the *x*-axis, beginning with very informal rectangular approximation methods and culminating with significantly more formal techniques. The actual connection bridging area approximation to the process of antidifferentiation will be investigated in Chapter 18.

This chapter is really low-tech compared to the chapters before and after it. After all, you spend almost the entire chapter either adding up the areas of little rectangles or plugging things mindlessly into formulas like the trapezoidal rule. Formal Riemann sums, the last topic in this chapter, are a little tricky, but they're not so bad once you get used to them. You may start wondering "Why the heck do I even want to approximate area? What does this have to do with anything?" It's just calculus's way of saying "In geometry, you could only find areas of a few different things—circles, rectangles, regular polygons, those sorts of things—but in calculus, you'll find the areas of things with irregular shapes."

Informal Riemann Sums

Left, Right, Midpoint, Upper, and Lower Sums

They're actually asking you to find the area BETWEEN the curve and the x-axis.

17.1 If n subdivisions of equal width are used to approximate the area beneath a curve on the x-interval $[a,b]$, calculate the width Δx of the rectangles.

The width of each rectangle is $\Delta x = \dfrac{b-a}{n}$.

Note: Problems 17.2–17.6 refer to the graph in Figure 17-1.

17.2 Approximate the shaded area using five rectangles of equal width and left Riemann sums.

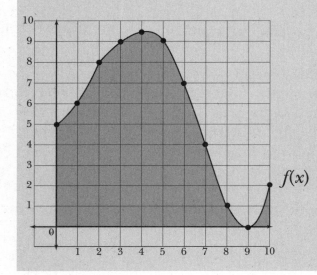

f(x)

Figure 17-1

The goal of Problems 17.2–17.6 is to approximate the shaded area between f(x) and the x-axis on the x-interval [0,10].

Calculate the width of the triangles using the formula from Problem 17.1.

$$\Delta x = \frac{10-0}{5} = \frac{10}{5} = 2$$

If each rectangle is 2 units wide, the rectangles will occupy the following intervals: [0,2], [2,4], [4,6], [6,8], and [8,10]. The height of each rectangle in a left Riemann sum is determined by the height of $f(x)$ at the *left* boundary of each interval, as illustrated by Figure 17-2.

So the rectangle on [0,2] has a height of f(0) = 5 (because x = 0 is the left boundary of the interval), the rectangle on [2,4] has a height of f(2) = 8, etc.

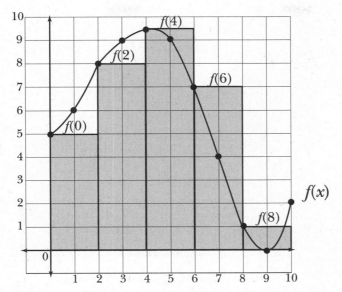

Figure 17-2 *The rectangles used to determine the left Riemann sum with n = 5 rectangles. The height of each rectangle is shown above it.*

The area of each rectangle is its width, 2, times its height. Calculate the sum of the areas of the rectangles.

$$2f(0) + 2f(2) + 2f(4) + 2f(6) + 2f(8) = 2[f(0) + f(2) + f(4) + f(6) + f(8)]$$

> Be-
> cause all the
> rectangles have
> the same width,
> you can factor it
> out.

Estimate the function values based on the graph in Figure 17-2.

$$2(5 + 8 + 9.5 + 7 + 1) = 2(30.5) = 61$$

The area beneath the curve is approximately 61 square units according to the left Riemann sum.

Note: Problems 17.2–17.6 refer to the graph in Figure 17-1.

17.3 Approximate the shaded area using five rectangles of equal width and right Riemann sums.

Because the interval and number of rectangles are the same as Problem 17.2, there is no need to recalculate the width of the rectangles: $\Delta x = 2$. The height of each rectangle in a right Riemann sum is dictated by the height of the function at the *right* boundary of each interval, as illustrated by Figure 17-3.

> Because
> the right
> boundary of
> the [0,2] interval
> is x = 2, that
> rectangle will have
> a height of f(2). The
> rectangle on [2,4]
> will have a height
> of f(4), and
> so on.

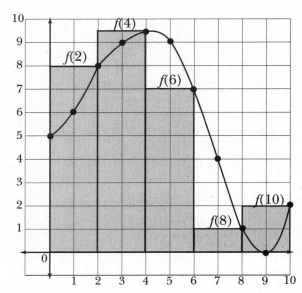

Figure 17-3 *The rectangles used to determine the right Riemann sum with*
n = 5 rectangles. The height of each rectangle is shown above it.

Calculate the sum of the areas of the rectangles.

$$2f(2)+2f(4)+2f(6)+2f(8)+2f(10)=2\big[f(2)+f(4)+f(6)+f(8)+f(10)\big]$$
$$=2\big[8+9.5+7+1+2\big]$$
$$=55$$

The area beneath the curve is approximately 55 square units, according to the
right Riemann sum.

Note: Problems 17.2–17.6 refer to the graph in Figure 17-1.

17.4 Approximate the shaded area using five rectangles of equal width and midpoint
Riemann sums.

Because the interval and number of rectangles are the same as Problems 17.2 and
17.3, there is no need to recalculate the width of the rectangles: $\Delta x = 2$. The height
of each rectangle in a midpoint Riemann sum is dictated by the height of the
function at the *midpoint* of each interval, as illustrated by Figure 17-4.

The
midpoint of the
interval [0,2] is x
= 1, so the rectangle
on that interval has a
height of f(1) = 6. The
heights of the rectangles
on [2,4], [4,6], [6,8], and
[8,10] are f(3) = 9, f(5) =
9, f(7) = 4, and f(9) =
0, respectively.

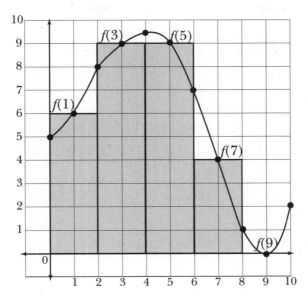

Figure 17-4 *The rectangles used to determine the midpoint Riemann sum with n = 5 rectangles. Note that the rectangle on [8,10] has a height of f(9) = 0, so its area is 0 as well.*

Calculate the sum of the areas of the rectangles.

$$2f(1) + 2f(3) + 2f(5) + 2f(7) + 2f(9) = 2\left[f(1) + f(3) + f(5) + f(7) + f(9)\right]$$
$$= 2[6 + 9 + 9 + 4 + 0]$$
$$= 56$$

The area beneath the curve is approximately 56 square units, according to the midpoint Riemann sum.

Note: Problems 17.2–17.6 refer to the graph in Figure 17-1.

17.5 Approximate the shaded area using 10 rectangles of equal width and upper Riemann sums.

According to the formula,

$$\Delta x = \frac{b - a}{n} = \frac{10 - 0}{10} = 1,$$

but you don't really need a formula to tell you that splitting something 10 units long into 10 sections means each section will measure 1 unit.

Because the number of rectangles is $n = 10$, $\Delta x = 1$. The height of each rectangle in an upper Riemann sum is the greatest function value in an interval. Therefore, the rectangle on the interval [0,1] has a height of $f(1) = 6$ (the right endpoint of the interval), whereas the rectangle on [7,8] will have a height of $f(7) = 4$ (the left endpoint of the interval). Note that the heights defining the rectangles in an upper Rieman sum need not occur at the endpoints of the interval, but in this problem they do.

Some textbooks call upper sum rectangles CIRCUMSCRIBED rectangles.

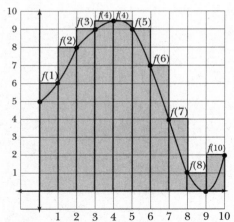

Figure 17-5   *The greatest function values for each interval occur at either the left or right endpoint, but that is not necessarily true for all functions. In fact, it only occurs when a function, like f(x), does not change direction between intervals.*

Calculate the sum of the areas of the rectangles.

$$1 \cdot f(1) + 1 \cdot f(2) + 1 \cdot f(3) + 1 \cdot f(4) + 1 \cdot f(4) + 1 \cdot f(5) + 1 \cdot f(6) + 1 \cdot f(7) + 1 \cdot f(8) + 1 \cdot f(10)$$
$$= 6 + 8 + 9 + 9.5 + 9.5 + 9 + 7 + 4 + 1 + 2$$
$$= 65$$

The shaded area measures approximately 65 square units, according to the upper Riemann sum.

> Lower Riemann sum rectangles are sometimes called "inscribed rectangles."

Note: Problems 17.2–17.6 refer to the graph in Figure 17-1.

17.6   Approximate the shaded area using 10 rectangles of equal width and lower Riemann sums.

The height of each rectangle in a lower Riemann sum is the smallest function value on the *x*-interval, as illustrated by Figure 17-6.

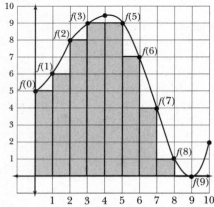

Figure 17-6   *The smallest function value over each interval defines the heights of the rectangles in a lower Riemann sum. Though these values occur at interval endpoints for this function, it is not necessarily true for all functions for the reasons given in Figure 17-5.*

Calculate the sum of the areas of the rectangles.

$1 \cdot f(0) + 1 \cdot f(1) + 1 \cdot f(2) + 1 \cdot f(3) + 1 \cdot f(5) + 1 \cdot f(6) + 1 \cdot f(7) + 1 \cdot f(8) + 1 \cdot f(9) + 1 \cdot f(9)$

$= 5 + 6 + 8 + 9 + 9 + 7 + 4 + 1 + 0 + 0$

$= 49$

The area beneath the curve is approximately 49 square units, according to the lower Riemann sum.

Note: Problems 17.7–17.11 refer to the area of the region bounded by the x-axis and the function $g(x) = x^2$ on the x-interval [0,5].

17.7 Estimate the area using a right Riemann sum with $n = 5$ rectangles of equal width.

Calculate the width of the rectangles using the formula from Problem 17.1.

$$\Delta x = \frac{b-a}{n} = \frac{5-0}{5} = \frac{5}{5} = 1$$

The rectangles will occupy these intervals on the x-axis: [0,1], [1,2], [2,3], [3,4], and [4,5]. The height of each rectangle will be determined by the function values of the right boundaries of each interval: $g(1)$, $g(2)$, $g(3)$, $g(4)$, and $g(5)$, respectively. Calculate the sum of the areas of the rectangles.

You can factor out the width in the first step.

$$\Delta x \left(g(1) + g(2) + g(3) + g(4) + g(5) \right) = 1 \left[\left(1^2\right) + \left(2^2\right) + \left(3^2\right) + \left(4^2\right) + \left(5^2\right) \right]$$
$$= 1(1 + 4 + 9 + 16 + 25)$$
$$= 55$$

According to the right Riemann sum, the area is approximately 55 square units.

Note: Problems 17.7–17.11 refer to the area of the region bounded by the x-axis and the function $g(x) = x$ on the x-interval [0,5].

17.8 Estimate the area using a left Riemann sum with $n = 5$ rectangles of equal width.

Use the same width from Problem 17.7: $\Delta x = 1$. When calculating left Riemann sums, the height of each rectangle is dictated by the height of the function at the left boundary of each rectangle. Calculate the sum of the areas of the rectangles.

$$\Delta x \left(g(0) + g(1) + g(2) + g(3) + g(4) \right) = 1 \left[0^2 + 1^2 + 2^2 + 3^2 + 4^2 \right]$$
$$= 1 + 4 + 9 + 16$$
$$= 30$$

According to the left Riemann sum, the area is approximately 30 square units.

17.9 Estimate the area using a midpoint Riemann sum with *n* = 4 rectangles of equal width.

Determine the width of the rectangles.

$$\Delta x = \frac{b-a}{n} = \frac{5-0}{4} = \frac{5}{4}$$

The number of rectangles has changed from Problems 17.7 and 17.8—make sure to calculate the new width.

The four rectangles will occupy these intervals: $\left[0,\frac{5}{4}\right], \left[\frac{5}{4},\frac{5}{2}\right], \left[\frac{5}{2},\frac{15}{4}\right], \left[\frac{15}{4},5\right]$.

Determine the midpoint of each interval by adding its boundaries and multiplying by $\frac{1}{2}$.

To get the interval boundaries, start with 0 and add $\Delta x = \frac{5}{4}$ four times.

$$\text{Midpoint of } \left[0,\frac{5}{4}\right] = \frac{1}{2}\left(0+\frac{5}{4}\right) = \frac{1}{2}\left(\frac{5}{4}\right) = \frac{5}{8}$$
$$\text{Midpoint of } \left[\frac{5}{4},\frac{5}{2}\right] = \frac{1}{2}\left(\frac{5}{4}+\frac{5}{2}\right) = \frac{1}{2}\left(\frac{5}{4}+\frac{10}{4}\right) = \frac{1}{2}\left(\frac{15}{4}\right) = \frac{15}{8}$$
$$\text{Midpoint of } \left[\frac{5}{2},\frac{15}{4}\right] = \frac{1}{2}\left(\frac{5}{2}+\frac{15}{4}\right) = \frac{1}{2}\left(\frac{10}{4}+\frac{15}{4}\right) = \frac{1}{2}\left(\frac{25}{4}\right) = \frac{25}{8}$$
$$\text{Midpoint of } \left[\frac{15}{4},5\right] = \frac{1}{2}\left(\frac{15}{4}+5\right) = \frac{1}{2}\left(\frac{15}{4}+\frac{20}{4}\right) = \frac{1}{2}\left(\frac{35}{4}\right) = \frac{35}{8}$$

Calculate the sum of the areas of the rectangles.

Plug each of the midpoints into g(x) = x².

$$\Delta x\left[g\left(\frac{5}{8}\right)+g\left(\frac{15}{8}\right)+g\left(\frac{25}{8}\right)+g\left(\frac{35}{8}\right)\right] = \frac{5}{4}\left(\frac{25}{64}+\frac{225}{64}+\frac{625}{64}+\frac{1,225}{64}\right)$$
$$= \frac{5}{4}\left(\frac{2,100}{64}\right)$$
$$= \frac{2,625}{64} \approx 41.016 \text{ square units}$$

17.10 Estimate the area using a lower Riemann sum with *n* = 6 rectangles of equal width.

Calculate the width of the rectangles.

$$\Delta x = \frac{b-a}{n} = \frac{5-0}{6} = \frac{5}{6}$$

The six rectangles will occupy these intervals: $\left[0,\frac{5}{6}\right]$, $\left[\frac{5}{6},\frac{5}{3}\right]$, $\left[\frac{5}{3},\frac{5}{2}\right]$, $\left[\frac{5}{2},\frac{10}{3}\right]$, $\left[\frac{10}{3},\frac{25}{6}\right]$, and $\left[\frac{25}{6},5\right]$. Because $g(x)$ is strictly increasing on $[0,5]$, the lowest function value in each interval occurs at the right boundary. Therefore, the lower Riemann sum is equivalent to the left Riemann sum for $f(x) = x^2$.

$$\frac{5}{6}\left(g(0)+g\left(\frac{5}{6}\right)+g\left(\frac{5}{3}\right)+g\left(\frac{5}{2}\right)+g\left(\frac{10}{3}\right)+g\left(\frac{25}{6}\right)\right)$$

$$=\frac{5}{6}\left(0+\frac{25}{36}+\frac{25}{9}+\frac{25}{4}+\frac{100}{9}+\frac{625}{36}\right)$$

$$=\frac{5}{6}\left(\frac{25+100+225+400+625}{36}\right)$$

$$=\frac{6{,}875}{216}\approx 31.829 \text{ square units}$$

> Because $g'(x) = 2x$ is positive for all x's between 0 and 5, $g(x)$ is increasing on the interval.

Note: Problems 17.7–17.11 refer to the area of the region bounded by the x-axis and the function $g(x) = x^2$ on the x-interval [0,5].

17.11 Estimate the area using an upper Riemann sum with $n = 6$ rectangles of equal width.

Follow the same method as Problem 17.10, but instead of the lesser of the function values, define the rectangle heights as the *greater* of the endpoints' function values.

$$\frac{5}{6}\left(g\left(\frac{5}{6}\right)+g\left(\frac{5}{3}\right)+g\left(\frac{5}{2}\right)+g\left(\frac{10}{3}\right)+g\left(\frac{25}{6}\right)+g(5)\right)$$

$$=\frac{5}{6}\left(\frac{25}{36}+\frac{25}{9}+\frac{25}{4}+\frac{100}{9}+\frac{625}{36}+25\right)$$

$$=\frac{5}{6}\left(\frac{25+100+225+400+625+900}{36}\right)$$

$$=\frac{11{,}375}{216}\approx 52.662 \text{ square units}$$

Note: Problems 17.12–17.16 refer to the area of the region bounded by $f(x) = 3x^2 + 1$ and the x-axis on the x-interval [0,6].

17.12 Estimate the area using a left Riemann sum with:
(a) $n = 3$ rectangles
(b) $n = 6$ rectangles

Which estimate more accurately approximates the correct area of 222 square units?

When $n = 3$ rectangles are used, $\Delta x = \dfrac{6-0}{3} = 2$; when $n = 6$ rectangles are used, $\Delta x = \dfrac{6-0}{6} = 1$.

(a) Left Riemann sum with $n = 3$ rectangles:

$\Delta x \big[f(0) + f(2) + f(4) \big]$

$= 2(1 + 13 + 49)$

$= 126$ square units

(b) Left Riemann sum with $n = 6$ rectangles:

$\Delta x \big[f(0) + f(1) + f(2) + f(3) + f(4) + f(5) \big]$

$= 1(1 + 4 + 13 + 28 + 49 + 76)$

$= 171$ square units

Using $n = 6$ rectangles produces a more accurate result.

> The more rectangles you use to calculate a Riemann sum, the more accurate your area estimate will be.

Note: Problems 17.12–17.16 refer to the area of the region bounded by $f(x) = 3x^2 + 1$ and the x-axis on the x-interval [0,6].

17.13 Why is the area estimate in Problem 17.12 significantly less than the actual area?

The graph of $f(x) = 3x^2 + 1$ is increasing on the interval $[0, \infty)$, so the function value at the right boundary of each interval is greater than that of the left boundary. Furthermore, the magnitude by the right endpoint is greater increases as x increases. Consider Figure 17-7, the graph of $f(x)$ and the $n = 6$ rectangles used to estimate the left Riemann sum.

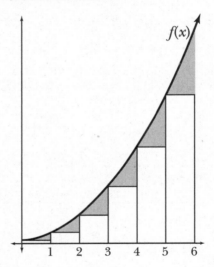

Figure 17-7 *The estimation error, represented by the shaded region of the graph, increases as x increases.*

Above each inscribed rectangle, a large region beneath the curve is omitted, causing the rectangles to underestimate the actual area beneath the curve.

Note: Problems 17.12–17.16 refer to the area of the region bounded by $f(x) = 3x^2 + 1$ and the x-axis on the x-interval [0,6].

17.14 Describe the accuracy of a right Riemann sum using $n = 6$ subintervals of equal width.

The greater the difference between a function's height at the left and right boundaries of each interval, the greater the estimation error when only the left or right Riemann sum is considered. Therefore, for exactly the opposite reasons presented in Problem 17.13, the right Riemann sum significantly overestimates the area. As Figure 17-8 demonstrates, the rectangles of the right Riemann sum now enclose far *more* than the area to be estimated.

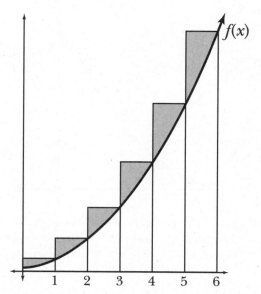

Figure 17-8 *The circumscribed rectangles representing the right (or upper) Riemann sum using n = 6 rectangles overestimate the area beneath f(x). The estimation error, represented by the shaded region on the graph, increases as f(x) increases.*

Note: Problems 17.12–17.16 refer to the area of the region bounded by $f(x) = 3x^2 + 1$ and the x-axis on the x-interval [0,6].

17.15 Why will the midpoint sum provide the most accurate estimate of the area when compared to the left and right Riemann sums using the same subintervals? Supplement your argument with a graph.

As explained in Problem 17.13, a left Riemann sum will underestimate the area beneath an increasing function. Similarly, using the highest function values on an interval (when calculating a right Riemann sum) overestimates the area. However, a midpoint sum produces a far more accurate approximation, because each rectangle both excludes area within the region and includes area outside of the region, as illustrated by Figure 17-9.

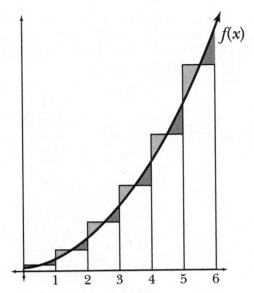

Figure 17-9 *Each rectangle excludes area beneath the curve (represented by the dark shaded regions) and includes area above the curve (represented by the light shaded regions).*

Although the dark and light areas on each interval in Figure 17-9 are not equal (and therefore won't cancel one another out to perfectly calculate the area beneath the curve), they serve to negate one another to some extent and provide a more accurate result than right, left, upper, or lower Riemann sums for $f(x) = 3x^2 + 1$. In fact, midpoint sums are the most accurate Riemann approximation for the vast majority of functions.

Note: Problems 17.12–17.16 refer to the area of the region bounded by $f(x) = 3x^2 + 1$ and the x-axis on the x-interval [0,6].

17.16 Verify the hypothesis presented in Problem 17.15 by estimating the area using a midpoint Riemann sum with $n = 6$ rectangles.

> The rectangles are located on the intervals [0,1], [1,2], [2,3], [3,4], [4,5], and [5,6].

The width of the rectangles is $\Delta x = \dfrac{6-0}{6} = 1$, and the midpoints of the intervals are $x = \dfrac{1}{2}, \dfrac{3}{2}, \dfrac{5}{2}, \dfrac{7}{2}, \dfrac{9}{2}$, and $\dfrac{11}{2}$. Calculate the sum of the areas of the rectangles.

$$\Delta x\left[f\left(\frac{1}{2}\right) + f\left(\frac{3}{2}\right) + f\left(\frac{5}{2}\right) + f\left(\frac{7}{2}\right) + f\left(\frac{9}{2}\right) + f\left(\frac{11}{2}\right)\right]$$

$$= 1\left[\frac{7}{4} + \frac{31}{4} + \frac{79}{4} + \frac{151}{4} + \frac{247}{4} + \frac{367}{4}\right]$$

$$= \frac{441}{2} \approx 220.5 \text{ square units}$$

The area estimate of 220.5 square units closely approximates the actual area of 222, especially considering that the left Riemann sum using the same number of rectangles is inaccurate by more than 50 square units (according to Problem 17.12).

Trapezoidal Rule
Similar to Riemann sums but much more accurate

17.17 Approximate the area defined in Problem 17.12 (the area of the region bounded by $f(x) = 3x^2 + 1$ and the x-axis on the x-interval $[0,6]$) using three trapezoids of equal width.

It is best to construct the trapezoids such that one of their nonparallel sides is horizontal and lies on the x-axis. The opposite side should connect the function values at the endpoints of the interval, as illustrated by Figure 17-10. ←

> You're probably used to trapezoids that have horizontal parallel sides. Just tip a traditional-looking trapezoid onto its left side and flatten that side out to get the trapezoids you'll use with the trapezoidal rule.

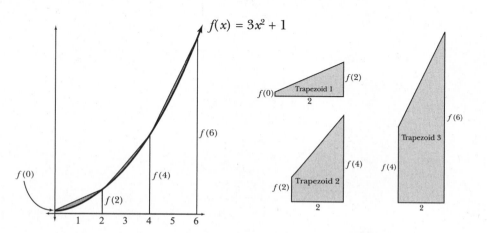

Figure 17-10   *Three trapezoids of height 2 are used to estimate the area between f(x) and the x-axis. In the left illustration, note that the shaded region represents the overestimation error. In the right portion of the illustration, the trapezoids are rendered separately to facilitate the calculation of their areas.*

The area of a trapezoid is $\frac{1}{2}h\left(b_1 + b_2\right)$, where b_1 and b_2 are the lengths of the bases and h represents the distance between them. Each of the trapezoids has the same height: $h = \Delta x = \dfrac{b-a}{n} = \dfrac{6-0}{3} = 2$.

> The BASES of a trapezoid are its parallel sides. That means the bases of the left trapezoid are f(0) and f(2), the bases of the middle trapezoid are f(2) and f(4), and the bases of the right trapezoid are f(4) and f(6).

Area of Trapezoid 1:

$\frac{1}{2}h\left(b_1 + b_2\right)$

$= \frac{1}{2}(2)\left(f(0) + f(2)\right)$

$= \frac{2}{2}\left[\left(3\cdot 0^2 + 1\right) + \left(3\cdot 2^2 + 1\right)\right]$

$= 1[1 + 13]$

$= 14$

Area of Trapezoid 2:

$\frac{1}{2}h\left(b_2 + b_3\right)$

$= \frac{1}{2}(2)\left(f(2) + f(4)\right)$

$= \frac{2}{2}\left[\left(3\cdot 2^2 + 1\right) + \left(3\cdot 4^2 + 1\right)\right]$

$= 1[13 + 49]$

$= 62$

Area of Trapezoid 3:

$\frac{1}{2}h\left(b_3 + b_4\right)$

$= \frac{1}{2}(2)\left(f(4) + f(6)\right)$

$= \frac{2}{2}\left[\left(3\cdot 4^2 + 1\right) + \left(3\cdot 6^2 + 1\right)\right]$

$= 1[49 + 109]$

$= 158$

The sum of the trapezoidal areas is $14 + 62 + 158 = 234$ square units. Note that this estimate is almost as accurate as the midpoint Riemann sum from Problem

17.16 despite using half as many geometric shapes ($n = 3$ trapezoids instead of $n = 6$ rectangles).

17.18 Construct the trapezoidal rule, which uses n trapezoids of equal width to approximate the area of the region bounded by a function $f(x)$ and the x-axis on the x-interval $[a,b]$.

Label the endpoints of the intervals as follows: $a = x_0$, $a + \Delta x = x_1$, $a + 2\Delta x = x_2$, $\cdots$, $b = x_n$.

> If each trapezoid is Δx wide, then at $x = a + \Delta x$, the first trapezoid ends and the second begins; at $x = a + 2\Delta x$, the second trapezoid ends and the third begins, etc.

Each of the trapezoids has area $\frac{1}{2} \cdot \Delta x \cdot (b_1 + b_2)$, where $\Delta x = \dfrac{b-a}{n}$. Factor $\frac{1}{2} \cdot \Delta x = \frac{1}{2}\left(\dfrac{b-a}{n}\right) = \dfrac{b-a}{2n}$ out of each term, leaving behind the sums of the bases of the trapezoids:

$$\frac{b-a}{2n}\left[\left(f(a)+f(x_1)\right)+\left(f(x_1)+f(x_2)\right)+\cdots+\left(f(x_{n-2})+f(x_{n-1})\right)+\left(f(x_{n-1})+f(b)\right)\right]$$

$$=\frac{b-a}{2n}\left[f(a)+2f(x_1)+2f(x_2)+\cdots+2f(x_{n-1})+f(b)\right]$$

> All of the function values have a coefficient of 2 except $x = a$ and $x = b$, because all of the bases represent the end of one trapezoid and the beginning of another except at the endpoints.

17.19 Apply the trapezoidal rule with $n = 5$ trapezoids to estimate the area of the shaded region in Figure 17-11. (Note that this is the same function $f(x)$ and region investigated in Problems 17.2–17.6.)

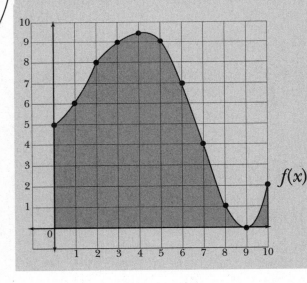

Figure 17-11

The graph of a continuous function $f(x)$.

Apply the trapezoidal rule formula from Problem 17.18.

$$\frac{b-a}{2n}\left[f(a)+2f(x_1)+2f(x_2)+2f(x_3)+2f(x_4)+f(b)\right]$$

Here, $a = 0$, $b = 10$, and $n = 5$; the bases are located at the same x-values as the interval endpoints: $a = 0$, $x_1 = 2$, $x_2 = 4$, $x_3 = 6$, $x_4 = 8$, and $b = 10$. Substitute these values into the formula.

$$\frac{10-0}{2(5)}\Big[f(0)+2f(2)+2f(4)+2f(6)+2f(8)+f(10)\Big]$$

$$=\frac{10}{10}(5+2\cdot8+2\cdot9.5+2\cdot7+2\cdot1+2)$$

$$=1(5+16+19+14+2+2)$$

$$=58 \text{ square units}$$

There's always one more function value in here than the number n. (Since n = 5, there are six terms in the brackets.) All of them are multiplied by 2 except the first and last ones, f(0) and f(10).

17.20 Problems 17.7–17.11 estimate the area of the region bounded by $g(x) = x^2$ and the x-axis on the x-interval [0,5]. Apply the trapezoidal rule using $n = 6$ trapezoids.

Apply the trapezoidal rule formula from Problem 17.18. Note that Problem 17.10 used the same number of intervals over the same interval, so use those boundaries and function values.

$$\frac{b-a}{2n}\Big[g(a)+2g(x_1)+2g(x_2)+2g(x_3)+2g(x_4)+2g(x_5)+g(b)\Big]$$

$$=\frac{5-0}{2(6)}\Big[g(0)+2g\Big(\frac{5}{6}\Big)+2g\Big(\frac{5}{3}\Big)+2g\Big(\frac{5}{2}\Big)+2g\Big(\frac{10}{3}\Big)+2g\Big(\frac{25}{6}\Big)+g(5)\Big]$$

$$=\frac{5}{12}\Big[0+2\Big(\frac{25}{36}\Big)+2\Big(\frac{25}{9}\Big)+2\Big(\frac{25}{4}\Big)+2\Big(\frac{100}{9}\Big)+2\Big(\frac{625}{36}\Big)+25\Big]$$

$$=\frac{5}{12}\Big(\frac{1{,}825}{18}\Big)$$

$$=\frac{9{,}125}{216}\approx 42.245$$

17.21 Estimate the area of the region bounded by $h(x) = \sin x$ and $y = 0$ on the x-interval $[0, \pi]$ using the trapezoidal rule with $n = 4$ trapezoids.

This is the equation of the x-axis.

Begin by calculating Δx.

$$\Delta x = \frac{b-a}{n} = \frac{\pi-0}{4} = \frac{\pi}{4}$$

Apply the trapezoidal rule formula.

$$\frac{b-a}{2n}\Big[f(a)+2f(x_1)+2f(x_2)+2f(x_3)+f(b)\Big]$$

$$=\frac{\pi-0}{2(4)}\Big[f(0)+2f\Big(\frac{\pi}{4}\Big)+2f\Big(\frac{\pi}{2}\Big)+2f\Big(\frac{3\pi}{4}\Big)+f(\pi)\Big]$$

$$=\frac{\pi}{8}\Big[\sin 0+2\sin\frac{\pi}{4}+2\sin\frac{\pi}{2}+2\sin\frac{3\pi}{4}+\sin\pi\Big]$$

Evaluate the sine function using the unit circle.

$$= \frac{\pi}{8}\left[0 + \left(2 \cdot \frac{\sqrt{2}}{2}\right) + (2 \cdot 1) + \left(2 \cdot \frac{\sqrt{2}}{2}\right) + 0\right]$$

$$= \frac{\pi}{8}\left[\sqrt{2} + 2 + \sqrt{2}\right]$$

$$= \frac{\pi\left(2 + 2\sqrt{2}\right)}{8}$$

Reduce the fraction by factoring 2 out of the numerator.

$$\frac{\pi \cdot \cancel{2}\left(1 + \sqrt{2}\right)}{\cancel{2} \cdot 4}$$

$$= \frac{\pi\left(1 + \sqrt{2}\right)}{4}$$

17.22 Estimate the area of the region bounded by $f(x) = \dfrac{1}{x}$, $y = 0$, $x = 1$ and $x = 11$ using six trapezoids of equal height. Report your answer accurate to three decimal places.

Apply the trapezoidal rule; begin by calculating Δx.

$$\Delta x = \frac{11 - 1}{6} = \frac{10}{6} = \frac{5}{3}$$

In order to make six trapezoids, you need seven boundaries: two endpoints for the whole interval ($x = 1$ and $x = 11$) and five equal steps (that are Δx units long) in between.

Add $\frac{5}{3}$ to the lower boundary five times to generate the boundaries of the trapezoids.

$$1 = \frac{3}{3}, \frac{3+5}{3} = \frac{8}{3}, \frac{8+5}{3} = \frac{13}{3}, \frac{13+5}{3} = \frac{18}{3} = 6, \frac{18+5}{3} = \frac{23}{3}, \frac{23+5}{3} = \frac{28}{3}, \frac{28+5}{3} = \frac{33}{3} = 11$$

Apply the trapezoidal rule.

$$\frac{11-1}{2(6)}\left[f(1) + 2f(x_1) + 2f(x_2) + 2f(x_3) + 2f(x_4) + 2f(x_5) + f(11)\right]$$

$$= \frac{10}{12}\left[f(1) + 2f\left(\frac{8}{3}\right) + 2f\left(\frac{13}{3}\right) + 2f(6) + 2f\left(\frac{23}{3}\right) + 2f\left(\frac{28}{3}\right) + f(11)\right]$$

$$= \frac{5}{6}\left[1 + 2 \cdot \frac{3}{8} + 2 \cdot \frac{3}{13} + 2 \cdot \frac{1}{6} + 2 \cdot \frac{3}{23} + 2 \cdot \frac{3}{28} + \frac{1}{11}\right]$$

$$= \frac{5}{6}\left[1 + \frac{3}{4} + \frac{6}{13} + \frac{1}{3} + \frac{6}{23} + \frac{3}{14} + \frac{1}{11}\right]$$

The common denominator of 3, 4, 11, 13, 23, and 28 is 276,276! Sure you don't want to add those fractions by hand?!

It is neither useful nor valuable to combine these fractions using a common denominator—use a graphing calculator to identify the decimal equivalent. The area is approximately 2.592 square units.

17.23 Prove that the trapezoidal rule approximation for the area beneath a continuous function $f(x)$ on the x-interval $[a,b]$ using n trapezoids is equal to the average of the right and left Riemann sums for the same area using n rectangles.

Figure 17-12 illustrates a function $f(x)$ split into n subintervals over $[a,b]$. Note that each interval has width $\Delta x = \dfrac{b-a}{n}$, creating the subintervals $[a,x_1]$, $[x_1,x_2]$, $[x_2,x_3]$, $\cdots$, $[x_{n-2}, x_{n-1}]$, and $[x_{n-1}, b]$.

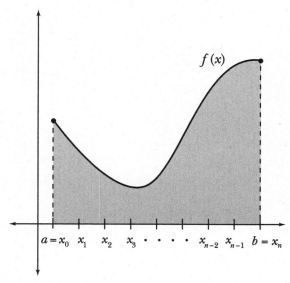

Figure 17-12 *Although $[a,b]$ is split into $n = 8$ subintervals, this diagram is labeled generically—any n subintervals would be labeled the same way along the x-axis.*

Calculate the left Riemann sum, S_L.

> Factor out the width of each rectangle like in Problem 17.7. Since you have no idea what the actual width is, leave the variables a, b, and n in there.

$$S_L = \frac{b-a}{n}\left(f(a)+f(x_1)+f(x_2)+f(x_3)+\cdots+f(x_{n-2})+f(x_{n-1})\right)$$

Note that each boundary is used except for $f(b)$, as that x-value does not represent the left boundary of any subinterval. Now calculate the right Riemann sum, S_R; this time only $f(a)$ will not appear in the formula, as it is not the right boundary of any subinterval.

$$S_R = \frac{b-a}{n}\left(f(x_1)+f(x_2)+f(x_3)+\cdots+f(x_{n-2})+f(x_{n-1})+f(b)\right)$$

The average of S_L and S_R, is the sum multiplied by $\dfrac{1}{2}$.

$$\frac{1}{2}\left(S_L+S_R\right)=\frac{1}{2}\left[\frac{b-a}{n}\left(f(a)+f(x_1)+\cdots+f(x_{n-2})+f(x_{n-1})\right)\right]+\frac{1}{2}\left[\frac{b-a}{n}\left(f(x_1)+f(x_2)+\cdots+f(x_{n-1})+f(b)\right)\right]$$

Every term in the sum contains $\dfrac{1}{2}\cdot\dfrac{b-a}{n}=\dfrac{b-a}{2n}$; factor out that expression.

$$=\frac{b-a}{2n}\left[\left(f(a)+f(x_1)+\cdots+f(x_{n-2})+f(x_{n-1})\right)+\left(f(x_1)+f(x_2)+\cdots+f(x_{n-1})+f(b)\right)\right]$$

When you're done averaging the right and left Riemann sums, you end up with the trapezoidal rule formula from Problem 17.18.

Notice that, apart from $f(a)$ and $f(b)$, every function value $f(x_1), f(x_2), f(x_3), \cdots,$ $f(x_{n-2}), f(x_{n-1})$ is repeated—appearing once in S_L and once in S_R. Add each pair: $f(x_1) + f(x_1) = 2f(x_1)$, $f(x_2) + f(x_2) = 2f(x_2)$, etc.

$$= \frac{b-a}{2n}\left[f(a) + 2f(x_1) + 2f(x_2) + 2f(x_3) + \cdots + 2f(x_{n-2}) + 2f(x_{n-1}) + f(b)\right]$$

17.24 A surveying company measures the distance between the northern and southern shores of a lake at fixed intervals 100 feet apart. Apply the trapezoidal rule to the data in Figure 17-13 in order to determine the approximate surface area of the lake.

So, x feet east of the left edge of the lake, $d(x)$ is the distance from the top shore of the lake to the bottom shore. For example, $d(200) = 289$ because 200 feet east of the left edge of the lake, the lake measures 289 feet across.

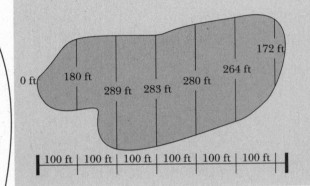

Figure 17-13

The distances between the shores of a 625-foot-long lake taken at regular 100-foot intervals. Note that the distance 0 at the leftmost measurement mark indicates the western boundary of the lake.

The lake's measurements are a function of the distance from its western shore, $d(x)$: $d(0) = 0$, $d(100) = 180$, $d(200) = 289$, $d(300) = 283$, $d(400) = 280$, $d(500) = 264$, and $d(600) = 172$.

Apply the trapezoidal rule to approximate the area beneath $d(x)$ on $[0,600]$ with $n = 6$ trapezoids.

Actually, the trapezoids on the far right and left aren't trapezoids at all—they're triangles because one of the bases is 0.

$$\frac{b-a}{2n}\left[d(a) + 2d(x_1) + 2d(x_2) + 2d(x_3) + 2d(x_4) + 2d(x_5) + d(b)\right]$$

$$= \frac{600 - 0}{2(6)}\left[0 + 2(180) + 2(289) + 2(283) + 2(280) + 2(264) + 172\right]$$

$$= 50[2,764]$$

$$= 138,200 \text{ ft}^2$$

17.25 The surveying company from Problem 17.24 is not entirely comfortable with its estimate, because that approximation neglects the portion of the lake less than 25 feet from its eastern boundary. To verify its previous estimate, seven trapezoids are used to approximate the surface area, but this time, the measurements are taken at unequal intervals, as illustrated by Figure 17-14. What is the approximate surface area of the lake according to this technique?

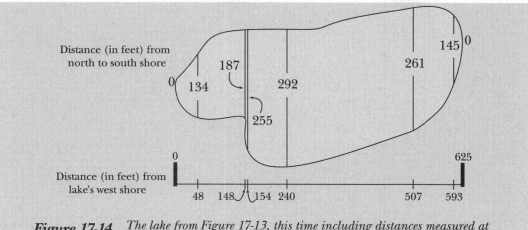

Figure 17-14 *The lake from Figure 17-13, this time including distances measured at unequal intervals.*

Though you are instructed to use trapezoids to approximate the area, you cannot actually apply the trapezoidal rule—it requires equal subintervals. Therefore, you simply find the area of each trapezoid individually. The leftmost two measurements, 0 and 134, are the bases of T_1 (the first trapezoid), 134 and 187 are the bases of T_2 (the second trapezoid), 187 and 255 are the bases of T_3, and so forth until T_7, the trapezoid with bases 145 and 0.

$$\text{Area of } T_1 + \text{Area of } T_2 + \text{Area of } T_3 + \text{Area of } T_4 + \text{Area of } T_5 + \text{Area of } T_6 + \text{Area of } T_7$$

$$= \frac{1}{2}h_1(0+134) + \frac{1}{2}h_2(134+187) + \frac{1}{2}h_3(187+255) + \frac{1}{2}h_4(255+292) + \frac{1}{2}h_5(292+261) + \frac{1}{2}h_6(261+145) + \frac{1}{2}h_7(145+0)$$

$$= \frac{1}{2}\left[h_1(134) + h_2(321) + h_3(442) + h_4(547) + h_5(553) + h_6(406) + h_7(145)\right]$$

Calculate the widths $(h_1, h_2, \cdots, h_7)$ of the trapezoids and substitute them into the expression.

$$= \frac{1}{2}\left[48(134) + 100(321) + 6(442) + 86(547) + 267(553) + 86(406) + 32(145)\right]$$

$$= \frac{1}{2}\left[6{,}432 + 32{,}100 + 2{,}652 + 47{,}042 + 147{,}651 + 34{,}916 + 4{,}640\right]$$

$$= \frac{1}{2}\left[275{,}433\right]$$

$$= 137{,}716.5 \text{ ft}^2$$

> The first trapezoid goes from 0 to 48 feet east of the west edge of the lake, so $h_1 = 48 - 0 = 48$. The second trapezoid goes from 48 to 148 feet east, so $h_2 = 148 - 48 = 100$, etc.

Note that this estimate of the lake's surface area is very close to the estimate from Problem 17.24.

17.26 Demonstrate that the exact area beneath a linear function can be calculated using a trapezoidal approximation.

Consider the linear function $f(x) = mx + b$ in Figure 17-15. The shaded area of the region bounded by $f(x)$ and the x-axis on the x-interval $[c,d]$ is, in fact, a trapezoid. Use the formula for the area of a trapezoid to calculate the shaded area.

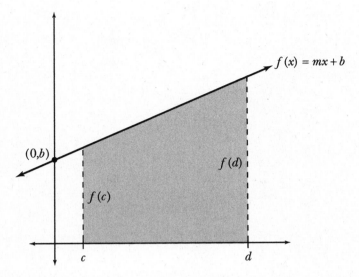

Figure 17-15   *The area beneath the linear function f(x) on the x-interval [c,d] is a trapezoid with bases f(c) and f(d) and height $\Delta x = d - c$.*

$$\text{area} = \frac{1}{2}\Delta x \left(b_1 + b_2 \right)$$

$$= \frac{1}{2}(d-c)\big(f(c) + f(d)\big)$$

Note that $f(c) = m(c) + b$ and $f(d) = m(d) + b$. Substitute these values into the area expression.

$$= \frac{1}{2}(d-c)\big[(mc+b) + (md+b)\big]$$

Not all areas between a linear function and the *x*-axis can be calculated using a single trapezoid. However, as Figure 17-16 demonstrates, those areas can be calculated using two triangles.

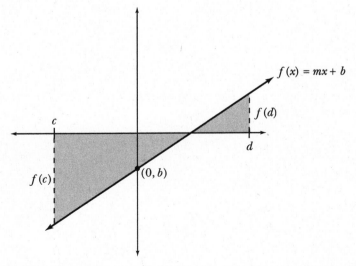

Figure 17-16   *If the exact area beneath a linear function cannot be calculated using a single trapezoid, it can be determined using two triangles.*

Simpson's Rule
Approximates area beneath curvy functions really well

17.27 Compare and contrast Simpson's rule and the trapezoidal rule.

Simpson's rule is an area approximation technique that uses equal subintervals, like the trapezoidal rule. However, rather than using straight lines to connect the interval endpoints, thereby constructing trapezoids to estimate the area, Simpson's rule uses parabolas. Therefore, Simpson's rule is able to exactly calculate the area beneath quadratic functions, whereas the trapezoidal rule is able only to exactly calculate the area beneath linear functions (as explained in Problem 17.26). Simpson's rule is also governed by one additional restriction: n, the number of subintervals used in the approximation, must be even.

According to Simpson's rule, the area between the continuous function $f(x)$ and the x-axis on the x-interval $[a,b]$ is approximately equal to the following:

$$\frac{b-a}{3n}\left(f(a)+4f(x_1)+2f(x_2)+4f(x_3)+2f(x_4)+\cdots+2f(x_{n-2})+4f(x_{n-1})+f(b)\right)$$

> In the Riemann sum formulas, you divide $b - a$ by n. In the trapezoidal rule formula, you divide $b - a$ by $2n$; and in the Simpson's rule formula, you divide $b - a$ by $3n$.

Like the trapezoidal rule, Simpson's rule attaches no coefficients to $f(a)$ and $f(b)$. However, Simpson's rule requires a coefficient of 4 before all odd subscripts of x and a coefficient of 2 before all even subscripts.

17.28 Problems 17.12–17.17 estimate the area of the region bounded by $f(x) = 3x^2 + 1$ and the x-axis on the x-interval $[0,6]$. Apply Simpson's rule with $n = 2$ subintervals to verify that the exact area of the region is 222 square units.

> You'll get the exact area because $f(x)$ is a quadratic function and Simpson's rule calculates area using parabolas (instead of trapezoids), which are quadratic functions.

Begin by calculating the widths of the subintervals.

$$\Delta x = \frac{b-a}{n} = \frac{6-0}{2} = 3$$

Apply Simpson's rule with $a = 0$, $b = 6$, $n = 2$, and $f(x) = 3x^2 + 1$.

$$\frac{b-a}{3n}\left[\left(f(a)+4f(x_1)+f(b)\right)\right]$$
$$=\frac{6-0}{3(2)}\left[f(0)+4f(3)+f(6)\right]$$
$$=\frac{6}{6}\left[(3\cdot 0^2+1)+4(3\cdot 3^2+1)+(3\cdot 6^2+1)\right]$$
$$=1\left[(0+1)+4(27+1)+(108+1)\right]$$
$$=1+112+109$$
$$=222$$

> When you have only $n = 2$ subintervals, don't switch back and forth between coefficients of 4 and 2 in the Simpson's rule formula. Just use one coefficient: 4.

17.29 Problem 17.20 estimates the area of the region bounded by $g(x) = x^2$ and the x-axis on the x-interval $[0,5]$ using $n = 6$ subintervals. Apply Simpson's rule to determine the exact area.

$$\frac{b-a}{3n}\left[g(a) + 4g(x_1) + 2g(x_2) + 4g(x_3) + 2g(x_4) + 4g(x_5) + g(b)\right]$$

$$= \frac{5-0}{3(6)}\left[g(0) + 4g\left(\frac{5}{6}\right) + 2g\left(\frac{5}{3}\right) + 4g\left(\frac{5}{2}\right) + 2g\left(\frac{10}{3}\right) + 4g\left(\frac{25}{6}\right) + g(5)\right]$$

$$= \frac{5}{18}\left[0 + 4\left(\frac{25}{36}\right) + 2\left(\frac{25}{9}\right) + 4\left(\frac{25}{4}\right) + 2\left(\frac{100}{9}\right) + 4\left(\frac{625}{36}\right) + 25\right]$$

$$= \frac{5}{18}[150]$$

$$= \frac{750}{18}$$

$$= \frac{125}{3}$$

17.30 Use Simpson's rule with $n = 4$ subintervals to estimate the area between the x-axis and the function $h(x) = \sin x$ (as defined in Problem 17.21) on the x-interval $[0, \pi]$.

The exact area beneath the curve is 2—it's worked out in Problem 18.20. Simpson's rule doesn't give you the exact area because sin x is a trigonometric, not a quadratic, function.

Use the subinterval width and function values from Problem 17.21 as you apply Simpson's rule.

$$\frac{b-a}{3n}\left[h(a) + 4h(x_1) + 2h(x_2) + 4h(x_3) + h(b)\right]$$

$$= \frac{\pi-0}{3(4)}\left[h(0) + 4h\left(\frac{\pi}{4}\right) + 2h\left(\frac{\pi}{2}\right) + 4h\left(\frac{3\pi}{4}\right) + h(\pi)\right]$$

$$= \frac{\pi}{12}\left(0 + 4\left(\frac{\sqrt{2}}{2}\right) + 2(1) + 4\left(\frac{\sqrt{2}}{2}\right) + 0\right)$$

$$= \frac{\pi}{12}\left(2\sqrt{2} + 2 + 2\sqrt{2}\right)$$

$$= \frac{\pi\left(2 + 4\sqrt{2}\right)}{12}$$

Factor 2 out of the numerator to reduce the fraction.

$$= \frac{\cancel{2}\pi\left(1 + 2\sqrt{2}\right)}{\cancel{2}\cdot 6}$$

$$= \frac{\pi\left(1 + 2\sqrt{2}\right)}{6}$$

This equals approximately 2.00456, which is very close to the exact answer of 2. By comparison, the trapezoidal rule estimate from Problem 17.21 was approximately 1.89612.

17.31 Use Simpson's rule with $n = 6$ subintervals to estimate the area between the x-axis and $f(x) = \dfrac{1}{x}$ (as defined in Problem 17.22) on the x-interval $[1,11]$. Provide an answer accurate to three decimal places.

The intervals and function values match those listed in Problem 17.22; apply them to Simpson's rule.

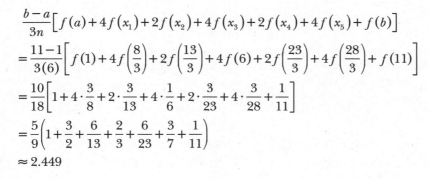

$$\frac{b-a}{3n}\Big[f(a)+4f(x_1)+2f(x_2)+4f(x_3)+2f(x_4)+4f(x_5)+f(b)\Big]$$

$$=\frac{11-1}{3(6)}\left[f(1)+4f\left(\frac{8}{3}\right)+2f\left(\frac{13}{3}\right)+4f(6)+2f\left(\frac{23}{3}\right)+4f\left(\frac{28}{3}\right)+f(11)\right]$$

$$=\frac{10}{18}\left[1+4\cdot\frac{3}{8}+2\cdot\frac{3}{13}+4\cdot\frac{1}{6}+2\cdot\frac{3}{23}+4\cdot\frac{3}{28}+\frac{1}{11}\right]$$

$$=\frac{5}{9}\left(1+\frac{3}{2}+\frac{6}{13}+\frac{2}{3}+\frac{6}{23}+\frac{3}{7}+\frac{1}{11}\right)$$

$$\approx 2.449$$

The actual area is ln 11 = 2.397895.... It's worked out in Problem 18.21.

Formal Riemann Sums

You'll want to poke your "i"s out

17.32 Describe how the Riemann sum $\displaystyle\sum_{i=1}^{n} f(c_i)\Delta x_i$ is used to calculate area in the coordinate plane when $f(x) \geq 0$.

This is just a variable i, not $i = \sqrt{-1}$.

This expression states that the approximate area beneath the continuous function $f(x)$ is equal to the sum of a series of rectangle areas. Each rectangle's area is equal to the product of its length, $f(c_i)$, and its width, Δx_i. For instance, the area of the third $(i = 3)$ rectangle is $f(c_3) \cdot \Delta x_3$, where Δx_3 is the distance between the endpoints of the interval on which the rectangle is constructed, and $f(c_3)$ is the height of the function $f(x)$ at some point $x = c_3$ on that interval.

17.33 Use a formal right Riemann sum with $n = 4$ subintervals of equal width to estimate the area of the region bounded by the x-axis and the positive, continuous function $f(x)$ on the x-interval $[a,b]$.

It doesn't really matter what point you pick on the interval, because you'll be using infinitely thin rectangles and all of the functions values will be roughly the same. (See Problem 17.37.)

Calculate the width of the rectangles using the formula from Problem 17.1.

$$\Delta x = \frac{b-a}{n}$$

Next, identify the boundaries of the four intervals. The leftmost interval has a left bound of $x = a$ and a right bound of $a + \Delta x$: $[a, a + \Delta x]$. The left boundary of the second interval matches the right boundary of the first interval $(a + \Delta x)$ and its right boundary is Δx to the right:

$$(a + \Delta x) + \Delta x = a + 2\Delta x$$

Therefore, the second interval is $[a + \Delta x, a + 2\Delta x]$. Similarly, the remaining intervals are $[a + 2\Delta x, a + 3\Delta x]$ and $[a + 3\Delta x, b]$.

$b = a + 4\Delta x$

Each rectangle of the right Riemann sum has a width of $\Delta x = \dfrac{b - a}{n}$ and a height equal to $f(c_i)$, if c_i is the right boundary of the ith interval. Calculate the sum of the areas of the rectangles.

The right interval boundaries are $a + \Delta x$, $a + 2\Delta x$, $a + 3\Delta x$, and $a + 4\Delta x$.

$$\Delta x \cdot f(a + \Delta x) + \Delta x \cdot f(a + 2\Delta x) + \Delta x \cdot f(a + 3\Delta x) + \Delta x \cdot f(a + 4\Delta x)$$
$$= \Delta x \big[f(a + \Delta x) + f(a + 2\Delta x) + f(a + 3\Delta x) + f(a + 4\Delta x) \big]$$

17.34 Use a formal right Riemann sum with n subintervals of equal width to estimate the area of the region bounded by the x-axis and the positive, continuous function $f(x)$ on the x-interval $[a,b]$.

Note that Problem 17.33 asks you to perform the same task, but with a specific number of subintervals ($n = 4$). This problem will therefore have a similar answer, but instead of 4 function values added parenthetically, there should be n:
$\Delta x \big[f(a + \Delta x) + f(a + 2\Delta x) + f(a + 3\Delta x) + \cdots + f(a + n\Delta x) \big]$.

First plug in $i = 1$ to get the area $\Delta x \cdot f(1\Delta x)$, then plug in $i = 2$ to get the area $\Delta x \cdot f(a + 2\Delta x)$, and so on, all the way up to $i = n$, the area $\Delta x \cdot f(a + n\Delta x)$. Once you've found all those rectangular areas, add them all up.

17.35 Write the solution to Problem 17.34 in sigma notation.

To indicate a sum of n terms, the limits of the summation are 1 and n. Use the variable i to indicate the individual rectangles: $\displaystyle\sum_{i=1}^{n} \big[\Delta x \cdot f(a + i\Delta x) \big]$.

17.36 In order to exactly calculate the area between a function $f(x)$ and the x-axis, an infinite number of rectangles must be used. Explain how to indicate this for Riemann sums written in sigma notation.

Apply the formula from Problem 17.35 and indicate an infinite number of rectangles by allowing n to approach infinity: $\displaystyle\lim_{n \to \infty} \sum_{i=1}^{n} \big[\Delta x \cdot f(a + i\Delta x) \big]$.

17.37 Justify the sufficiency of a right Riemann sum (or in fact any other Riemann sum) to exactly calculate a finite area using an infinite number of rectangles.

No matter what $b - a$ is, if you divide it by an infinitely gigantic number n, the result is basically 0. (See the note on Problem 10.28.)

In order for an infinite number of rectangles to fit in a finite space, the width of the rectangles, $\Delta x = \dfrac{b - a}{n}$, must approach 0. If the rectangles have an infinitely small width, the heights of the function beneath which you are approximating the area will not change significantly on the subintervals.

Therefore, all of the function values on each subinterval are essentially the same, and the right, left, midpoint, or any other Riemann sum will produce the exact same value, the precise area between the function and the *x*-axis.

17.38 Identify the values of the following expressions: $\sum_{i=1}^{n} k$, $\sum_{i=1}^{n} i$, $\sum_{i=1}^{n} i^2$, and $\sum_{i=1}^{n} i^3$.

Calculating area with formal Riemann sums often requires you to know the values of these sums. It is helpful to memorize their values rather than to generate them when they are needed. In fact, generating them is both trivial and irrelevant for the scope of this exercise, and is therefore omitted. ←

$$\sum_{i=1}^{n} k = kn, \quad \sum_{i=1}^{n} i = \frac{n^2 + n}{2}, \quad \sum_{i=1}^{n} i^2 = \frac{2n^3 + 3n^2 + n}{6}, \quad \text{and} \quad \sum_{i=1}^{n} i^3 = \frac{n^4 + 2n^3 + n^2}{4}$$

Generating these is MUCH HARDER than just memorizing them, so don't even bother figuring it out unless you have a lot of time on your hands. The first one's easy, but the rest of 'em? Not so much.

17.39 According to Problem 17.29, the exact area of the region bounded by $g(x) = x^2$ and the *x*-axis on the *x*-interval [0,5] is equal to $\frac{125}{3}$ square units. Verify this value using a formal right Riemann sum (i.e., using infinite number of rectangles).

Problem 17.36 states that the right Riemann sum using an infinite number of rectangles is equal to $\lim_{n \to \infty} \sum_{i=1}^{n} [\Delta x \cdot g(a + i\Delta x)]$. To apply the formula, begin by calculating Δx.

$$\Delta x = \frac{b-a}{n} = \frac{5-0}{n} = \frac{5}{n}$$

Substitute $\Delta x = \frac{5}{n}$ into the formal Riemann sum formula. Note that $g(a + i\Delta x) = g(i\Delta x)$ because $a = 0$.

$$\lim_{n \to \infty} \sum_{i=1}^{n} [\Delta x \cdot g(i\Delta x)] = \lim_{n \to \infty} \sum_{i=1}^{n} \left[\frac{5}{n} \cdot g\left(\frac{5i}{n}\right) \right]$$

According to a summation property, $\sum a \cdot g(x) = a \sum g(x)$. ←

In this problem, 5/n is a constant (i changes during the problem but n doesn't). Pull it outside the summation symbol.

$$= \lim_{n \to \infty} \left[\frac{5}{n} \cdot \sum_{i=1}^{n} g\left(\frac{5i}{n}\right) \right]$$

Substitute $\frac{5i}{n}$ into $g(x) = x^2$.

$$= \lim_{n \to \infty} \left[\frac{5}{n} \cdot \sum_{i=1}^{n} \left(\frac{5i}{n}\right)^2 \right]$$

$$= \lim_{n \to \infty} \left[\frac{5}{n} \cdot \sum_{i=1}^{n} \left(\frac{25i^2}{n^2}\right) \right]$$

Again apply the summation property $\sum a \cdot g(x) = a \sum g(x)$ to remove the constant $\dfrac{25}{n^2}$ from within the summation.

$$= \lim_{n \to \infty} \left[\frac{5}{n} \cdot \frac{25}{n^2} \cdot \sum_{i=1}^{n} i^2 \right]$$

$$= \lim_{n \to \infty} \left[\frac{125}{n^3} \cdot \sum_{i=1}^{n} i^2 \right]$$

Apply the $\sum_{i=1}^{n} i^2$ formula from Problem 17.38.

$$= \lim_{n \to \infty} \left[\frac{125}{n^3} \cdot \frac{2n^3 + 3n^2 + n}{6} \right]$$

$$= \lim_{n \to \infty} \left[\frac{250n^3}{6n^3} + \frac{375n^2}{6n^3} + \frac{125n}{6n^3} \right]$$

$$= \lim_{n \to \infty} \frac{250n^3}{6n^3} + \lim_{n \to \infty} \frac{375n^2}{6n^3} + \lim_{n \to \infty} \frac{125n}{6n^3}$$

L'Hôpital's Rule is covered in Problems 14.27–14.34.

Evaluate each limit either by comparing the degrees of the numerator and denominator or using L'Hôpital's Rule.

$$= \frac{250}{6} + 0 + 0 = \frac{125}{3}$$

17.40 According to Problem 17.28, the exact area of the region bounded by $f(x) = 3x^2 + 1$ and the x-axis on the x-interval $[0,6]$ is equal to 222 square units. Verify this value by calculating the same area using a right Riemann sum and an infinite number of rectangles.

Apply the techniques described in Problem 17.39. Begin by calculating Δx.

$$\Delta x = \frac{b-a}{n} = \frac{6-0}{n} = \frac{6}{n} \quad \text{and} \quad f(a + i\Delta x) = f(i\Delta x), \text{ because } a = 0$$

Substitute Δx into the formal right Riemann sum formula.

$$\lim_{n \to \infty} \sum_{i=1}^{n} \left[\Delta x \cdot f(i\Delta x) \right] = \lim_{n \to \infty} \sum_{i=1}^{n} \left[\frac{6}{n} \cdot f\left(\frac{6i}{n} \right) \right]$$

Apply the summation property that states $\sum a \cdot g(x) = a \sum g(x)$ and evaluate $f\left(\dfrac{6i}{n} \right)$.

$$= \lim_{n \to \infty} \left(\frac{6}{n} \cdot \sum_{i=1}^{n} \left[3\left(\frac{6i}{n} \right)^2 + 1 \right] \right)$$

$$= \lim_{n \to \infty} \left(\frac{6}{n} \cdot \sum_{i=1}^{n} \left[3 \cdot \frac{36i^2}{n^2} + 1 \right] \right)$$

$$= \lim_{n \to \infty} \left(\frac{6}{n} \cdot \sum_{i=1}^{n} \left[\frac{108i^2}{n^2} + 1 \right] \right)$$

According to a summation property, $\sum(a+b) = \sum a + \sum b$. ←

$$= \lim_{n\to\infty}\left[\frac{6}{n}\left(\sum_{i=1}^{n}\frac{108i^2}{n^2} + \sum_{i=1}^{n}1\right)\right]$$

Distribute $\frac{6}{n}$ through the parentheses.

> You're allowed to split the sum inside sigma into the sum of two separate sigma expressions.

$$= \lim_{n\to\infty}\left(\frac{6}{n}\sum_{i=1}^{n}\frac{108i^2}{n^2} + \frac{6}{n}\sum_{i=1}^{n}1\right)$$

$$= \lim_{n\to\infty}\left(\frac{6}{n}\cdot\frac{108}{n^2}\sum_{i=1}^{n}i^2 + \frac{6}{n}\sum_{i=1}^{n}1\right)$$

> You can pull the constant $\frac{108}{n^2}$ outside the sigma sign now.

According to Problem 17.38, $\sum_{i=1}^{n}i^2 = \frac{2n^3 + 3n^2 + n}{6}$ and $\sum_{i=1}^{n}1 = (1)n = n$.

$$= \lim_{n\to\infty}\left[\frac{648}{n^3}\left(\frac{2n^3 + 3n^2 + n}{6}\right) + \frac{6}{n}(n)\right]$$

$$= \lim_{n\to\infty}\left[\frac{648}{6n^3}\left(2n^3 + 3n^2 + n\right) + \frac{6\cancel{n}}{\cancel{n}}\right]$$

$$= \lim_{n\to\infty}\left[\frac{108}{n^3}\left(2n^3 + 3n^2 + n\right) + 6\right]$$

$$= \lim_{n\to\infty}\left[\frac{216n^3}{n^3} + \frac{324n^2}{n^3} + \frac{108n}{n^3} + 6\right]$$

$$= \lim_{n\to\infty}\frac{216n^3}{n^3} + \lim_{n\to\infty}\frac{324n^2}{n^3} + \lim_{n\to\infty}\frac{108n}{n^3} + \lim_{n\to\infty}6$$

Evaluate the first three limits by comparing the degrees of the numerator and denominator or using L'Hôpital's Rule. Note that $\lim_{n\to\infty}6 = 6$.

$$= 216 + 0 + 0 + 6$$

$$= 222$$

17.41 Calculate the exact area of the region bounded by $j(x) = x^3 + x$ and the x-axis on the x-interval $[-1,0]$ using a formal right Riemann sum.

Determine the value of Δx.

$$\Delta x = \frac{0 - (-1)}{n} = \frac{0 + 1}{n} = \frac{1}{n}$$

Substitute $\Delta x = \frac{1}{n}$ into the formal right Riemann sum formula.

$$\lim_{n\to\infty}\sum_{i=1}^{n}\left[\Delta x \cdot j(a + i\Delta x)\right] = \lim_{n\to\infty}\sum_{i=1}^{n}\left[\frac{1}{n}\cdot j\left(-1 + \frac{i}{n}\right)\right]$$

$$= \lim_{n\to\infty}\frac{1}{n}\sum_{i=1}^{n}j\left(-1 + \frac{i}{n}\right)$$

Evaluate $j\left(-1 + \dfrac{i}{n}\right)$.

$$= \lim_{n \to \infty} \frac{1}{n} \sum_{i=1}^{n} \left[\left(-1 + \frac{i}{n}\right)^3 + \left(-1 + \frac{i}{n}\right) \right]$$

$$= \lim_{n \to \infty} \frac{1}{n} \sum_{i=1}^{n} \left(-1 + \frac{3i}{n} - \frac{3i^2}{n^2} + \frac{i^3}{n^3} - 1 + \frac{i}{n} \right)$$

$$= \lim_{n \to \infty} \frac{1}{n} \sum_{i=1}^{n} \left(\frac{i^3}{n^3} - \frac{3i^2}{n^2} + \frac{4i}{n} - 2 \right)$$

$$= \lim_{n \to \infty} \frac{1}{n} \left(\sum_{i=1}^{n} \frac{i^3}{n^3} - \sum_{i=1}^{n} \frac{3i^2}{n^2} + \sum_{i=1}^{n} \frac{4i}{n} - \sum_{i=1}^{n} 2 \right)$$

$$\lim_{n \to \infty} \frac{1}{n} \left(\frac{1}{n^3} \sum_{i=1}^{n} i^3 - \frac{3}{n^2} \sum_{i=1}^{n} i^2 + \frac{4}{n} \sum_{i=1}^{n} i - \sum_{i=1}^{n} 2 \right)$$

> When you split the big sigma into four smaller sigmas, make sure you distribute that $1/n$ to both of them.

Replace $\displaystyle\sum_{i=1}^{n} i^3$, $\displaystyle\sum_{i=1}^{n} i^2$, and $\displaystyle\sum_{i=1}^{n} i$ with the formulas from Problem 17.38.

$$= \lim_{n \to \infty} \frac{1}{n} \left(\frac{1}{n^3} \cdot \frac{n^4 + 2n^3 + n^2}{4} - \frac{3}{n^2} \cdot \frac{2n^3 + 3n^2 + n}{6} + \frac{4}{n} \cdot \frac{n^2 + n}{2} - 2n \right)$$

$$= \lim_{n \to \infty} \left(\frac{n^4 + 2n^3 + n^2}{4n^4} - \frac{6n^3 + 9n^2 + 3n}{6n^3} + \frac{4n^2 + 4n}{2n^2} - \frac{2n}{n} \right)$$

$$= \frac{1}{4} - 1 + 2 - 2$$

$$= -\frac{3}{4}$$

> $j(x)$ is the only function in this chapter that's below the x-axis.

Because the graph of $j(x)$ is negative on the x-interval $(-1,0)$, the region bounded by $j(x)$ and the x-axis has a *signed area* of $-\dfrac{3}{4}$. Signed area reports both the area of the region and whether that area appears above or below the x-axis (much like velocity reports an object's speed as well as its direction via its sign). However, area should be reported using positive units, so the area of the region is $\dfrac{3}{4}$.

Chapter 18
INTEGRATION

Now the derivative's not the answer, it's the question

In this chapter, you will apply the fundamental theorem of calculus, which uses the antiderivative of a function in order to exactly calculate the area bounded by that function and the *x*-axis. Antidifferentiation requires a wider variety of techniques and algorithms than differentiation, so Chapters 18–24 are dedicated to thoroughly exploring antidifferentiation techniques and their applications. At this, the onset of said exploration, the power rule for integration, trigonometric, logarithmic, and exponential function antiderivatives, and variable substitution are discussed.

This chapter is almost a clone of Chapter 12, except instead of learning basic differentiation techniques, you'll learn basic integration (or antidifferentiation) techniques. Chapter 11 ended with the "hard way" to compute derivatives (the difference quotient)—sort of like Chapter 17, which ended with the "hard way" to calculate exact area (formal Riemann sums). Right smack at the beginning of Chapter 12, the "easy way" to take derivatives showed up, the power rule for differentiation. This chapter starts with the "easy way" to find ANTIderivatives, the power rule for integration. It then moves on to basic integration concepts like the fundamental theorem of calculus and integration methods like u-substitution.

Power Rule for Integration
ADD 1 to the exponent and DIVIDE by the new power

In other words, coefficients can be pulled outside the integral sign. Just integrate what's left and multiply by the coefficient when you're done. You do the same thing with derivatives— to differentiate $10x^3$, just differentiate x^3 to get $3x^2$ and multiply that by 10 to get $30x^2$.

18.1 Given a and n are real numbers, determine the antiderivative: $\int ax^n\,dx$.

According to a property of integrals, $\int af(x)\,dx = a\int f(x)\,dx$.

$$\int ax^n\,dx = a\int x^n\,dx$$

The power rule for integration states that $\int x^n\,dx = \dfrac{x^{n+1}}{n+1} + C$. In other words, increase the exponent by 1 and divide the term by the new exponent; the result is the antiderivative. Additionally, you must add a constant of integration "$+C$" to the expression to indicate the possible presence of a constant. Notice that differentiating this expression with respect to x will return the original integrand.

$$\frac{d}{dx}\left(\frac{x^{n+1}}{n+1} + C\right) = \frac{d}{dx}\left(\frac{1}{n+1}x^{n+1} + C\right) = \frac{1}{n+1}(n+1)x^{(n+1)-1} = \frac{n+1}{n+1}x^{n+1-1} = x^n$$

The INTEGRAND is the expression inside the integral (between the integral sign and the "dx").

Because the derivative of *any* constant is 0, there is no way to determine what constant, if any, was eliminated by differentiating, so "$+C$" is used as a generic placeholder. Therefore, $a\int x^n\,dx = \dfrac{ax^{n+1}}{n+1} + C$.

18.2 Integrate the expression: $\int x^8\,dx$.

Add 1 to the exponent ($8 + 1 = 9$) and divide the expression by the new exponent:
$\int x^8\,dx = \dfrac{x^{8+1}}{8+1} + C = \dfrac{x^9}{9} + C$.

18.3 Integrate the expression: $\int (6x^2 + 9x)\,dx$.

Even though there are two integrals here, you need only to use one C. Each integral has a constant of integration, but those two unknown constants add up to some other unknown constant you label C.

According to a property of integrals, $\int [f(x) \pm g(x)]\,dx = \int f(x)\,dx \pm \int g(x)\,dx$: The integral of a sum or difference is equal to the sum or difference of the individual integrals.

$$\int (6x^2 + 9x)\,dx = \int 6x^2\,dx + \int 9x\,dx$$
$$= 6\int x^2\,dx + 9\int x\,dx$$

Apply the power rule for integration.

$$= 6 \cdot \frac{x^{2+1}}{2+1} + 9 \cdot \frac{x^{1+1}}{1+1} + C$$
$$= \frac{6x^3}{3} + \frac{9x^2}{2} + C$$
$$= 2x^3 + \frac{9}{2}x^2 + C$$

18.4 Integrate the expression: $\int \frac{dx}{x}$.

Note that the integrand can be rewritten as $\int \frac{1}{x} dx$ —moving dx out of the expression to be integrated does not affect the result, as it is still multiplied by $\frac{1}{x}$ in both cases. However, the power rule for integration doesn't apply because it gives an undefined result.

$$\int \frac{1}{x} dx = \int x^{-1} dx = \frac{x^{-1+1}}{-1+1} + C = \frac{x^0}{0} + C$$

Division by 0 is mathematically invalid, so an alternative to the power rule for integration is needed in order to integrate. Recall that $\frac{d}{dx}(\ln x) = \frac{1}{x}$. Therefore, $\int \frac{1}{x} dx = \ln|x| + C$. The absolute value signs are required because the domain of $y = \ln x$ is $(0,\infty)$, the positive real numbers.

18.5 Integrate the expression: $\int \sqrt{x^3}\, dx$.

Rewrite the radical expression using rational exponents (as discussed in Problem 2.10).

$$\int \sqrt{x^3}\, dx = \int x^{3/2} dx$$

Apply the power rule for integration.

$$= \frac{x^{(3/2)+1}}{3/2+1} + C$$

$$= \frac{x^{3/2+2/2}}{3/2+2/2} + C$$

$$= \frac{x^{5/2}}{5/2} + C$$

Eliminate the complex fraction.

$$= \frac{2}{5} x^{5/2} + C$$

Dividing $x^{5/2}$ by 5/2 is the same as multiplying $x^{5/2}$ by 2/5.

18.6 Integrate the expression: $\int \left(9x^4 + 7\sqrt{x} - 5\sqrt[8]{x^3}\right) dx$.

By applying an integral property, you can split the integrand into three separate integrals, move the coefficient of each outside its integral, and rewrite the radical expressions using rational exponents.

$$\int \left(9x^4 + 7\sqrt{x} - 5\sqrt[8]{x^3}\right) dx = \int 9x^4 dx + \int 7\sqrt{x}\, dx + \int \left(-5\sqrt[8]{x^3}\, dx\right)$$

$$= 9\int x^4 dx + 7\int \sqrt{x}\, dx - 5\int \sqrt[8]{x^3}\, dx$$

$$= 9\int x^4 dx + 7\int x^{1/2} dx - 5\int x^{3/8} dx$$

Apply the power rule for integration.

Use common denominators: $(3/8) + 1 = 3/8 + 8/8 = 11/8$.

$$= 9 \cdot \frac{x^{4+1}}{4+1} + 7 \cdot \frac{x^{(1/2)+1}}{(1/2)+1} - 5 \cdot \frac{x^{(3/8)+1}}{(3/8)+1} + C$$

$$= 9 \cdot \frac{x^5}{5} + 7 \cdot \frac{x^{3/2}}{3/2} - 5 \cdot \frac{x^{11/8}}{11/8} + C$$

$$= \frac{9}{5}x^5 + 7\left(\frac{2}{3}\right)x^{3/2} - 5\left(\frac{8}{11}\right)x^{11/8} + C$$

$$= \frac{9}{5}x^5 + \frac{14}{3}x^{3/2} - \frac{40}{11}x^{11/8} + C$$

18.7 Integrate the expression: $\int \frac{5x - 6x^7}{x^4} \, dx$.

You can only split a fraction up into two smaller fractions if there's a sum in the NUMERATOR. You can't turn $\frac{a}{b+c}$ into $\frac{a}{b} + \frac{a}{c}$.

Note that $\frac{f(x) + g(x)}{h(x)} = \frac{f(x)}{h(x)} + \frac{g(x)}{h(x)}$, so rewrite the integrand as two separate integrals.

$$\int \frac{5x - 6x^7}{x^4} \, dx = \int \frac{5x}{x^4} \, dx - \int \frac{6x^7}{x^4} \, dx$$

$$= \int 5x^{1-4} dx - \int 6x^{7-4} dx$$

$$= 5 \int x^{-3} dx - 6 \int x^3 dx$$

Apply the power rule for integration.

Even though the exponent is negative, you still add 1 to it and divide by the new power.

$$= 5\left(\frac{x^{-3+1}}{-3+1}\right) - 6\left(\frac{x^{3+1}}{3+1}\right) + C$$

$$= 5\left(\frac{x^{-2}}{-2}\right) - 6\left(\frac{x^4}{4}\right) + C$$

$$= -\frac{5}{2x^2} - \frac{3}{2}x^4 + C$$

18.8 Integrate the expression: $\int \left(\frac{x^3 - x^4}{x^4}\right) dx$.

Rewrite the rational integrand as two separate integrals, as explained in Problem 18.7.

$$\int \left(\frac{x^3 - x^4}{x^4}\right) dx = \int \frac{x^3}{x^4} \, dx - \int \frac{x^4}{x^4} \, dx$$

$$= \int x^{3-4} dx - \int 1 \, dx$$

$$= \int x^{-1} dx - \int dx$$

According to Problem 18.4, the antiderivative of $\frac{1}{x}$ is $\ln |x| + C$; the antiderivative of 1 (which can be written $1 dx$ or simply dx) is x, because the derivative of x is 1.

$$= \ln |x| - x + C$$

18.9 Integrate the expression: $\int x^2 \left(5 - \sqrt{x}\right) dx$.

Distribute x^2 before applying the power rule for integration.

$$\int x^2 \left(5 - \sqrt{x}\right) dx = \int \left(5x^2 - x^2 \cdot x^{1/2}\right) dx$$
$$= \int \left(5x^2 - x^{5/2}\right) dx$$
$$= 5 \int x^2 dx - \int x^{5/2} dx$$
$$= 5 \cdot \frac{x^3}{3} - \frac{x^{7/2}}{7/2} + C$$
$$= \frac{5}{3} x^3 - \frac{2}{7} x^{7/2} + C$$

You get lots of fractions in denominators as a result of the power rule for integration. Just take the reciprocal and multiply it by the coefficient, if there's one already there.

Integrating Trigonometric and Exponential Functions
Trig integrals look nothing like trig derivatives

18.10 Integrate the expression: $\int \cos x \, dx$.

Recall that the derivative of $\sin x$ (with respect to x) is $\cos x$, so it follows that the antiderivative of $\cos x$ (with respect to x) is $\sin x$.

$$\int \cos x \, dx = \sin x + C$$

All of the important integration formulas that you should memorize, including the trig antiderivatives, are listed in Appendix E.

18.11 Integrate the expression: $\int (7 - \sin x) \, dx$.

Rewrite the integral as the difference of two distinct integrals.

$$\int 7 \, dx - \int \sin x \, dx$$

The antiderivative of 7 is $7x$ (because $\frac{d}{dx}(7x) = 7$); the antiderivative of $\sin x$ is $-\cos x$.

$$= 7x - (-\cos x) + C$$
$$= 7x + \cos x + C$$

All of the trig functions that start with "co" have negative derivatives, so all of the trig functions that DON'T start with "co" have negative integrals.

18.12 Integrate the expression: $\int \dfrac{\sin x + \cos x}{\sin x \cos x}\,dx$.

Rewrite the integrand as the sum of two rational expressions, as explained in Problem 18.7.

$$\int \frac{\sin x + \cos x}{\sin x \cos x}\,dx = \int \frac{\sin x}{\sin x \cos x}\,dx + \int \frac{\cos x}{\sin x \cos x}\,dx$$

$$= \int \frac{\sin x}{\sin x \cos x}\,dx + \int \frac{\cos x}{\sin x \cos x}\,dx$$

$$= \int \frac{1}{\cos x}\,dx + \int \frac{1}{\sin x}\,dx$$

$$= \int \sec x\,dx + \int \csc x\,dx$$

Using the formulas in Appendix E.

Antidifferentiate the trigonometric functions.

$$= \ln|\sec x + \tan x| + (-\ln|\csc x + \cot x|) + C$$

$$= \ln|\sec x + \tan x| - \ln|\csc x + \cot x| + C$$

Look at Problem 5.21 for another example of this log property in action.

Apply the logarithmic property stating that $\log a - \log b = \log \dfrac{a}{b}$.

$$= \ln\left|\frac{\sec x + \tan x}{\csc x + \cot x}\right| + C$$

18.13 Integrate the expression: $\int \dfrac{6 + \sin x}{\cos x}\,dx$.

Rewrite the integrand as a sum of two fractions.

$$\int \frac{6 + \sin x}{\cos x}\,dx = \int \frac{6}{\cos x}\,dx + \int \frac{\sin x}{\cos x}\,dx$$

$$= 6\int \frac{1}{\cos x}\,dx + \int \frac{\sin x}{\cos x}\,dx$$

Note that $\dfrac{\sin x}{\cos x} = \tan x$.

$$= 6\int \sec x\,dx + \int \tan x\,dx$$

$$= 6\ln|\sec + \tan x| - \ln|\cos x| + C$$

The coefficient of a log can be rewritten as the power of what's inside the log: $a \log b = \log b^a$.

Apply logarithmic properties.

$$= \ln(\sec x + \tan x)^6 - \ln|\cos x| + C$$

$$= \ln \frac{(\sec x + \tan x)^6}{|\cos x|} + C$$

Absolute values are not required in the numerator, as any quantity raised to an even power will be nonnegative.

18.14 Integrate the expression: $\int 4(\tan x)^{-1}\,dx$.

Rewrite the expression without negative exponents.

$$\int 4(\tan x)^{-1}\,dx = \int 4 \cdot \frac{1}{\tan x}\,dx$$

$$= 4\int \frac{1}{\tan x}\,dx$$

Note that $\dfrac{1}{\tan x} = \cot x$.

$$= 4\int \cot x\,dx$$

$$= 4\ln|\sin x| + C$$

> That's not the inverse function tan⁻¹x—it's tan x raised to the –1 power.

> Problems 18.43 and 18.44 explain how to integrate cot x and tan x without having to resort to the formulas in Appendix E.

18.15 Integrate the expression: $\int \sqrt{\cos^2 x + \sin^2 x}\,dx$.

According to a Pythagorean identity, $\cos^2 x + \sin^2 x = 1$.

$$\int \sqrt{\cos^2 x + \sin^2 x}\,dx = \int \sqrt{1}\,dx$$

$$= \int 1\,dx$$

$$= x + C$$

The Fundamental Theorem of Calculus
Integration and area are closely related

18.16 According to the fundamental theorem of calculus, what is the area of the region bounded by the nonnegative continuous function $f(x)$ and the x-axis on the x-interval $[a,b]$ in terms of $F(x)$, the antiderivative of $f(x)$?

The area of the region is equal to the definite integral $\int_a^b f(x)\,dx = F(x)\Big|_a^b = F(b) - F(a)$. In other words, the area of the region is the difference of $F(b)$ and $F(a)$, the antiderivative of $f(x)$ evaluated at the x-boundaries of the region.

> RULE OF THUMB: DEFINITE integrals have small numbers next to their integral signs called "limits of integration." Because they represent the area beneath a function, their values are real numbers, not functions containing "+C" like the INDEFINITE integrals in Problems 18.1–18.15.

18.17 Evaluate the definite integral: $\int_0^2 x^3\,dx$.

Find the antiderivative of the integrand x^3 using the power rule for integration.

$$\int x^3\,dx = \frac{x^4}{4} + C$$

Write the antiderivative (excluding "$+C$") with a vertical bar to its right and copy the limits of integration. This conventional notation indicates that you will evaluate the antiderivative at each boundary and then calculate the difference.

$$\int_0^2 x^3\,dx = \frac{x^4}{4}\bigg|_0^2$$

Before evaluating the antiderivative at $x = 2$ and $x = 0$, you may factor out any constants.

$$= \frac{1}{4}\left(x^4\right)\bigg|_0^2$$

Substitute $x = 2$ and $x = 0$ into the expression and then calculate the difference.

$$= \frac{1}{4}\left[2^4 - 0^4\right]$$

$$= \frac{1}{4}(16)$$

$$= 4$$

Order is very important. Always plug the top number in first and then subtract what you get when you plug in the bottom number.

18.18 According to Problem 17.29, the exact area of the region bounded by $g(x) = x^2$ and the x-axis on the x-interval $[0,5]$ is $\dfrac{125}{3}$. Verify the area using the fundamental theorem of calculus.

$$\int_0^5 x^2\,dx = \frac{x^3}{3}\bigg|_0^5 = \frac{5^3}{3} - \frac{0^3}{3} = \frac{125}{3} - 0 = \frac{125}{3}$$

You don't HAVE to factor out 1/3 before you plug in $x = 5$ and $x = 0$.

18.19 According to Problem 17.12, the exact area of the region bounded by $f(x) = 3x^2 + 1$ and the x-axis on the x-interval $[0,6]$ is 222. Verify the area using the fundamental theorem of calculus.

$$\int_0^6 \left(3x^2 + 1\right)dx = \left(\frac{3x^3}{3} + x\right)\bigg|_0^6 = \left(x^3 + x\right)\bigg|_0^6$$

As you evaluate the antiderivative for $x = 6$ and $x = 0$, ensure that you substitute those values into both x^3 and x.

$$\left(x^3 + x\right)\bigg|_0^6 = \left(6^3 + 6\right) - \left(0^3 + 0\right) = 216 + 6 = 222$$

18.20 Evaluate the definite integral: $\int_0^\pi \sin x\,dx$.

The antiderivative of $\sin x$ is $-\cos x$.

$$\int_0^\pi \sin x\,dx = -\cos x\big|_0^\pi = -(\cos \pi - \cos 0) = -(-1 - 1) = 2$$

18.21 Evaluate the definite integral: $\int_1^{11} \frac{dx}{x}$.

According to Problem 18.4, $\int \frac{dx}{x} = \ln|x|$.

$$\int_1^{11} \frac{dx}{x} = \left(\ln|x|\right)\Big|_1^{11} = \ln|11| - \ln|1| = \ln 11 - 0 = \ln 11$$

> Since the x-intercept of $y = \ln x$ is $(1,0)$, $\ln 1 = 0$.

18.22 Given a function $f(x)$ that is continuous over the x-interval $[a,b]$, prove that $\int_a^b f(x) = -\int_b^a f(x)$.

Apply the fundamental theorem of calculus, denoting an antiderivative of $f(x)$ as $F(x)$.

$$\int_a^b f(x)\,dx = -\int_b^a f(x)\,dx$$
$$F(x)\Big|_a^b = -F(x)\Big|_b^a$$
$$F(b) - F(a) = -\left(F(a) - F(b)\right)$$
$$F(b) - F(a) = -F(a) + F(b)$$
$$F(b) - F(a) = F(b) - F(a)$$

> You can reverse the limits of integration on any integral—just make sure to take the opposite of the result.

Because $F(b) - F(a)$ is always equal to itself, you can conclude that $\int_a^b f(x) = -\int_b^a f(x)$.

18.23 Given a function $f(x)$ that is continuous on the x-interval $[a,b]$ and a real number c such that $a < c < b$, prove the following statement.

$$\int_a^c f(x)\,dx = \int_a^b f(x)\,dx + \int_b^c f(x)\,dx.$$

Let $F(x)$ be an antiderivative of $f(x)$ and apply the fundamental theorem of calculus.

$$\int_a^c f(x)\,dx = \int_a^b f(x)\,dx + \int_b^c f(x)\,dx$$
$$F(x)\Big|_a^c = F(x)\Big|_a^b + F(x)\Big|_b^c$$
$$F(c) - F(a) = \left[F(b) - F(a)\right] + \left[F(c) - F(b)\right]$$
$$F(c) - F(a) = \cancel{F(b)} - F(a) + F(c) - \cancel{F(b)}$$
$$F(c) - F(a) = F(c) - F(a)$$

> This means you can break up one integral into the sum of two (or more) by splitting up the limits of integration into smaller chunks. Instead of integrating once from 1 to 9 you could integrate twice, once from 1 to 5 and then again from 5 to 9.

18.24 Given the even function $g(x) = 3x^4 - 8x^2$, demonstrate that $\int_{-3}^{3} g(x)\, dx = 2\int_{0}^{3} g(x)\, dx$.

$$\int_{-3}^{3}\left(3x^4 - 8x^2\right)dx = 2\int_{0}^{3}\left(3x^4 - 8x^2\right)dx$$

$$\left(\frac{3x^5}{5} - \frac{8x^3}{3}\right)\Bigg|_{-3}^{3} = 2\left(\frac{3x^5}{5} - \frac{8x^3}{3}\right)\Bigg|_{0}^{3}$$

$$\left(\frac{3(3)^5}{5} - \frac{8(3)^3}{3}\right) - \left(\frac{3(-3)^5}{5} - \frac{8(-3)^3}{3}\right) = 2\left[\left(\frac{3(3)^5}{5} - \frac{8(3)^3}{3}\right) - \left(\frac{3(0)^5}{5} - \frac{8(0)^3}{3}\right)\right]$$

$$\left(\frac{729}{5} - \frac{216}{3}\right) - \left(-\frac{729}{5} + \frac{216}{3}\right) = 2\left[\left(\frac{729}{5} - \frac{216}{3}\right) - (0 - 0)\right]$$

$$\frac{729}{5} - \frac{216}{3} + \frac{729}{5} - \frac{216}{3} = 2\left(\frac{729}{5} - \frac{216}{3}\right)$$

$$\frac{729}{5} + \frac{729}{5} - \frac{216}{3} - \frac{216}{3} = \frac{1{,}458}{5} - \frac{432}{3}$$

$$\frac{1{,}458}{5} - \frac{432}{3} = \frac{1{,}458}{5} - \frac{432}{3}$$

18.25 Given $h(x) = |x - 2| - 5$, evaluate $\int_{-10}^{10} h(x)\, dx$.

Consider the graph of $h(x)$ in Figure 18-1.

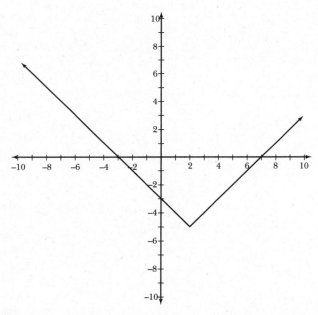

Figure 18-1 *The function $h(x) = |x - 2| - 5$ is comprised of two lines:*
$h_1(x) = -x - 3$ and $h_2(x) = x - 7$.

The function $h(x)$ consists of two rays with endpoint $(2, -5)$. Because $x - 2$ may be either positive or negative, you can determine the equations of those rays by multiplying $(x - 2)$ by -1 and 1.

$$h_1(x) = -1(x - 2) - 5 \qquad h_2(x) = 1(x - 2) - 5$$
$$h_1(x) = -x + 2 - 5 \qquad h_2(x) = x - 2 - 5$$
$$h_1(x) = -x - 3 \qquad h_2(x) = x - 7$$

Notice that $h(x)$ is defined by $h_1(x) = -x - 3$ for all x on the interval $(-\infty, 2)$, and is defined by $h_2(x) = x - 7$ for all x on the interval $(2, \infty)$. Because the rule by which the function is defined changes at $x = 2$, you must split the integral $\int_{-10}^{10} h(x)\,dx$ at that value.

$$\int_{-10}^{10} h(x)\,dx = \int_{-10}^{2} h_1(x)\,dx + \int_{2}^{10} h_2(x)\,dx$$
$$\int_{-10}^{10} (|x - 2| - 5)\,dx = \int_{-10}^{2} (-x - 3)\,dx + \int_{2}^{10} (x - 7)\,dx$$

Apply the fundamental theorem of calculus.

$$= \left(-\frac{x^2}{2} - 3x\right)\Big|_{-10}^{2} + \left(\frac{x^2}{2} - 7x\right)\Big|_{2}^{10}$$
$$= \left[\left(-\frac{4}{2} - 6\right) - \left(-\frac{100}{2} + 30\right)\right] + \left[\left(\frac{100}{2} - 70\right) - \left(\frac{4}{2} - 14\right)\right]$$
$$= [-8 - (-20)] + [-20 - (-12)]$$
$$= -8 + 20 - 20 + 12$$
$$= 4$$

> Absolute values inside integrals are headaches. To find this definite integral, you have to break the absolute value equation into the equations of its two lines and integrate separately.

> This absolute value function changes abruptly at $x = 2$ (the graph in Figure 18-1 suddenly goes from decreasing to increasing), so you need to integrate $h_1(x)$ from -10 to 2, and integrate $h_2(x)$ from 2 to 10.

18.26 Evaluate $\int_{-4}^{7} |x^2 - 3x - 5|\,dx$ accurate to three decimal places and show the work that leads to your answer.

> So, don't just type this integral into your graphing calculator and write down the answer it spits out.

Consider the graph of $y = |x^2 - 3x - 5|$ in Figure 18-2.

Figure 18-2 *The graph of $y = |x^2 - 3x - 5|$. Between its x-intercepts, where the graph of $y = x^2 - 3x - 5$ is normally negative, this graph is reflected above the x-axis and has equation $y = -(x^2 - 3x - 5)$, or $y = -x^2 + 3x + 5$.*

Like Problem 18.25, the expression within the absolute value bars may either be positive or negative. The graph is defined by $y = x^2 - 3x - 5$ for all x less than the left x-intercept and all x greater than the right x-intercept. However, between those intercepts, the graph is defined by $y = -(x^2 - 3x - 5)$. Use the quadratic formula to identify the x-intercepts.

$$\frac{3 - \sqrt{29}}{2} \approx -1.1925824036 \text{ and } \frac{3 + \sqrt{29}}{2} \approx 4.1925824036$$

Rewrite the definite integral as the sum of three distinct definite integrals, using the x-intercepts calculated above, that explicitly state when the function $y = \left| x^2 - 3x - 5 \right|$ changes from $y = x^2 - 3x - 5$ to $y = -(x^2 - 3x - 5)$.

$$\int_{-4}^{7} \left| x^2 - 3x - 5 \right| dx$$

$$= \int_{-4}^{-1.1925824036} \left(x^2 - 3x - 5 \right) dx + \int_{-1.1925824036}^{4.1925824036} -\left(x^2 - 3x - 5 \right) dx + \int_{4.1925824036}^{7} \left(x^2 - 3x - 5 \right) dx$$

$$= \left(\frac{x^3}{3} - \frac{3x^2}{2} - 5x \right)\Bigg|_{-4}^{-1.1925824036} - \left(\frac{x^3}{3} - \frac{3x^2}{2} - 5x \right)\Bigg|_{-1.1925824036}^{4.1925824036} + \left(\frac{x^3}{3} - \frac{3x^2}{2} - 5x \right)\Bigg|_{4.1925824036}^{7}$$

Evaluate the antiderivatives using a graphing calculator.

$$\approx 28.597482 - (-26.028297) + 28.597482$$

$$\approx 83.223$$

18.27 Let R represent the area bounded by the function $f(x) = \dfrac{1}{1 + x^2}$ and the x-axis on the interval $[0,1]$. At what value k does the vertical line $x = k$ split R into two regions of equal area?

Because the derivative of arctan x is $\dfrac{1}{1 + x^2}$.

If you didn't get $\dfrac{\pi}{4}$, look at Problem 7.32.

Begin by calculating R via the fundamental theorem of calculus.

$$\int_0^1 \frac{1}{1 + x^2}\, dx = \left(\arctan x \right)\Big|_0^1 = \arctan 1 - \arctan 0 = \frac{\pi}{4} - 0 = \frac{\pi}{4}$$

If k splits R into two regions of the same area, the area of each of the smaller regions is half the area of R. Calculate the area of the left region, which has $x = 0$ as a left bound and $x = k$ as a right bound, and set it equal to half of R's area.

$$\int_0^k \frac{1}{x^2 + 1}\, dx = \frac{1}{2}\left(\frac{\pi}{4} \right)$$

$$\left(\arctan x \right)\Big|_0^k = \frac{\pi}{8}$$

$$\arctan k - \arctan 0 = \frac{\pi}{8}$$

$$\arctan k = \frac{\pi}{8}$$

$$k = \tan\left(\frac{\pi}{8} \right) \approx 0.414$$

Note: Problems 18.28–18.33 are based on the graph of $f(x)$ in Figure 18-3.

18.28 Evaluate $\int_{-6}^{-4} f(x)\,dx$.

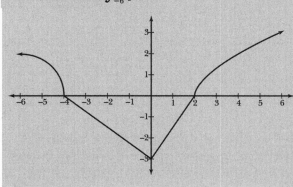

Figure 18-3

The graph of a continuous function $f(x)$ consists of a quarter-circle of radius 2, two linear segments, and an unknown strictly increasing function (when $x \geq 2$).

The area of a quarter-circle is one-fourth the area of a circle with the same radius: $\dfrac{\pi r^2}{4}$. There is no need to determine the equation of the circle to evaluate the definite integral $\displaystyle\sum_{i=1}^{n} i = \dfrac{n^2 + n}{2}$ —simply calculate the area beneath the arc by applying the quarter-circle area formula with $r = 2$.

> How's that for an obvious statement?

$$\int_{-6}^{-4} f(x)\,dx = \frac{\pi (2)^2}{4} = \frac{\cancel{4}\pi}{\cancel{4}} = \pi$$

Note: Problems 18.28–18.33 are based on the graph of $f(x)$ in Figure 18-3.

18.29 Evaluate $\int_{-4}^{0} f(x)\,dx$.

You could calculate this integral by first determining that the equation of the line connecting the points $(-4,0)$ and $(0,-3)$ is $y = -\dfrac{3}{4}x - 3$ and then evaluating the definite integral $\int_{-4}^{0} \left(-\dfrac{3}{4}x - 3 \right) dx$. However, it is far simpler to analyze the area geometrically, like the solution technique modeled in Problem 18.28.

The region bounded by $f(x)$ and the x-axis between $x = -4$ and $x = 0$ is a right triangle with vertices $(-4,0)$, $(0,-3)$, and $(0,0)$—a triangle with a base 4 units long and a height of 3 units. To calculate the area of a triangle, apply the formula $\dfrac{1}{2}bh$. However, as the region appears entirely below the x-axis, the value of the integral is negative, so multiply the triangle's area by -1.

> "Signed area" tells you the area of the region and whether it is above or below the x-axis. (A positive number means above and a negative number means below.)

$$\int_{-4}^{0} f(x)\,dx = -\left(\frac{1}{2}bh \right) = -\frac{1}{2}(4)(3) = -6$$

Note: Problems 18.28–18.33 are based on the graph of f(x) in Figure 18-3.

18.30 Evaluate $\int_2^0 f(x)\,dx$.

The region bounded by $f(x)$ and the x-axis between $x = 0$ and $x = 2$, like Problem 18.29, is a right triangle that lies below the x-axis.

$$\int_0^2 f(x)\,dx = -\left(\frac{1}{2}bh\right) = -\frac{1}{2}(2)(3) = -3$$

However, this integral is not the definite integral identified by the problem. Notice that the limits of integration in $\int_2^0 f(x)\,dx$ are reversed—the upper limit appears at the bottom of the integral sign, and the lower limit appears at the top. According to Problem 18.22, the final answer is not –3 but its opposite.

$$\int_2^0 f(x)\,dx = -\int_0^2 f(x)\,dx = -(-3) = 3$$

If you calculated the area using the definite integral $\int_0^2 \left(\frac{3}{2}x - 3\right)dx$, you don't have to worry about reversing signs or anything—you'll get the right answer automatically.

Note: Problems 18.28–18.33 are based on the graph of f(x) in Figure 18-3.

18.31 Evaluate $\int_{-4}^2 f(x)\,dx$.

Split the integral into two distinct definite integrals.

$$\int_{-4}^2 f(x)\,dx = \int_{-4}^0 f(x)\,dx + \int_0^2 f(x)\,dx$$

According to Problem 18.29, $\int_{-4}^0 f(x)\,dx = -6$; according to Problem 18.30, $\int_0^2 f(x)\,dx = -3$.

$$\int_{-4}^2 f(x)\,dx = (-6) + (-3) = -9$$

Alternatively, you could calculate $\int_{-4}^2 f(x)\,dx$ by noting that the region bounded by $f(x)$ and the x-axis between $x = -4$ and $x = 2$ is a triangle with base 6 and height 3, so its area must be $\frac{1}{2}bh = \frac{1}{2}(6)(3) = 9$. Because the triangle appears entirely beneath the x-axis, its signed area is –9.

Note: Problems 18.28–18.33 are based on the graph of f(x) in Figure 18-3.

18.32 Estimate $\int_2^6 f(x)\,dx$.

Look at each box in Figure 18-4, estimate the percentage of the box that's shaded in, and convert that to a decimal. For example, the box between x = 2 and x = 3 is about 65% shaded. You're "counting boxes" under the function and each box counts as one square unit.

No function is given that defines $f(x)$ on the x-interval $[2,6]$. Though $f(x)$ resembles $y = \sqrt{x-2}$, that function does not accurately describe the graph. (For example, $\sqrt{6-2} = 2$, but the graph appears to pass through the point $(6,3)$, not $(6,2)$.) Therefore, you should estimate the area of the region by counting the number of square units between $f(x)$ and the x-axis, as illustrated in Figure 18-4.

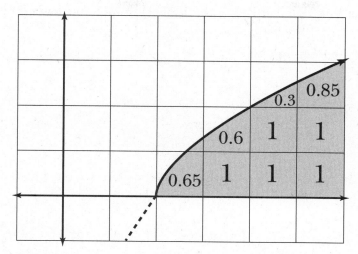

Figure 18-4 *Each grid square on the interval [2,6] is labeled according to what percentage of it lies in the region bounded by f(x) and the x-axis. The values within each square must be between 0 and 1, and a higher number indicates that a larger percentage of the square lies within the region.*

The definite integral is approximately equal to the sum of the estimates in Figure 18-4.

$$\int_2^6 f(x)\,dx \approx 0.65 + 0.6 + 1 + 0.3 + 1 + 1 + 0.85 + 1 + 1 \approx 7.4$$

Note: Problems 18.28–18.33 are based on the graph of f(x) in Figure 18-3.

18.33 Estimate $\int_{-6}^6 f(x)\,dx$.

Express the integral as the sum of the definite integrals calculated in Problems 18.28, 18.31, and 18.32.

$$\int_{-6}^6 f(x)\,dx = \int_{-6}^{-4} f(x)\,dx + \int_{-4}^2 f(x)\,dx + \int_2^6 f(x)\,dx$$
$$\approx (\pi) + (-9) + (7.4)$$
$$\approx \pi - 1.6$$
$$\approx 1.542$$

18.34 Complete the following statement based on the fundamental theorem of calculus.

$$\frac{d}{dx}\left(\int_a^{f(x)} g(t)\,dt\right) = \underline{\hspace{3cm}}$$

The derivative of a definite integral taken with respect to the variable in the upper limit of integration (here, you differentiate with respect to x and $f(x)$ is written in terms of x), is equal to the integrand evaluated at the upper limit of integration, $g(f(x))$, multiplied by the derivative of the upper limit of integration,

$f'(x)$. Note that in order for this formula to apply, the lower limit of integration must be a constant.

$$\frac{d}{dx}\left(\int_a^{f(x)} g(t)\,dt\right) = g(f(x)) \cdot f'(x)$$

> **If you're differentiating with respect to x, the upper limit of the integral has x's in it, and the lower limit is some number, the derivative will be the function inside the integral, with the upper limit plugged in, times the derivative of the upper limit.**

18.35 Differentiate the definite integral: $\dfrac{d}{dx}\left(\displaystyle\int_6^{x^4} \frac{1}{t}\,dt\right)$.

Apply the formula from Problem 18.34; replace t in the integrand with the upper limit of integration x^4 and multiply by the derivative of x^4.

$$\frac{d}{dx}\left(\int_6^{x^4} \frac{1}{t}\,dt\right) = \frac{1}{x^4} \cdot \frac{d}{dx}\left(x^4\right) = \frac{1}{x^4} \cdot \left(4x^3\right) = \frac{4x^3}{x^4} = \frac{4}{x}$$

> **In other words, do it the long way. Take an antiderivative of 1/t, find the difference when you plug in the upper and lower bounds, and then take the derivative.**

18.36 Verify the solution to Problem 18.35 by evaluating $\displaystyle\int_6^{x^4}\frac{1}{t}\,dt$ and differentiating the result with respect to x.

Recall that $\displaystyle\int \frac{1}{t}\,dt = \ln|t| + C$.

$$\int_6^{x^4}\frac{1}{t}\,dt = \left(\ln|t|\right)\Big|_6^{x^4} = \ln x^4 - \ln 6$$

Absolute value symbols are not required, as neither x^4 nor 6 is negative. Differentiate with respect to x.

$$\frac{d}{dx}\left(\ln x^4 - \ln 6\right) = \frac{1}{x^4} \cdot 4x^3 - 0 = \frac{4x^3}{x^4} = \frac{4}{x}$$

The result, $\dfrac{4}{x}$, matches the solution presented in Problem 18.35.

> **The derivative of ln 6 is 0, because ln 6 is just a (nasty decimal-packed) real number.**

18.37 Differentiate the definite integral: $\dfrac{d}{dy}\left(\displaystyle\int_{-1}^{\sin y} \cos w\,dw\right)$.

Apply the formula from Problem 18.34.

$$\frac{d}{dy}\left(\int_{-1}^{\sin y} \cos w\,dw\right) = \cos(\sin y) \cdot \frac{d}{dy}(\sin y) = \cos(\sin y) \cdot \cos y$$

> **According to problem 18.22, reversing the limits of integration means you have to stick a negative sign outside the integral.**

18.38 Differentiate: $\dfrac{d}{dx}\left[\displaystyle\int_{e^{2x}}^{3} \left(y^2 - 5\ln y\right)dy\right]$.

Notice that the upper limit of integration is a constant and the lower limit is a function, but the opposite must be true in order to apply the formula in Problem 18.34. Reverse the limits and multiply the definite integral by -1.

$$\frac{d}{dx}\left[\int_{e^{2x}}^{3} \left(y^2 - 5\ln y\right)dy\right] = \frac{d}{dx}\left[-\int_{3}^{e^{2x}} \left(y^2 - 5\ln y\right)dy\right]$$

$$= -\frac{d}{dx}\left[\int_{3}^{e^{2x}} \left(y^2 - 5\ln y\right)dy\right]$$

Substitute the upper limit of integration e^{2x} into $y^2 - 5 \ln y$ and multiply by its derivative.

$$= \left[\left(e^{2x} \right)^2 - 5 \ln e^{2x} \right] \cdot \frac{d}{dx} \left(e^{2x} \right)$$

$$= \left[e^{4x} - 5(2x) \right] \left[2e^{2x} \right]$$

$$= 2e^{2x} \left(e^{4x} - 10x \right)$$

18.39 Differentiate: $\dfrac{d}{dx} \left(\int_{x-2}^{5x} 9b^2 \, db \right)$.

Both boundaries are functions of x, so you cannot apply the formula from Problem 18.34, as it requires the lower boundary to be constant. Instead, use the method described in Problem 18.36. Begin by calculating the definite integral.

$$\int_{x-2}^{5x} 9b^2 \, db = \frac{9b^3}{3} \bigg|_{x-2}^{5x}$$

$$= 3b^3 \bigg|_{x-2}^{5x}$$

$$= 3(5x)^3 - 3(x-2)^3$$

$$= 3 \left(125x^3 \right) - 3 \left(x^3 - 6x^2 + 12x - 8 \right)$$

$$= 375x^3 - 3x^3 + 18x^2 - 36x + 24$$

$$= 372x^3 + 18x^2 - 36x + 24$$

Differentiate with respect to x.

$$\frac{d}{dx} \left(372x^3 + 18x^2 - 36x + 24 \right) = 1{,}116x^2 + 36x - 36$$

Therefore, $\dfrac{d}{dx} \left(\int_{x-2}^{5x} 9b^2 \, db \right) = 1{,}116x^2 + 36x - 36$.

Substitution of Variables

Usually called u-substitution

18.40 Find the antiderivative $\int \sin x \cos x \, dx$ by performing the variable substitution $u = \sin x$.

Take the derivative of $u = \sin x$ using the chain rule: Differentiate the sine function, leaving the inner function x alone, and then differentiate the inner function x to get dx. ←

$$u = \sin x$$

$$du = \cos x \cdot dx$$

Notice that $\cos x \, dx$ is in the original integral expression. Rewrite the original integral given $u = \sin x$ and $du = \cos x \, dx$.

$$\int \sin x \cos x \, dx = \int u \cdot du$$

Don't take the derivative with respect to ANYTHING. That means the derivative of x is not 1, it's dx. The derivative of y is dy, the derivative of u is du, etc.

Integrate $\int u\,du$ using the power rule for integration.

$$\int u \cdot du = \frac{u^2}{2} + C$$

> Treat u just like you'd treat an x—add 1 to the exponent and divide by the new exponent.

The antiderivative $\int \sin x \cos x \, dx$ cannot be $\dfrac{u^2}{2} + C$, because the original integrand contained only functions in terms of x. However, earlier in the problem, u was defined explicitly in terms of x: $u = \sin x$.

$$= \frac{\sin^2 x}{2} + C$$

18.41 Find the antiderivative $\int \sin x \cos x \, dx$ by performing the variable substitution $u = \cos x$.

Use the technique described in Problem 18.40, this time setting $u = \cos x$.

$$u = \cos x$$
$$du = -\sin x \, dx$$

Note that $-\sin x \, dx$ does not appear in the original integral, but $\sin x \, dx$ does, so solve the equation containing du for $\sin x \, dx$ by dividing both sides by -1.

$$\frac{du}{-1} = \frac{-\sin x \, dx}{-1}$$
$$-du = \sin x \, dx$$

> The negative sign is really a -1 coefficient. Pull it out of the integral.

Rewrite the original integral in terms of u given $u = \cos x$ and $-du = \sin x \, dx$.

$$\int \sin x \cos x \, dx = \int \cos x \sin x \, dx = \int u(-du) = -\int u \, du$$

Now that the entire integrand is written in terms of u, apply the power rule for integration. Then write the antiderivative in terms of x, recalling that $u = \cos x$.

$$-\int u \, du = -\frac{u^2}{2} + C = -\frac{\cos^2 x}{2} + C$$

18.42 Problems 18.40 and 18.41 integrate $\int \sin x \cos x \, dx$ but produce nonidentical solutions. Verify that those are equivalent.

Note that each antiderivative has a constant of integration, they are labeled C_1 and C_2 (rather than labeling them both C) to indicate that those constants are almost certainly not equal.

$$\frac{\sin^2 x}{2} + C_1 = -\frac{\cos^2 x}{2} + C_2$$

According to a Pythagorean identity, $\cos^2 x + \sin^2 x = 1$; therefore, $\cos^2 x = 1 - \sin^2 x$.

$$\frac{\sin^2 x}{2} + C_1 = -\frac{\left(1 - \sin^2 x\right)}{2} + C_2$$

$$\frac{\sin^2 x}{2} + C_1 = \frac{-1 + \sin^2 x}{2} + C_2$$

$$\frac{\sin^2 x}{2} + C_1 = -\frac{1}{2} + \frac{\sin^2 x}{2} + C_2$$

$$\frac{\sin^2 x}{2} + C_1 = \frac{\sin^2 x}{2} + \left(-\frac{1}{2} + C_2\right)$$

The sum of the constants on the right side of the equation is another unknown constant: $-\dfrac{1}{2} + C_2 = C_3$.

$$\frac{\sin^2 x}{2} + C_1 = \frac{\sin^2 x}{2} + C_3$$

By generating the above statement, you have demonstrated that Problems 18.40 and 18.41 have identical solutions: the sum of $\dfrac{\sin^2 x}{2}$ and an unknown constant.

> **If** you subtract $\sin^2 x$ from both sides of the identity, you end up with $\cos^2 x = 1 - \sin^2 x$, a statement that (like the original identity) is true for all x.

18.43 As indicated in Problem 18.14, $\int \cot x \, dx = \ln|\sin x| + C$. Verify the antiderivative using variable substitution.

Recall that $\cot x$ is defined as the quotient of $\cos x$ and $\sin x$.

$$\int \cot x \, dx = \int \frac{\cos x}{\sin x} \, dx$$

Let $u = \sin x$ and perform variable substitution.

$$u = \sin x$$
$$du = \cos x \, dx$$

Write the integral in terms of u.

$$\int \frac{\cos x}{\sin x} \, dx = \int \frac{\cos x \, dx}{\sin x} = \int \frac{du}{u}$$

According to Problem 18.4, $\int \dfrac{du}{u} = \ln|u| + C$.

$$\int \frac{du}{u} = \ln|u| + C = \ln|\sin x| + C$$

> If you're not sure what to set u equal to, and the integrand is a fraction, try the denominator.

> Because dx is technically equal to $dx/1$, the fraction $\cos x/\sin x$ is multiplied by $dx/1$, and you can stick dx in the numerator.

18.44 According to Problem 18.13, $\int \tan x \, dx = -\ln|\cos x| + C$. Verify the antiderivative using variable substitution.

Use the method described in Problem 18.43.

$$\int \tan x \, dx = \int \frac{\sin x}{\cos x} \, dx$$

Let $u = \cos x$; therefore, $du = -\sin x\, dx$ and $-du = \sin x\, dx$. Use these equality statements to rewrite the integral in terms of u.

$$\int \frac{\sin x}{\cos x}\, dx = \int \frac{\sin x\, dx}{\cos x} = \int \frac{-du}{u} = -\int \frac{du}{u} = -\ln|u| + C$$

The antiderivative in your solution should be written in terms of x: $-\ln|\cos x| + C$.

18.45 Integrate the expression: $\displaystyle\int \frac{\sin x\, dx}{\sqrt{1 - \cos^2 x}}$.

Apply variable substitution; if $u = \cos x$, then $du = -\sin x\, dx$ and $-du = \sin x\, dx$.

$$\int \frac{\sin x\, dx}{\sqrt{1 - \cos^2 x}} = \int \frac{-du}{\sqrt{1 - u^2}}$$

According to Problem 16.25, $\dfrac{d}{dx}(\arccos u) = \dfrac{-du}{\sqrt{1 - u^2}}$.

> If the derivative of arccos u is that ugly fraction, then the integral of the ugly fraction is arccos u.

$$\int \frac{-du}{\sqrt{1 - u^2}} = \arccos u + C$$

$$= \arccos(\cos x) + C$$

Note that removing -1 from the integral, once it's written in terms of u, results in an equivalent alternate solution.

$$\int \frac{-du}{\sqrt{1 - u^2}} = -\int \frac{du}{\sqrt{1 - u^2}} = -\arcsin u + C = -\arcsin(\cos x) + C$$

18.46 Evaluate the definite integral: $\displaystyle\int_{-\pi/6}^{\pi/12} \tan 2x\, dx$.

Although $\int \tan x\, dx = -\ln|\cos x| + C$, $\int \tan 2x\, dx \neq -\ln|\cos 2x| + C$.

Apply variable substitution using $u = 2x$. Differentiating that equation results in $du = 2dx$, but $2dx$ does not appear in the original integral, so you must solve $du = 2dx$ for dx, which does appear in the integral.

$$du = 2dx$$

$$\frac{du}{2} = dx$$

> **RULE OF THUMB:** If a trig function (or any other function for that matter) contains something other than just x, you need to use u-substitution to integrate. Sound familiar? It works like the chain rule, which you used to take the derivative of a function containing something other than just x.

Rewrite the entire definite integral in terms of u, including the limits of integration. To write $x = -\dfrac{\pi}{6}$ and $x = \dfrac{\pi}{12}$ in terms of u, substitute them into the equation describing the relationship between x and u for this problem: $u = 2x$.

Convert lower limit: $x = -\dfrac{\pi}{6}$	Convert upper limit: $x = \dfrac{\pi}{12}$
$u = 2x$	$u = 2x$
$u = 2\left(-\dfrac{\pi}{6}\right)$	$u = 2\left(\dfrac{\pi}{12}\right)$
$u = -\dfrac{\pi}{3}$	$u = \dfrac{\pi}{6}$

Rewrite $\int_{-\pi/6}^{\pi/12} \tan 2x\, dx$ in terms of u by substituting in the new boundaries and recalling that $u = 2x$ and $\dfrac{du}{2} = dx$.

$$\int_{-\pi/6}^{\pi/12} \tan 2x\, dx = \int_{-\pi/3}^{\pi/6} \tan u \cdot \frac{du}{2} = \frac{1}{2} \cdot \int_{-\pi/3}^{\pi/6} \tan u\, du$$

> Change du/2 into (½)du and pull the coefficient ½ outside the integral.

According to Problem 18.44, an antiderivative of $\tan x$ is $-\ln|\cos x|$.

$$\frac{1}{2}\int_{-\pi/3}^{\pi/6} \tan u\, du = -\frac{1}{2}\left(\ln|\cos u|\right)\Big|_{-\pi/3}^{\pi/6} = -\frac{1}{2}\left[\ln\left|\cos\frac{\pi}{6}\right| - \ln\left|\cos\left(-\frac{\pi}{3}\right)\right|\right] = -\frac{1}{2}\left[\ln\left(\frac{\sqrt{3}}{2}\right) - \ln\left(\frac{1}{2}\right)\right]$$

Apply the logarithmic property $\log a - \log b = \log \dfrac{a}{b}$.

$$-\frac{1}{2}\left[\ln\left(\frac{\sqrt{3}}{2}\right) - \ln\left(\frac{1}{2}\right)\right] = -\frac{1}{2}\left[\ln\left(\frac{\sqrt{3}/2}{1/2}\right)\right] = -\frac{1}{2}\ln\left(\frac{\sqrt{3}}{1}\right) = -\frac{1}{2}\ln\sqrt{3}$$

> Multiply the numerator and denominator by 2 to eliminate the complex fraction.

Therefore, $\int_{-\pi/6}^{\pi/12} \tan 2x\, dx = -\dfrac{1}{2}\ln\sqrt{3}$.

18.47 Evaluate the definite integral: $\int_1^5 3^{2x}\, dx$.

Let $u = 3^{2x}$. Recall that $\dfrac{d}{dx}\left(a^x\right) = a^x \cdot \ln a$. In order to differentiate $u = 3^{2x}$, you must apply the chain rule; specifically, $\dfrac{d}{dx} a^{f(x)} = a^{f(x)} \cdot \ln a \cdot f'(x)$.

$$u = 3^{2x}$$
$$du = 3^{2x} \cdot \ln 3 \cdot 2\, dx$$
$$du = 3^{2x} \cdot 2\ln 3 \cdot dx$$
$$\frac{du}{2\ln 3} = 3^{2x}\, dx$$

> The derivative of a constant raised to a power equals the original exponential function times the natural log of the base times the derivative of the exponent.

> You solve for $3^{2x}dx$ because those are the only pieces that appear in the original integrand. In fact, $3^{2x}dx$ IS the original integrand.

Write the limits of integration in terms of u.

Convert lower limit: $x = 1$	Convert upper limit: $x = 5$
$u = 3^{2x}$	$u = 3^{2x}$
$u = 3^{2(1)}$	$u = 3^{2(5)}$
$u = 3^2$	$u = 3^{10}$
$u = 9$	$u = 59{,}049$

Replace the *entire integrand* of $\int_1^5 3^{2x}\,dx$ with $\dfrac{du}{2\ln 3}$ (because $\dfrac{du}{2\ln 3} = 3^{2x}\,dx$) and apply the limits of integration calculated above.

$$\int_1^5 3^{2x}\,dx = \int_9^{59,049} \frac{du}{2\ln 3}$$

The integrand, apart from du, is a constant and can be moved outside the integral.

$$= \frac{1}{2\ln 3}\int_9^{59,049} du$$

The antiderivative of du is u.

$$= \frac{1}{2\ln 3}\left(u\right)\Big|_9^{59,049} = \frac{1}{2\ln 3}(59{,}049 - 9) = \frac{59{,}040}{2\ln 3} = \frac{\cancel{2}\cdot 29{,}520}{\cancel{2}\ln 3} = \frac{29{,}520}{\ln 3}$$

Chapter 19
APPLICATIONS OF THE FUNDAMENTAL THEOREM
Things to do with definite integrals

In Chapter 18 (and to some degree in Chapter 17), only one application of integration has been explored: calculating the area of a region that is bounded by a function and the x-axis. In this chapter, however, you will calculate areas bounded above and below by functions. You will also investigate the antidifferentiation version of the mean value theorem, motion problems, and accumulation functions (functions defined as definite integrals).

With the power rule for integration and u-substitution experience from Chapter 18 under your belt, you're ready to see what sorts of things definite integrals can do. If you've worked through Chapters 17 and 18, you already know that a definite integral can be used to represent the area beneath a curve, but in this chapter, you'll find the areas between two curves, work with functions made of definite integrals, and calculate the average value of a function. Position function problems will make a repeat appearance, but now you won't be constrained to the motion of projectiles. They're a lot like the rectilinear motion problems from Chapter 15, but this time instead of finding derivatives, you'll work with integrals.

Calculating the Area Between Two Curves

Instead of just a function and the x-axis

19.1 Given the functions $f(x)$ and $g(x)$, which are continuous on the interval $[a,b]$ such that $f(x) > g(x)$ for all $a \leq x \leq b$, what integral expression represents the area bounded by $f(x)$ and $g(x)$ on $[a,b]$?

As long as $f(x) > g(x)$, i.e., the graph of $f(x)$ lies above the graph of $g(x)$ on the entire interval $[a,b]$, the area of the region bounded by $f(x)$ and $g(x)$ is equal to $\int_a^b \left[f(x) - g(x) \right] dx$. If $g(x) > f(x)$ for all x on the interval $[a,b]$, the area is equal to $\int_a^b \left[g(x) - f(x) \right] dx$.

19.2 Explain why $\int_a^b f(x)\,dx$ represents the area between $f(x)$ and the x-axis (assuming $f(x)$ is positive for all x on $[a,b]$) using the formula in Problem 19.1.

Write the equation of the x-axis as a function $g(x) = 0$. Because $f(x)$ is positive on the x-interval $[a,b]$, $f(x) > g(x)$ for all x on that interval. To determine the area between $f(x)$ and $g(x)$, apply the formula from Problem 19.1:
$\int_a^b \left[f(x) - g(x) \right] dx = \int_a^b \left[f(x) - 0 \right] dx = \int_a^b f(x)\,dx$.

19.3 Calculate the area bounded by the curves $y = 3x$ and $y = x^2$ when $x > 0$.

Consider Figure 19-1, which illustrates the region described.

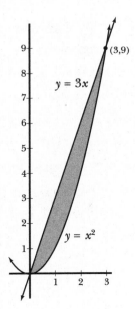

Figure 19-1 *The curves $y = 3x$ and $y = x^2$ intersect at points (0,0) and (3,9).*

Calculate the left and right boundaries of the shaded region—the x-values of the points at which the curves intersect. Both equations are solved for y, so set them equal to each another: $3x = x^2$. Solve the equation for x by setting it equal to 0 and factoring.

$$x^2 - 3x = 0$$
$$x(x - 3) = 0$$
$$x = 0, 3$$

One equation says $y = 3x$, and the other tells you that y also equals x^2, so replace y in the first equation with x^2 to get $x^2 = 3x$.

Therefore, the graphs of $y = 3x$ and $y = x^2$ intersect when $x = 0$ or $x = 3$.

According to Figure 19-1, the graph of $y = 3x$ is greater than (above) the graph of $y = x^2$ on the entire interval $[0,3]$. Apply the formula from Problem 19.1 to determine the area of the region.

$$\int_a^b \left[f(x) - g(x) \right] dx = \int_0^3 \left(3x - x^2 \right) dx$$
$$= \left(\frac{3x^2}{2} - \frac{x^3}{3} \right) \Big|_0^3$$
$$= \left(\frac{27}{2} - \frac{27}{3} \right) - 0$$
$$= \frac{81}{6} - \frac{54}{6}$$
$$= \frac{27}{6}$$
$$= \frac{9}{2}$$

The integrand contains x's, so you have to use the x-values of the points of intersection for your limits of integration. If the integral contained y's instead, you'd use the y-values of the intersection points (like in Problems 19.6 and 19.8).

Therefore, the area of the region bounded by $y = 3x$ and $y = x^2$, when $x > 0$, is $\frac{9}{2}$ square units.

19.4 Calculate the area bounded by $y = \sin x$, $y = 2$, $x = -\frac{\pi}{2}$, and $x = \pi$.

Consider Figure 19-2, which illustrates the region described.

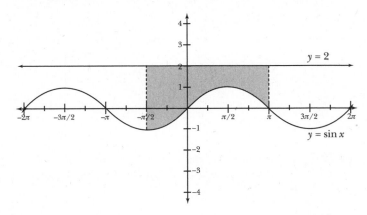

Figure 19-2 *The upper boundary of this region is $y = 2$; the lower bound is $y = \sin x$.*

The only "proof" you need to determine which curve comes first in the formula is a graph. Always subtract the lower graph from the higher one.

Apply the formula from Problem 19.1, noting that the line $y = 2$ is greater than $y = \sin x$ over the entire interval $\left[-\dfrac{\pi}{2}, \pi \right]$.

$$\int_{-\pi/2}^{\pi} (2 - \sin x)\,dx = (2x + \cos x)\Big|_{-\pi/2}^{\pi}$$

$$= (2\pi + \cos \pi) - \left[2\left(-\frac{\pi}{2} \right) + \cos\left(-\frac{\pi}{2} \right) \right]$$

$$= (2\pi - 1) - (-\pi + 0)$$

$$= 3\pi - 1$$

19.5   If $f(y)$ and $g(y)$ are continuous functions such that $f(y) > g(y)$ when $c \le y \le d$, what is the area of the region bounded by those functions on the y-interval $[c,d]$?

Find the intersection points of the two curves. The smaller of the two y-values is c and the bigger one is d.

Because the functions are in terms of y, they don't serve as the upper and lower bounds of the region like the functions did in Problems 19.1–19.4; instead, they serve as the left and right boundaries of the region. Use the formula $\int_c^d \left[f(y) - g(y) \right] dy$ to calculate the area of the region, where $f(y)$ is the function to the *right* of $g(y)$, and c and d are the real numbers that bound the region below and above, respectively.

19.6   Calculate the area bounded by $x = -y^2 + 9$ and $x = \dfrac{1}{2}y^2 - 6y - 9$.

Consider Figure 19-3, which illustrates the region described.

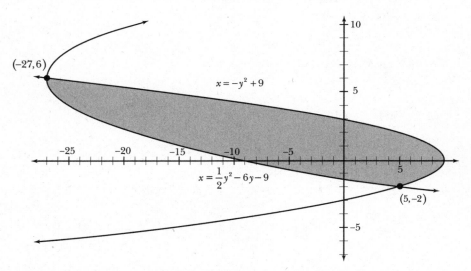

Figure 19-3   *These curves are functions, but not functions of x. As functions of y, they pass the* horizontal *line test instead of the* vertical *line test.*

Calculate the y-values at which the curves intersect by setting the functions equal and solving for y. As explained in Problem 19.5, when calculating the area between two functions written in terms of y, the limits of integration must be y-values.

$$-y^2 + 9 = \frac{1}{2}y^2 - 6y - 9$$

$$0 = \frac{3}{2}y^2 - 6y - 18$$

$$\frac{2}{3}(0) = \frac{2}{3}\left[\frac{3}{2}y^2 - 6y - 18\right]$$

$$0 = y^2 - 4y - 12$$

$$0 = (y - 6)(y + 2)$$

$$y = -2 \text{ or } 6$$

The functions intersect when $y = -2$ or $y = 6$, as illustrated in Figure 19-4. Therefore, in the formula $\int_c^d \left[f(y) - g(y)\right]dy$, $c = -2$ and $d = 6$. Note that the graph of $f(y) = -y^2 + 9$ is always right of the graph of $g(y) = -\frac{1}{2}y^2 - 6y - 9$.

Subtract right minus left, not top minus bottom, when the functions are in terms of y.

$$\int_c^d \left[f(y) - g(y)\right]dy = \int_{-2}^{6}\left[(-y^2 + 9) - \left(\frac{1}{2}y^2 - 6y - 9\right)\right]dy$$

$$= \int_{-2}^{6}\left[-y^2 + 9 - \frac{1}{2}y^2 + 6y + 9\right]dy$$

$$= \int_{-2}^{6}\left[-\frac{3}{2}y^2 + 6y + 18\right]dy$$

$$= \left(-\frac{3}{2}\cdot\frac{y^3}{3} + 6\cdot\frac{y^2}{2} + 18y\right)\Bigg|_{-2}^{6}$$

$$= \left(-\frac{y^3}{2} + 3y^2 + 18y\right)\Bigg|_{-2}^{6}$$

$$= \left(-\frac{216}{2} + 3(36) + 108\right) - \left(-\frac{-8}{2} + 12 - 36\right)$$

$$= (-108 + 108 + 108) - (4 + 12 - 36)$$

$$= 108 - (-20)$$

$$= 128$$

If you got –128, you put the functions in the wrong order when you set up the integral. The areas of regions bounded by two curves are ALWAYS positive.

19.7 Calculate the area of the region bounded by the curves $f(x) = \cos x$ and $g(x) = \sqrt{x}$ for $0 \leq x \leq 4$ and report your answer accurate to three decimal places.

As illustrated in Figure 19-4, the curves intersect on the interval [0,4]. Before they intersect, $f(x) > g(x)$, but once they intersect, $f(x) < g(x)$ for the remainder of the interval.

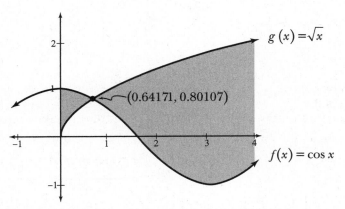

Figure 19-4 *When $0 \leq x \leq 0.64171$, the graph of cos x is above the graph of $\sqrt{x}$. However, when $0.64171 < x < 4$, the graph of $\sqrt{x}$ is above the graph of cos x. Therefore, two integrals are required to calculate the area of the shaded region.*

Use a graphing calculator to determine the x-value of the point at which $f(x)$ and $g(x)$ intersect.

> Set this equation equal to 0 to get the function $h(x) = \cos x - \sqrt{x}$. Graph h(x) on your calculator and find the root of the function to get 0.6417... .

$$\cos x = \sqrt{x}$$
$$\cos x - \sqrt{x} = 0$$
$$x \approx 0.641714370873$$

Use two integrals to calculate the area of the region, one that describes the interval on which $f(x) > g(x)$, and one that describes the interval on which $g(x) > f(x)$.

$$\int_0^{0.641714370873} \left(\cos x - \sqrt{x}\right) dx + \int_{0.641714370873}^4 \left(\sqrt{x} - \cos x\right) dx$$

Evaluate the definite integrals using a graphing calculator.

$$\approx 0.2558639 + 6.3459997 \approx 6.602$$

19.8 Calculate the area bounded by the graphs of $x - y = 3$ and $x = y^2 - y$.

You must first decide whether to write the linear equation in terms of x (by solving it for y), or vice versa. Because the quadratic equation is already written explicitly in terms of y (and cannot easily be solved for y to rewrite the equation in terms of x), solve the linear equation for x as well.

$$x - y = 3$$
$$x = y + 3$$

Determine the y-values at which $x = y + 3$ and $x = y^2 - y$ intersect.

$$y + 3 = y^2 - y$$
$$0 = y^2 - 2y - 3$$
$$0 = (y - 3)(y + 1)$$
$$y = -1 \text{ or } 3$$

The region bounded by the curves is pictured in Figure 19-5. Notice that the graph of $x = y + 3$ is always positioned to the right of the graph of $x = y^2 - y$ when $-1 \le y \le 3$.

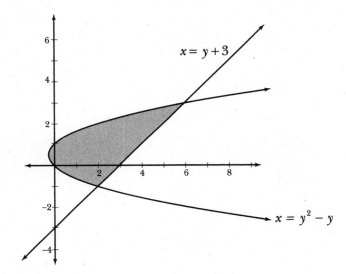

Figure 19-5 *The region bounded by $x = y + 3$ and $x = y^2 - y$. Note that the graph of $x = y + 3$ is the same as the graph of $y = x - 3$, which is slightly easier to graph, because it is in slope-intercept form.*

Calculate the area using the formula from Problem 19.5.

$$\int_c^d \left[f(y) - g(y) \right] dy = \int_{-1}^{3} \left[(y+3) - (y^2 - y) \right] dy$$
$$= \int_{-1}^{3} \left(-y^2 + 2y + 3 \right) dy$$
$$= \left(-\frac{y^3}{3} + 2 \cdot \frac{y^2}{2} + 3y \right) \bigg|_{-1}^{3}$$
$$= \left(-\frac{y^3}{3} + y^2 + 3y \right) \bigg|_{-1}^{3}$$
$$= (-9 + 9 + 9) - \left(\frac{1}{3} + 1 - 3 \right)$$
$$= \frac{32}{3}$$

The Mean Value Theorem for Integration
Make a rectangle that matches the area beneath a curve

19.9 State the mean value theorem for integration.

Given a function $f(x)$ that is continuous over the interval $[a,b]$, there exists a c such that $a \leq c \leq b$ for which $\int_a^b f(x)\,dx = f(c)(b-a)$.

19.10 Explain the geometric implications of the mean value theorem for integration.

> The area of the darker, rectangular region is (width)(length) = $f(c)(b-a)$, the right side of the mean value theorem for integration.

The mean value theorem for integration states that the area of the region bounded by $f(x)$ and the x-axis on the x-interval $[a,b]$ is exactly equal to the area of a rectangle with length $b-a$ and width $f(c)$, if you find the correct value(s) of c between a and b. Consider Figure 19-6. According to the mean value theorem for integration, the lightly shaded region beneath $f(x)$ has the same area as the darker region, a rectangle with length $b-a$ and width $f(c)$.

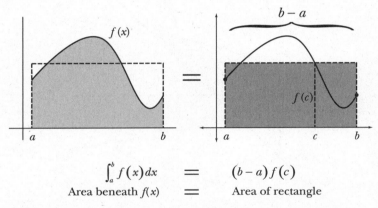

$$\int_a^b f(x)\,dx \quad = \quad (b-a)f(c)$$
$$\text{Area beneath } f(x) \quad = \quad \text{Area of rectangle}$$

Figure 19-6 *The mean value theorem for integration guarantees that there exists some $x = c$ between a and b such that the rectangle on $[a,b]$ with height $f(c)$ has the same area as the region bounded by $f(x)$ and the x-axis on $[a,b]$.*

19.11 What is the average value of a continuous function $f(x)$ over the closed interval $[a,b]$?

The average value is $f(c)$, as described by Problem 19.10. Solve the equation of the mean value theorem for integration for $f(c)$ to generate a formula that calculates the average value of $f(x)$.

$$\int_a^b f(x)\,dx = (b-a) \cdot f(c)$$
$$\frac{\int_a^b f(x)\,dx}{b-a} = \frac{(b-a) \cdot f(c)}{b-a}$$
$$f(c) = \frac{\int_a^b f(x)\,dx}{b-a}$$

The average value formula may also be written as a product rather than a fraction:

$$f(c) = \frac{1}{b-a} \int_a^b f(x)\,dx.$$

Note: Problems 19.12–19.13 refer to the function $f(x) = x^2$.

19.12 Calculate the average value of $f(x)$ between $x = 0$ and $x = 4$.

Apply the average value formula generated in Problem 19.11.

$$\frac{1}{b-a} \int_a^b f(x)\,dx = \frac{1}{4-0} \int_0^4 x^2\,dx = \frac{1}{4}\left(\frac{x^3}{3}\right)\Big|_0^4 = \frac{1}{4}\left[\frac{64}{3} - 0\right] = \frac{16}{3}$$

Some people try to find the average value by averaging the function values. Since f(0) = 0 and f(4) = 16, they'd add 0 + 16 = 16 and divide by 2 (16 ÷ 2 = 8) to get an average value of 8. As you can see, that does NOT always give you the right answer!

Note: Problems 19.12–19.13 refer to the function $f(x) = x^2$.

19.13 At what value c on the x-interval $[0,4]$ does $f(x)$ satisfy the mean value theorem for integration?

According to Problem 19.12, the average value of $f(x) = x^2$ on $[0,4]$ is $\frac{16}{3}$.
According to the mean value theorem for integration, there exists some c such that $0 \leq c \leq 4$ and $f(c) = \frac{16}{3}$. Substitute $x = c$ into $f(x)$.

In other words, what should you plug into f(x) to get the average value from Problem 19.12?

$$f(c) = c^2$$
$$\frac{16}{3} = c^2$$

Solve for c.

$$\pm\sqrt{\frac{16}{3}} = \sqrt{c^2}$$
$$c = \pm\frac{4}{\sqrt{3}} = \pm\frac{4\sqrt{3}}{3}$$

Recall that $0 \leq c \leq 4$, so $-\frac{4\sqrt{3}}{3}$ is not a valid value for c; therefore, $c = \frac{4\sqrt{3}}{3}$.

Note: Problems 19.14–19.15 refer to the function $g(x) = e^{2x}$.

19.14 Calculate the average value of $g(x)$ on the x-interval $[-1,1]$.

Apply the average value formula from Problem 19.11.

$$\frac{1}{b-a} \int_a^b g(x)\,dx = \frac{1}{1-(-1)} \int_{-1}^1 e^{2x}\,dx$$
$$= \frac{1}{2} \int_{-1}^1 e^{2x}\,dx$$

Integrate the expression using substitution of variables: $u = e^{2x}$ and $du = 2e^{2x}dx$. Therefore, $\dfrac{du}{2} = e^{2x}dx$. Translate the limits of integration in terms of u by substituting them into $u = e^{2x}$.

Exponential functions feel different than any other u-substitution. When you substitute, there's usually no u in the integrand, because du contains most, if not all, of the original expression.

$$\text{Lower limit } (x = -1): u = e^{2x} = e^{2(-1)} = e^{-2}$$
$$\text{Upper limit } (x = 1): \quad u = e^{2x} = e^{2(1)} = e^{2}$$

Rewrite the definite integral in terms of u.

$$\frac{1}{2}\int_{-1}^{1} e^{2x}dx = \frac{1}{2}\int_{e^{-2}}^{e^{2}} \frac{du}{2}$$
$$= \frac{1}{4}\int_{e^{-2}}^{e^{2}} du$$
$$= \frac{1}{4}(u)\Big|_{e^{-2}}^{e^{2}}$$
$$= \frac{1}{4}\left(e^{2} - e^{-2}\right)$$

Eliminate the negative exponent in the solution.

By the way, you can write that last line in terms of the hyperbolic sine function: $\dfrac{\sinh(2)}{2}$. If you have no idea what the heck that "h" is doing attached to "sin," don't sweat it.

$$= \frac{1}{4}\left(e^{2} - \frac{1}{e^{2}}\right)$$
$$= \frac{1}{4}\left(\frac{e^{4} - 1}{e^{2}}\right)$$
$$= \frac{e^{4} - 1}{4e^{2}}$$

Note: Problems 19.14–19.15 refer to the function $g(x) = e^{2x}$.

19.15 At what value c on the x-interval $[-1,1]$ does $g(x)$ satisfy the mean value theorem for integration?

According to Problem 19.14, the average value of $g(x)$ on the interval $[-1,1]$ is $\dfrac{e^{4} - 1}{4e^{2}}$. Therefore, there exists a value c such that $-1 \le c \le 1$ and $g(c) = \dfrac{e^{4} - 1}{4e^{2}}$. Substitute c and $g(c)$ into $g(x)$.

$$g(x) = e^{2x}$$
$$g(c) = e^{2c}$$
$$\frac{e^{4} - 1}{4e^{2}} = e^{2c}$$

Take the natural logarithm of both sides of the equation to solve for c.

$$\ln\left(\frac{e^{4} - 1}{4e^{2}}\right) = \ln\left(e^{2c}\right)$$
$$\ln\left(\frac{e^{4} - 1}{4e^{2}}\right) = 2c$$
$$\frac{1}{2}\ln\left(\frac{e^{4} - 1}{4e^{2}}\right) = c$$

Expand and simplify the logarithmic expression. ⟵

$$\ln\left(e^4 - 1\right) - \ln\left(4e^2\right) = 2c$$

$$\ln\left(e^4 - 1\right) - \left(\ln 4 + \ln e^2\right) = 2c$$

$$\ln\left(e^4 - 1\right) - \ln 4 - \ln e^2 = 2c$$

$$\ln\left(e^4 - 1\right) - \ln 4 - 2 = 2c$$

$$\frac{\ln\left(e^4 - 1\right) - \ln 4 - 2}{2} = c$$

> If you need practice with this, check out Problems 5.25–5.27. Remember that ln $e^a = a$, only the power is left because ln and e cancel each other out.

19.16 Calculate the average value of $f(x) = \dfrac{1}{3x}$ over the x-interval $\left[\dfrac{1}{2}, 2\right]$.

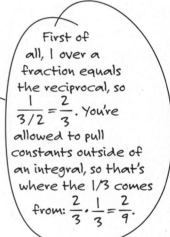

Apply the average value formula.

$$\frac{1}{b-a}\int_a^b f(x)\,dx = \frac{1}{2 - (1/2)}\int_{1/2}^2 \frac{1}{3x}\,dx$$

$$= \frac{1}{3/2} \cdot \frac{1}{3}\int_{1/2}^2 \frac{dx}{x}$$

$$= \frac{2}{9}\int_{1/2}^2 \frac{dx}{x}$$

> First of all, 1 over a fraction equals the reciprocal, so $\dfrac{1}{3/2} = \dfrac{2}{3}$. You're allowed to pull constants outside of an integral, so that's where the 1/3 comes from: $\dfrac{2}{3} \cdot \dfrac{1}{3} = \dfrac{2}{9}$.

According to Problem 18.4, $\int \dfrac{dx}{x} = \ln|x| + C$.

$$= \frac{2}{9}\left(\ln|x|\right)\Big|_{1/2}^2$$

$$= \frac{2}{9}\left(\ln 2 - \ln\frac{1}{2}\right)$$

According to a logarithmic property, $a \log b = \log b^a$. Therefore,

$$-1 \cdot \ln\frac{1}{2} = \ln\left(\frac{1}{2}\right)^{-1} = \ln 2.$$

$$= \frac{2}{9}(\ln 2 + \ln 2)$$

$$= \frac{2}{9}(2\ln 2)$$

$$= \frac{4\ln 2}{9}$$

Apply the logarithmic property $a \log b = \log b^a$ again: $4 \ln 2 = \ln 2^4 = \ln 16$.
Therefore, the average value of $f(x) = \dfrac{1}{3x}$ on the interval $\left[\dfrac{1}{2}, 2\right]$ is $\dfrac{\ln 16}{9}$.

Note: In Problems 19.17–19.19, assume that h(x) is a continuous function over the interval [–4,5]. If a and b are fixed real numbers such that –4 < a < b < 5, the following definite integral statements are true.

$$\int_{-4}^{b} h(x) = -22, \quad \int_{a}^{5} h(x) = 13, \quad and \quad \int_{5}^{-4} h(x) = 10$$

19.17 Calculate the average value of $h(x)$ over the x-interval $[-4,5]$.

Apply the average value formula from Problem 19.11.

$$\frac{1}{5-(-4)}\int_{-4}^{5} h(x)\,dx = \frac{1}{9}\int_{-4}^{5} h(x)\,dx$$

Although you are not given the value of $\int_{-4}^{5} h(x)\,dx$, you are given $\int_{5}^{-4} h(x) = 10$.

$$\frac{1}{9}\int_{-4}^{5} h(x)\,dx = \frac{1}{9}(-10) = -\frac{10}{9}$$

If you reverse the limits of integration (switch –4 and 5 next to the integral sign), you have to multiply its value by –1. That's why –10 suddenly turns into 10.

Note: Problems 19.17–19.19 refer to h(x), a, b, and the definite integrals described in Problem 19.17.

19.18 Calculate the average value of $h(x)$ over the interval $[-4,b]$.

Apply the average value formula.

$$\frac{1}{b-(-4)}\int_{-4}^{b} h(x)\,dx = \frac{1}{b+4}\int_{-4}^{b} h(x)\,dx$$

According to the information given, $\int_{-4}^{b} h(x) = -22$.

$$= \frac{1}{b+4}(-22)$$

$$= -\frac{22}{b+4}$$

The average value of $h(x)$ over $[-4,b]$ is $-\dfrac{22}{b+4}$.

Note: Problems 19.17–19.19 refer to h(x), a, b, and the definite integrals described in Problem 19.17.

19.19 Calculate the average value of $h(x)$ over the interval $[a,b]$.

As the value of $\int_{a}^{b} h(x)\,dx$ is not explicitly given, you must calculate it before applying the average value formula. Because $-4 < a < b < 5$, you can expand $\int_{-4}^{5} h(x)$ into three definite integrals.

$$\int_{-4}^{5} h(x)\,dx = \int_{-4}^{a} h(x)\,dx + \int_{a}^{b} h(x)\,dxdx + \int_{b}^{5} h(x)\,dx$$

Add $\int_a^b h(x)\,dx$ to both sides of the equation (indicated below by the underlined expression).

$$\int_{-4}^5 h(x)\,dx + \underline{\int_a^b h(x)\,dx} = \int_{-4}^a h(x)\,dx + \int_a^b h(x)\,dx + \underline{\int_a^b h(x)\,dx} + \int_b^5 h(x)\,dx$$

Notice that $\int_{-4}^a h(x)\,dx + \int_a^b h(x)\,dx = \int_{-4}^b h(x)\,dx$ and $\int_a^b h(x)\,dx + \int_b^5 h(x)\,dx = \int_a^5 h(x)\,dx$.

$$\int_{-4}^5 h(x)\,dx + \int_a^b h(x)\,dx = \int_{-4}^b h(x)\,dx + \int_a^5 h(x)\,dx$$

Substitute the known values of the definite integrals into the equation and solve for $\int_a^b h(x)\,dx$.

$$-10 + \int_a^b h(x)\,dx = -22 + 13$$

$$-10 + \int_a^b h(x)\,dx = -9$$

$$\int_a^b h(x)\,dx = 1$$

Now that you have determined the value of $\int_a^b h(x)\,dx$, calculate the average value of $h(x)$ on the interval $[a,b]$.

$$\frac{1}{b-a}\int_a^b h(x)\,dx = \frac{1}{b-a}(1) = \frac{1}{b-a}$$

> *This is the tricky part that you may not have thought of. Adding $\int_a^b h(x)\,dx$ creates two definite integrals with known values on the right side of the equation.*

Note: Problems 19.20–19.22 refer to the position equation $s(t)$, a sprinter's distance from the starting line during the first 4 seconds of a race (measured in meters after t seconds have elapsed), as defined below.

$$s(t) = -\frac{8t}{t-5}$$

19.20 Use $s(t)$ to determine the average velocity of the runner during the first four seconds of the race.

The average velocity of the runner is the average rate of change of position on the interval, which equals the slope of the secant line connecting points $(0, s(0))$ and $(4, s(4))$.

$$v_{\text{avg}} = \frac{s(4) - s(0)}{4 - 0}$$

$$= \frac{-\dfrac{8(4)}{4-5} - \left(-\dfrac{8(0)}{0-5}\right)}{4}$$

$$= \frac{-\dfrac{32}{-1} - 0}{4}$$

$$= 8 \text{ meters/second}$$

> *The slope of the tangent line to the graph of position represents instantaneous velocity, and the slope of the secant line to position represents average velocity. Look at Problem 15.5 if you forget how to calculate a secant slope.*

Note: Problems 19.20–19.22 refer to the position equation s(t), a sprinter's distance from the starting line during the first 4 seconds of a race (measured in meters after t seconds have elapsed), which is defined in Problem 19.20.

19.21 Identify the function $v(t)$ that models the velocity of the runner during the first four seconds of the race.

Given a position equation $s(t)$, the velocity equation is the derivative with respect to t. Apply the quotient rule to differentiate.

$$v(t) = \frac{d}{dt}\left(-\frac{8t}{t-5}\right)$$

$$= -\left[\frac{(t-5)(8)-(8t)(1)}{(t-5)^2}\right]$$

$$= -\frac{8t-40-8t}{(t-5)^2}$$

$$= \frac{40}{(t-5)^2}$$

Note: Problems 19.20–19.22 refer to the position equation s(t), a sprinter's distance from the starting line during the first 4 seconds of a race (measured in meters after t seconds have elapsed), which is defined in Problem 19.20.

19.22 Calculate the average value of $v(t)$ from Problem 19.21 to demonstrate that the average rate of change of $s(t)$ is equal to the average value of $v(t)$.

Apply the average value formula to $v(t)$, such that $a = 0$ and $b = 4$.

$$\frac{1}{4-0}\int_0^4 v(t)\,dt$$

$$= \frac{1}{4}\int_0^4 \frac{40}{(t-5)^2}\,dt$$

Remove the constant from the integrand and apply variable substitution.

$$= \frac{40}{4}\int_0^4 \frac{dt}{(t-5)^2}$$

$$= 10\int_{-5}^{-1} u^{-2}\,du$$

$$= 10\left(\frac{u^{-1}}{-1}\right)\Big|_{-5}^{-1}$$

$$= -10\left(\frac{1}{u}\right)\Big|_{-5}^{-1}$$

$$= -10\left(-1-\left(-\frac{1}{5}\right)\right)$$

$$= -10\left(-\frac{4}{5}\right)$$

$$= 8 \text{ meters/second}$$

Set $u = t - 5$ and you'll get $du = dt$. That changes the denominator to u^2 and the numerator to du. Plug the limits $t = 0$ and $t = 4$ into $u = t - 5$ to get new integral limits of -5 and -1.

According to Problem 19.20, the average rate of change of $s(t)$ on the t-interval $[0,4]$ is 8 meters/second, which is equal to the average vale of $v(t)$ on $[0,4]$.

19.23 Assume $f(x)$ is a continuous function and the chart below represents a selection of its function values. Estimate the average value of $f(x)$ on the interval $[-3,7]$ using the trapezoidal rule.

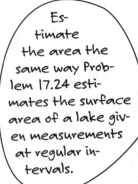

Es-timate the area the same way Problem 17.24 esti-mates the surface area of a lake giv-en measurements at regular in-tervals.

x	-3	-1	1	3	5	7
$f(x)$	6	8	9	4	-1	-5

Divide the interval $[-3,7]$ into five equal subintervals, each of width $\Delta x = 2$: $[-3,-1]$, $[-1,1]$, $[1,3]$, $[3,5]$, and $[5,7]$.

$$\int_a^b f(x)\,dx \approx \frac{b-a}{2n}\left[f(a)+2f(x_1)+2f(x_2)+2f(x_3)+2f(x_4)+f(b)\right]$$

$$\int_{-3}^7 f(x)\,dx \approx \frac{7-(-3)}{2(5)}\left[f(-3)+2f(-1)+2f(1)+2f(3)+2f(5)+f(7)\right]$$

$$\int_{-3}^7 f(x)\,dx \approx \frac{10}{10}\left[6+2(8)+2(9)+2(4)+2(-1)+(-5)\right]$$

$$\int_{-3}^7 f(x)\,dx \approx 41$$

Substitute this approximation of $\int_{-3}^7 f(x)\,dx$ into the average value formula.

$$\frac{1}{b-a}\int_a^b f(x)\,dx = \frac{1}{7-(-3)}\int_{-3}^7 f(x)\,dx$$

$$\approx \frac{41}{10}$$

19.24 Approximate the average value of $g(x)$, as graphed in Figure 19-7, over the interval $[-2,3]$. Show the work that leads to your answer.

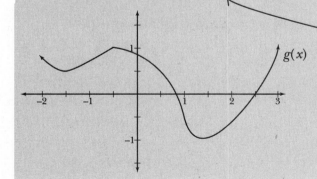

Figure 19-7

The graph of a continuous function $g(x)$.

$g(x)$

In other words, use a formula to get your answer—don't just look at the graph and say "Looks like the answer is around 1."

In order to approximate the average value of $g(x)$, you must first estimate $\int_{-2}^3 g(x)\,dx$. Use the method of Problem 18.32, counting the number of squares (formed by grid lines) between $g(x)$ and the x-axis, as illustrated by Figure 19-8.

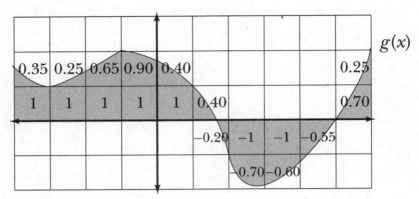

Figure 19-8 *The number in each grid square represents the approximate percentage of the square that is occupied by the shaded region bounded by f(x) and the x-axis (expressed as a decimal). Note that area below the x-axis is considered negative signed area.*

So, add up all 19 of the numbers in Figure 19-8 and multiply the sum by 1/4 (the actual area of one square).

Unlike Problem 18.32, each grid mark has length $\frac{1}{2}$, so each square has area $\frac{1}{2} \cdot \frac{1}{2} = \frac{1}{4}$, rather than $1 \cdot 1 = 1$. Each term in the sum below represents the sum of the values in each "column" in Figure 19-8.

$$\int_{-2}^{3} g(x)\,dx \approx \frac{1}{4}(1.35 + 1.25 + 1.65 + 1.9 + 1.4 + 0.2 - 1.7 - 1.6 - 0.55 + 0.95)$$

$$\approx 0.25(4.85)$$

$$\approx 1.2125$$

Apply the average value formula.

$$\frac{1}{b-a}\int_{a}^{b} g(x)\,dx = \frac{1}{3-(-2)}\int_{-2}^{3} g(x)\,dx$$

$$\approx \frac{1}{5}(1.2125)$$

$$\approx .2425$$

Accumulation Functions and Accumulated Change

Integrals with x limits and "real life" uses for integration

Note: Problems 19.25–19.30 refer to the function $f(x) = \int_{2}^{x} h(t)\,dt$, given the graph of h(t) in Figure 19-9.

19.25 Evaluate $f(2)$.

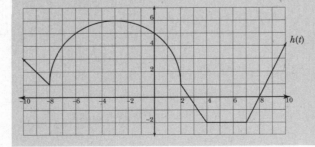

Figure 19-9

The graph of h(t) consists of a semicircle and four linear segments of differing slope.

Substitute $x = 2$ into $f(x)$: $f(2) = \int_2^2 h(t)\,dt$. According to the property of definite integrals that states $\int_a^a f(x)\,dx = 0$, $f(2) = 0$.

> There's no area under the curve if you start and stop measuring at the same x-value.

Note: Problems 19.25–19.30 refer to the function $f(x) = \int_2^x h(t)\,dt$, given the graph of $h(t)$ in Figure 19-9.

19.26 Evaluate $f(8)$.

Substitute $x = 8$ into $f(x)$: $f(8) = \int_2^8 h(t)\,dt$. The function value at $x = 8$ is defined as the area of the region bounded by $h(t)$ and the x-axis between $x = 2$ and $x = 8$, as illustrated by Figure 19-10.

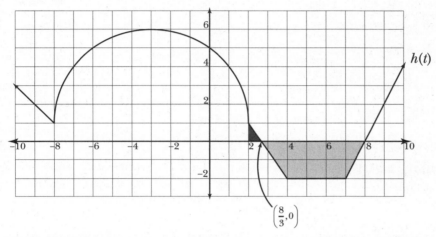

Figure 19-10   *The area bounded by h(t) and the x-axis consists of a right triangle with a positive signed area (dark shaded region) and a trapezoid with a negative signed area (light shaded region).*

The line segment connecting $(2,1)$ and $(4,-2)$ has slope $\dfrac{-2-1}{4-2} = -\dfrac{3}{2}$. Use the point-slope formula to get the equation of the line: $y = -\dfrac{3}{2}x + 4$. Substitute $y = 0$ into the equation and solve for x to calculate the x-intercept.

> Plug $m = -3/2$, $x_1 = 2$, and $y_1 = 1$ into $y - y_1 = m(x - x_1)$ and solve for y.

$$0 = -\frac{3}{2}x + 4$$

$$\left(\frac{2}{3}\right)\frac{3}{2}x = 4\left(\frac{2}{3}\right)$$

$$x = \frac{8}{3}$$

Calculate $\int_2^8 h(t)\,dt$ by combining the areas of the shaded regions in Figure 19-10; subtract the area of the trapezoid from the area of the triangle to account for its position below the x-axis.

$$\int_2^8 h(t)\,dt = \text{area of right triangle} - \text{area of trapezoid}$$

> Subtract the endpoints of the segments to get their lengths.

The right triangle has base $b = \frac{8}{3} - 2 = \frac{2}{3}$ and height $h_A = 1$. The trapezoid has bases of length $b_1 = 7 - 4 = 3$ and $b_2 = 8 - \frac{8}{3} = \frac{16}{3}$; its height is $h_B = 2$.

$$\int_2^8 h(t)\,dt = \left[\frac{1}{2}\cdot b \cdot h_A\right] - \left[\frac{1}{2}\cdot h_B \cdot (b_1 + b_2)\right]$$

$$= \left[\frac{1}{2}\cdot\frac{2}{3}\cdot 1\right] - \left[\frac{1}{2}\cdot 2 \cdot \left(3 + \frac{16}{3}\right)\right]$$

$$= \frac{1}{3} - \frac{25}{3}$$

$$= -8$$

Note: Problems 19.25–19.30 refer to the function $f(x) = \int_2^x h(t)\,dt$, given the graph of h(t) in Figure 19-9.

19.27 Evaluate $f(-8)$.

Substitute $x = -8$ into $f(x)$: $f(-8) = \int_2^{-8} h(t)\,dt$. Reverse the limits of integration so that the lesser of the two is the lower limit; according to Problem 18.22, this requires you to multiply the integral by –1.

$$f(-8) = -\int_{-8}^2 h(t)\,dt$$

Note that $\int_{-8}^2 h(t)\,dt$ is equivalent to the sum of two areas: a semicircle with radius 5 and a rectangle with length 10 and width 1, as illustrated by Figure 19-11. Add the areas of those regions to evaluate the definite integral.

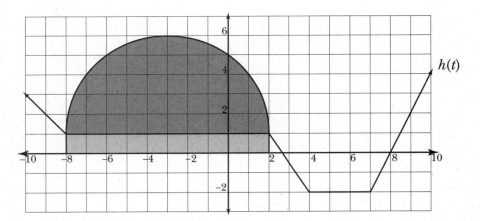

Figure 19-11 *The area bounded by h(t) and the x-axis on the interval [–8,2] consists of a semicircle (dark shaded region) and a rectangle (light shaded region).*

$$\int_{-8}^{2} h(t)\,dt = \text{area of semicircle} + \text{area of rectangle}$$

$$= \left(\frac{1}{2}\pi r^2\right) + (l \cdot w)$$

$$= \frac{1}{2}\pi(5)^2 + (10 \cdot 1)$$

$$= \frac{25\pi}{2} + 10$$

$$= \frac{25\pi + 20}{2}$$

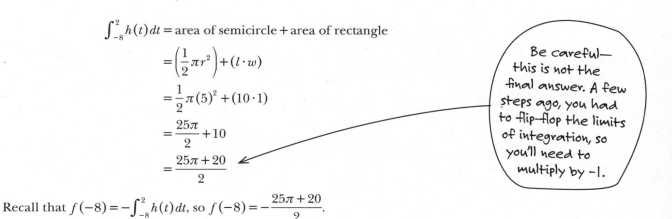

Be careful—this is not the final answer. A few steps ago, you had to flip-flop the limits of integration, so you'll need to multiply by -1.

Recall that $f(-8) = -\int_{-8}^{2} h(t)\,dt$, so $f(-8) = -\frac{25\pi + 20}{2}$.

Note: Problems 19.25–19.30 refer to the function $f(x) = \int_{2}^{x} h(t)\,dt$, given the graph of $h(t)$ in Figure 19-9.

19.28 Evaluate $f(-10)$.

Substitute $x = -10$ into $f(x)$ and reverse the limits of integration.

$$f(-10) = \int_{2}^{-10} h(t)\,dt$$

$$f(-10) = -\int_{-10}^{2} h(t)\,dt$$

Rewrite $\int_{-10}^{2} h(t)\,dt$ as a sum of two definite integrals.

This integral is calculated in Problem 19.27.

$$f(-10) = -\left[\int_{-10}^{-8} h(t)\,dt + \int_{-8}^{2} h(t)\,dt\right]$$

Note that $\int_{-10}^{-8} h(t)\,dt$ equals the area of a trapezoid with bases of length 1 and 3 and height 2.

$$\int_{-10}^{-8} h(t)\,dt = \frac{1}{2}h(b_1 + b_2) = \frac{1}{2}(2)(1+3) = 4$$

According to Problem 19.27, $\int_{-8}^{2} h(t)\,dt = -\frac{25\pi + 20}{2}$.

$$f(-10) = -\left[\int_{-10}^{-8} h(t)\,dt + \int_{-8}^{2} h(t)\,dt\right]$$

$$= -\left[4 - \frac{25\pi + 20}{2}\right]$$

$$= -\left[\frac{8}{2} + \frac{-25\pi - 20}{2}\right]$$

$$= -\left[\frac{-25\pi - 12}{2}\right]$$

$$= \frac{25\pi + 12}{2}$$

Distribute this -1 through the numerator, then change this - sign into a + sign.

Note: Problems 19.25–19.30 refer to the function $f(x) = \int_2^x h(t)\,dt$, given the graph of $h(t)$ in Figure 19-9.

19.29 Graph $f'(x)$.

According to Problem 18.34, $\dfrac{d}{dx}\left(\int_a^{f(x)} g(t)\,dt\right) = g(f(x)) \cdot f'(x)$.

$$\frac{d}{dx}\left(\int_2^x h(t)\,dt\right) = h(x) \cdot \frac{d}{dx}(x)$$
$$= h(x) \cdot 1$$
$$= h(x)$$

Therefore, the graph of $f'(x)$, pictured in Figure 19-12, is equivalent to the graph of $h(t)$.

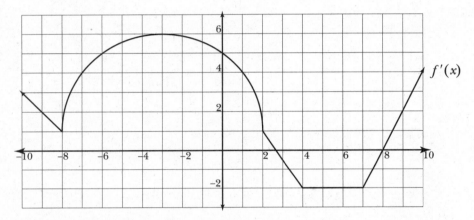

Figure 19-12   *The graph of $h(t)$ is also the graph of $f'(x)$.*

Note: Problems 19.25–19.30 refer to the function $f(x) = \int_2^x h(t)\,dt$, given the graph of $h(t)$ in Figure 19-9.

19.30 Rank the following values from least to greatest: $f'(-3), f'(0), f'(2),$ and $f'(10)$.

According to Problem 19.29, $f'(x) = h(x)$. Evaluate each derivative by determining the height of $h(t)$ in Figure 9-12 at each x-value.

$$f'(-3) = 6 \qquad f'(0) = 5 \qquad f'(2) = 1 \qquad f'(10) = 4$$

Therefore, $f'(2) < f'(10) < f'(0) < f'(-3)$.

The letter in the function doesn't matter: h(x) and h(t) have the same value no matter what you call the independent variable.

Note: Problems 19.31–19.35 discuss a particle moving back and forth along the x-axis with velocity $v(t) = -t^2 + 9t - 20$ (measured in meters per second after t seconds have elapsed) and an initial position 3 feet left of the origin.

19.31 Identify the function $s(t)$ that models the position of the particle at time t with respect to the origin.

> This means $s(0) = -3$. If the particle is left of the origin, its position is considered negative.

The derivative of a position function is its velocity function. Therefore, the antiderivative of the velocity function is the position function.

$$s(t) = \int v(t)\,dt$$
$$= \int (-t^2 + 9t - 20)\,dt$$
$$= -\frac{t^3}{3} + \frac{9t^2}{2} - 20t + C$$

According to the information given, $s(0) = -3$. Use this initial condition to calculate C.

$$s(0) = -\frac{(0)^3}{3} + \frac{9(0)^2}{2} - 20(0) + C$$
$$-3 = C$$

> When you know one of the values of an antiderivative, you can actually figure out what C equals. Substitute $t = 0$ into the function, and plug in $s(0) = -3$.

Therefore, $s(t) = -\dfrac{t^3}{3} + \dfrac{9t^2}{2} - 20t - 3$.

Note: Problems 19.31–19.35 discuss a particle moving back and forth along the x-axis with velocity $v(t) = -t^2 + 9t - 20$ (measured in meters per second after t seconds have elapsed) and an initial position 3 feet left of the origin.

19.32 Calculate the total displacement of the particle from $t = 0$ to $t = 4$.

> DISPLACEMENT means "You're in DIS PLACE now, and you were in DAT PLACE before." It compares starting and ending positions of the object— that's it.

Displacement of the particle on the time interval $[a,b]$ is defined as $s(b) - s(a)$; in this problem, $a = 0$ and $b = 4$.

$$s(b) - s(a) = s(4) - s(0)$$
$$= \left[-\frac{4^3}{3} + \frac{9(4)^2}{2} - 20(4) - 3 \right] - (-3)$$
$$= -\frac{64}{3} + \frac{144}{2} - 80 - 3 + 3$$
$$= -\frac{64}{3} + 72 - 80$$
$$= \frac{-64 - 24}{3}$$
$$= -\frac{88}{3}$$

The particle is $\dfrac{88}{3}$ meters left of where it began at $t = 4$ seconds.

> That doesn't necessarily mean it traveled a total of 88/3 meters, though. If the particle changes direction at all, it'll travel farther than the displacement suggests.

Note: Problems 19.31–19.35 discuss a particle moving back and forth along the x-axis with velocity $v(t) = -t^2 + 9t - 20$ (measured in meters per second after t seconds have elapsed) and an initial position 3 feet left of the origin.

19.33 Determine the total distance traveled by the particle on the time interval [0,4].

You must first identify any *t*-values at which the particle changes direction, indicated by a sign change in the velocity function. Use the technique demonstrated in Problem 15.20, which begins by setting the velocity function equal to 0 and solving the equation to locate critical numbers.

$$-t^2 + 9t - 20 = 0$$
$$-1\left(-t^2 + 9t - 20\right) = -1(0)$$
$$t^2 - 9t + 20 = 0$$
$$(t-4)(t-5) = 0$$
$$t = 4, 5$$

> The particle does change direction at t = 4, where its velocity is 0. Even though the particle has momentarily stopped moving, it hasn't actually begun traveling the other way yet.

The particle changes direction at $t = 4$ seconds and again at $t = 5$ seconds, because $v(t)$ changes sign at both of those critical numbers. Neither critical number affects the distance traveled by the particle from $t = 0$ to $t = 4$, because the particle travels in only one direction during that time. Therefore, its displacement and distance traveled are equivalent: $\dfrac{88}{3}$ meters (according to Problem 19.32).

Note: Problems 19.31–19.35 discuss a particle moving back and forth along the x-axis with velocity $v(t) = -t^2 + 9t - 20$ (measured in meters per second after t seconds have elapsed) and an initial position 3 feet left of the origin.

19.34 Determine the displacement of the particle on the time interval [3,7].

As explained in Problem 19.32, the displacement of the particle is the difference of its positions at the endpoints of the specified *t*-interval: $s(7) - s(3)$.

$$s(7) - s(3) = \left[-\frac{7^3}{3} + \frac{9(7)^2}{2} - 20(7) - 3\right] - \left[-\frac{3^3}{3} + \frac{9(3)^2}{2} - 20(3) - 3\right]$$
$$= \left[-\frac{343}{3} + \frac{441}{2} - 140 - 3\right] - \left[-9 + \frac{81}{2} - 60 - 3\right]$$
$$= \left[\frac{-686 + 1,323 - 840 - 18}{6}\right] - \left[\frac{-18 + 81 - 120 - 6}{2}\right]$$
$$= -\frac{221}{6} + \frac{63}{2}$$
$$= -\frac{32}{6}$$
$$= -\frac{16}{3}$$

At $t = 7$ seconds, the particle is $\dfrac{16}{3}$ meters left of its position when $t = 3$.

Note: Problems 19.31–19.35 discuss a particle moving back and forth along the x-axis with velocity $v(t) = -t^2 + 9t - 20$ (measured in meters per second after t seconds have elapsed) and an initial position 3 feet left of the origin.

19.35 What is the total distance traveled by the particle on the time interval [3,7]?

According to Problem 19.33, the particle changes direction at $t = 4$ and $t = 5$. Definite integrals of the velocity function calculate the total distance traveled, as long as the particle travels only one direction between the limits of integration. However, if the particle is moving left, the definite integral will be negative. Therefore, to calculate the total distance traveled by the particle, take the absolute value of each integral and add the results.

$$\text{distance traveled} = \left|\int_3^4 v(t)\,dt\right| + \left|\int_4^5 v(t)\,dt\right| + \left|\int_5^7 v(t)\,dt\right|$$

$$= \left|\left(-\frac{t^3}{3} + \frac{9t^2}{2} - 20t\right)\Big|_3^4\right| + \left|\left(-\frac{t^3}{3} + \frac{9t^2}{2} - 20t\right)\Big|_4^5\right| + \left|\left(-\frac{t^3}{3} + \frac{9t^2}{2} - 20t\right)\Big|_5^7\right|$$

$$= |s(4) - s(3)| + |s(5) - s(4)| + |s(7) - s(5)|$$

$$= \left|-\frac{5}{6}\right| + \left|\frac{1}{6}\right| + \left|-\frac{14}{3}\right|$$

$$= \frac{17}{3}$$

> You don't use $C = -3$ when you're working with definite integrals—just integrate each term of $v(t)$ using the power rule for integration.

The particle travels a total distance of $\frac{17}{3}$ meters between $t = 3$ and $t = 7$. The particle travels left for most of that distance $\left(\frac{5}{6} + \frac{14}{3} = \frac{11}{2}\text{ meters}\right)$, traveling right only $\frac{1}{6}$ of a meter, between $t = 4$ and $t = 5$ seconds.

19.36 Seven hours after a community water tank is filled, monitoring equipment reports that water is leaking from the tank at a rate of $l(t) = \sqrt{\frac{t+4}{3}} - \frac{1}{t+1}$ gallons per hour (where t is the number of hours elapsed since the tank was last filled). Calculate the total amount of water that leaked out of the tank during those seven hours.

The definite integral $\int_a^b l(t)\,dt$ calculates the total volume of water that leaked out of the tank between $t = a$ and $t = b$. Unlike Problems 19.33 and 19.35, there is no need to identify critical points or split the definite integral because $l(t) > 0$ for all t. Evaluate $\int_0^7 l(t)\,dt$.

> If $l(t)$ is the rate that the water's leaking out, a negative $l(t)$ means water is leaking back in, and that doesn't make any sense.

$$\int_0^7 \left(\sqrt{\frac{t+4}{3}} - \frac{1}{t+1}\right)dt = \int_0^7 \frac{\sqrt{t+4}}{\sqrt{3}}\,dt - \int_0^7 \frac{dt}{t+1}$$

$$= \frac{1}{\sqrt{3}}\int_0^7 (t+4)^{1/2}\,dt - \int_0^7 \frac{dt}{t+1}$$

Apply variable substitution. In the first integral, $u = t + 4$ and $du = dt$. In the second integral, $v = t + 1$ and $dv = dt$. As you write the integrands in terms of u and v, remember to write the limits of integration in terms of u and v as well.

> You don't always have to use u as the variable. Actually, it's better to use two different variables (like u and v in this problem) if you're working on more than one integral at a time.

$$= \frac{1}{\sqrt{3}} \int_4^{11} u^{1/2} du - \int_1^8 \frac{dv}{v}$$

$$= \frac{1}{\sqrt{3}} \cdot \frac{2}{3} \left(u^{3/2}\right)\Big|_4^{11} - \ln|v| \Big\|_1^8$$

$$= \frac{2}{3\sqrt{3}} \left(11^{3/2} - 4^{3/2}\right) - \left(\ln 8 - \ln 1\right)$$

Note that $4^{3/2} = \left(\sqrt{4}\right)^3 = 8$ and $\ln 1 = 0$.

$$= \frac{2}{3\sqrt{3}} \left(11^{3/2} - 8\right) - \ln 8$$

$$= \frac{2\sqrt{3}}{9} \left(11^{3/2} - 8\right) - \ln 8$$

$$\approx 8.884 \text{ gallons}$$

Chapter 20
INTEGRATING RATIONAL EXPRESSIONS

What to do when there's a fraction inside the integral

Minor differences in a rational integrand require vastly different solution methods. For instance, integrating a prime quadratic denominator paired with a constant numerator typically requires you to complete the square. However, if the same denominator is paired with a linear numerator, variable substitution may be required. Furthermore, a numerator with degree two or higher will likely involve long division before you are able to integrate the expression. All of these methods are explored in detail. The chapter culminates with the integration by partial fraction decomposition technique, a powerful and rigorous tool used to express an integrand as the sum of fractions whose denominators are factors of the original rational expression.

Derivatives are much easier than integrals. Want to differentiate a product? Use the product rule. Need the derivative of a fraction? The quotient rule always works. Not so with integrals, though. Different fractions, even if they're only very slightly different, will have completely different looking solutions that require you to use completely different techniques. That's why this entire chapter is spent on the quagmire that is integrating fractions. Watch out for little tricks along the way such as adding and subtracting the same thing inside an integral (like −x + x), which is really the same thing as adding 0. Two of the methods you'll learn (separation and partial fractions) work by converting one big fraction into a bunch of small fractions added together, but using partial fractions is much more complicated than separation. It's not all THAT hard, but there are a lot of steps, which means there are a lot of places to make mistakes.

Separation

Make one big ugly fraction into smaller, less ugly ones

If there's addition or subtraction in the numerator, you can write each term over a copy of the denominator and add (or subtract) those fractions. You can't split apart addition or subtraction in the denominator like that, though.

20.1 Demonstrate that $\dfrac{a+b}{c} = \dfrac{a}{c} + \dfrac{b}{c}$ but $\dfrac{a}{b+c} \neq \dfrac{a}{b} + \dfrac{a}{c}$.

Notice that $\dfrac{a}{c} + \dfrac{b}{c}$ have common denominator c. You may combine the numerators of such fractions: $\dfrac{a}{c} + \dfrac{b}{c} = \dfrac{a+b}{c}$. However, in order to combine the fractions in the expression $\dfrac{a}{b} + \dfrac{a}{c}$, you must first rewrite the sum using the common denominator bc.

$$\frac{a}{b}\left(\frac{c}{c}\right) + \frac{a}{c}\left(\frac{b}{b}\right) = \frac{ac+ab}{bc}$$

Therefore, $\dfrac{a}{b+c} \neq \dfrac{a}{b} + \dfrac{a}{c}$; instead, $\dfrac{a}{b} + \dfrac{a}{c} = \dfrac{ac+ab}{bc}$.

20.2 Integrate the expression: $\int \dfrac{x^9 - 3x^4}{x^6}\,dx$.

Separate the rational expression into two rational expressions with a common denominator.

$$\int \frac{x^9 - 3x^4}{x^6}\,dx = \int \left(\frac{x^9}{x^6} - \frac{3x^4}{x^6}\,dx\right)$$

$$= \int \frac{x^9}{x^6}\,dx - 3\int \frac{x^4}{x^6}\,dx$$

$$= \int x^3\,dx - 3\int x^{-2}\,dx$$

Apply the power rule for integration.

$$= \frac{x^4}{4} - 3\cdot\frac{x^{-1}}{-1} + C$$

$$= \frac{x^4}{4} + \frac{3}{x} + C$$

20.3 Integrate the expression: $\int \dfrac{\sqrt{x}+1}{x}\,dx$.

Separate the expression into two indefinite integrals containing integrands with common denominators.

$$\int \frac{\sqrt{x}+1}{x}\,dx = \int \frac{\sqrt{x}}{x}\,dx + \int \frac{1}{x}\,dx$$

$$= \int \frac{x^{1/2}}{x}\,dx + \int \frac{dx}{x}$$

$$= \int x^{-1/2}\,dx + \int \frac{dx}{x}$$

$$= \frac{x^{1/2}}{1/2} + \ln|x| + C$$

$$= 2x^{1/2} + \ln|x| + C$$

> Dividing by ½ is the same as multiplying by 2 (just like dividing by 2 is the same as multiplying by ½).

20.4 Integrate the expression: $\int \frac{\cos 2x}{\cos^2 x}\,dx$.

Recall that $\cos 2x = 2\cos^2 x - 1$. Use this identity to rewrite the numerator and separate the integrand into the difference of two fractions.

$$\int \frac{\cos 2x}{\cos^2 x}\,dx = \int \frac{2\cos^2 x - 1}{\cos^2 x}\,dx$$

$$= \int \frac{2\cos^2 x}{\cos^2 x}\,dx - \int \frac{1}{\cos^2 x}\,dx$$

$$= \int 2\,dx - \int \sec^2 x\,dx$$

$$= 2x - \tan x + C$$

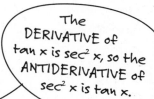

> The DERIVATIVE of $\tan x$ is $\sec^2 x$, so the ANTIDERIVATIVE of $\sec^2 x$ is $\tan x$.

20.5 Integrate the expression: $\int \frac{e^{4x} - 2}{e^{4x}}\,dx$.

Rewrite the integrand as the difference of two rational expressions.

$$\int \frac{e^{4x} - 2}{e^{4x}}\,dx = \int \frac{e^{4x}}{e^{4x}}\,dx - \int \frac{2}{e^{4x}}\,dx$$

$$= \int 1 \cdot dx - 2\int e^{-4x}\,dx$$

Integrate $\int e^{-4x}\,dx$ using variable substitution: $u = e^{4x}$ and $du = -4e^{4x}\,dx$, so $-\dfrac{du}{4} = e^{-4x}\,dx$.

$$= \int 1 \cdot dx - 2\int -\frac{du}{4}$$

$$= \int dx + \frac{1}{2}\int du$$

$$= x + \frac{1}{2}u + C$$

$$= x + \frac{1}{2e^{4x}} + C$$

20.6 Integrate the expression: $\int \dfrac{-2\,dx}{e^x - 2}$.

Add and subtract e^x in the numerator.

$$\int \frac{-2}{e^x - 2}\,dx = \int \frac{-2 + e^x - e^x}{e^x - 2}\,dx$$

Separate the first two terms of the numerator from the third term, creating two indefinite integrals.

$$= \int \frac{-2 + e^x}{e^x - 2}\,dx + \int \frac{-e^x\,dx}{e^x - 2}$$

$$= \int \frac{e^x \cancel{- 2}}{e^x \cancel{- 2}}\,dx - \int \frac{e^x\,dx}{e^x - 2}$$

$$= \int dx - \int \frac{e^x\,dx}{e^x - 2}$$

Apply variable substitution to the remaining rational integrand: $u = e^x - 2$ and $du = e^x\,dx$.

$$= \int dx - \int \frac{du}{u}$$

$$= x - \ln|u| + C$$

$$= x - \ln\left|e^x - 2\right| + C$$

Leaving –2 inside the integral (instead of pulling it out in front of the integral sign) will actually save you a step or two in this problem, but either way you'll get the same final answer.

20.7 Integrate the expression: $\int \dfrac{\sqrt{x-4} + x}{x - 4}\,dx$.

Rewrite the integrand as the sum of two rational expressions.

$$\int \frac{\sqrt{x-4} + x}{x-4}\,dx = \int \frac{\sqrt{x-4}}{x-4}\,dx + \int \frac{x}{x-4}\,dx$$

$$= \int \frac{(x-4)^{1/2}}{(x-4)^{1}}\,dx + \int \frac{x}{x-4}\,dx$$

$$= \int (x-4)^{-1/2}\,dx + \int \frac{x}{x-4}\,dx$$

Apply the same variable substitution to both integrals: $u = x - 4$ and $du = dx$.

$$= \int u^{-1/2}\,du + \int \frac{u+4}{u}\,du$$

$$= \int u^{-1/2}\,du + \int \frac{\cancel{u}}{\cancel{u}}\,du + \int \frac{4}{u}\,du$$

$$= \frac{u^{1/2}}{1/2} + u + 4\ln|u| + C$$

$$= 2\sqrt{x-4} + (x-4) + 4\ln|x-4| + C$$

$$= 2\sqrt{x-4} + x + 4\ln|x-4| + C - 4$$

If $u = x - 4$, you can add 4 to both sides to get $u + 4 = x$.

Note that $C-4$ is another arbitrary constant. To indicate that the new arbitrary constant is different than C as it appeared in the preceding steps, you can use a new constant, such as K, but it is common to continue to refer to the arbitrary constant as C, even if the value of C may change throughout the problem. Also note that, according to the logarithmic property that states $a \log b = \log b^a$, $4 \ln |x-4| = \ln (x-4)^4$.

$$= 2\sqrt{x-4} + x + \ln(x-4)^4 + C$$

Long Division
Divide before you integrate

20.8 Under what circumstances is it beneficial to perform long division on a rational integrand?

If the numerator and denominator of the rational expression are polynomials such that the degree of the numerator is greater than or equal to the degree of the denominator, it is often beneficial to perform long (or synthetic) division before integrating.

20.9 According to Problem 20.2, $\int \dfrac{x^9 - 3x^4}{x^6}\, dx = \dfrac{x^4}{4} + \dfrac{3}{x} + C$. Verify the antiderivative by performing long division on the integrand before integrating.

As the degree of the numerator is greater than or equal to the degree of the denominator ($9 \geq 6$), you can apply long division.

$$x^6 \overline{)\; x^9 + 0x^8 + 0x^7 + 0x^6 + 0x^5 - 3x^4 + 0x^3 + 0x^2 + 0x + 0}$$

quotient $x^3 - 3x^{-2}$

$$-x^9$$
$$-3x^4 + 0x^3 + 0x^2$$
$$+3x^4$$
$$0$$

> If the DENOMINATOR'S degree is larger, you can try to integrate using partial fractions, which is covered at the end of this chapter.

> If you need to review long division of polynomials, look at Problems 2.16 and 2.17.

Rewrite the rational integrand as the quotient that results from long division.

$$\int \frac{x^9 - 3x^4}{x^6}\, dx = \int \left(x^3 - 3x^{-2}\right) dx$$
$$= \int x^3\, dx - 3\int x^{-2}\, dx$$
$$= \frac{x^4}{4} - 3 \cdot \frac{x^{-1}}{-1} + C$$
$$= \frac{x^4}{4} + \frac{3}{x} + C$$

20.10 Integrate the expression: $\int \dfrac{2x^3 - 13x^2 - 57x + 108}{2x - 3}\, dx$.

Perform polynomial long division.

$$
\begin{array}{r}
x^2\ -5x\ -36 \\
2x - 3\overline{)\ 2x^3 - 13x^2 - 57x + 108} \\
\underline{-2x^3 + 3x^2} \\
-10x^2 - 57x \\
\underline{+10x^2 - 15x} \\
-72x + 108 \\
\underline{+72x - 108} \\
0
\end{array}
$$

Even though Problems 20.9 and 20.10 divide evenly, long division still helps if there's a remainder, as you'll see in Problems 20.11–20.13.

Rewrite the rational integrand as the quotient that results from long division.

$$\int \frac{2x^3 - 13x^2 - 57x + 108}{2x - 3}\, dx = \int \left(x^2 - 5x - 36\right) dx$$

$$= \frac{x^3}{3} - \frac{5x^2}{2} - 36x + C$$

20.11 Integrate the expression: $\int \dfrac{x^3 - 8x^2 + 3x + 16}{x^2 - 2}\, dx$.

Perform polynomial long division.

$$
\begin{array}{r}
x - 8 \\
x^2 - 0x - 2\overline{)\ x^3 - 8x^2 + 3x + 16} \\
\underline{-x^3 + 0x^2 + 2x} \\
-8x^2 + 5x + 16 \\
\underline{+8x^2 - 0x - 16} \\
5x + 0
\end{array}
$$

The polynomials in front and inside of the division symbols need to have all their x's. Because $x^2 - 2$ is missing an x term, fill in the space with 0x.

Note that the remainder is $5x$; therefore, $\dfrac{x^3 - 8x^2 + 3x - 4}{x^2 - 2} = x - 8 + \dfrac{5x}{x^2 - 2}$.

$$\int \frac{x^3 - 8x^2 + 3x - 4}{x^2 - 2}\, dx = \int \left(x - 8 + \frac{5x}{x^2 - 2}\right) dx$$

$$= \int x\, dx - \int 8\, dx + \int \frac{5x}{x^2 - 2}\, dx$$

$$= \int x\, dx - 8\int dx + 5\int \frac{x}{x^2 - 2}\, dx$$

The rightmost integral requires variable substitution: $u = x^2 - 2$ and $du = 2x\,dx$, so $\dfrac{du}{2} = x\,dx$.

$$= \int x\,dx - 8\int dx + 5 \cdot \frac{1}{2}\int \frac{du}{u}$$

$$= \frac{x^2}{2} - 8x + \frac{5}{2}\ln|u| + C$$

$$= \frac{x^2}{2} - 8x + \frac{5}{2}\ln\left|x^2 - 2\right| + C$$

20.12 Integrate the expression: $\displaystyle\int \frac{4x+7}{3x-1}\,dx$.

Perform long division.

$$
\begin{array}{r}
\frac{4}{3} \\
3x-1{\overline{\smash{\big)}\,4x+\ \ 7}} \\
\underline{-4x+\ \ \frac{4}{3}} \\
\frac{25}{3}
\end{array}
$$

> Put 4/3 above the division symbol because $3x\left(\dfrac{4}{3}\right) = \dfrac{12x}{3} = 4x$. It answers the question "3x times what equals 4x?"

Therefore, $\dfrac{4x+7}{3x-1} = \dfrac{4}{3} + \dfrac{25/3}{3x-1} = \dfrac{4}{3} + \dfrac{25}{3(3x-1)}$.

$$\int \frac{4x+7}{3x-1}\,dx = \int\left[\frac{4}{3} + \frac{25}{3(3x-1)}\right]dx$$

$$= \int \frac{4}{3}\,dx + \frac{25}{3}\int \frac{dx}{3x-1}$$

Variable substitution is required to antidifferentiate $\displaystyle\int \frac{dx}{3x-1}$: $u = 3x - 1$ and $du = 3\,dx$, so $\dfrac{du}{3} = dx$.

$$= \int \frac{4}{3}\,dx + \frac{25}{3}\cdot\frac{1}{3}\int \frac{du}{u}$$

$$= \frac{4}{3}x + \frac{25}{9}\ln|u| + C$$

$$= \frac{4}{3}x + \frac{25}{9}\ln|3x-1| + C$$

> You could also have done this problem using separation, but the u-substitution is a little trickier:
> $$\int \frac{4x+7}{3x-1}\,dx = \int \frac{4x}{3x-1}\,dx +$$
> $$\int \frac{7}{3x-1}\,dx.$$

20.13 Integrate the expression: $\int \dfrac{x^5 - 6x^4 + 3x^3 - x - 1}{x+2}\,dx$.

> If you can't remember how synthetic division works, look at Problems 2.18 and 2.19.

Because the denominator is a linear binomial with leading coefficient 1, synthetic division is preferable to long division.

$$
\begin{array}{r|rrrrrr}
-2 & 1 & -6 & 3 & 0 & -1 & -1 \\
   &   & -2 & 16 & -38 & 76 & -150 \\
\hline
   & 1 & -8 & 19 & -38 & 75 & -151
\end{array}
$$

Rewrite the integrand using the quotient resulting from synthetic division.

$$\int \frac{x^5 - 6x^4 + 3x^3 - x - 1}{x+2}\,dx = \int \left(x^4 - 8x^3 + 19x^2 - 38x + 75 - \frac{151}{x+2} \right)dx$$

$$= \int x^4\,dx - 8\int x^3\,dx + 19\int x^2\,dx - 38\int x\,dx + 75\int dx - 151\int \frac{dx}{x+2}$$

$$= \frac{x^5}{5} - \frac{8x^4}{4} + \frac{19x^3}{3} - \frac{38x^2}{2} + 75x - 151\cdot\ln|x+2| + C$$

$$\frac{x^5}{5} - 2x^4 + \frac{19x^3}{3} - 19x^2 + 75x - 151\cdot\ln|x+2| + C$$

Applying Inverse Trigonometric Functions
Very useful, but only in certain circumstances

20.14 Explain why $\int \dfrac{x\,dx}{1+x^2}$ can be integrated using variable substitution but $\int \dfrac{dx}{1+x^2}$ cannot.

> Problem 20.14 says you can't integrate this with u-substitution, but that doesn't mean that (A) you can't integrate it at all, or (B) u-substitution won't be involved. You CAN integrate it, and usually u-substitution IS an important step, even if it's not the only step.

If $u = x^2 + 1$, the shared denominator, then $du = 2x\,dx$ and $\dfrac{du}{2} = x\,dx$. Because the derivative contains "$x\,dx$," the numerator of the fraction must contain this quantity as well, in order to perform variable substitution. Whereas $\int \dfrac{x\,dx}{1+x^2}$ does contain the required x (as well as the dx, which appears in both integrals), $\int \dfrac{dx}{1+x^2}$ does not.

20.15 Integrate the expression: $\int \dfrac{dx}{1+x^2}$.

According to Problem 16.26, $\dfrac{d}{du}(\arctan u) = \dfrac{1}{1+u^2}\cdot\dfrac{du}{dx}$. If $u = x$ in that

rational expression, it becomes $\dfrac{1}{1+(x)^2}\cdot\dfrac{d}{dx}(x) = \dfrac{1}{1+x^2}\cdot 1$. Therefore,

$\int \dfrac{dx}{1+x^2} = \arctan x + C$.

More generally, any integral of the form $\int \dfrac{du}{a^2 + u^2}$, where u is a function of x and a is a real number, has an antiderivative of $\dfrac{1}{a}\arctan\left(\dfrac{u}{a}\right) + C$.

20.16 Integrate the expression: $\int \dfrac{6xdx}{x^4 + 16}$.

This integral has form $\int \dfrac{du}{a^2 + u^2}$, if $a = 4$ (because $4^2 = 16$) and $u = x^2$ (because $(x^2)^2 = x^4$). To integrate, apply variable substitution: $u = x^2$ and $du = 2x\,dx$, so $\dfrac{du}{2} = x\,dx$.

$$\int \frac{6x\,dx}{x^4 + 16} = 6\int \frac{x\,dx}{x^4 + 16}$$

$$= 6\int \frac{du/2}{u^2 + a^2}$$

$$= 6 \cdot \frac{1}{2}\int \frac{du}{u^2 + a^2}$$

According to Problem 20.15, $\int \dfrac{du}{a^2 + u^2} = \dfrac{1}{a}\arctan\left(\dfrac{u}{a}\right) + C$.

$$= 3 \cdot \frac{1}{a}\arctan\left(\frac{u}{a}\right) + C$$

$$= \frac{3}{4}\arctan\left(\frac{x^2}{4}\right) + C$$

The u's inside the inverse trig integrals should remind you to use u-substitution if u equals something other than just x.

The order in the denominator doesn't matter—u² + a² and a² + u² are equivalent, just like 5 + 3 and 3 + 5.

20.17 Integrate the expression: $\int \dfrac{e^x dx}{\sqrt{5 - e^{2x}}}$.

Note that $\int \dfrac{du}{\sqrt{a^2 - u^2}} = \arcsin\left(\dfrac{u}{a}\right) + C$.

Let $a = \sqrt{5}$ (because $a^2 = \left(\sqrt{5}\right)^2 = 5$) and $u = e^x$ (because $u^2 = (e^x)^2 = e^{2x}$). Perform variable substitution: if $u = e^x$, then $du = e^x\,dx$.

$$\int \frac{e^x dx}{\sqrt{5 - e^{2x}}} = \int \frac{du}{\sqrt{a^2 - u^2}}$$

$$= \arcsin\left(\frac{u}{a}\right) + C$$

$$= \arcsin\left(\frac{e^x}{\sqrt{5}}\right) + C$$

Rationalize the denominator to get the equivalent solution $\arcsin\left(\dfrac{e^x \sqrt{5}}{5}\right) + C$.

This formula (which looks a lot like the arcsin x derivative formula from Problem 16.24) doesn't have 1/a out front like the arctan x formula in Problem 20.15.

20.18 Integrate the expression: $\int \dfrac{dx}{x\sqrt{1-(\ln x)^2}}$.

Use the arcsine antiderivative from Problem 20.17 again, because inside the radical you've got a number minus a variable ($a^2 - u^2$). In Problem 20.19, you deal with $u^2 - a^2$ inside the radical.

Apply the inverse trigonometric antiderivative $\int \dfrac{du}{\sqrt{a^2-u^2}} = \arcsin\left(\dfrac{u}{a}\right) + C$ with $a = 1$, $u = \ln x$, and $du = \dfrac{dx}{x}$.

$$\int \frac{dx}{x\sqrt{1-(\ln x)^2}} = \int \frac{1}{\sqrt{1-(\ln x)^2}} \cdot \frac{dx}{x}$$

$$= \int \frac{1}{\sqrt{a^2-u^2}} \cdot du$$

$$= \arcsin\left(\frac{u}{a}\right) + C$$

$$= \arcsin(\ln x) + C$$

The extra x in front of the radical sign in the denominator of the original problem disappears when you replace dx/x with du.

20.19 Integrate the expression: $\int \dfrac{\cos 3x \, dx}{\sin 3x \sqrt{\sin^2(3x) - 2}}$.

Apply the arcsecant antiderivative formula: $\int \dfrac{du}{u\sqrt{u^2 - a^2}} = \dfrac{1}{a}\operatorname{arcsec}\left(\dfrac{|u|}{a}\right) + C$. Let $a = \sqrt{2}$ and $u = \sin 3x$. Therefore, $du = 3\cos 3x \, dx$ and $\dfrac{du}{3} = \cos 3x \, dx$.

$$\int \frac{\cos 3x \, dx}{\sin 3x\sqrt{\sin^2(3x)-2}} = \int \frac{du/3}{u\sqrt{u^2-a^2}}$$

$$= \frac{1}{3}\int \frac{du}{u\sqrt{u^2-a^2}}$$

$$= \frac{1}{3} \cdot \frac{1}{a}\operatorname{arcsec}\left(\frac{|u|}{a}\right) + C$$

$$= \frac{1}{3\sqrt{2}}\operatorname{arcsec}\left(\frac{|\sin(3x)|}{\sqrt{2}}\right) + C$$

Rationalize the expressions to get the equivalent solution

$$\frac{\sqrt{2}}{6}\operatorname{arcsec}\left(\frac{\sqrt{2}\,|\sin(3x)|}{2}\right) + C.$$

20.20 Integrate the expression: $\int \dfrac{x \, dx}{\left(\arctan x^2 + x^4 \arctan x^2\right)\sqrt{\arctan^2(x^2) - 25}}$.

The word "parenthetical" just means "the stuff inside the parentheses."

Factor $\arctan x^2$ out of the parenthetical quantity in the denominator.

$$\int \frac{x \, dx}{\arctan x^2 \left(1 + x^4\right)\sqrt{\arctan^2(x^2) - 25}}$$

Apply the arcsecant trigonometric antiderivative such that $a = 5$, because $a^2 = 5^2 = 25$, and $u = \arctan x^2$, because $u^2 = (\arctan x^2)^2 = \arctan^2 (x^2)$. Differentiate $u = \arctan x^2$ to determine du.

$$u = \arctan x^2$$

$$du = \frac{1}{1 + \left(x^2\right)^2} \cdot \frac{d}{dx}\left(x^2\right)$$

$$du = \frac{1}{1 + x^4} \cdot 2x\, dx$$

$$\frac{du}{2} = \frac{x\, dx}{1 + x^4}$$

> The $x\, dx$ in the numerator and the $1 + x^4$ in the denominator are replaced by $du/2$.

$$\int \frac{x\, dx}{\arctan x^2 \left(1 + x^4\right)\sqrt{\arctan^2 \left(x^2\right) - 25}} = \int \frac{du/2}{u\sqrt{u^2 - a^2}}$$

$$= \frac{1}{2} \cdot \frac{1}{a} \operatorname{arcsec}\left(\frac{|u|}{a}\right) + C$$

$$= \frac{1}{2} \cdot \frac{1}{5} \operatorname{arcsec}\left(\frac{\left|\arctan x^2\right|}{5}\right) + C$$

> Check out the graph of $y = \tan^{-1} x$ on your calculator—it's above the x-axis when $x > 0$.

Note that $\arctan x^2 > 0$ for all x, so the absolute values are unnecessary.

$$= \frac{1}{10} \operatorname{arcsec}\left(\frac{\arctan x^2}{5}\right) + C$$

Completing the Square

For quadratics down below and no variables up top

20.21 Integrate the expression: $\int \dfrac{dx}{(4x - 1)^2 + 9}$.

Apply the inverse tangent antiderivative from Problem 20.15, such that $a = 3$ and $u = 4x - 1$. Therefore, $du = 4\, dx$ and $\dfrac{du}{4} = dx$.

$$\int \frac{dx}{(4x - 1)^2 + 9} = \int \frac{du/4}{u^2 + a^2}$$

$$= \frac{1}{4} \cdot \frac{1}{a} \cdot \arctan\left(\frac{u}{a}\right) + C$$

$$= \frac{1}{4}\left(\frac{1}{3}\right)\arctan\left(\frac{4x - 1}{3}\right) + C$$

$$= \frac{1}{12}\arctan\left(\frac{4x - 1}{3}\right) + C$$

20.22 Complete the square in the denominator of the integrand: $\int \dfrac{dx}{x^2 - 8x + 20}$.

> Just like in Problem 20.6, adding and subtracting something at the same time really means you're adding 0, which won't change the value of the integral.

Compute the square of one-half the x-coefficient: $\left(-8 \cdot \dfrac{1}{2}\right)^2 = 16$. Add and subtract this value from the denominator.

$$\int \frac{dx}{x^2 - 8x + 20} = \int \frac{dx}{x^2 - 8x + 20 + \mathbf{16} - \mathbf{16}}$$

Reorder the terms in the denominator so that the 16 is grouped with the x-terms and -16 is grouped with the constant.

$$= \int \frac{dx}{\left(x^2 - 8x + 16\right) + (20 - 16)}$$

Factor the trinomial, which (as a result of the above arithmetic manipulation) will be a perfect square. Combine the constants as well.

$$= \int \frac{dx}{(x-4)^2 + 4}$$

20.23 Integrate the expression $\int \dfrac{dx}{(x-4)^2 + 4}$, generated by Problem 20.22.

Integrate using the method described by Problem 20.21, setting $a = 2$, $u = x - 4$, and $du = dx$.

$$\int \frac{dx}{(x-4)^2 + 4} = \int \frac{du}{u^2 + a^2}$$

$$= \frac{1}{a}\arctan\left(\frac{u}{a}\right) + C$$

$$= \frac{1}{2}\arctan\left(\frac{x-4}{2}\right) + C$$

20.24 Integrate $\int \dfrac{x\,dx}{x^2 - 8x + 20}$.

> The numerator already has an x—it just needs a −4. If you put a −4 in there, you have to put 4 in there at the same time (because −4 + 4 = 0, and adding 0 won't change anything.)

Attempt to integrate using variable substitution: set $u = x^2 - 8x + 20$ and differentiate.

$$du = (2x - 8)\,dx$$
$$du = 2(x - 4)\,dx$$
$$\frac{du}{2} = (x - 4)\,dx$$

In order to apply the variable substitution technique, the numerator must be $(x - 4)\,dx$ instead of $x\,dx$. Add and subtract 4 in the numerator.

$$\int \frac{x\,dx}{x^2 - 8x + 20} = \int \frac{x - 4 + 4}{x^2 - 8x + 20}\,dx$$

Split the expression into the sum of two integrals, such that $x - 4$ (the expression required for variable substitution) is one of the numerators.

$$= \int \frac{x-4}{x^2 - 8x + 20}\, dx + \int \frac{4\, dx}{x^2 - 8x + 20}$$

$$= \int \frac{x-4}{x^2 - 8x + 20}\, dx + 4\int \frac{dx}{x^2 - 8x + 20}$$

Apply variable substitution to the first integral: $u = x^2 - 8x + 20$ and $\dfrac{du}{2} = (x-4)\, dx$, as calculated above. The second integral is calculated in Problem 20.23.

$$= \left[\int \frac{du/2}{u} \right] + \left[4\int \frac{dx}{x^2 - 8x + 20} \right]$$

$$= \left[\frac{1}{2}\ln|u| \right] + \left[4 \cdot \frac{1}{2}\arctan\left(\frac{x-4}{2} \right) \right] + C$$

$$= \frac{1}{2}\ln\left(x^2 - 8x + 20 \right) + 2\arctan\left(\frac{x-4}{2} \right) + C$$

> Every point on the graph of $y = x^2 - 8x + 20$ is above the x-axis, so no matter what x you plug in, you'll get out a positive number. That means you can drop the absolute value bars.

20.25 Integrate the expression: $\int \dfrac{dx}{2x^2 - 4x + 14}$.

In order to complete the square, the x^2-term must have a coefficient of 1.

$$\int \frac{dx}{2x^2 - 4x + 14} = \int \frac{dx}{2\left(x^2 - 2x + 7 \right)}$$

$$= \frac{1}{2}\int \frac{dx}{x^2 - 2x + 7}$$

> If it doesn't, factor the coefficient of x^2 out of EVERY term in the denominator.

Complete the square in the denominator and apply the inverse tangent antiderivative.

$$= \frac{1}{2}\int \frac{dx}{x^2 - 2x + 1 + 7 - 1}$$

$$= \frac{1}{2}\int \frac{dx}{\left(x^2 - 2x + 1 \right) + (7 - 1)}$$

$$= \frac{1}{2}\int \frac{dx}{\left(x - 1 \right)^2 + 6}$$

Let $a = \sqrt{6}$, $u = x - 1$, and $du = dx$.

$$= \frac{1}{2}\int \frac{du}{u^2 + a^2}$$

$$= \frac{1}{2} \cdot \frac{1}{\sqrt{6}}\arctan\left(\frac{x-1}{\sqrt{6}} \right) + C$$

Rationalize the expression.

$$\frac{\sqrt{6}}{12}\arctan\left(\frac{\sqrt{6}\,(x-1)}{6} \right) + C$$

20.26 Integrate the expression: $\int \dfrac{5\,dx}{3x^2 - 2x + 13}$.

The leading coefficient of the quadratic must be 1, so factor 3 out of each term in the denominator.

$$\int \frac{5\,dx}{3x^2 - 2x + 13} = \int \frac{5\,dx}{3\left[x^2 - (2/3)x + 13/3\right]}$$

$$= \frac{5}{3}\int \frac{dx}{x^2 - (2/3)x + 13/3}$$

> **Get ready, because this is going to get ugly. To "factor" 3 out of 2, divide 2 by 3: 2/3. Same thing with 13: 13/3.**

Compute the square of half of the x-coefficient.

$$\left[\left(-\frac{2}{3}\right)\left(\frac{1}{2}\right)\right]^2 = \left(-\frac{1}{3}\right)^2 = \frac{1}{9}$$

$$= \frac{5}{3}\int \frac{dx}{\left[x^2 - (2/3)x + 1/9\right] + (13/3 - 1/9)}$$

$$= \frac{5}{3}\int \frac{dx}{(x - 1/3)^2 + 38/9}$$

> **The number inside the perfect square will always be half the x-coefficient, so the factored form of this quadratic is $\left(x - \dfrac{1}{3}\right)^2$.**

Apply the inverse trigonometric antiderivative formula such that $\sqrt{\dfrac{38}{9}} = \dfrac{\sqrt{38}}{3} = a$, $u = x - \dfrac{1}{3}$, and $du = dx$.

$$= \frac{5}{3}\int \frac{du}{u^2 + a^2}$$

$$= \frac{5}{3}\cdot\frac{1}{a}\arctan\left(\frac{u}{a}\right) + C$$

$$= \frac{5}{3}\cdot\frac{1}{\sqrt{38}/3}\arctan\left(\frac{x - 1/3}{\sqrt{38}/3}\right) + C$$

Eliminate the complex fractions by multiplying their numerators and denominators by $\dfrac{3}{\sqrt{38}}$.

$$= \frac{5}{3}\cdot\frac{3}{\sqrt{38}}\arctan\left[\frac{3}{\sqrt{38}}\left(x - \frac{1}{3}\right)\right] + C$$

$$= \frac{5}{\sqrt{38}}\arctan\left(\frac{3x - 1}{\sqrt{38}}\right) + C$$

Rationalize the expression to get the equivalent solution

$$\frac{5\sqrt{38}}{38}\arctan\left[\frac{\sqrt{38}(3x - 1)}{38}\right] + C.$$

Partial Fractions
A fancy way to break down big fractions

20.27 What is the final goal of partial fraction decomposition?

The ultimate goal of partial fraction decomposition is to rewrite a single fraction as a sum of fractions whose denominators are factors (or powers of factors) of the original fraction.

> Problem 20.32 explains what this weird little qualifying statement means. Until then, don't worry about it.

20.28 Perform partial fraction decomposition on the rational expression: $\dfrac{2x-3}{x^2-25}$.

Factor the denominator.

$$\frac{2x-3}{x^2-25}=\frac{2x-3}{(x+5)(x-5)}$$

The goal of partial fraction decomposition is to rewrite the expression as a sum of fractions whose denominators are the factors of x^2-25.

$$\frac{2x-3}{(x+5)(x-5)}=\frac{A}{x+5}+\frac{B}{x-5}$$

> The denominators $x+5$ and $x-5$ have degree 1. The numerators have a degree that's exactly one smaller (0). In other words, the A and B values are real numbers.

Eliminate all of the fractions in the equation by multiplying every term by the least common denominator $(x+5)(x-5)$.

$$\left[\frac{(x+5)(x-5)}{1}\right]\left[\frac{2x-3}{(x+5)(x-5)}\right]=\left[\frac{(x+5)(x-5)}{1}\right]\left[\frac{A}{x+5}+\frac{B}{x-5}\right]$$

$$\frac{(x+5)(x-5)(2x-3)}{(x+5)(x-5)}=\frac{(x+5)(x-5)A}{x+5}+\frac{(x+5)(x-5)B}{x-5}$$

$$2x-3=(x-5)A+(x+5)B$$

Distribute A and B.

$$2x-3=Ax-5A+Bx+5B$$

Group like terms and factor x out of the variable terms.

$$2x-3=Ax+Bx-5A+5B$$
$$2x-3=x(A+B)+(-5A+5B)$$

If the expressions on the right and left sides of the equation are equal, their coefficients must be equal. The x-coefficient on the left side of the equation is 2, so the x-coefficient on the right side $(A+B)$ must equal 2 as well: $A+B=2$. Similarly, the constant on the left (-3) must equal the constant on the right $(-5A+5B)$: $-5A+5B=-3$. In order to identify A and B, you must solve the system of equations.

$$\begin{cases} A+B=2 \\ -5A+5B=-3 \end{cases}$$

The system is quickly solved using substitution.

$$-5A + 5B = -3$$
$$-5(2 - B) + 5B = -3$$
$$-10 + 5B + 5B = -3$$
$$10B = 7$$
$$B = \frac{7}{10}$$

Solve the first equation for A to get A = 2 – B and substitute that into the second equation.

Substitute B into either equation of the system to calculate A.

$$A + B = 2$$
$$A + \frac{7}{10} = 2$$
$$A = \frac{20}{10} - \frac{7}{10}$$
$$A = \frac{13}{10}$$

Substitute the values of A and B into the original decomposition equation.

$$\frac{2x - 3}{(x + 5)(x - 5)} = \frac{A}{x + 5} + \frac{B}{x - 5} = \frac{13/10}{x + 5} + \frac{7/10}{x - 5} = \frac{13}{10(x + 5)} + \frac{7}{10(x - 5)}$$

20.29 Verify the partial fraction decomposition from Problem 20.28 by demonstrating that $\dfrac{2x - 3}{(x + 5)(x - 5)} = \dfrac{13}{10(x + 5)} + \dfrac{7}{10(x - 5)}$.

Can't remember how to find the least common denominator? Look at Problems 3.1 and 3.2.

Eliminate the fractions by multiplying by the least common denominator: $10(x + 5)(x - 5)$.

$$\left[\frac{10(x - 5)(x + 5)}{1}\right]\left[\frac{2x - 3}{(x + 5)(x - 5)}\right] = \left[\frac{10(x - 5)(x + 5)}{1}\right]\left[\frac{13}{10(x + 5)}\right] + \left[\frac{10(x - 5)(x + 5)}{1}\right]\left[\frac{7}{10(x - 5)}\right]$$

$$\frac{10\cancel{(x + 5)}\cancel{(x - 5)}(2x - 3)}{\cancel{(x + 5)}\cancel{(x - 5)}} = \frac{\cancel{10} \cdot 13\cancel{(x + 5)}(x - 5)}{\cancel{10}\cancel{(x + 5)}} + \frac{\cancel{10} \cdot 7\cancel{(x - 5)}(x + 5)}{\cancel{10}\cancel{(x - 5)}}$$

$$10(2x - 3) = 13(x - 5) + 7(x + 5)$$
$$20x - 30 = 13x - 65 + 7x + 35$$
$$20x - 30 = 20x - 30$$

This is a true statement, so the original statement—equating the fraction to its partial fraction decomposition—is true as well.

20.30 Integrate the expression: $\int \dfrac{2x-3}{x^2-25}\,dx$, from Problem 20.28.

According to Problem 20.28, $\dfrac{2x-3}{x^2-25} = \dfrac{13}{10(x+5)} + \dfrac{7}{10(x-5)}$.

$$\int \frac{2x-3}{x^2-25}\,dx = \int \frac{13\,dx}{10(x+4)} + \int \frac{7\,dx}{10(x-5)}$$

$$= \frac{13}{10}\int \frac{dx}{x+4} + \frac{7}{10}\int \frac{dx}{x-5}$$

Integrate each expression using variable substitution: $u = x+4$, $v = x-5$, $du = dx$, and $dv = dx$.

$$= \frac{13}{10}\int \frac{du}{u} + \frac{7}{10}\int \frac{dv}{v}$$

$$= \frac{13}{10}\ln|u| + \frac{7}{10}\ln|v| + C$$

$$= \frac{13}{10}\ln|x+4| + \frac{7}{10}\ln|x-5| + C$$

> Even though it's not a huge deal, don't use u for two different denominators in the same problem. Use v for the second one to avoid confusion.

There are alternate ways to write the solution if logarithmic properties are applied, but the above solution is preferable, as it clearly identifies the partial fractions from which it is derived and is unencumbered by unwieldy rational exponents.

20.31 Integrate the expression: $\int \dfrac{11x-15}{4x^2-3x}\,dx$.

Factor the denominator to get $\int \dfrac{11x-15}{x(4x-3)}\,dx$. Perform partial fraction decomposition on the integrand, as described by Problem 20.28.

$$\frac{11x-15}{x(4x-3)} = \frac{A}{x} + \frac{B}{4x-3}$$

$$\frac{x(4x-3)}{1}\left[\frac{11x-15}{x(4x-3)}\right] = \frac{x(4x-3)}{1}\left[\frac{A}{x} + \frac{B}{4x-3}\right]$$

$$11x-15 = A(4x-3) + Bx$$

$$11x-15 = 4Ax - 3A + Bx$$

$$11x-15 = x(4A+B) - 3A$$

Set the x-coefficients on both sides of the equation equal, and do the same for the constants on both sides of the equation. This generates a system of equations.

$$\begin{cases} 4A+B = 11 \\ -3A = -15 \end{cases}$$

Solve the second equation to get $A = 5$. Substitute $A = 5$ into the first equation to calculate B.

$$4A + B = 11$$
$$4(5) + B = 11$$
$$B = -9$$

Rewrite the integrand according to its partial fraction decomposition.

$$\int \frac{11x - 15}{4x^2 - 3x} dx = \int \frac{A}{x} dx + \int \frac{B}{4x - 3} dx$$
$$= \int \frac{5}{x} dx + \int \frac{-9}{4x - 3} dx$$
$$= 5\ln|x| - \frac{9}{4}\ln|4x - 3| + C$$

> Integrate using variable substitution: $u = 4x - 3$ and $du = 4\,dx$, so $du/4 = dx$. That's where the denominator 4 comes from in the next step.

20.32 Perform partial fraction decomposition: $\dfrac{x}{4x^2 + 4x + 1}$.

Factor the denominator to get $(2x + 1)(2x + 1) = (2x + 1)^2$; $2x + 1$ is a repeated factor of the quadratic. Therefore, the partial fraction decomposition must include all natural number exponents of the repeated factor, up to and including its original exponent (i.e., powers from 1 to n if the factor is raised to the n power).

> The whole factor is raised to a power, so you have to include both $(2x + 1)^2$ and $2x + 1$ as decomposition denominators. If it had factored into $(2x + 1)^4$, you'd have had to include $(2x + 1)^4$, $(2x + 1)^3$, $(2x + 1)^2$, and $2x + 1$. Keep subtracting 1 from the power until you get to 1.

$$\frac{x}{(2x+1)^2} = \frac{A}{2x+1} + \frac{B}{(2x+1)^2}$$
$$\frac{(2x+1)^2}{1}\left[\frac{x}{(2x+1)^2}\right] = \frac{(2x+1)^2}{1}\left[\frac{A}{2x+1} + \frac{B}{(2x+1)^2}\right]$$
$$x = A(2x+1) + B$$
$$x = 2Ax + A + B$$

As both x-coefficients must be equal, $2A = 1$, so $A = \dfrac{1}{2}$. The constants on both sides of the equation must be equal as well, so $A + B = 0$. Therefore, $B = -\dfrac{1}{2}$. Substitute these values into the partial fraction decomposition equation.

> Substitute $A = ½$ into $A + B = 0$ to get $½ + B = 0$. Solve for B: $B = 0 - ½ = ½$.

$$\frac{x}{4x^2 + 4x + 1} = \frac{1/2}{2x+1} + \frac{-1/2}{(2x+1)^2}$$
$$= \frac{1}{2(2x+1)} - \frac{1}{2(2x+1)^2}$$

20.33 Integrate the expression $\int \dfrac{x}{4x^2+4x+1}\,dx$, from Problem 20.32.

According to Problem 20.32, $\dfrac{x}{4x^2+4x+1} = \dfrac{1}{2(2x+1)} - \dfrac{1}{2(2x+1)^2}$.

$$\int \frac{x}{4x^2+4x+1}\,dx = \int \frac{dx}{2(2x+1)} - \int \frac{dx}{2(2x+1)^2}$$

$$= \frac{1}{2}\int \frac{dx}{2x+1} - \frac{1}{2}\int \frac{dx}{(2x+1)^2}$$

Integrate both expressions using variable substitution: $u = 2x + 1$ and $du = 2\,dx$, so $\dfrac{du}{2} = dx$.

$$= \frac{1}{2}\int \frac{du/2}{u} - \frac{1}{2}\int \frac{du/2}{u^2}$$

$$= \frac{1}{2}\cdot\frac{1}{2}\int \frac{du}{u} - \frac{1}{2}\cdot\frac{1}{2}\int u^{-2}\,du$$

$$= \frac{1}{4}\ln|u| - \frac{1}{4}\cdot\frac{u^{-1}}{-1} + C$$

$$= \frac{1}{4}\ln|2x+1| + \frac{1}{4}\cdot\frac{1}{2x+1} + C$$

$$= \frac{\ln|2x+1|}{4} + \frac{1}{8x+4} + C$$

20.34 Perform partial fraction decomposition: $\dfrac{2x^3 - 5x^2 + 6x - 3}{x^4 + 3x^2}$.

Factor the denominator.

$$\frac{2x^3 - 5x^2 + 6x - 3}{x^2\left(x^2 + 3\right)}$$

Unlike the preceding partial fractions exercises, the factor $x^2 + 3$ requires the linear numerator $Ax + B$ rather than a constant numerator. Note that x^2 is considered a repeated factor, so you must include a denominator of x^1 as well as a denominator of x^2.

$$\frac{2x^3 - 5x^2 + 6x - 3}{x^2\left(x^2 + 3\right)} = \frac{A}{x} + \frac{B}{x^2} + \frac{Cx + D}{x^2 + 3}$$

$$\frac{x^2\left(x^2+3\right)}{1}\left[\frac{2x^3 - 5x^2 + 6x - 3}{x^2\left(x^2+3\right)}\right] = \frac{x^2\left(x^2+3\right)}{1}\left[\frac{A}{x} + \frac{B}{x^2} + \frac{Cx + D}{x^2 + 3}\right]$$

$$2x^3 - 5x^2 + 6x - 3 = Ax\left(x^2 + 3\right) + B\left(x^2 + 3\right) + x^2\left(Cx + D\right)$$

$$2x^3 - 5x^2 + 6x - 3 = Ax^3 + 3Ax + Bx^2 + 3B + Cx^3 + Dx^2$$

$$2x^3 - 5x^2 + 6x - 3 = (A + C)x^3 + (B + D)x^2 + 3Ax + 3B$$

RULE OF THUMB: The numerator of the mini-fractions have to be one degree less than the denominator. If the denominator is $x + 3$ (degree 1), the numerator is $Ax^0 = A$. If the denominator is $x^2 + 3$, the numerator is $Ax + B$ (or $Cx + D$, the letters don't matter). If the denominator were $x^3 + 3$, you'd use the numerator of $Ax^2 + Bx + C$.

Match the coefficients of the terms on both sides of the equations; you can immediately calculate A and B.

x^3 coefficients	x^2 coefficients	x coefficients	constants
$A+C=2$	$B+D=-5$	$3A=6$ $A=2$	$3B=-3$ $B=-1$

Substitute $A=2$ into the equation $A+C=2$ to determine that $C=0$. Substitute $B=-1$ into the equation $B+D=-5$ to determine that $D=-4$. Substitute A, B, C, and D into the partial fraction decomposition equation.

$$\frac{2x^3-5x^2+6x-3}{x^4+3x^2}=\frac{A}{x}+\frac{B}{x^2}+\frac{Cx+D}{x^2+3}$$
$$=\frac{2}{x}+\frac{-1}{x^2}+\frac{(0)x+(-4)}{x^2+3}$$
$$=\frac{2}{x}-\frac{1}{x^2}-\frac{4}{x^2+3}$$

20.35 Integrate the expression: $\int\dfrac{2x^3-5x^2+6x-3}{x^4+3x^2}\,dx$, from Problem 20.34.

According to Problem 20.34, $\dfrac{2x^3-5x^2+6x-3}{x^4+3x^2}=\dfrac{2}{x}-\dfrac{1}{x^2}-\dfrac{4}{x^2+3}$.

$$\int\frac{2x^3-5x^2+6x-3}{x^4+3x^2}\,dx=\int\frac{2}{x}\,dx-\int\frac{1}{x^2}\,dx-\int\frac{4}{x^2+3}\,dx$$
$$=2\int\frac{dx}{x}-\int x^{-2}\,dx-4\int\frac{dx}{x^2+3}$$
$$=2\ln|x|-\frac{x^{-1}}{-1}-4\left(\frac{x}{\sqrt{3}}\right)\arctan\left(\frac{x}{\sqrt{3}}\right)+C$$
$$=2\ln|x|+\frac{1}{x}-\frac{4x}{\sqrt{3}}\arctan\left(\frac{x}{\sqrt{3}}\right)+C$$

Use the arctangent antiderivative here. Set $u=x$, $du=dx$, and $a=\sqrt{3}$. That way, $\int\dfrac{du}{u^2+a^2}=\dfrac{1}{\sqrt{3}}\arctan\left(\dfrac{x}{\sqrt{3}}\right)+C$.

Rationalize the expression to get an equivalent solution:

$$2\ln|x|+\frac{1}{x}-\frac{4x\sqrt{3}}{3}\arctan\left(\frac{x\sqrt{3}}{3}\right)+C.$$

Chapter 21
ADVANCED INTEGRATION TECHNIQUES

Even more ways to find integrals—must be your birthday

Integration by parts, the first topic investigated in this chapter, is an extremely useful tool, the importance of which cannot be overestimated. By comparison, the remainder of the techniques discussed in this chapter (trigonometric substitution and improper integration), are less universally useful. Like most integration techniques, they are exceedingly useful in very specific situations, but integration by parts is an excellent "tool of last resort" when other integration techniques fail.

It's just what you always wanted—another grab bag of integration methods. Of the three topics in the chapter, you will use the first one a heck of a lot more than the others. You can even use it to integrate f(x) = ln x, an antiderivative that has been conspicuously absent until now. (In fact, you'll integrate ln x in Problem 3.) Trig powers and trig substitution are only handy if you've got an integral with (surprise, surprise) trig functions in it that is defying all of your integration efforts. The last topic, improper integration, explains how to handle definite integrals that either (A) have infinity (instead of a real number) as a limit of integration, or (B) have a vertical asymptote cutting through the interval you're integrating over.

Integration by Parts
It's like the product rule, but for integrals

21.1 The method of integration by parts states that $\int u\,dv = uv - \int v\,du$, given differentiable functions u and v. Generate this formula by differentiating the product uv.

> You have no idea how u and v are defined (using x's? using y's?) so the derivatives of u and v are very vague: du and dv.

Because u and v are functions, you must apply the product rule.

$$d(uv) = u \cdot dv + v \cdot du$$

Integrate both sides of the equation.

$$\int d(uv) = \int (u \cdot dv + v \cdot du)$$
$$\int d(uv) = \int u\,dv + \int v\,du$$

Note that $\int d(uv) = uv$, because the antiderivative of a function's derivative equals the function itself.

> You don't even have to include "+C" because the original function, uv, has no constant. The integral and the derivative basically cancel each another out.

$$uv = \int u\,dv + \int v\,du$$

Solve the equation for $\int u\,dv$.

$$uv - \int v\,du = \int u\,dv$$

21.2 Integrate using the parts method: $\int xe^x\,dx$.

> Leave off the +C while you integrate, or things could get messy. Just remember to attach +C to your final answers when you're dealing with indefinite integrals.

Rewrite the integral as $\int u\,dv$. This requires you to define u and dv using parts of the integrand (hence the name "integration by parts"). It is most helpful to set u equal to an easily differentiable quantity and dv equal to an easily integrable quantity: $u = x$ and $dv = e^x\,dx$. Therefore, $du = dx$ and $v = \int dv = \int e^x\,dx = e^x$. Substitute these values into the formula from Problem 21.1.

$$\int u\,dv = uv - \int v\,du$$
$$\int xe^x\,dx = xe^x - \int e^x\,dx$$
$$= xe^x - e^x + C$$

21.3 Integrate the expression: $\int \ln x\,dx$.

> When you set up integration by parts, the u and dv you pick have to multiply together to get the original integral. In this case, u = ln x and dv = dx. So, u dv = ln x dx, the original integral.

Integrate by parts, setting $u = \ln x$ and $dv = dx$. Differentiate u and integrate dv to determine du and v: $du = d(\ln x) = \dfrac{dx}{x}$ and $v = \int dv = \int dx = x$.

$$\int u\,dv = uv - \int v\,du$$
$$\int \ln x\,dx = (\ln x)(x) - \int (x)\left(\frac{dx}{x}\right)$$
$$= x\ln x - \int dx$$
$$= x\ln x - x + C$$

21.4 Integrate: $\int x^2 \sin x \, dx$.

Neither x^2 nor $\sin x$ is difficult to differentiate—this allows you additional freedom to choose u and dv. Given this choice, however, you should set u equal to the function that will eventually equal zero if it is differentiated repeatedly. Therefore, $u = x^2$, $dv = \sin x \, dx$, $du = d(x^2) = 2x \, dx$, and $v = \int dv = \int \sin x \, dx = -\cos x$. Substitute these values into the integration by parts formula.

$$\int u \, dv = uv - \int v \, du$$
$$\int x^2 \sin x \, dx = \left(x^2\right)(-\cos x) - \int (-\cos x)(2x \, dx)$$
$$= -x^2 \cos x + 2\int x \cos x \, dx$$

> The derivative of x^2 is $2x$, the derivative of $2x$ is 2, and the derivative of 2 is 0. Since x^2 eventually has a derivative of 0, set $u = x^2$.

Integrating $\int x \cos x \, dx$ again requires integration by parts. Set $u = x$ and $dv = \cos x \, dx$; it follows that $du = dx$ and $v = \sin x$.

$$\int u \, dv = uv - \int v \, du$$
$$\int x \cos x \, dx = x \sin x - \int \sin x \, dx$$
$$= x \sin x - (-\cos x) + C$$
$$= x \sin x + \cos x + C$$

This is not the final answer. Rather, this is the value of $\int x \cos x \, dx$, part of the original attempt to integrate by parts. Substitute this antiderivative into the original integration by parts formula.

$$\int x^2 \sin x \, dx = -x^2 \cos x + 2\int x \cos x \, dx$$
$$= -x^2 \cos x + 2(x \sin x + \cos x + C)$$
$$= -x^2 \cos x + 2x \sin x + 2\cos x + C$$

> Technically, C gets multiplied by 2 and then added to another arbitrary constant (you integrate twice in this problem, which will give you two constants), but all of that stuff results in some unknown number, and it's easiest just to keep writing that number as C.

21.5 According to Problem 21.2, $\int xe^x \, dx = xe^x - e^x + C$. Verify this antiderivative using the integration by parts tabular method.

The *tabular* method consists of a *table* with three columns. The first column contains u and its subsequent derivatives, the second column contains dv and its subsequent integrals, and the final column alternates between $+1$ and -1 (always beginning with $+1$).

Use the same values for u and dv from Problem 21.2. Find consecutive derivatives of u until the derivative equals 0 and list them vertically in the left column. (The final number in the left column must be 0.) Fill the second column with a corresponding number of antiderivatives. The right column should always have one more row in it than the other two columns.

> Antiderivatives of dv, that is.

u	dv	± 1
x	e^x	$+1$
1	e^x	-1
0	e^x	$+1$
		-1

Multiply each term in the left column with the other values along a downward diagonal, as illustrated by Figure 21-1.

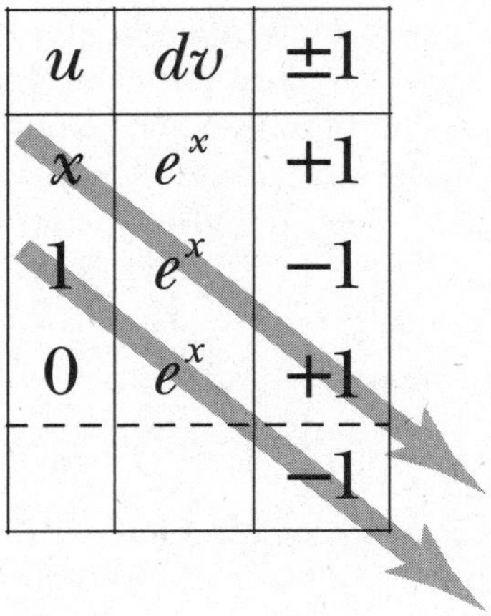

u	dv	± 1
x	e^x	$+1$
1	e^x	-1
0	e^x	$+1$
		-1

Figure 21-1 *Starting with the first term in the u column, move down and to the right, following the paths indicated by the arrows. Find the product of the terms along the path. Then, move to the next term in the u column (1) and multiply along a similar path. (There is no need to begin a path at 0, as the product will be 0.) Finally, add the products together.*

Multiply along the paths in Figure 21-1 and add the results.

$$(x)(e^x)(1)+(1)(e^x)(-1)=xe^x-e^x$$

Therefore, $\int xe^x\,dx=xe^x-e^x+C$, which verifies the solution to Problem 21.2.

Don't forget to add "+C" to the end because it's an indefinite integral. To practice using integration by parts with DEFINITE integrals, check out Problems 22.11, 22.12, and 22.31 through 22.33.

21.6 According to Problem 21.4, $\int x^2\sin x\,dx=-x^2\cos x+2x\sin x+2\cos x+C$. Verify this antiderivative using the integration by parts tabular method.

Construct a table (as explained in Problem 21.5) using the values of u and dv defined by Problem 21.4: $u=x^2$ and $dv=\sin x\,dx$.

u	dv	± 1
x^2	$\sin x$	$+1$
$2x$	$-\cos x$	-1
2	$-\sin x$	$+1$
0	$\cos x$	-1
		$+1$

Draw diagonal paths (as illustrated in Figure 21-2), calculate the product of the terms on each, and compute the sum of those products.

u	dv	± 1
x^2	$\sin x$	$+1$
$2x$	$-\cos x$	-1
2	$-\sin x$	$+1$
0	$\cos x$	-1
		$+1$

Figure 21-2 *There are three nonzero terms in the u-column, so products must be calculated along three paths (as indicated by the arrows in the diagram).*

$$\int x^2 \sin x \, dx = \left(x^2\right)(-\cos x)(+1) + (2x)(-\sin x)(-1) + (2)(\cos x)(1) + C$$

$$= -x^2 \cos x + 2x \sin x + 2\cos x + C$$

21.7 Integrate the expression: $\int e^x \sin x \, dx$.

Integrate by parts, setting $u = \sin x$ and $dv = e^x \, dx$. Accordingly, $du = \cos x \, dx$ and $v = e^x$.

$$\int u \, dv = uv - \int v \, du$$

$$\int e^x \sin x \, dx = e^x \sin x - \int e^x \cos x \, dx$$

Integrating $\int e^x \cos x \, dx$ also requires integration by parts. $\longleftarrow$

$$\int e^x \sin x \, dx = e^x \sin x - \int e^x \cos x \, dx$$

$$\int e^x \sin x \, dx = e^x \sin x - \left(e^x \cos x + \int e^x \sin x \, dx\right)$$

$$\int e^x \sin x \, dx = e^x \sin x - e^x \cos x - \int e^x \sin x \, dx$$

Add $\int e^x \sin x \, dx$ to both sides of the equation.

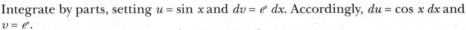

$$\int e^x \sin x \, dx + \int e^x \sin x \, dx = e^x \sin x - e^x \cos x - \int e^x \sin x \, dx + \int e^x \sin x \, dx$$

$$2\int e^x \sin x \, dx = e^x (\sin x - \cos x)$$

$$\int e^x \sin x \, dx = \frac{e^x (\sin x - \cos x)}{2} + C$$

This time, set $u = \cos x$ and $dv = e^x \, dx$. That means $du = -\sin x \, dx$ and $v = e^x$. Integration by parts tells you that
$$\int e^x \cos x \, dx =$$
$$e^x \cos x + \int e^x \sin x \, dx.$$

Plug this thing into the original integration by parts formula.

Trigonometric Substitution

Using identities and little right triangle diagrams

21.8 Compute the definite integral: $\int_{\pi/4}^{\pi/3} \tan^2 x \, dx$.

According to a Pythagorean trigonometric identity, $1 + \tan^2 x = \sec^2 x$. Therefore, $\tan^2 x = \sec^2 x - 1$.

$$\int_{\pi/4}^{\pi/3} \tan^2 x \, dx = \int_{\pi/4}^{\pi/3} \left(\sec^2 x - 1\right) dx$$

$$= \int_{\pi/4}^{\pi/3} \sec^2 x \, dx - \int_{\pi/4}^{\pi/3} 1 \, dx$$

$$= \left(\tan x \Big|_{\pi/4}^{\pi/3}\right) - \left(x \Big|_{\pi/4}^{\pi/3}\right)$$

$$= \left(\tan \frac{\pi}{3} - \tan \frac{\pi}{4}\right) - \left(\frac{\pi}{3} - \frac{\pi}{4}\right)$$

Evaluate the tangent by rewriting it in terms of sine and cosine.

$$= \left(\frac{\sin \pi/3}{\cos \pi/3} - \frac{\sin \pi/4}{\cos \pi/4}\right) - \left(\frac{\pi}{3} - \frac{\pi}{4}\right)$$

$$= \left(\frac{\sqrt{3}/2}{1/2} - \frac{\sqrt{2}/2}{\sqrt{2}/2}\right) - \left(\frac{4\pi}{12} - \frac{3\pi}{12}\right)$$

$$= \sqrt{3} - 1 - \frac{\pi}{12}$$

21.9 Integrate the expression: $\int \cos^3 x \sin^2 x \, dx$.

Given an integral containing $\cos^a x$ and $\sin^b x$, where a and b are natural numbers, a is odd, and b is even, rewrite $\cos^a x$ as $(\cos x)(\cos^{a-1} x)$.

$$\int \cos^3 x \sin^2 x \, dx = \int \cos x \cdot \cos^2 x \cdot \sin^2 x \, dx$$

Apply the Pythagorean identity $\cos^2 x = 1 - \sin^2 x$ to rewrite the now even-powered trigonometric expression.

$$= \int \cos x \left(1 - \sin^2 x\right) \sin^2 x \, dx$$

$$= \int \cos x \sin^2 x \left(1 - \sin^2 x\right) dx$$

Distribute $\cos x \sin^2 x$.

$$= \int \left(\cos x \sin^2 x - \cos x \sin^4 x\right) dx$$

$$= \int \cos x \sin^2 x \, dx - \int \cos x \sin^4 x \, dx$$

Integrating using variable substitution: $u = \sin x$ and $du = \cos x \, dx$.

$$= \int u^2 \, du - \int u^4 \, du$$

$$= \frac{u^3}{3} - \frac{u^5}{5} + C$$

$$= \frac{1}{3} \sin^3 x - \frac{1}{5} \sin^5 x + C$$

> If the integral has cos x and sin x raised to powers and only ONE of those powers is odd, here's what to do: take a single cos x or sin x out of the odd power. In this case, cos³ x is the odd power, so rewrite it as (cos x)(cos² x).

21.10 Integrate: $\int \cos^8 x \sin^3 x \, dx$.

Rewrite $\sin^3 x$ as $(\sin x)(\sin^2 x)$.

This problem is very similar to Problem 21.9, except this time $\sin x$ is raised to an odd power and $\cos x$ is raised to an even power. Do the same things to $\sin^3 x$ that Problem 21.9 did to $\cos^3 x$.

$$\int \cos^8 x \sin^3 x \, dx = \int \cos^8 x \cdot \sin^2 x \cdot \sin x \, dx$$
$$= \int \cos^8 x \left(1 - \cos^2 x\right) \sin x \, dx$$
$$= \int \cos^8 x \sin x \left(1 - \cos^2 x\right) dx$$
$$= \int \cos^8 x \sin x \, dx - \int \cos^{10} x \sin x \, dx$$

Use variable substitution to integrate both expressions: $u = \cos x$ and $du = -\sin x \, dx$.

$$= -\int u^8 du - \left(-\int u^{10} du\right)$$
$$= -\frac{u^9}{9} + \frac{u^{11}}{11} + C$$
$$= -\frac{1}{9}\cos^9 x + \frac{1}{11}\cos^{11} x + C$$

21.11 Integrate the expression: $\int \sin^2 x \, dx$.

According to a double angle identity, $\cos 2x = 1 - 2\sin^2 x$. Therefore, $\sin^2 x = \dfrac{1 - \cos 2x}{2}$.

$$\int \sin^2 x \, dx = \int \frac{1 - \cos 2x}{2} \, dx$$
$$= \frac{1}{2}\int (1 - \cos 2x) \, dx$$
$$= \frac{1}{2}\int 1 \, dx - \frac{1}{2}\int \cos 2x \, dx$$

Power-reducing identities like this one are useful when you've got cosine or sine raised to even powers, either all by themselves (like in $\int \cos^4 x \, dx$) or multiplied together (like in $\int \sin^2 x \cos^4 x \, dx$).

Use variable substitution to integrate $\int \cos 2x \, dx$: $u = 2x$ and $du = 2dx$, so $\dfrac{du}{2} = dx$.

$$= \frac{1}{2}\int dx - \frac{1}{2}\cdot\frac{1}{2}\int \cos u \, du$$
$$= \frac{1}{2}x - \frac{1}{4}\sin u + C$$
$$= \frac{1}{2}x - \frac{1}{4}\sin 2x + C$$

21.12 Integrate: $\int \sin^3 x \, dx$.

This is the same way Problem 21.10 starts. Technically, cos x is raised to an even power in this integral: $(\cos x)^0 = 1$, so $(\cos x)^0 (\sin^3 x) = \sin^3 x$.

Rewrite $\sin^3 x$ as $(\sin x)(\sin^2 x)$.

$$\int \sin^3 x \, dx = \int \sin x \cdot \sin^2 x \, dx$$
$$= \int \sin x \left(1 - \cos^2 x\right) dx$$
$$= \int \sin x \, dx - \int \sin x \cos^2 x \, dx$$

Use variable substitution to integrate $\int \sin x \cos^2 x \, dx$: $u = \cos x$ and $du = -\sin x \, dx$.

$$= \int \sin x \, dx - \left(-\int u^2 \, du\right)$$
$$= -\cos x + \frac{u^3}{3} + C$$
$$= -\cos x + \frac{1}{3}\cos^3 x + C$$

21.13 Integrate: $\int \sin^4 x \, dx$.

Apply the $\sin^2 x$ power-reducing formula from Problem 21.11: $\sin^2 x = \dfrac{1 - \cos 2x}{2}$.

$$\int \sin^4 x \, dx = \int \left(\sin^2 x\right)^2 dx$$
$$= \int \left(\frac{1 - \cos 2x}{2}\right)^2 dx$$
$$= \int \left(\frac{1 - 2\cos 2x + \cos^2 2x}{4}\right) dx$$
$$= \frac{1}{4}\left(\int 1 \, dx - 2\int \cos 2x \, dx + \int \cos^2 2x \, dx\right)$$

Remember, cosine or sine by itself inside an integral and raised to an even power usually requires the power-reducing formulas.

Apply the power-reducing formula $\cos^2 \theta = \dfrac{1 + \cos 2\theta}{2}$ to integrate $\int \cos^2 2x \, dx$.

$$= \frac{1}{4}\left[\int 1 \, dx - 2\int \cos 2x \, dx + \int \frac{1 + \cos 2(2x)}{2} \, dx\right]$$
$$= \frac{1}{4}\left[\int 1 \, dx - 2\int \cos 2x \, dx + \frac{1}{2}\int (1 + \cos 4x) \, dx\right]$$
$$= \frac{1}{4}\left[\int 1 \, dx - 2\int \cos 2x \, dx + \frac{1}{2}\left(\int dx + \int \cos 4x \, dx\right)\right]$$

Use variable substitution to integrate the trigonometric integrals: $u = 2x$ and $\dfrac{du}{2} = dx$.

$$= \frac{1}{4}\left[x - 2 \cdot \frac{1}{2}\sin 2x + \frac{1}{2}\left(x + \frac{1}{4}\sin 4x \right) \right] + C$$

$$= \frac{1}{4}\left[x - \sin 2x + \frac{1}{2}x + \frac{1}{8}\sin 4x \right] + C$$

$$= \frac{1}{4}\left[\frac{3}{2}x - \sin 2x + \frac{1}{8}\sin 4x \right] + C$$

$$= \frac{3}{8}x - \frac{1}{4}\sin 2x + \frac{1}{32}\sin 4x + C$$

21.14 Integrate using a product-to-sum identity: $\int \cos x \cos 2x \, dx$.

> There are three product-to-sum formulas in Appendix C. This one is used only to rewrite the product of two cosines.

Apply the identity $\cos A \cdot \cos B = \dfrac{\cos(A - B) + \cos(A + B)}{2}$ such that $A = x$ and $B = 2x$.

$$\cos x \cos 2x = \frac{\cos(x - 2x) + \cos(x + 2x)}{2}$$

$$= \frac{\cos(-x) + \cos(3x)}{2}$$

$$= \frac{\cos x + \cos 3x}{2}$$

> Cosine is an even function, and all even functions have the property $f(-x) = f(x)$. That's why you drop the negative sign in the next step.

Substitute $\cos x \cos 2x = \dfrac{\cos x + \cos 3x}{2}$ into the integrand.

$$\int \cos x \cos 2x \, dx = \int \frac{\cos x + \cos 3x}{2} \, dx$$

$$= \frac{1}{2}\int \cos x \, dx + \frac{1}{2}\int \cos 3x \, dx$$

Use variable substitution to integrate $\int \cos 3x \, dx$.

$$= \frac{1}{2}\sin x + \frac{1}{2} \cdot \frac{1}{3}\sin 3x + C$$

$$= \frac{1}{2}\sin x + \frac{1}{6}\sin 3x + C$$

21.15 Identify an alternate solution to Problem 21.14 by applying a trigonometric identity to integrate $\int \cos x \cos 2x \, dx$.

Apply the double angle identity $\cos 2x = 1 - 2\sin^2 x$.

$$\int \cos x \cos 2x \, dx = \int \cos x \left(1 - 2\sin^2 x \right) dx$$

$$= \int \cos x \, dx - 2\int \cos x \sin^2 x \, dx$$

Apply variable substitution to the second integral ($u = \sin x$ and $du = \cos x\, dx$).

$$= \int \cos x\, dx - 2 \int u^2\, du$$

$$= \sin x - 2 \cdot \frac{u^3}{3} + C$$

$$= \sin x - \frac{2}{3} \sin^3 x + C$$

21.16 Integrate the expression: $\int 7x \sin x^2 \sin 4x^2\, dx$.

Apply the product-to-sum formula $\sin A \sin B = \dfrac{\cos(A-B) - \cos(A+B)}{2}$ to rewrite $\sin x^2 \sin(4x^2)$.

$$\int 7x \sin x^2 \sin 4x^2\, dx = \int 7x \left[\frac{\cos(x^2 - 4x^2) - \cos(x^2 + 4x^2)}{2} \right] dx$$

$$= \frac{7}{2} \int x \left[\cos(-3x^2) - \cos 5x^2 \right] dx$$

$$= \frac{7}{2} \int x \left[\cos 3x^2 - \cos 5x^2 \right] dx$$

$$= \frac{7}{2} \int x \cos 3x^2\, dx - \frac{7}{2} \int x \cos 5x^2\, dx$$

In the first integral, $u = 3x^2$ and $du/6 = x\, dx$. In the second integral, $v = 5x^2$ and $dv/10 = x\, dx$.

Apply variable substitution to integrate.

$$= \frac{7}{2} \cdot \frac{1}{6} \int \cos u\, du - \frac{7}{2} \cdot \frac{1}{10} \int \cos v\, dv$$

$$= \frac{7}{12} \sin 3x^2 - \frac{7}{20} \sin 5x^2 + C$$

21.17 Integrate the expression $\int \sec^2 x \tan x\, dx$ using the variable substitution $u = \sec x$.

Expand the $\sec^2 x$ factor in the integrand.

$$\int \sec^2 x \tan x\, dx = \int \sec x \cdot \sec x \cdot \tan x\, dx$$

Differentiate $u = \sec x$ to get $du = \sec x \tan x\, dx$.

$$= \int u\, du$$

$$= \frac{u^2}{2} + C$$

$$= \frac{1}{2} \sec^2 x + C$$

21.18 Use the variable substitution $u = \tan x$ to integrate $\int \sec^2 x \tan x \, dx$ and verify the solution to Problem 21.17.

If $u = \tan x$, then $du = \sec^2 x \, dx$.

$$\int \sec^2 x \tan x \, dx = \int u \, du$$

$$= \frac{u^2}{2} + C$$

$$= \frac{1}{2} \tan^2 x + C$$

Set $\frac{1}{2} \tan^2 x + C$ equal to $\frac{1}{2} \sec^2 x + C$ and prove that the expressions are equivalent. Multiply by 2 to eliminate the fractions. (Note that C is an arbitrary constant, so $2C = C$ for the purposes of the proof.) ←

> 2 times some unknown number is some other unknown number.

$$2\left[\frac{1}{2} \tan^2 x + C\right] = 2\left[\frac{1}{2} \sec^2 x + C\right]$$

$$\tan^2 x + C = \sec^2 x + C$$

Apply the trigonometric identity $\tan^2 x + 1 = \sec^2 x$.

$$\tan^2 x + C = (1 + \tan^2 x) + C$$

Note that $1 + C = C$ for the reasons described above. ←

> Some unknown number plus 1 equals some other unknown number. I have no idea why talking like this is fun but, oddly, it is.

$$\tan^2 x + C = \tan^2 x + C$$

Because this statement is true, the statement $\frac{1}{2} \tan^2 x + C = \frac{1}{2} \sec^2 x + C$ is true as well, which verifies that the solutions to Problems 21.17 and 21.18 are equivalent.

21.19 Integrate the expression: $\int \sec^6 x \tan^4 x \, dx$.

If an integrand consists of the product $\sec^a x \tan^b x$, such that where a and b are natural numbers and a is even, rewrite the integrand as $\left(\sec^2 x\right)\left(\sec^2 x\right)^{(a-2)/2}$. Keep a single $\sec^2 x$ factor separate and rewrite the rest of it as $(\sec^2 x)^n$, where n is an even number. For example, you'd change $\sec^{12} x$ into $(\sec^2 x)(\sec^{10} x)$ and then change $\sec^{10} x$ into $(\sec^2 x)^5$.

$$\int \sec^6 x \tan^4 x \, dx = \int \sec^2 x \cdot \sec^4 x \cdot \tan^4 x \, dx$$

$$= \int \sec^2 x \left(\sec^2 x\right)^2 \tan^4 x \, dx$$

Apply the Pythagorean identity $1 + \tan^2 x = \sec^2 x$ to the $\sec^2 x$ factor that is raised to the second power.

$$= \int \sec^2 x \left(1 + \tan^2 x\right)^2 \tan^4 x \, dx$$

$$= \int \sec^2 x \tan^4 x \left(1 + 2\tan^2 x + \tan^4 x\right) dx$$

$$= \int \sec^2 x \tan^4 x \, dx + 2\int \sec^2 x \tan^6 x \, dx + \int \sec^2 x \tan^8 x \, dx$$

Perform variable substitution using $u = \tan x$ and $du = \sec^2 x\, dx$.

$$= \int u^4\, du + 2\int u^6\, du + \int u^8\, du$$

$$= \frac{u^5}{5} + 2 \cdot \frac{u^7}{7} + \frac{u^9}{9} + C$$

$$= \frac{1}{5}\tan^5 x + \frac{2}{7}\tan^7 x + \frac{1}{9}\tan^9 x + C$$

21.20 Integrate the expression: $\int \sec^7 x \tan^5 x\, dx$.

If an integrand consists of the product $\sec^a x \tan^b x$, such that a and b are natural numbers and b is odd, rewrite the integrand as $(\sec x \tan x)(\sec^{a-1} x \tan^{b-1} x)$. Note that $b - 1$ will be even (because b is odd). Then rewrite $\tan^{b-1} x$ as $(\sec^2 x - 1)^{(b-1)/2}$.

$$\int \sec^7 x \tan^5 x\, dx = \int (\sec x \tan x)\sec^6 x \tan^4 x\, dx$$

$$= \int (\sec x \tan x)\sec^6 x \left(\sec^2 x - 1\right)^2 dx$$

$$= \int (\sec x \tan x)\sec^6 x \left(\sec^4 x - 2\sec^2 x + 1\right) dx$$

$$= \int (\sec x \tan x)\left(\sec^{10} x - 2\sec^8 x + \sec^6 x\right) dx$$

Distribute $\sec x \tan x$ through the quantity, and separate each term of the integrand into individual integrals.

$$= \int \sec^{10} x (\sec x \tan x)\, dx - 2\int \sec^8 x (\sec x \tan x)\, dx + \int \sec^6 x (\sec x \tan x)\, dx$$

Perform variable substitution: $u = \sec x$ and $du = \sec x \tan x\, dx$.

$$= \int u^{10}\, du - 2\int u^8\, du + \int u^6\, du$$

$$= \frac{u^{11}}{11} - 2 \cdot \frac{u^9}{9} + \frac{u^7}{7} + C$$

$$= \frac{1}{11}\sec^{11} x - \frac{2}{9}\sec^9 x + \frac{1}{7}\sec^7 x + C$$

21.21 Given $\sin \theta = \dfrac{x}{4}$, draw a right triangle to determine the value of $\cos \theta$, assuming $0 < \theta \le \dfrac{\pi}{2}$.

Recall that the sine ratio relates the side opposite an angle to the hypotenuse of the right triangle. If $\sin \theta = \dfrac{x}{4}$, the side opposite θ is x and the hypotenuse of the right triangle is 4. According to the Pythagorean theorem, the remaining side of the triangle has length $\sqrt{16 - x^2}$.

In Problem 21.19, you held $\sec^2 x$, rewrote the rest of the secants as $(\sec^2 x)$ to some even power, and then used a Pythagorean identity. In this problem, you hold out $\sec x \tan x$, rewrite the rest of the tangents as $(\tan^2 x)$ to some even power, and then use the Pythagorean identity.

This means you should draw the right triangle in the first quadrant, which means you don't have to worry about the signs of the sides like in Problems 7.26 and 7.27.

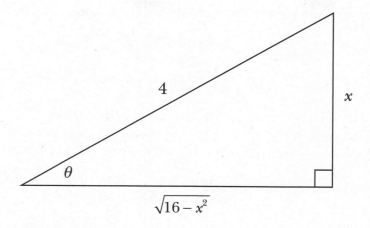

Figure 21-3 *Construct this right triangle to illustrate an acute angle θ with opposite side x and hypotenuse 4; you can then use it to evaluate other trigonometric functions of θ.*

Recall that the cosine of an angle is the quotient of its adjacent side and the hypotenuse of the triangle. Therefore, $\cos\theta = \dfrac{\sqrt{16-x^2}}{4}$.

21.22 Integrate the expression: $\int \sqrt{16-x^2}\,dx$.

This integral requires the technique of trigonometric substitution. The integrand is of the form $\sqrt{a^2-u^2}$, where $a=4$ and $u=x$. Given radicals of this form, make the trigonometric substitution $u = a\sin\theta$. Substitute u and a into the equation.

$$x = 4\sin\theta$$

Divide both sides of the equation by 4 to get $\dfrac{x}{4} = \sin\theta$, which is the equation from Problem 21.21. Differentiate $x = 4\sin\theta$.

$$dx = 4\cos\theta\,d\theta$$

Now to address the integral $\int \sqrt{16-x^2}\,dx$. Rewrite the integrand as a trigonometric function based on the triangle in Figure 21-3. According to Problem 21.21, $\cos\theta = \dfrac{\sqrt{16-x^2}}{4}$, so $4\cos\theta = \sqrt{16-x^2}$. (Note that the cosine ratio is chosen so that the constant leg of the triangle, 4, is used rather than side x, to reduce the number of variables involved.)

$$\int \sqrt{16-x^2}\,dx = \int 4\cos\theta\,dx$$

Replace dx using the derivative identified above: $dx = 4\cos\theta\,d\theta$.

$$= \int (4\cos\theta)(4\cos\theta\,d\theta)$$
$$= 16\int \cos^2\theta\,d\theta$$

Integration by trig substitution (Problems 21.22-21.29) helps you integrate things like $\sqrt{x^2-u^2}$, $\sqrt{u^2-x^2}$, and $\sqrt{u^2+a^2}$ containing function u and real number a.

So, if it's a trig substitution problem and you see $\sqrt{a^2-u^2}$, the first thing you write is $u = a\sin\theta$.

Integrate using a power-reducing formula (as demonstrated by Problem 21.11).

$$= 16 \int \frac{1 + \cos 2\theta}{2} \, d\theta$$

$$= \frac{16}{2} \left[\int 1 \, d\theta + \int \cos 2\theta \, d\theta \right]$$

$$= 8\theta + 8 \cdot \frac{1}{2} \sin 2\theta + C$$

> **This is important because you can only use the triangle in Figure 21-3 to figure out single angle trig functions (like $\cos \theta$ and $\tan \theta$), not double angles like $\sin 2\theta$.**

Apply the double angle trigonometric identity $\sin 2\theta = 2 \sin \theta \cos \theta$.

$$= 8\theta + 4(2 \sin \theta \cos \theta) + C$$

$$= 8\theta + 8 \sin \theta \cos \theta + C$$

You must now replace θ, $\sin \theta$, and $\cos \theta$ with expressions written in terms of x. Recall that $x = 4 \sin \theta$. Solve for θ using the inverse sine function.

$$x = 4 \sin \theta$$

$$\frac{x}{4} = \sin \theta$$

$$\arcsin\left(\frac{x}{4}\right) = \theta$$

Substitute the values of $\sin \theta$ and $\cos \theta$ identified above: $\frac{x}{4} = \sin \theta$ and $\cos \theta = \frac{\sqrt{16 - x^2}}{4}$. Substitute all of these values into the antiderivative.

$$8\theta + 8 \sin \theta \cos \theta + C = 8 \arcsin\left(\frac{x}{4}\right) + 8\left(\frac{x}{4}\right)\left(\frac{\sqrt{16 - x^2}}{4}\right) + C$$

$$= 8 \arcsin\left(\frac{x}{4}\right) + \frac{1}{2} x \sqrt{16 - x^2} + C$$

21.23 Given $\sec \theta = \dfrac{x}{\sqrt{3}}$, draw a right triangle to determine the value of $\sin \theta$, assuming $0 < \theta \leq \dfrac{\pi}{2}$.

If $\sec \theta = \dfrac{x}{\sqrt{3}}$ then $\cos \theta = \dfrac{\sqrt{3}}{x}$, so the side adjacent to θ has length $\sqrt{3}$, and the hypotenuse has length x, as illustrated by Figure 21-4.

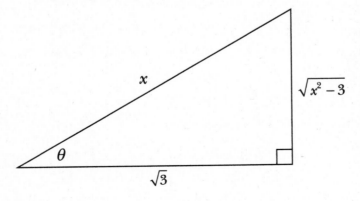

Figure 21-4

The cosine of an angle is the quotient of its adjacent side and the hypotenuse, so $\cos \theta = \dfrac{\sqrt{3}}{x}$. According to the Pythagorean theorem, the opposite side has length $\sqrt{x^2 - 3}$.

According to Figure 21-4, $\sin\theta = \dfrac{\sqrt{x^2-3}}{x}$, the quotient of the side opposite θ and the hypotenuse of the triangle.

21.24 Integrate: $\int \dfrac{\sqrt{x^2-3}}{x}\,dx$.

This integral contains $\sqrt{u^2-a^2}$ (where $u=x$ and $a=\sqrt{3}$), and requires the trigonometric substitution $u=a\sec\theta$.

$$x = \sqrt{3}\sec\theta$$

$$\frac{x}{\sqrt{3}} = \sec\theta \leftarrow$$

> Once you get this equation, you can draw the right triangle in Figure 21-4—that's the information Problem 21.23 gave you about the triangle.

Note that (not coincidentally), the triangle in Figure 21-4 is based upon the equation $\sec\theta = \dfrac{x}{\sqrt{3}}$. Differentiate $x=\sqrt{3}\sec\theta$.

$$dx = \sqrt{3}\sec\theta\tan\theta\,d\theta$$

Rewrite the integrand using this derivative and the conclusion drawn by Problem 21.23: $\sin\theta = \dfrac{\sqrt{x^2-3}}{x}$.

$$\int \frac{\sqrt{x^2-3}}{x}\,dx = \int \sin\theta\left(\sqrt{3}\sec\theta\tan\theta\,d\theta\right)$$

$$= \sqrt{3}\int \sin\theta\cdot\frac{1}{\cos\theta}\cdot\tan\theta\,d\theta$$

$$= \sqrt{3}\int \frac{\sin\theta}{\cos\theta}\tan\theta\,d\theta$$

$$= \sqrt{3}\int \tan^2\theta\,d\theta$$

Apply the Pythagorean trigonometric identity $\tan^2\theta = \sec^2\theta - 1$.

$$= \sqrt{3}\int\left(\sec^2\theta-1\right)d\theta$$

$$= \sqrt{3}\left[\int\sec^2\theta\,d\theta - \int 1\,d\theta\right]$$

$$= \sqrt{3}\left[\tan\theta-\theta\right]+C$$

> Because that's the side opposite θ divided by the side adjacent to θ.

According to Figure 21-4, $\tan\theta = \dfrac{\sqrt{x^2-3}}{\sqrt{3}}$. Solve the equation $x=\sqrt{3}\sec\theta$ for θ to find an expression equivalent to θ.

$$x = \sqrt{3}\sec\theta$$

$$\frac{x}{\sqrt{3}} = \sec\theta$$

$$\operatorname{arcsec}\left(\frac{x}{\sqrt{3}}\right) = \theta$$

Write the antiderivative $\sqrt{3}\left[\tan\theta - \theta\right] + C$ in terms of x.

$$\sqrt{3}\left[\tan\theta - \theta\right] + C = \sqrt{3}\left[\frac{\sqrt{x^2 - 3}}{\sqrt{3}} - \text{arcsec}\left(\frac{x}{\sqrt{3}}\right)\right] + C$$

$$= \sqrt{x^2 - 3} - \sqrt{3}\,\text{arcsec}\left(\frac{x}{\sqrt{3}}\right) + C$$

Rationalize the expression.

$$\sqrt{x^2 - 3} - \sqrt{3}\,\text{arcsec}\left(\frac{x\sqrt{3}}{3}\right) + C$$

21.25 Construct an equivalent solution for Problem 21.24 by replacing the inverse trigonometric function.

The expression $\text{arcsec}\left(\frac{x}{\sqrt{3}}\right)$ is generated by solving the equation $\sec\theta = \frac{x}{\sqrt{3}}$ for θ. Based on Figure 21-4, you can define θ in terms of any other trigonometric ratio; for instance, $\tan\theta = \frac{\sqrt{x^2 - 3}}{\sqrt{3}}$ and $\sin\theta = \frac{\sqrt{x^2 - 3}}{x}$. Solve the tangent equation for θ.

> Of course, you could use the sine function (or any of the other trig functions for that matter).

$$\theta = \arctan\left(\frac{\sqrt{x^2 - 3}}{\sqrt{3}}\right)$$

This value of θ is as valid as $\theta = \text{arcsec}\left(\frac{x}{\sqrt{3}}\right)$ from Problem 21.24, so substituting it for θ in the antiderivative $\sqrt{3}\left[\tan\theta - \theta\right] + C$ results in an equivalent solution.

$$\sqrt{3}\left[\tan\theta - \theta\right] + C = \sqrt{3}\left[\frac{\sqrt{x^2 - 3}}{\sqrt{3}} - \arctan\left(\frac{\sqrt{x^2 - 3}}{\sqrt{3}}\right)\right] + C$$

$$= \sqrt{x^2 - 3} - \sqrt{3}\arctan\left(\frac{\sqrt{x^2 - 3}}{\sqrt{3}}\right) + C$$

21.26 Given $\tan\theta = \frac{x\sqrt{5}}{\sqrt{6}}$, draw a right triangle to determine the value of $\sec\theta$, assuming $0 < \theta \leq \frac{\pi}{2}$.

Recall that the tangent of an angle is the quotient of its opposite and adjacent sides. Therefore, the side opposite θ has length $x\sqrt{5}$ and the side adjacent to θ has length $\sqrt{6}$, as illustrated in Figure 21-5.

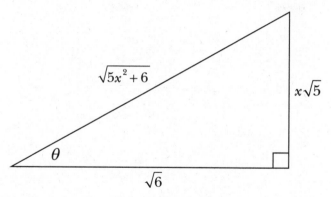

Figure 21-5 *Apply the Pythagorean theorem to calculate the length of the missing side—in this case the hypotenuse.*

> Because that's the side adjacent to θ divided by the side opposite θ.

According to Figure 21-5, $\cos\theta = \dfrac{\sqrt{6}}{\sqrt{5x^2+6}}$, so $\sec\theta = \dfrac{\sqrt{5x^2+6}}{\sqrt{6}}$.

21.27 Integrate: $\displaystyle\int \frac{x^3\,dx}{\sqrt{5x^2+6}}$.

This integral contains $\sqrt{u^2+a^2}$ and requires the trigonometric substitution $u = a\tan\theta$, where $u = \sqrt{5x^2} = x\sqrt{5}$ and $a = \sqrt{6}$.

$$u = a\tan\theta$$
$$x\sqrt{5} = \sqrt{6}\tan\theta$$
$$\frac{x\sqrt{5}}{\sqrt{6}} = \tan\theta$$

> Here are the three substitutions you need to know: $\sqrt{a^2-u^2}$ uses $u = a\sin\theta$, $\sqrt{u^2-a^2}$ uses $u = a\sec\theta$, and $\sqrt{u^2+a^2}$ uses $u = a\tan\theta$.

This trigonometric expression is illustrated by Figure 21-5. Solve for x and differentiate the equation.

$$x = \frac{\sqrt{6}}{\sqrt{5}}\tan\theta$$
$$dx = \frac{\sqrt{6}}{\sqrt{5}}\sec^2\theta\,d\theta$$

According to Problem 21.26, $\sec\theta = \dfrac{\sqrt{5x^2+6}}{\sqrt{6}}$, so $\sqrt{6}\sec\theta = \sqrt{5x^2+6}$. Substitute this into the original integrand.

$$\int \frac{x^3\,dx}{\sqrt{5x^2+6}} = \int \frac{x^3\,dx}{\sqrt{6}\sec\theta}$$

Also substitute the values of x and dx generated above: $x = \dfrac{\sqrt{6}}{\sqrt{5}}\tan\theta$ and $dx = \dfrac{\sqrt{6}}{\sqrt{5}}\sec^2\theta\,d\theta$.

Dividing by $\sqrt{6}\sec\theta$ is the same as multiplying by $\dfrac{1}{\sqrt{6}\sec\theta}$.

$$= \int \frac{\left[\left(\frac{\sqrt{6}}{\sqrt{5}}\tan\theta\right)^3\left(\frac{\sqrt{6}}{\sqrt{5}}\sec^2\theta\,d\theta\right)\right]}{\sqrt{6}\sec\theta}$$

$$= \int \left(\frac{6\sqrt{6}}{5\sqrt{5}}\tan^3\theta\right)\left(\frac{\sqrt{6}}{\sqrt{5}}\sec^2\theta\,d\theta\right)\cdot\frac{1}{\sqrt{6}\sec\theta}$$

$$= \int \frac{6\sqrt{6}\,\sqrt{6}\,\tan^3\theta\,\sec\theta\,\sec\theta\,d\theta}{5\sqrt{5}\sqrt{5}\,\sqrt{6}\,\sec\theta}$$

$$= \int \frac{6\sqrt{6}\tan^3\theta\sec\theta\,d\theta}{5(5)}$$

$$= \frac{6\sqrt{6}}{25}\int \tan^3\theta\sec\theta\,d\theta$$

The integrand is a product of $\tan\theta$ and $\sec\theta$ with the tangent function raised to an odd power; use the method described in Problem 21.20 to integrate.

$$= \frac{6\sqrt{6}}{25}\int \sec\theta\tan\theta\left(\tan^2\theta\right)d\theta$$

$$= \frac{6\sqrt{6}}{25}\int \sec\theta\tan\theta\left(\sec^2\theta-1\right)d\theta$$

$$= \frac{6\sqrt{6}}{25}\left[\int \sec^2\theta\sec\theta\tan\theta\,d\theta - \int \sec\theta\tan\theta\,d\theta\right]$$

The antiderivative of $\sec\theta\tan\theta$ is $\sec\theta$. Use variable substitution on the left integral: $u=\sec\theta$ and $du=\sec\theta\tan\theta\,d\theta$.

$$= \frac{6\sqrt{6}}{25}\left[\int u^2\,du - \int \sec\theta\tan\theta\,d\theta\right]$$

$$= \frac{6\sqrt{6}}{25}\left[\frac{1}{3}\sec^3\theta-\sec\theta\right]+C$$

Rewrite $\left(\sqrt{5x^2+6}\right)^3$ as $\left(\sqrt{5x^2+6}\right)^2\left(\sqrt{5x^2+6}\right)$. The square root and the square in the left factor cancel out, and you don't have to worry about absolute value signs because $5x^2+6$ is always positive.

Recall that $\sec\theta=\dfrac{\sqrt{5x^2+6}}{\sqrt{6}}$.

$$= \frac{6\sqrt{6}}{25}\left[\frac{1}{3}\left(\frac{\sqrt{5x^2+6}}{\sqrt{6}}\right)^3-\frac{\sqrt{5x^2+6}}{\sqrt{6}}\right]+C$$

$$= \frac{6\sqrt{6}}{25}\left[\frac{1}{3}\cdot\frac{(5x^2+6)\sqrt{5x^2+6}}{6\sqrt{6}}-\frac{\sqrt{5x^2+6}}{\sqrt{6}}\right]+C$$

$$= \frac{6\sqrt{6}}{25}\left[\frac{(5x^2+6)\sqrt{5x^2+6}}{18\sqrt{6}}-\frac{\sqrt{5x^2+6}}{\sqrt{6}}\right]+C$$

Multiply the right fraction inside the brackets by $\dfrac{18}{18}$ in order to establish common denominators and combine the fractions.

$$= \frac{6\sqrt{6}}{25}\left[\frac{\left(5x^2+6\right)\sqrt{5x^2+6}}{18\sqrt{6}} - \frac{18\sqrt{5x^2+6}}{18\sqrt{6}}\right] + C$$

$$= \frac{6\sqrt{6}}{25}\left[\frac{\left(5x^2+6\right)\sqrt{5x^2+6} - 18\sqrt{5x^2+6}}{18\sqrt{6}}\right] + C$$

Because $\left(5x^2+6\right)\sqrt{5x^2+6}$ and $-18\sqrt{5x^2+6}$ contain common radicals (i.e., their radicands are equal), you can combine their coefficients: $5x^2 + 6 - 18 = 5x^2 - 12$.

$$= \frac{\cancel{6}\,\cancel{\sqrt{6}}}{25}\left[\frac{\left(5x^2-12\right)\sqrt{5x^2+6}}{\cancel{6}\cdot 3\cancel{\sqrt{6}}}\right] + C$$

$$= \frac{\left(5x^2-12\right)\sqrt{5x^2+6}}{75} + C$$

21.28 According to Problem 20.15, $\displaystyle\int \frac{dx}{\sqrt{1-x^2}} = \arcsin x + C$. Verify the antiderivative using trigonometric substitution.

The denominator has form $\sqrt{a^2 - u^2}$, where $a = 1$ and $u = x$. Use the method outlined in Problem 21.22, applying the trigonometric substitution $x = 1(\sin\theta)$. Differentiate the equation to get $dx = \cos\theta\,d\theta$.

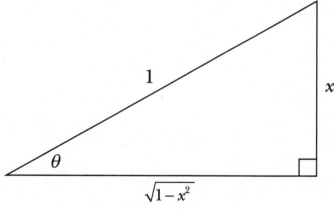

Figure 21-6 *If $x = \sin\theta$ then $\sin\theta = \dfrac{x}{1}$. Therefore, the side opposite θ is x and the hypotenuse of the right triangle is 1.*

Because that's the adjacent side $\left(\sqrt{1-x^2}\right)$ divided by the hypotenuse (1).

According to Figure 21-6, $\cos\theta = \sqrt{1-x^2}$. Replace the radical expression and dx in the original integral with functions written in terms of θ.

$$\int \frac{dx}{\sqrt{1-x^2}} = \int \frac{\cos\theta\, d\theta}{\cos\theta}$$
$$= \int d\theta$$
$$= \theta + C$$

Rewrite the antiderivative in terms of x. If $\sin\theta = x$, then $\theta = \arcsin x$.

$$= \arcsin x + C$$

21.29 Integrate: $\int \dfrac{dx}{\sqrt{x^2 - 8x - 9}}$.

Complete the square in the denominator.

$$\int \frac{dx}{\sqrt{x^2 - 8x - 9}} = \int \frac{dx}{\sqrt{x^2 - 8x + 16 - 9 - 16}}$$
$$= \int \frac{dx}{\sqrt{(x-4)^2 - 25}}$$

Use the equation $u = a\sec\theta$. Substitute u and a to get $x - 4 = 5\sec\theta$ and solve for x.

The denominator has form $\sqrt{u^2 - a^2}$, where $u = x - 4$ and $a = 5$. Use the technique described in Problem 21.24, applying the trigonometric substitution $x = 5\sec\theta + 4$. Differentiate the equation to get $dx = 5\sec\theta\tan\theta\,d\theta$.

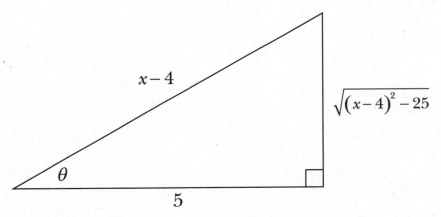

Figure 21-7 *Solving $x = 5\sec\theta + 4$ for $\sec\theta$ yields the trigonometric ratio $\dfrac{x-4}{5} = \sec\theta$. Therefore, $\cos\theta = \dfrac{5}{x-4}$ — the side adjacent to θ has length 5 and the hypotenuse of the triangle has length $x - 4$.*

According to Figure 21-7, $\tan\theta = \dfrac{\sqrt{(x-4)^2-25}}{5}$; therefore, $5\tan\theta = \sqrt{(x-4)^2-25}$. Rewrite the original integral in terms of θ.

$$\int \frac{dx}{\sqrt{(x-4)^2-25}} = \int \frac{5\sec\theta\tan\theta\,d\theta}{5\tan\theta}$$

$$= \int \sec\theta\,d\theta$$

$$= \ln|\sec\theta + \tan\theta| + C$$

Recall that $\sec\theta = \dfrac{x-4}{5}$ and $\tan\theta = \dfrac{\sqrt{(x-4)^2-25}}{5}$.

$$= \ln\left|\frac{x-4}{5} + \frac{\sqrt{(x-4)^2-25}}{5}\right| + C$$

It is customary to factor $\dfrac{1}{5}$ out of the expression and expand the perfect square into its original form.

$$= \ln\frac{1}{5}\left|x-4+\sqrt{x^2-8x-9}\right| + C$$

Expand the expression.

$$= \ln\frac{1}{5} + \ln\left|x-4+\sqrt{x^2-8x-9}\right| + C$$

> The log of something multiplied equals two logs added together: log ab = log a + log b.

Note that $\ln\dfrac{1}{5}$ is a constant, so $\ln\dfrac{1}{5}+C$ is merely another arbitrary constant C. More rigorous treatment of the arbitrary constant is neither required nor necessary.

$$= \ln\left|x-4+\sqrt{x^2-8x-9}\right| + C$$

Improper Integrals
Integrating despite asymptotes and infinite boundaries

21.30 What characteristics classify an integral as "improper"?

Two conditions characterize the vast majority of improper integrals: discontinuity of the integrand on the closed interval defined by the limits of integration or the presence of an infinite limit of integration.

> One of the limits of integration (or some number between them) can't be plugged into the function

21.31 When is an improper integral considered "divergent"?

As discussed in Problem 21.32, improper integrals are evaluated by means of a limit. If that limit does not equal a real number, then the corresponding integral is described as "divergent."

> Definite integrals need real number boundaries, so if one of the limits is ∞ or −∞, the integral is automatically improper.

21.32 Determine whether or not $\int_0^\infty \sin\theta\, d\theta$ converges; if it does, evaluate the integral.

The definite integral $\int_0^\infty \sin\theta\, d\theta$ is an improper integral because its upper limit of integration is infinite. In order to evaluate the integral, replace the infinite limit with constant a and find the limit of the integral as a approaches infinity.

$$\int_0^\infty \sin\theta\, d\theta = \lim_{a\to\infty}\left(\int_0^a \sin\theta\, d\theta\right)$$

$$= \lim_{a\to\infty}\left(-\cos\theta\big|_0^a\right)$$

$$= \lim_{a\to\infty}\left(-\cos a - (-\cos 0)\right)$$

$$= \lim_{a\to\infty}\left(-\cos a + 1\right)$$

$$= -\lim_{a\to\infty}\cos a + \lim_{a\to\infty}1$$

$$= -\lim_{a\to\infty}\cos a + 1$$

Unfortunately, $\lim_{a\to\infty}(-\cos a + 1)$ does not exist, because $y = \cos\theta$ oscillates infinitely as a approaches infinity. The graph is periodic, and cycles through the interval $-1 \leq \cos a \leq 1$, never approaching a single finite value. Therefore, $\int_0^\infty \sin\theta\, d\theta$ diverges.

21.33 Determine whether or not $\int_0^{10} \ln x\, dx$ converges; if it does, evaluate the integral.

Although $y = \ln x$ is continuous on the x-interval $(0,\infty)$, it is not defined when $x = 0$. Replace the invalid limit of integration with constant a and evaluate the limit as a approaches 0 from the right.

> You have to approach from the right, because $y = \ln x$ isn't defined to the left of $x = 0$ (you can't take the natural log of negative numbers). The little plus sign in the limit is just for notation—it won't affect your work.

$$\int_0^{10} \ln x\, dx = \lim_{a\to 0^+}\left(\int_a^{10} \ln x\, dx\right)$$

According to Problem 21.3, the antiderivative of $\ln x$ is $x\ln x - x$.

$$= \lim_{a\to 0^+}\left[(x\ln x - x)\big|_a^{10}\right]$$

$$= \lim_{a\to 0^+}\left[(10\ln 10 - 10) - (a\ln a - a)\right]$$

$$= \lim_{a\to 0^+}(10\ln 10 - 10) - \lim_{a\to 0^+}(a\ln a) + \lim_{a\to 0^+}a$$

Whereas $\lim_{a\to 0^+}(10\ln 10 - 10) = 10\ln 10 - 10$ (the limit of any constant as a approaches any value is equal to the constant) and $\lim_{a\to 0^+}a = 0$ (by substitution) you must apply L'Hôpital's Rule to evaluate $\lim_{a\to 0^+}(a\ln a)$.

> L'Hôpital's Rule only works on fractions, so rewrite $a\ln a$—multiplying by a is the same thing as dividing by the reciprocal of a.

$$\lim_{a\to 0^+}(a\ln a) = \lim_{a\to 0^+}\frac{\ln a}{1/a}$$

$$= \lim_{a\to 0^+}\frac{\ln a}{a^{-1}}$$

Differentiate the numerator and denominator individually.

$$= \lim_{a\to 0^+}\frac{1/a}{-1/a^2}$$

Eliminate the complex fraction by multiplying the numerator and denominator by $-a^2$.

$$\lim_{a \to 0^+}(a \ln a) = \lim_{a \to 0^+}\left[\frac{\frac{1}{a}\left(\frac{-a^2}{1}\right)}{-\frac{1}{a^2}\left(-\frac{a^2}{1}\right)}\right]$$
$$= \lim_{a \to 0^+}(-a)$$
$$= 0$$

Substitute the values $\lim_{a \to 0^+}(10\ln 10 - 10) = 10\ln 10 - 10$, $\lim_{a \to 0^+} a = 0$, and $\lim_{a \to 0^+}(a \ln a) = 0$ into the antiderivative.

$$\int_0^{10} \ln x \, dx = \lim_{a \to 0^+}(10\ln 10 - 10) - \lim_{a \to 0^+}(a \ln a) + \lim_{a \to 0^+} a$$
$$= 10\ln 10 - 10 - 0 + 0$$
$$= 10(\ln 10 - 1)$$

21.34 Determine whether or not $\int_1^\infty \frac{dx}{x}$ converges; if it does, evaluate the integral.

The integral is improper due to the infinite upper integration limit. Replace ∞ with a and evaluate the limit of the integral as $a \to \infty$.

$$\int_1^\infty \frac{dx}{x} = \lim_{a \to \infty}\left(\int_1^a \frac{dx}{x}\right)$$
$$= \lim_{a \to \infty}\left(\ln|x|\big|_1^a\right)$$
$$= \lim_{a \to \infty}(\ln|a| - \ln 1)$$
$$= \lim_{a \to \infty}\ln|a| - \lim_{a \to \infty} 0$$
$$= \lim_{a \to \infty}\ln|a|$$

The graph of $y = \ln x$ has an x-intercept of $(1,0)$, so $\ln 1 = 0$.

The bigger the x, the bigger the natural log. The graph of $y = \ln x$ doesn't have a vertical asymptote, so it increases without bound as x approaches infinity (even if it does so kind of slowly).

Because $y = \ln a$ increases without bound as a approaches infinity, the integral diverges.

21.35 Determine whether or not $\int_{-\infty}^{-1} \frac{dx}{x^2}$ converges; if it does, evaluate the integral.

The integral is improper due to the infinite lower integration limit. Replace $-\infty$ with a and evaluate the limit as a approaches $-\infty$.

$$\int_{-\infty}^{-1}\frac{dx}{x^2} = \lim_{a\to-\infty}\left(\int_{a}^{-1}x^{-2}dx\right)$$

$$= \lim_{a\to-\infty}\left(-\frac{1}{x}\bigg|_{a}^{-1}\right)$$

$$= \lim_{a\to-\infty}\left(1+\frac{1}{a}\right)$$

$$= \lim_{a\to-\infty}1 + \lim_{a\to-\infty}\frac{1}{a}$$

$$= 1+0$$

$$= 1$$

21.36 Determine whether or not $\int_{0}^{2}\frac{x^3-1}{x-1}dx$ converges; if it does, evaluate the integral.

This integral is not improper because of its limits of integration, but because the integrand is discontinuous at $x = 1$. Split the integral at the location of the discontinuity.

$$\int_{0}^{2}\frac{x^3-1}{x-1}dx = \int_{0}^{1}\frac{x^3-1}{x-1}dx + \int_{1}^{2}\frac{x^3-1}{x-1}dx$$

Replace the invalid limits in the integrals with constants, and evaluate the limits of each integral as the constant approaches 1 from the appropriate direction (i.e., from the left in the left interval and from the right in the right integral).

$$= \lim_{a\to1^-}\left(\int_{0}^{a}\frac{x^3-1}{x-1}dx\right) + \lim_{a\to1^+}\left(\int_{a}^{2}\frac{x^3-1}{x-1}dx\right)$$

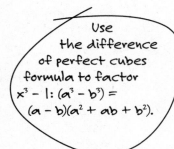
Use the difference of perfect cubes formula to factor $x^3 - 1$: $(a^3 - b^3) = (a - b)(a^2 + ab + b^2)$.

Factor the rational expression: $\dfrac{x^3-1}{x-1} = \dfrac{(x-1)(x^2+x+1)}{x-1}$. As long as $x \neq 1$, you can eliminate the common factor $x - 1$ in the numerator and denominator: $\dfrac{x^3-1}{x-1} = x^2+x+1$. Because you are calculating the limit as a *approaches* (but does not equal) 1, it is safe to assume that $x \neq 1$.

$$= \lim_{a\to1^-}\left[\int_{0}^{a}\left(x^2+x+1\right)dx\right] + \lim_{a\to1^+}\left[\int_{a}^{2}\left(x^2+x+1\right)dx\right]$$

$$= \lim_{a\to1^-}\left[\left(\frac{x^3}{3}+\frac{x^2}{2}+x\right)\bigg|_{0}^{a}\right] + \lim_{a\to1^+}\left[\left(\frac{x^3}{3}+\frac{x^2}{2}+x\right)\bigg|_{a}^{2}\right]$$

$$= \lim_{a\to1^-}\left(\frac{a^3}{3}+\frac{a^2}{2}+a-0\right) + \lim_{a\to1^+}\left[\left(\frac{8}{3}+2+2\right)-\left(\frac{a^3}{3}+\frac{a^2}{2}+a\right)\right]$$

$$= \left(\frac{1}{3}+\frac{1}{2}+1\right) + \left[\left(\frac{20}{3}\right)-\left(\frac{1}{3}+\frac{1}{2}+1\right)\right]$$

$$= \frac{20}{3}$$

21.37 Determine whether or not $\int_{-\infty}^{\infty} \dfrac{dx}{1+x^2}$ converges; if it does, evaluate the integral.

There are no x-values for which the integrand is discontinuous; however, neither the upper nor the lower limit of integration is finite. You must split the integral into the sum of two integrals and address each invalid boundary separately. ← Because the integrand is defined for all real x, you can choose to split the integrals at any real number. To simplify the ensuing calculations, it is advisable to use $x = 0$.

If there are two bad boundaries, you'll have to take care of them one at a time, using one limit on each of the two integrals.

$$\int_{-\infty}^{\infty} \frac{dx}{1+x^2} = \int_{-\infty}^{0} \frac{dx}{1+x^2} + \int_{0}^{\infty} \frac{dx}{1+x^2}$$

$$= \lim_{a \to -\infty} \left(\int_{a}^{0} \frac{dx}{1+x^2} \right) + \lim_{b \to \infty} \left(\int_{0}^{b} \frac{dx}{1+x^2} \right)$$

According to Problem 20.15, $\int \dfrac{dx}{1+x^2} = \arctan x + C$.

$$= \lim_{a \to -\infty} \left(\arctan x \Big|_{a}^{0} \right) + \lim_{b \to \infty} \left(\arctan x \Big|_{0}^{b} \right)$$

$$= \lim_{a \to -\infty} \left(\arctan 0 - \arctan a \right) + \lim_{b \to \infty} \left(\arctan b - \arctan 0 \right)$$

$$= \lim_{a \to -\infty} \arctan 0 - \lim_{a \to -\infty} \arctan a + \lim_{b \to \infty} \arctan b - \lim_{b \to \infty} \arctan 0$$

Because $\lim\limits_{x \to \pi/2} \tan x = \infty$, $\lim\limits_{x \to \infty} \arctan x = \dfrac{\pi}{2}$. Similarly, $\lim\limits_{x \to -\infty} \arctan x = -\dfrac{\pi}{2}$ because $\lim\limits_{x \to -\pi/2} \tan x = -\infty$.

$$= 0 - \left(-\frac{\pi}{2} \right) + \left(\frac{\pi}{2} \right) - 0$$

$$= \pi$$

Chapter 22
CROSS-SECTIONAL AND ROTATIONAL VOLUME

Please put on your 3-D glasses at this time

This chapter modifies the technique of calculating the two-dimensional area of a region (originally introduced in Chapters 18 and 19) to calculate the volume of three-dimensional solids. As Riemann sums added infinitely many areas of rectangular regions to determine the area of a region, the methods herein calculate infinitely many cross-sectional areas of a solid to exactly calculate its volume.

Even though all of the problems in this chapter deal with three-dimensional figures, none of the formulas you'll use look much different than the ones from Chapter 18 and 19. That's because there are lots of sneaky ways to calculate the volume of a solid just by looking at its cross-sections—the layers you'd get if you shaved the solid down with that big silver meat slicer they have at the deli that turns giant boulders of roast beef into thin, sandwich-ready slices. The first set of problems deals with solids that "grow upward" from a region on the coordinate plane, and the rest of the problems deal with solids created by swinging some region of the coordinate plane around an axis of rotation.

Volume of a Solid with Known Cross-Sections

Cut the solid into pieces and measure those instead

22.1 If a three-dimensional solid has cross-sections perpendicular to the *x*-axis along the interval [*a,b*] whose areas are modeled by the function $A(x)$, what is the volume of the solid?

The volume of the solid will be $\int_a^b A(x)\,dx$. This formula states that the three-dimensional volume of a solid can be determined by slicing the solid into infinitely many, infinitely thin cross-sectional slices, determining the volumes of the cross-sections, and then calculating the sum.

Finding the formula for the cross-section's area is the trickiest part. Once you figure that out, you just integrate like you've been doing since Chapter 17.

Note: Problems 22.2–22.4 refer to a region B in the coordinate plane that is bounded by $y = \sqrt{x}$, $y = 0$, *and* $x = 4$.

22.2 Calculate the volume of the solid with base *B* that has square cross-sections perpendicular to the *x*-axis, as illustrated in Figure 22-1.

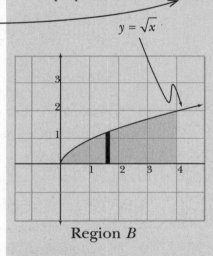

Region *B*

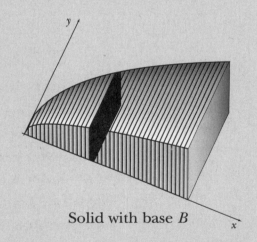

Solid with base *B*

Figure 22-1 *The base of the solid (region B) determines the size of the solid's cross-sections. The darkened rectangle on the left graph corresponds to the darkened square cross-section on the right.*

> **If the cross-sections are perpendicular to the x-axis, then the formula and the integral you'll use should both be in terms of x.**

> **It's a good idea to draw a dark line or rectangle like this, one that's perpendicular to the correct axis and stretches across the region.**

Consider the graph of Region *B* in Figure 22-1. The problem states that the solid has square cross-sections, so visualize a square lying along the darkened rectangle on the left, perpendicular to the *x*-axis and the printed page, growing upwards from this book rather than lying flat against it. The sides of a square all have the same length, so the square is as tall as the darkened rectangle is long. By calculating the length of the rectangle, you are also determining the length of the sides of the square cross-section.

The rectangle is bounded above by $y = \sqrt{x}$ and below by $y = 0$. Therefore, the length of the darkened rectangle in Figure 22-1 is $\sqrt{x} - 0 = \sqrt{x}$, the

upper boundary of the region minus the lower boundary. The square cross-sections have side length $s = f(x) = \sqrt{x}$, so the area $A(x)$ of those squares is s^2:
$A(x) = s^2 = (\sqrt{x})^2 = x$.

Apply the formula from Problem 22.1 to determine the volume of the solid.

$$\int_a^b A(x)\,dx = \int_0^4 x\,dx$$
$$= \frac{x^2}{2}\Big|_0^4$$
$$= \frac{16}{2} - 0$$
$$= 8$$

Note: Problems 22.2–22.4 refer to a region B in the coordinate plane that is bounded by $y = \sqrt{x}$, $y = 0$, and $x = 4$.

22.3 Calculate the volume of the solid with base B that has rectangular cross-sections of height 3 that are perpendicular to the y-axis.

The region is the same as Problem 22.2, as is the orientation of the cross-sections (perpendicular to the x-axis), so you will use the representative length calculated in Problem 22.2: $\sqrt{x} - 0 = \sqrt{x}$. In this problem, that length represents one dimension of the rectangle; the other dimension of the rectangle is stated explicitly—the rectangles have a fixed height of 3.

> "Representative lengths" are the lengths of the darkened rectangles like the one drawn on Figure 22-1.

Construct a function that represents the area $A(x)$ of the cross-sections; the area of a rectangle is the product of its length and width.

$$A(x) = \text{length} \cdot \text{width} = \sqrt{x} \cdot 3$$

Apply the formula from Problem 22.1 to calculate the volume of the solid. Use the integration limits from Problem 22.2, as the region and orientation of the cross-sections are the same.

$$\int_a^b A(x)\,dx = \int_0^4 3\sqrt{x}\,dx$$
$$= 3\int_0^4 x^{1/2}\,dx$$
$$= 3 \cdot \frac{x^{3/2}}{3/2}\Big|_0^4$$
$$= 3 \cdot \frac{2}{3} \cdot (x^{3/2})\Big|_0^4$$
$$= 2\left[4^{3/2} - 0\right]$$
$$= 2(8)$$
$$= 16$$

> $4^{3/2} = (4^{1/2})^3 =$
> $(\sqrt{4})^3 = 2^3 = 8$

22.4 Calculate the volume of the solid with base B that has square cross-sections perpendicular to the y-axis.

> **RULE OF THUMB:** If the cross-sections are perpendicular to the x-axis, solve the equations for y. If they're perpendicular to the y-axis, solve the equations for x.

If the cross-sections are perpendicular to the y-axis, you should draw the representative length horizontally instead of vertically, as illustrated by Figure 22-2. Furthermore, you must express the functions in terms of y instead of x. To do so, solve the equation $y = \sqrt{x}$ for x.

$$y^2 = \left(\sqrt{x}\right)^2$$
$$x = y^2$$

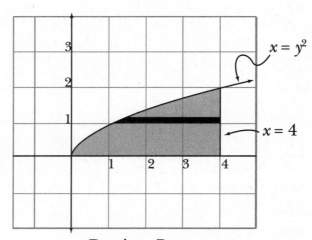

Region B

Figure 22-2 *The shaded region B is identical to the region defined in Problem 22.2, but this solid has square cross-sections perpendicular to the y-axis. The right boundary of the region is x = 4 and the left boundary is x = y².*

> Because all of the functions are in terms of y this time, the integral boundaries should be as well. Describe where the solid is along the y-axis—from y = 0 to y = 2.

To calculate a horizontal representative length, find the difference between the right and left boundaries: $4 - y^2$. The area of each square cross-section is $A(y) = s^2 = (4 - y^2)^2 = 16 - 8y^2 + y^4$. Calculate the volume of the solid.

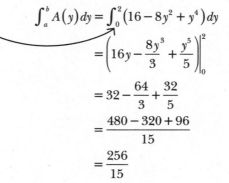

$$\int_a^b A(y)\,dy = \int_0^2 \left(16 - 8y^2 + y^4\right)dy$$
$$= \left(16y - \frac{8y^3}{3} + \frac{y^5}{5}\right)\Bigg|_0^2$$
$$= 32 - \frac{64}{3} + \frac{32}{5}$$
$$= \frac{480 - 320 + 96}{15}$$
$$= \frac{256}{15}$$

Note that the solution differs from the solution to Problem 22.2. Using cross-sections with different orientations usually changes the shape and volume of the ensuing solid, even when the solid has the same two-dimensional base.

Note: Problems 22.5–22.6 refer to region R, which is bounded by the circle $x^2 + y^2 = 25$.

22.5 Calculate the volume of the solid with base R that has semicircular cross-sections perpendicular to the y-axis.

Because the cross-sections are perpendicular to the y-axis, the cross-sectional area, the function must be written in terms of y. Solve the equation of the circle for x.

$$x^2 + y^2 = 25$$
$$x^2 = 25 - y^2$$
$$x = \pm\sqrt{25 - y^2}$$

The circular equation can be expressed using two functions written in terms of y: $x = \sqrt{25 - y^2}$ and $x = -\sqrt{25 - y^2}$, as illustrated in Figure 22-3.

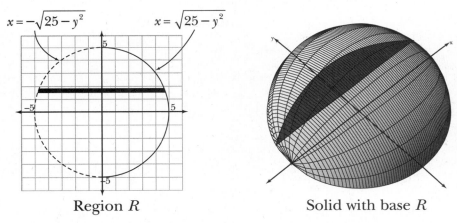

$$x = -\sqrt{25 - y^2} \qquad x = \sqrt{25 - y^2}$$

Region R Solid with base R

Figure 22-3 *The darkened representative length (in the left graph) is horizontal because the cross-sections are perpendicular to the y-axis. The darkened cross-section in the right graph may not look semicircular, but that is due only to the isometric angle at which the graph is rendered.*

Calculate the representative length in Figure 22-3 by subtracting the left boundary of the region from the right boundary.

$$\sqrt{25 - y^2} - \left(-\sqrt{25 - y^2}\right) = \sqrt{25 - y^2} + \sqrt{25 - y^2} = 2\sqrt{25 - y^2}$$

According to Figure 22-3, this length represents the diameter of the semicircular cross-section. Divide it by 2 to determine the radius of the semicircle.

$$r = \frac{2\sqrt{25 - y^2}}{2} = \sqrt{25 - y^2}$$

Construct the cross-sectional area function, $A(y)$, by substituting the radius into the formula for the area of a semicircle.

You don't need to use absolute values to simplify this, because the radius of the semicircle will always be positive.

$$A(y) = \frac{\pi r^2}{2} = \frac{\pi \left(\sqrt{25 - y^2}\right)^2}{2} = \frac{\pi}{2}\left(25 - y^2\right)$$

Calculate the volume of the solid.

$$\int_a^b A(y)\,dy = \frac{\pi}{2}\int_{-5}^5 \left(25 - y^2\right)dy$$

$$= \frac{\pi}{2}\left(25y - \frac{y^3}{3}\right)\Bigg|_{-5}^5$$

$$= \frac{\pi}{2}\left[\left(125 - \frac{125}{3}\right) - \left(-125 + \frac{125}{3}\right)\right]$$

$$= \frac{\pi}{2}\left[\left(\frac{250}{3}\right) - \left(-\frac{250}{3}\right)\right]$$

$$= \frac{\pi}{2}\left(\frac{500}{3}\right)$$

$$= \frac{250\pi}{3}$$

This solid is a sphere of radius 5 cut in half. The volume of that sphere is

$$\frac{4}{3}\pi r^3 = \frac{4(125)\pi}{3} = \frac{500\pi}{3},$$

and the answer to Problem 22.5 is half of that.

Note: Problems 22.5–22.6 refer to region R, which is bounded by the circle $x^2 + y^2 = 25$.

22.6 Calculate the volume of the solid with base R that has semicircular cross-sections perpendicular to the x-axis. Are the volumes of this solids described Problems 22.5 and 22.6 equal? Why or why not?

In order to calculate the representative length for this solid, you need to solve the equation of the circle for y: $y = \pm\sqrt{25 - x^2}$. Therefore, the upper boundary of the circle is $y = \sqrt{25 - x^2}$, and the lower boundary is $y = -\sqrt{25 - x^2}$. Compute the representative length by subtracting the lower from the upper boundary:

$\sqrt{25 - x^2} - \left(-\sqrt{25 - x^2}\right) = \sqrt{25 - x^2} + \sqrt{25 - x^2} = 2\sqrt{25 - x^2}$. This length represents the diameter of the semicircle (as it did in Problem 22.6), so

$r = \frac{2\sqrt{25 - x^2}}{2} = \sqrt{25 - x^2}$. Use the formula for the area of a semicircle to construct $A(x)$.

Like Problem 22.6, this solid is a hemisphere (a sphere sliced in half) of radius 5, just generated in a slightly different way. No matter how you make that half-sphere, it has the same volume.

$$A(x) = \frac{\pi r^2}{2} = \frac{\pi\left(\sqrt{25 - x^2}\right)^2}{2} = \frac{\pi}{2}\left(25 - x^2\right)$$

The upper and lower x-boundaries of integration mirror the upper and lower y-boundaries from Problem 22.6 when you apply the volume formula.

$$\int_a^b A(x)\,dx = \frac{\pi}{2}\int_{-5}^5 \left(25 - x^2\right)dx$$

The only difference between this integral and the integral in Problem 22.6 is the variable used (x instead of y). The label used to represent a variable does not affect the value of an integral, so $\frac{\pi}{2}\int_{-5}^5 \left(25 - x^2\right)dx = \frac{\pi}{2}\int_{-5}^5 \left(25 - y^2\right)dy = \frac{250\pi}{3}$.

Note: Problems 22.7–22.8 refer to region G, which is bounded by the graphs of y = −cos x and y = sin $\frac{x}{2}$ on the interval $[0,\pi]$.

22.7 Calculate the volume of the solid with base G whose cross-sections are equilateral triangles perpendicular to the x-axis.

Consider the darkened representative length illustrated in the left graph of Figure 22-4. It is bounded above by $y = \sin \frac{x}{2}$ and below by $y = -\cos x$. Therefore, it has length $\sin\frac{x}{2} - (-\cos x) = \sin\frac{x}{2} + \cos x$.

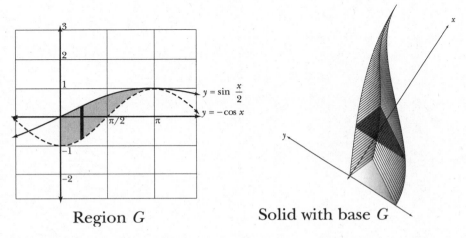

Region G Solid with base G

Figure 22-4 *The darkened rectangle in the left graph represents one side of the equilateral triangle cross-section.*

In order to construct $A(x)$, the area of an equilateral triangle with side length s, apply the area formula of an equilateral triangle: $\frac{s^2\sqrt{3}}{4}$, as illustrated in Figure 22-5.

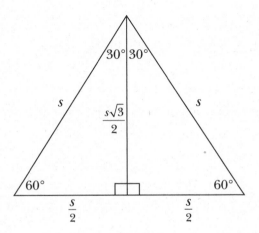

$$\text{area} = \frac{1}{2}b\cdot h$$
$$= \frac{1}{2}(s)\left(\frac{s\sqrt{3}}{2}\right)$$
$$= \frac{s^2\sqrt{3}}{4}$$

Figure 22-5

To find the height of an equilateral triangle, divide it into two congruent right triangles and use the 30°-60°-90° right triangle theorem from geometry, which states that the leg opposite the 30° angle is half the length of the hypotenuse, and the leg opposite the 60° angle is $\sqrt{3}$ times the length of the other leg.

Substitute $s = \sin\dfrac{x}{2} + \cos x$ into the area formula to generate $A(x)$.

$$A(x) = \frac{s^2 \sqrt{3}}{4} = \frac{\sqrt{3}}{4}\left(\sin\frac{x}{2} + \cos x\right)^2 = \frac{\sqrt{3}}{4}\left(\sin^2\frac{x}{2} + 2\cos x \sin\frac{x}{2} + \cos^2 x\right)$$

Calculate the volume of the solid.

$$\int_a^b A(x)\,dx = \frac{\sqrt{3}}{4}\int_0^\pi \left(\sin^2\frac{x}{2} + 2\cos x \sin\frac{x}{2} + \cos^2 x\right)dx$$

$$= \frac{\sqrt{3}}{4}\int_0^\pi \sin^2\frac{x}{2}\,dx + \frac{2\sqrt{3}}{4}\int_0^\pi \cos x \sin\frac{x}{2}\,dx + \frac{\sqrt{3}}{4}\int_0^\pi \cos^2 x\,dx$$

> **Integrate the first thing using the power-reducing formula from Problem 21.11 along with variable substitution: $u = x/2$ and $2\,du = dx$. Use a product-to-sum formula (like in Problems 21.15 and 21.16) on the second integral, and then it's back to power-reduction formulas for the third integral.**

Integrate using the techniques from Chapter 21.

$$= \frac{\sqrt{3}}{4}\int_0^\pi \left(\frac{1}{2} - \frac{\cos x}{2}\right)dx + \frac{2\sqrt{3}}{4}\int_0^\pi \left(\frac{\sin(3x/2)}{2} - \frac{\sin(x/2)}{2}\right)dx + \frac{\sqrt{3}}{4}\int_0^\pi \left(\frac{1}{2} + \frac{\cos 2x}{2}\right)dx$$

$$= \frac{\sqrt{3}}{8}(x - \sin x)\Big|_0^\pi + \frac{2\sqrt{3}}{4 \cdot 2}\left(-\frac{2}{3}\cos\frac{3x}{2} + 2\cos\frac{x}{2}\right)\Big|_0^\pi + \frac{\sqrt{3}}{8}\left(x + \frac{1}{2}\sin 2x\right)\Big|_0^\pi$$

$$= \frac{\sqrt{3}}{8}[\pi] + \frac{\sqrt{3}}{4}\left[0 - \left(-\frac{2}{3} + 2\right)\right] + \frac{\sqrt{3}}{8}[\pi]$$

$$= \frac{\pi\sqrt{3}}{8} - \frac{4\sqrt{3}}{4 \cdot 3} + \frac{\pi\sqrt{3}}{8}$$

$$= \frac{6\pi\sqrt{3} - 8\sqrt{3}}{24}$$

$$= \frac{2(3\pi - 4)\sqrt{3}}{2 \cdot 12}$$

$$= \frac{(3\pi - 4)\sqrt{3}}{12}$$

Note: Problems 22.7–22.8 refer to region G, which is bounded by the graphs of $y = -\cos x$ and $y = \sin\dfrac{x}{2}$ on the interval $[0,\pi]$.

22.8 Determine the volume of a solid with base G whose cross-sections are semi-ellipses perpendicular to the x-axis with a fixed height of 2.

> **Just like it didn't matter which has the length and width of the rectangle in Problem 22.3. You're just going to multiply them together, and in multiplication, the order doesn't matter.**

The area of an ellipse is πab, where a is half the length of the major axis and b is half the length of the minor axis (for the purposes of this problem, it doesn't matter which is which). Therefore, a semiellipse has area $\dfrac{\pi ab}{2} = \dfrac{\pi}{2}ab$.

You are given one dimension of the ellipse—the fixed height of 2 is the distance from the center to the endpoint of one of the axes, so let $a = 2$. The representative length $\sin\dfrac{x}{2} + \cos x$, as calculated in Problem 22.7, is the remaining axis of the ellipse. Divide it by 2, because b is *half* the length of the axis: $b = \dfrac{1}{2}\left(\sin\dfrac{x}{2} + \cos x\right)$.

Generate the cross-sectional area function $A(x)$ by substituting a and b into the formula for the area of a semiellipse: $\frac{\pi}{2}ab$.

$$A(x) = \frac{\pi}{2}ab$$

$$= \frac{\pi}{2}(2)\left[\frac{1}{2}\left(\sin\frac{x}{2} + \cos x\right)\right]$$

$$= \frac{\pi}{2}\left(\sin\frac{x}{2} + \cos x\right)$$

Calculate the volume of the solid.

$$\int_0^\pi \frac{\pi}{2}\left(\sin\frac{x}{2} + \cos x\right) = \frac{\pi}{2}\left[\left(-2\cos\frac{x}{2} + \sin x\right)\Big|_0^\pi\right]$$

$$= \frac{\pi}{2}[0 - (-2)]$$

$$= \pi$$

Disc Method

Circles are the easiest possible cross-sections

22.9 Determine a general formula to calculate the volume of a solid whose circular cross-sections are perpendicular to the x-axis from $x = a$ to $x = b$.

According to Problem 22.1, the volume of the solid is $\int_a^b A(x)\,dx$, where $A(x)$ is the area of the a cross-section. Because the cross-sections are circular, $A(x) = \pi[r(x)]^2$ given radius $r(x)$. Substitute this into the volume formula from Problem 22.1.

$$\int_a^b A(x)\,dx = \int_a^b \pi[r(x)]^2\,dx = \pi\int_a^b [r(x)]^2\,dx \longleftarrow$$

This formula is referred to as the disc method and is used to calculate the volume of rotational solids. If the cross-sections of the solid are perpendicular to the y-axis, use the formula $\pi\int_a^b [r(y)]^2\,dy \longleftarrow$

> *If you're rotating around a horizontal line, this is the disc method formula. Make sure you draw a vertical representative radius and put everything (including boundaries) in terms of x.*

22.10 Rotate the region bounded by the graphs of $y = e^x$, $y = 0$, $x = -1$, and $x = 1$ about the x-axis and calculate the volume of the resulting solid.

Figure 22-6 illustrates the region described and the three-dimensional solid created by rotating that shaded region about the x-axis.

> *When you rotate around a vertical line, use this formula and draw a horizontal representative radius and put everything in terms of y.*

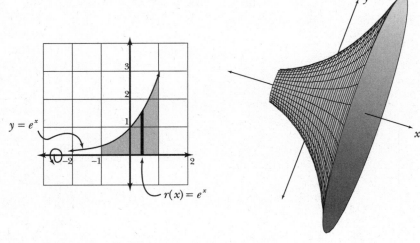

It doesn't matter where you draw the representative radius. All that matters is that it goes from the upper boundary to the lower boundary of the region.

Figure 22-6 *If the darkened representative radius in the left diagram is rotated about the x-axis, it creates a circular cross-section of the solid, the radius of which is the length is the segment itself.*

The darkened length in Figure 22-6 is a representative radius of the solid. Calculate its length by subtracting the lower boundary ($y = 0$) from the upper boundary ($y = e^x$): $r(x) = e^x - 0 = e^x$.

Plug $r(x)$ into the disc method formula from Problem 22.9.

$$\pi \int_a^b [r(x)]^2 \, dx = \pi \int_{-1}^1 (e^x)^2 \, dx$$
$$= \pi \int_{-1}^1 e^{2x} \, dx$$
$$= \frac{\pi}{2} \int_{-2}^2 e^u \, du$$
$$= \frac{\pi}{2} e^u \Big|_{-2}^2$$
$$= \frac{\pi}{2}\left(e^2 - e^{-2}\right)$$
$$= \frac{\pi}{2}\left(e^2 - \frac{1}{e^2}\right)$$
$$= \frac{\pi\left(e^4 - 1\right)}{2e^2}$$

Make the substitution $u = 2x$ and $du/2 = dx$. Don't forget to change the boundaries by plugging $x = -1$ and $x = 1$ into $u = 2x$.

22.11 Rotate the region in the first quadrant bounded by $y = e^x$, $x = 0$, and $y = e$ about the y-axis and calculate the volume of the resulting solid.

When rotating about a vertical axis, use a horizontal representative radius that extends from the left to the right boundary of the region (as illustrated in

Figure 22-7). You must also ensure that the left and right boundary equations are expressed in terms of *y*, so solve $y = e^x$ for *x*.

$$y = e^x$$
$$\ln y = \ln e^x$$
$$\ln y = x$$

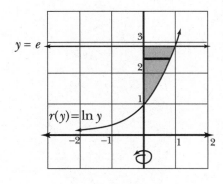

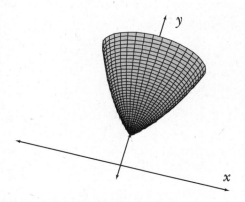

Just like in the last section, a horizontal representative length means you need to solve the boundary equations for x (to put them in terms of y) and use y-boundaries on the integral.

Figure 22-7 *Rotating the shaded region at left results in the solid on the right. Note that the representative radius is written r(y), as it must be expressed in terms of y.*

Find the length of $r(y)$ in Figure 22-7 by subtracting the left boundary ($x = 0$) from the right boundary ($x = \ln y$) of the region: $r(y) = \ln y - 0 = \ln y$. Substitute this expression into the disc method formula, using $a = 1$ and $b = e$ as the limits of integration (because they define, respectively, the lower and upper *y*-boundaries of the region).

$$\pi \int_a^b \left[r(y) \right]^2 dy = \pi \int_1^e \left(\ln y \right)^2 dy$$

Apply integration by parts: $u = (\ln y)^2$ and $dv = dy$, so $du = \dfrac{2 \ln y}{y} dy$ and $v = y$.

$$\pi \int_1^e \left(\ln y \right)^2 dy = \pi \left[y \left(\ln y \right)^2 - \int \frac{\cancel{y} \cdot 2 \ln y}{\cancel{y}} dy \right]_1^e$$

$$= \pi \left[y \left(\ln y \right)^2 - 2 \int \ln y \, dy \right]_1^e$$

According to Problem 21.3, $\int \ln y \, dy = y \ln y - y$.

$$= \pi \left[y \left(\ln y \right)^2 - 2 \left(y \ln y - y \right) \right]_1^e$$
$$= \pi \left[e \left(\ln e \right)^2 - 2 \left(e \ln e - e \right) \right] - \left[1 (\ln 1)^2 - 2 (1 \ln 1 - 1) \right]$$
$$= \pi \left[e - 2 (e - e) \right] - \left[1 \cdot 0 - 2 (0 - 1) \right]$$
$$= \pi (e - 2)$$

22.12 Rotate the region bounded by $y = x \sin x$ and the x-axis between $x = 0$ and $x = \pi$ about the x-axis and calculate the volume of the resulting solid.

Consider Figure 22-8, which illustrates the region, the representative radius, and the three-dimensional solid of rotation.

RULE OF THUMB: Radii in the disc method are perpendicular to the axis of rotation. Vertical radii, like vertical lines, are described using x's.

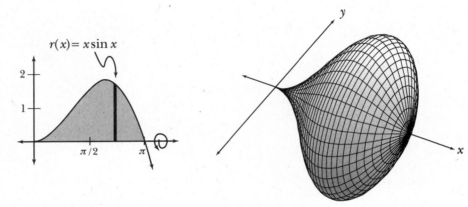

Figure 22-8 *Using the disc method to calculate the volume generated by rotating a region about the x-axis requires boundary functions in terms of x, limits of integration in terms of x, and a vertical representative radius.*

Find $r(x)$ by subtracting the lower boundary from the upper boundary: $r(x) = x \sin x - 0 = x \sin x$.

$$\pi \int_a^b \left[r(x) \right]^2 dx = \pi \int_0^\pi \left(x \sin x \right)^2 dx$$

$$= \pi \int_0^\pi x^2 \sin^2 x \, dx$$

Use the integration by parts tabular method, as described in Problems 21.5 and 21.6, to find the antiderivative.

This integral is calculated in Problem 21.11.

u	dv	± 1
x^2	$\sin^2 x$	$+1$
$2x$	$\dfrac{1}{2}x - \dfrac{1}{4}\sin 2x$	-1
2	$\dfrac{x^2}{4} + \dfrac{1}{8}\cos 2x$	$+1$
0	$\dfrac{x^3}{12} + \dfrac{1}{16}\sin 2x$	-1
		$+1$

$$\pi \int_0^\pi x^2 \sin^2 x\, dx = \pi \left[x^2 \left(\frac{1}{2}x - \frac{1}{4}\sin 2x \right) - 2x \left(\frac{x^2}{4} + \frac{1}{8}\cos 2x \right) + 2 \left(\frac{x^3}{12} + \frac{1}{16}\sin 2x \right) \right]\Big|_0^\pi$$

$$= \pi \left[\left(\frac{x^3}{2} - \frac{x^2 \sin 2x}{4} \right) - \left(\frac{x^3}{2} + \frac{x \cos 2x}{4} \right) + \left(\frac{x^3}{6} + \frac{\sin 2x}{8} \right) \right]\Big|_0^\pi$$

$$= \pi \left[\left(\frac{\pi^3}{2} - 0 - \frac{\pi^3}{2} - \frac{\pi}{4} + \frac{\pi^3}{6} + 0 \right) - \left(0 - 0 + 0 \right) \right]$$

$$= \pi \left(-\frac{\pi}{4} + \frac{\pi^3}{6} \right)$$

$$= \frac{\pi \left(2\pi^3 - 3\pi \right)}{12}$$

$$= \frac{\pi^2 \left(2\pi^2 - 3 \right)}{12}$$

22.13 Rotate the region bounded by the graphs of $y = x^3 - x + 1$ and $y = 1$ about the line $y = 1$ and calculate the volume of the resulting solid.

As illustrated by Figure 22-9, the upper and lower boundaries reverse at $x = 0$. Therefore, you must calculate two separate volumes, one for the region on the x-interval $[-1,0]$, and one for the region on the x-interval $[0,1]$.

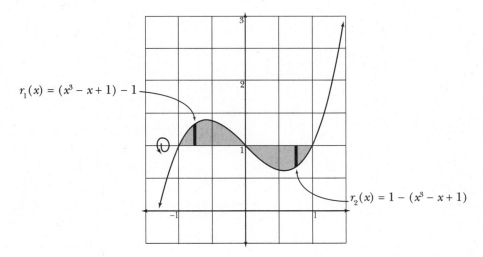

Figure 22-9 *Because the boundaries for the two regions are reversed, you must use different representative radii for each region, denoted $r_1(x)$ and $r_2(x)$. Though the bounding functions are the same ($y = x^3 - x + 1$ and $y = 1$), each serves once as an upper boundary and once as a lower boundary.*

Apply the disc method twice, using the representative radii identified in Figure 22-9; add the results.

Because $(x^3 - x)$ and $(-x^3 + x)$ are opposites, their squares are equal, just like a number and its opposite have equal squares: $(-3)^2 = 3^2 = 9$.

$$\pi \int_{-1}^{0} \left[x^3 - x + 1 - 1 \right]^2 dx + \pi \int_{0}^{1} \left[1 - \left(x^3 - x + 1 \right) \right]^2 dx$$

$$= \pi \int_{-1}^{0} \left(x^3 - x \right)^2 dx + \pi \int_{0}^{1} \left(-x^3 + x \right)^2 dx$$

$$= \pi \int_{-1}^{0} \left(x^6 - 2x^4 + x^2 \right) dx + \pi \int_{0}^{1} \left(x^6 - 2x^4 + x^2 \right) dx$$

$$= \pi \left(\frac{x^7}{7} - \frac{2x^5}{5} + \frac{x^3}{3} \right) \Bigg|_{-1}^{0} + \pi \left(\frac{x^7}{7} - \frac{2x^5}{5} + \frac{x^3}{3} \right) \Bigg|_{0}^{1}$$

$$= \pi \left[(0 - 0 + 0) - \left(-\frac{1}{7} + \frac{2}{5} - \frac{1}{3} \right) \right] + \pi \left[\left(\frac{1}{7} - \frac{2}{5} + \frac{1}{3} \right) - (0 - 0 + 0) \right]$$

$$= \pi \left(\frac{8}{105} \right) + \pi \left(\frac{8}{105} \right)$$

$$= \frac{16\pi}{105}$$

22.14 Rotate the region bounded by the graphs of $x = y^2 - 6y + 4$ and $x = -4$ about the line $x = -4$ and calculate the volume of the resulting solid.

Consider Figure 22-10, which illustrates the region and its representative radius.

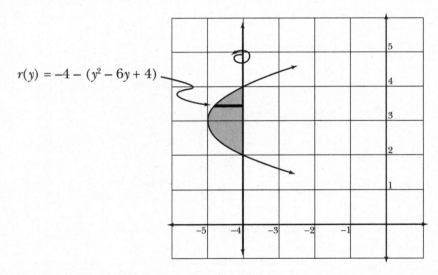

$r(y) = -4 - (y^2 - 6y + 4)$

Figure 22-10 *Rotating about a vertical axis requires a horizontal representative radius, whose length r(y) is equal to the right boundary (x = −4) minus the left boundary (x = y² − 6y + 4).*

Substitute $r(y)$ into the disc method formula, using $y = 2$ and $y = 4$ as the lower and upper limits of integration, respectively.

$$\pi \int_2^4 \left[-4 - \left(y^2 - 6y + 4 \right) \right]^2 dy$$

$$= \pi \int_2^4 \left[-y^2 + 6y - 8 \right]^2 dy$$

$$= \pi \int_2^4 \left(y^4 - 12y^3 + 52y^2 - 96y + 64 \right) dy$$

$$= \pi \left(\frac{y^5}{5} - 3y^4 + \frac{52y^3}{3} - 48y^2 + 64y \right) \Bigg|_2^4$$

$$= \pi \left(\frac{512}{15} - \frac{496}{15} \right)$$

$$= \frac{16\pi}{15}$$

> Those are the lowest and highest values reached by the shaded region in Figure 22-10.

22.15 Use the disc method to prove that the volume of a right circular cylinder with radius R and height h is $\pi R^2 h$. (Assume R and h are positive real numbers.)

Consider the rectangular region in the first quadrant of the coordinate plane bounded by $x = 0$, $x = R$, $y = 0$, and $y = h$, as illustrated by Figure 22.11. If this region is rotated about the y-axis, it produces a right circular cylinder with radius R and height h.

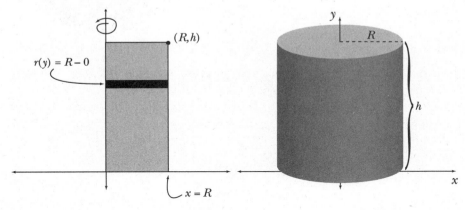

Figure 22-11 *Because you are rotating the shaded rectangular region about a vertical axis, you must use a horizontal representative radius, r(y); its length is the right boundary (x = R) minus the left boundary (x = 0). The limits of integration must be in terms of y as well: y = 0 and y = h.*

Apply the disc method.

> R^2 can be pulled out of the integral because it's a constant.

$$\pi \int_a^b \left[r(y) \right]^2 dy = \pi \int_0^h (R)^2 dy$$
$$= \pi R^2 \int_0^h dy$$
$$= \pi R^2 (y) \Big|_0^h$$
$$= \pi R^2 [h - 0]$$
$$= \pi R^2 h$$

22.16 Use the disc method to prove that the volume of a sphere with radius R is $\frac{4}{3}\pi R^3$. (Assume R is a positive real number.)

A circle centered at the origin with radius R has equation $x^2 + y^2 = R^2$. Solve the equation for y to determine the equation of the semicircle pictured in Figure 22-12.

> Is it me, or does that sphere look EXACTLY like the Death Star from Star Wars? The force (mass times acceleration) is strong in this one.

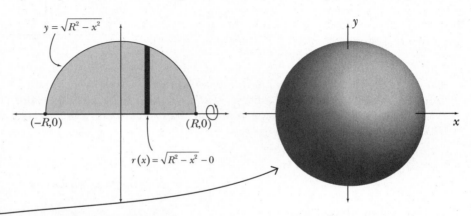

Figure 22-12   *Rotating the region defined by the x-axis and the semicircle with radius R generates a sphere with radius R.*

Apply the disc method, writing the representative radius and limits of integration in terms of x.

> Because R is a constant, $\int R^2 dx = R^2 \int dx = R^2 x$
> $\int R^2 dx \neq \frac{R^3}{3}$ because you are integrating with respect to x, not R.

$$\pi \int_a^b \left[r(x) \right]^2 dx = \pi \int_{-R}^R \left[\sqrt{R^2 - x^2} \right]^2 dx$$
$$= \pi \int_{-R}^R (R^2 - x^2) dx$$
$$= \pi \left(R^2 x - \frac{x^3}{3} \right) \Bigg|_{-R}^R$$
$$= \pi \left[\left(R^3 - \frac{R^3}{3} \right) - \left(-R^3 - \frac{-R^3}{3} \right) \right]$$
$$= \pi \left[\left(\frac{3R^3 - R^3}{3} \right) - \left(\frac{-3R^3 + R^3}{3} \right) \right]$$
$$= \pi \left[\frac{2R^3}{3} + \frac{2R^3}{3} \right]$$
$$= \frac{4\pi R^3}{3}$$

22.17 Use the disc method to prove that the volume of a right circular cone with radius R and height h is $\frac{1}{3}\pi R^2 h$. (Assume R and h are nonzero real numbers.)

Consider the first quadrant region bounded by $x = 0$, $y = 0$, $y = h$, and the segment with endpoints $(0,0)$ and (R,h), as illustrated by Figure 22-13. Rotating this region about the y-axis generates a right circular cone with height h and radius R.

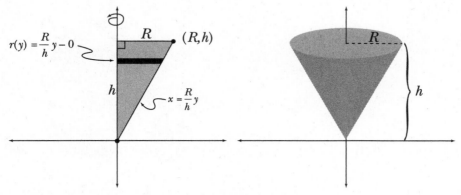

Figure 22-13   _The linear equation serving as the region's right boundary must be solved for x, because the disc method integrand and limits of integration must be written in terms of y when revolving around a vertical axis._

The equation of the line connecting the origin and the point (R,h) is $y = \frac{h}{R}x$. Solve the equation for y.

$$y = \frac{h}{R}x$$

$$\frac{R}{h}y = x$$

The slope of the line is
$$\frac{h-0}{R-0} = \frac{h}{R}$$
and the y-intercept is 0. Plug those into slope-intercept form (y = mx + b) to get the equation:
$$y = \frac{h}{R}x.$$

Determine $r(y)$, the length of the representative radius: $\frac{R}{h}y - 0 = \frac{R}{h}y$, and apply the disc method.

$$\pi \int_a^b \left[r\left(y \right) \right]^2 dy = \pi \int_0^h \left(\frac{R}{h}y \right)^2 dy$$

$$= \pi \cdot \frac{R^2}{h^2} \int_0^h y^2 \, dy$$

$$= \frac{\pi R^2}{h^2} \left(\frac{y^3}{3} \right) \Big|_0^h$$

$$= \frac{\pi R^2}{h^2} \left(\frac{h^3}{3} \right)$$

$$= \frac{\pi R^2 h}{3}$$

Washer Method
Find volumes even if the "solids" aren't solid

Note: Problems 22.18–22.19 refer to region K in Figure 22-14.

22.18 Explain why the disc method cannot be used to determine the volume of the solid generated when *K* is rotated about the line *y = c* in Figure 22-14.

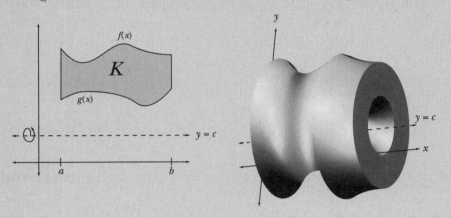

> *In other words, f(x) is always above g(x).*

Figure 22-14 *Region K is bounded by f(x), g(x), x = a, and x = b. Notice that f(x) ≥ g(x) for all x on the interval [a,b].*

> *In Problems 22.10–22.17 the axis you rotated around was also one of the boundaries of the region. When that's not true, you have to use either the washer method or the shell method.*

Notice that a gap separates the lower boundary, *g(x)*, from the axis of rotation, *x = c*. This gap is rotated about the line *y = c* as well, resulting in a hollow cavity within the rotational solid, which is clearly visible in Figure 22-14. The disc method can be used only to calculate the volume of a solid with circular cross-sections, which is not true of the rotational solid in Figure 22-14.

Note: Problems 22.18–22.19 refer to region K in Figure 22-14.

22.19 Use the washer method to construct a definite integral representing the volume of the rotational solid in Figure 22-14.

> *Find the lengths of R(x) and r(x) the same way you did with the disc method—the upper boundary of each radius minus its lower boundary.*

The washer method, like the disc method, uses representative radii that are perpendicular to the axis of rotation. However, it uses two radii: the outer radius (which extends from the axis of rotation to the outer boundary of the region) and the inner radius (which extends from the axis of rotation to the inner boundary of the region). In Figure 22-15, *R(x)* represents the outer radius and *r(x)* represents the inner radius.

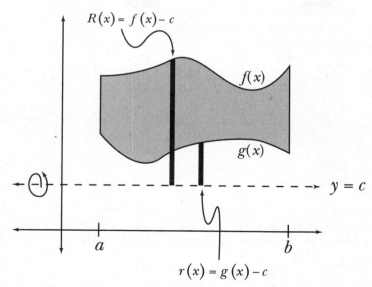

Figure 22-15 *Both the inner and outer radii extend from the axis of rotation to an edge of the region. The inner radius extends to the edge that is closer to the rotational axis and the outer radius extends to the edge that is farther away.*

According to the washer method, the volume of the rotational solid is $\pi \int_a^b \left([R(x)]^2 - [r(x)]^2 \right) dx$. Substitute the values of the radii determined in Figure 22-15.

$$\pi \int_a^b \left([f(x) - c]^2 - [g(x) - c]^2 \right) dx$$

22.20 Use the washer method to construct a definite integral representing the volume of the rotational solid in Figure 22-16.

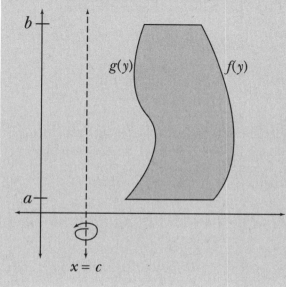

Figure 22-16

Subtract c from f(y) and g(y), respectively, to calculate the lengths of R(y) and r(y). In other words, subtract the left boundary from the right boundary when given horizontal radii.

This region is similar to the region pictured in Figure 22-14, but because the region is rotated about a vertical axis, the functions and limits of integration must be written in terms of y.

$$\pi\int_a^b\left(\left[R(y)\right]^2-\left[r(y)\right]^2\right)dy=\pi\int_a^b\left(\left[f(y)-c\right]^2-\left[g(y)-c\right]^2\right)dy$$

Note: Problems 22.21–22.22 refer to the region M in the first quadrant that is bounded by the graphs of y = sin x, y = cos x, and x = 0.

22.21 Determine the volume of the solid generated by rotating M about the x-axis.

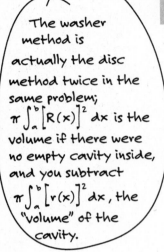

The washer method is actually the disc method twice in the same problem; $\pi\int_a^b\left[R(x)\right]^2dx$ is the volume if there were no empty cavity inside, and you subtract $\pi\int_a^b\left[r(x)\right]^2dx$, the "volume" of the cavity.

Consider Figure 22-17, which identifies the region and the outer and inner radii required to apply the washer method. Determine $R(x)$ and $r(x)$ by subtracting the lower boundary of each radius from its upper boundary.

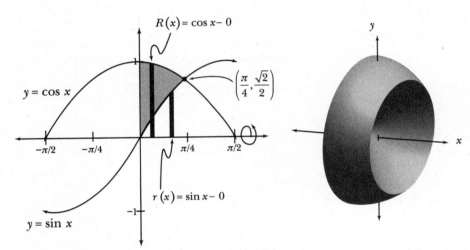

Figure 22-17 *Note that the outer radius R(x) extends from the axis of rotation (y = 0) to y = cos x (the region boundary that's farther away from the x-axis), and the inner radius r(x) extends from the axis of rotation to y = sin x (the region boundary closer to the x-axis).*

Substitute $R(x)$ and $r(x)$ into the washer method formula.

$$\pi\int_a^b\left(\left[R(x)\right]^2-\left[r(x)\right]^2\right)dx=\pi\int_a^b\left[(\cos x)^2-(\sin x)^2\right]dx$$
$$=\pi\int_a^b\left(\cos^2 x-\sin^2 x\right)dx$$

The region's left limit of integration is $a=0$. To determine the right limit b, you must identify the x-value at which $\cos x$ and $\sin x$ intersect. Set the functions equal and solve for x.

Cosine and sine intersect at lots of places, not just at one point, but you're looking for the spot that marks the right edge of the shaded area in Figure 22-17.

$$\cos x=\sin x$$
$$\frac{\cos x}{\cos x}=\frac{\sin x}{\cos x}$$
$$1=\tan x$$
$$x=\frac{\pi}{4}$$

Substitute a and b into the integral.

$$= \pi \int_0^{\pi/4} \left(\cos^2 x - \sin^2 x \right) dx$$

According to a double angle identity, $\cos 2x = \cos^2 x - \sin^2 x$.

$$= \pi \int_0^{\pi/4} \cos 2x \, dx$$

Apply variable substitution: $u = 2x$ and $\dfrac{du}{2} = dx$.

$$= \pi \cdot \frac{1}{2} \int_0^{\pi/2} \cos u \, du$$

$$= \frac{\pi}{2} (\sin u) \Big|_0^{\pi/2}$$

$$= \frac{\pi}{2} \left(\sin \frac{\pi}{2} - \sin 0 \right)$$

$$= \frac{\pi}{2}$$

> Don't forget to change the x-boundaries into u-boundaries by plugging each of them into $u = 2x$.

Note: Problems 22.21–22.22 refer to the region M in the first quadrant that is bounded by the graphs of $y = \sin x$, $y = \cos x$, and $x = 0$.

22.22 Calculate the volume of the solid generated by rotating M about the line $y = -1$.

Like Problem 22.21, region M is rotated about a horizontal axis. However, $R(x)$ and $r(x)$ are different due to the different axis of rotation, as illustrated by Figure 22-18.

> In this problem, $R(x) = \cos x - (-1) = \cos x + 1$ and $r(x) = \sin x + 1$.

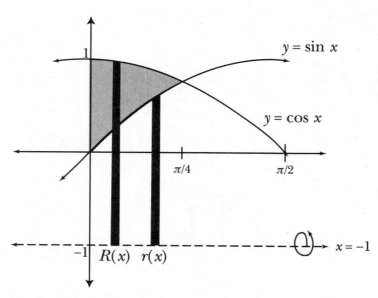

Figure 22-18 *The outer and inner radii are each exactly one unit longer than the radii in Problem 22.21, because the axis of rotation is exactly one unit further away from the region.*

Apply the washer method using the limits of integration from Problem 22.21.

$$= \pi \int_0^{\pi/4} \left[(\cos x + 1)^2 - (\sin x + 1)^2 \right] dx$$

$$= \pi \int_0^{\pi/4} \left[\cos^2 x + 2\cos x + 1 - \left(\sin^2 x + 2\sin x + 1 \right) \right] dx$$

$$= \pi \int_0^{\pi/4} \left[\cos^2 x + 2\cos x + \cancel{1} - \sin^2 x - 2\sin x \cancel{-1} \right] dx$$

Regroup the integrand into two integrals.

$$= \pi \int_0^{\pi/4} \left(\cos^2 x - \sin^2 x \right) dx + 2\pi \int_0^{\pi/4} \left(\cos x - \sin x \right) dx$$

According to Problem 22.21, $\pi \int_0^{\pi/4} \left(\cos^2 x - \sin^2 x \right) dx = \dfrac{\pi}{2}$.

$$= \frac{\pi}{2} + 2\pi \int_0^{\pi/4} \left(\cos x - \sin x \right) dx$$

$$= \frac{\pi}{2} + 2\pi \left[\sin x + \cos x \right]_0^{\pi/4}$$

$$= \frac{\pi}{2} + 2\pi \left[\left(\sin \frac{\pi}{4} + \cos \frac{\pi}{4} \right) - \left(\sin 0 + \cos 0 \right) \right]$$

$$= \frac{\pi}{2} + 2\pi\sqrt{2} - 2\pi$$

$$= \frac{4\pi\sqrt{2} - 3\pi}{2}$$

$$= \frac{\pi\left(4\sqrt{2} - 3 \right)}{2}$$

> This isn't the ONLY acceptable answer. It would have been perfectly fine to stop simplifying two steps ago, but I suppose this does look a little more compact since it's a single fraction.

Note: Problems 22.23–22.24 refer to region J, which is bounded by the graphs of $y = \dfrac{x}{2}$ and $y = \sqrt{x}$.

22.23 Calculate the volume of the solid generated by rotating *J* about the *x*-axis.

Consider Figure 22-19, which illustrates the region and the radii necessary to apply the washer method. To determine the points of intersection, set the functions equal and solve for *x*.

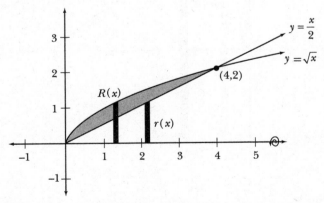

Figure 22-19 *Determine R(x) and r(x) by subtracting the top boundary of each radius from its bottom boundary: $R(x) = \sqrt{x} - 0 = \sqrt{x}$ and $r(x) = \dfrac{x}{2} - 0 = \dfrac{x}{2}$.*

$$\frac{x}{2} = \sqrt{x}$$

$$\left(\frac{x}{2}\right)^2 = \left(\sqrt{x}\right)^2$$

$$\frac{x^2}{4} = x$$

$$x^2 = 4x$$

$$x^2 - 4x = 0$$

$$x(x-4) = 0$$

$$x = 0 \quad \text{or} \quad x = 4$$

This time, the points of intersection mark the left and right edges of the region, so they become the limits on the washer method definite integral.

These x-values represent the limits of integration for the washer method: $a = 0$ and $b = 4$.

$$= \pi \int_0^4 \left[\left(\sqrt{x}\right)^2 - \left(\frac{x}{2}\right)^2\right] dx$$

$$= \pi \int_0^4 \left(x - \frac{x^2}{4}\right) dx$$

$$= \pi \left[\frac{x^2}{2} - \frac{x^3}{12}\right]_0^4$$

$$= \pi \left(8 - \frac{16}{3}\right)$$

$$= \frac{8\pi}{3}$$

Note: Problems 22.23–22.24 refer to region J, which is bounded by the graphs of $y = \dfrac{x}{2}$ and $y = \sqrt{x}$.

22.24 Calculate the volume of the solid generated by rotating J about the y-axis.

Because the rotational axis is vertical, the radii (and therefore the boundary functions) must be written in terms of y. Solve the equations for x to get $x = 2y$ and $x = y^2$. Figure 22-20 illustrates region J and the horizontal outer and inner radii, written in terms of y.

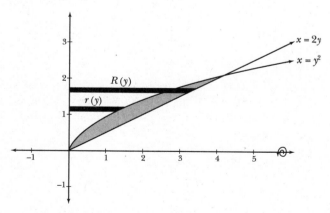

Figure 22-20   *Find the lengths of $R(y)$ and $r(y)$ by subtracting the left boundaries from the right boundaries: $R(y) = 2y - 0$ and $r(y) = y^2 - 0$.*

Apply the washer method.

Problem 22.23 says the graphs intersect at $x = 0$ and $x = 4$. Plug those x's into $y = x/2$ to get y-values: $0/2 = 0$ and $4/2 = 2$. That means $a = 0$ and $b = 2$ in the washer method formula.

$$\pi \int_a^b \left(\left[R(y) \right]^2 - \left[r(y) \right]^2 \right) dy = \pi \int_0^2 \left[(2y)^2 - (y^2)^2 \right] dy$$

$$= \pi \int_0^2 (4y^2 - y^4) \, dy$$

$$= \pi \left(\frac{4y^3}{3} - \frac{y^5}{5} \right) \Bigg|_0^2$$

$$= \pi \left(\frac{32}{3} - \frac{32}{5} \right)$$

$$= \pi \left(\frac{160 - 96}{15} \right)$$

$$= \frac{64\pi}{15}$$

Note: Problems 22.25–22.26 refer to region A, which is bounded by the graphs of $y = 3 - x^2$ and the x-axis.

22.25 Calculate the volume of the solid generated by rotating A about the line $y = -2$.

Graph the region, the axis of rotation, the outer radius, and the inner radius (as illustrated by Figure 22-21) and calculate the lengths of the radii.

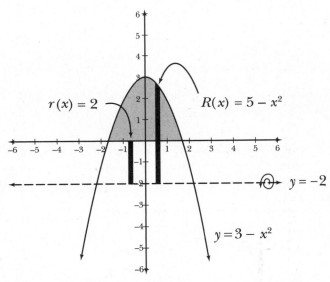

Figure 22-21 *The inner radius extends from the x-axis ($y = 0$) to the axis of rotation ($y = -2$), so $r(x) = 0 - (-2) = 2$. The outer radius extends from the curve $y = 3 - x^2$ to the axis of rotation, so $R(x) = (3 - x^2) - (-2) = 5 - x^2$.*

Determine the limits of integration by setting the boundaries of the region equal and solving for x.

$$3 - x^2 = 0$$
$$x^2 = 3$$
$$x = \pm\sqrt{3}$$

Substitute $R(x)$, $r(x)$, $a = -\sqrt{3}$, and $b = \sqrt{3}$ into the washer method formula.

$$\pi\int_a^b\left([R(x)]^2 - [r(x)]^2\right)dx = \pi\int_{-\sqrt{3}}^{\sqrt{3}}\left[(5-x^2)^2 - (2)^2\right]dx$$

$$= \pi\int_{-\sqrt{3}}^{\sqrt{3}}\left(25 - 10x^2 + x^4 - 4\right)dx$$

$$= \pi\int_{-\sqrt{3}}^{\sqrt{3}}\left(x^4 - 10x^2 + 21\right)dx$$

$$= \pi\left(\frac{x^5}{5} - \frac{10x^3}{3} + 21x\right)\Bigg|_{-\sqrt{3}}^{\sqrt{3}}$$

$$= \pi\left[\left(\frac{9\sqrt{3}}{5} - 10\sqrt{3} + 21\sqrt{3}\right) - \left(-\frac{9\sqrt{3}}{5} + 10\sqrt{3} - 21\sqrt{3}\right)\right]$$

$$= \pi\left(\frac{18\sqrt{3} - 100\sqrt{3} + 210\sqrt{3}}{5}\right)$$

$$= \frac{128\pi\sqrt{3}}{5}$$

Note: Problems 22.25–22.26 refer to region A, which is bounded by the graphs of $y = 3 - x^2$ and the x-axis.

22.26 Calculate the volume of the solid generated by rotating A about the line $y = 3$.

Consider Figure 22-22, which identifies the region to be rotated and the radii necessary to apply the washer method.

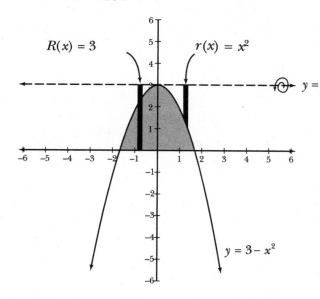

$R(x) = 3$ $r(x) = x^2$

$y = 3$

$y = 3 - x^2$

Figure 22-22

The inner radius extends from the axis of rotation ($y = 3$) to the inner boundary of the region ($y = 3 - x^2$). The outer radius extends from $y = 3$ to $y = 0$.

Construct expressions for $R(x)$ and $r(x)$.

$$R(x) = 3 - 0 = 3$$
$$r(x) = 3 - (3 - x^2) = x^2$$

> The region didn't change—only the line you're rotating about is different than Problem 2.25. That changes $R(x)$ and $r(x)$ but not a and b.

Apply the washer method using the same limits of integration as Problem 2.25.

$$\pi \int_a^b \left([R(x)]^2 - [r(x)]^2 \right) dx = \pi \int_{-\sqrt{3}}^{\sqrt{3}} \left[(3)^2 - (x^2)^2 \right] dx$$

$$= \pi \int_{-\sqrt{3}}^{\sqrt{3}} (9 - x^4) \, dx$$

$$= \pi \left(9x - \frac{x^5}{5} \right) \Bigg|_{-\sqrt{3}}^{\sqrt{3}}$$

$$= \pi \left[\left(9\sqrt{3} - \frac{9\sqrt{3}}{5} \right) - \left(-9\sqrt{3} + \frac{9\sqrt{3}}{5} \right) \right]$$

$$= \pi \left(18\sqrt{3} - \frac{18\sqrt{3}}{5} \right)$$

$$= \frac{\pi \left(90\sqrt{3} - 18\sqrt{3} \right)}{5}$$

$$= \frac{72\pi\sqrt{3}}{5}$$

Note: Problems 22.27–22.28 refer to the region L, which is bounded by the graphs of $x = \dfrac{1}{y}$, $x = 0$, $y = \dfrac{1}{2}$, and $y = 4$.

22.27 Calculate the volume of the solid generated by rotating L about the line $x = -4$.

Consider Figure 22-23, which illustrates the region and the radii required to apply the washer method.

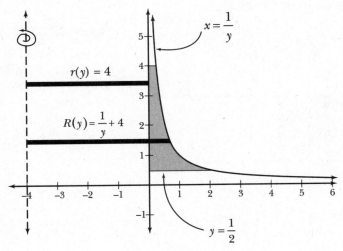

Figure 22-23

The graph of $x = \dfrac{1}{y}$ is identical to the graph of $y = \dfrac{1}{x}$. Solve the first equation for y or the second equation for x to generate the other equation.

Determine the lengths of the radii.

$$R(y) = \frac{1}{y} - (-4) = \frac{1}{y} + 4$$

$$r(y) = 0 - (-4) = 4$$

Apply the washer method.

$$\pi \int_a^b \left(\left[R(y) \right]^2 - \left[r(y) \right]^2 \right) dy = \pi \int_{1/2}^4 \left[\left(\frac{1}{y} + 4 \right)^2 - (4)^2 \right] dy$$

$$= \pi \int_{1/2}^4 \left(\frac{1}{y^2} + \frac{8}{y} + 16 - 16 \right) dy$$

$$= \pi \int_{1/2}^4 \left(y^{-2} + 8y^{-1} \right) dy$$

$$= \pi \left(-\frac{1}{y} + 8 \ln |y| \right) \Big|_{1/2}^4$$

$$= \pi \left[\left(-\frac{1}{4} + 8 \ln 4 \right) - \left(-2 + 8 \ln \frac{1}{2} \right) \right]$$

$$= \pi \left[\left(2 - \frac{1}{4} \right) + 8 \left(\ln 4 - \ln \frac{1}{2} \right) \right]$$

> The boundaries of the integral are a = ½ and b = 4, because you need to use y limits of integration with horizontal radii.

Apply the logarithmic property $\log a - \log b = \log \frac{a}{b}$.

$$= \pi \left[\frac{7}{4} + 8 \left(\ln \frac{4}{1/2} \right) \right]$$

$$= \pi \left[\frac{7}{4} + 8 \ln 8 \right]$$

Note: Problems 22.27–22.28 refer to the region L, which is bounded by the graphs of $x = \frac{1}{y}$, $x = 0$, $y = \frac{1}{2}$, and $y = 4$.

22.28 Calculate the volume of the solid generated by rotating L about the line $x = 6$.

> You're still rotating around a vertical axis (like Problem 22.26) so the limits of integration are the same and you'll still use horizontal radii, but those radii will be totally different.

The axis of rotation $x = 6$ is to the right of the region—unlike the axis of rotation $x = -4$ in Problem 22.26, which was left of the region. Therefore, the region boundary that was once closer to the rotational axis is now farther away, and vice versa. The radii changes are reflected in Figure 2-24.

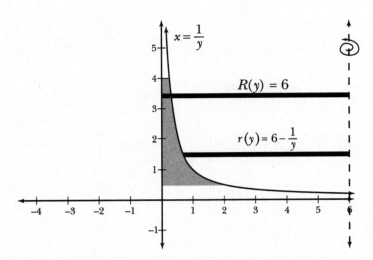

Figure 22-24 *The right boundary of each radius is now x = 6, so the length of each radius is 6 minus its left boundary.*

Apply the washer method.

$$\pi\int_a^b\left(\left[R(y)\right]^2-\left[r(y)\right]^2\right)dy=\pi\int_{1/2}^4\left[(6)^2-\left(6-\frac{1}{y}\right)^2\right]dy$$

$$=\pi\int_{1/2}^4\left[36-\left(36-\frac{12}{y}+\frac{1}{y^2}\right)\right]dy$$

$$=\pi\int_{1/2}^4\left(\frac{12}{y}-y^{-2}\right)dy$$

$$=\pi\left(12\ln|y|+\frac{1}{y}\right)\Bigg|_{1/2}^4$$

$$=\pi\left[\left(12\ln 4+\frac{1}{4}\right)-\left(12\ln\frac{1}{2}+2\right)\right]$$

$$=\pi\left[-\frac{7}{4}+12\left(\ln 4-\ln\frac{1}{2}\right)\right]$$

Apply the logarithmic property $\log a-\log b=\log\dfrac{a}{b}$.

$$=\pi\left[12(\ln 8)-\frac{7}{4}\right]$$

Shell Method
Something to fall back on when the washer method fails

22.29 If the region pictured in Figure 22-25 is rotated about the line $x = c$, the shell method calculates the volume of the resulting solid according to the formula $2\pi \int_a^b d(x)h(x)\,dx$. Explain how to determine the values of a, b, $d(x)$, and $h(x)$.

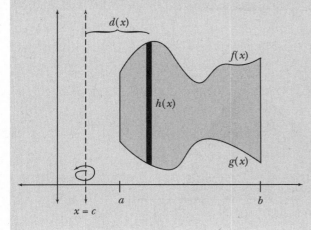

Figure 22-25

In order to calculate the volume of this rotational solid, the shell method requires only one representative radius.

Unlike the disc and washer methods, the shell method uses a representative radius *parallel* to the axis of rotation, rather than perpendicular to the axis. Because the region in Figure 22-25 is rotated about a vertical line, so you must therefore use a vertical representative radius. A vertical radius must be written in terms of x (as must the integrand and its boundaries). Notice that the boundaries of the region along the x-axis are $x = a$ and $x = b$.

Determine the length of the representative radius $h(x)$ in Figure 22-25 by subtracting its bottom boundary from its top boundary: $h(x) = f(x) - g(x)$. The remaining variable expression in the shell method formula, $d(x)$, describes the distance from the axis of rotation to $h(x)$. Calculate the length of $d(x)$ by subtracting the left boundary from the right boundary. While the rotational axis has an explicit location along the x-axis ($x = c$), $h(x)$ does not, so describe its position along the x-axis generically, as "x." In Figure 22-25, $d(x)$ is equal to the x-position of $h(x)$ minus the x-position of the axis of rotation: $d(x) = x - c$.

> This is true of ALL the volume formulas from this chapter. Put things in terms of x when you use vertical radii, and use y's when you're working with horizontal radii.

> Unless the radius is horizontal, in which case you refer to the radius as "y."

Note: Problems 22.30–22.32 refer to region H, which is bounded by the graphs of y = x sin x, y = 0, x = 0, and x = π.

22.30 Problem 22.12 uses the disc method to determine the volume generated when H is rotated about the x-axis. Explain why the washer method cannot be applied to determine the volume if H is rotated about the y-axis.

If the region were rotated about the y-axis, the washer method would require the use of horizontal radii. Therefore, the function $y = x \sin x$ would need to be rewritten in terms of y, but to do so would require solving the equation for x, and that is not possible.

22.31 Use the shell method to calculate the volume of the solid generated by rotating *H* about the *y*-axis.

> The fancy name for the shell method is "the method of cylindrical shells."

Consider Figure 22-26, which contains the region to be rotated and a graphical representation of $d(x)$ and $h(x)$, the expressions needed to apply the shell method formula.

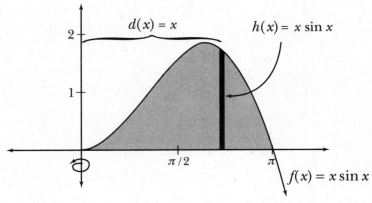

Figure 22-26 *Using the shell method to calculate the volume of a region rotated about the y-axis (a vertical line) requires a vertical representative radius h(x) and the boundaries of the region along the x-axis: a = 0 and b = π.*

> Use integration by parts: $u = x^2$, $du = 2x\,dx$, $dv = \sin x\,dx$, and $v = -\cos x$.

The radius $h(x)$ appears to the right of the rotational axis $(y = 0)$, at position x. Find $d(x)$, the difference of the x-positions of $h(x)$ and the rotational axis: $d(x) = x - 0 = x$. Next, define $h(x)$, the difference of the region's upper boundary $(y = x \sin x)$ and its lower boundary $(y = 0)$: $h(x) = x \sin x - 0 = x \sin x$. Substitute $d(x)$, $h(x)$, $a = 0$, and $b = \pi$ into the shell method formula.

$$2\pi \int_a^b d(x) h(x)\,dx = 2\pi \int_0^\pi x \cdot x \sin x\,dx$$
$$= 2\pi \int_0^\pi x^2 \sin x\,dx$$
$$= 2\pi \left(-x^2 \cos x + 2x \sin x + 2\cos x\right)\Big|_0^\pi$$
$$= 2\pi \left[\left(-\pi^2 \cos \pi + 2\pi \sin \pi + 2\cos \pi\right) - \left(0 + 0 + 2\right)\right]$$
$$= 2\pi \left(\pi^2 + 0 - 2 - 2\right)$$
$$= 2\pi \left(\pi^2 - 4\right)$$

22.32 Find the volume generated when *H* is rotated about the line $x = \dfrac{5\pi}{4}$.

Except for $d(x)$, this problem is nearly identical to Problem 22.31, as illustrated by Figure 22-27.

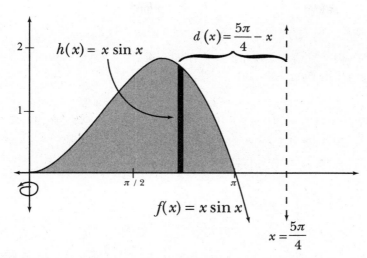

Figure 22-27 *Recall that d(x) represents the horizontal distance between the*
axis of rotation $x = \dfrac{5\pi}{4}$ and the radius h(x).

To calculate $d(x)$, first note the relative positions of the radius and rotational axis—$d(x)$ equals the difference of their x-positions: $d(x) = \dfrac{5\pi}{4} - x$. Apply the shell method.

$$2\pi \int_a^b d(x)h(x)\,dx = 2\pi \int_0^\pi \left(\frac{5\pi}{4} - x\right)(x\sin x)\,dx$$

Integrate by parts: $u = \dfrac{5\pi}{4} - x$, $dv = x\sin x\,dx$, $du = -dx$, and $v = -x\cos x + \sin x$.

$$= 2\pi\left[\left(\frac{5\pi}{4} - x\right)(-x\cos x + \sin x) - \int(-x\cos x + \sin x)(-dx)\right]_0^\pi$$

$$= 2\pi\left[\left(\frac{5\pi}{4} - x\right)(-x\cos x + \sin x) - \int x\cos x\,dx + \int \sin x\,dx\right]_0^\pi$$

Use integration by parts to determine that $\int x\cos x\,dx = x\sin x + \cos x$.

$$= 2\pi\left[\left(\frac{5\pi}{4} - x\right)(-x\cos x + \sin x) - (x\sin x + \cos x) - \cos x\right]_0^\pi$$

$$= 2\pi\left[\left(\frac{5\pi}{4} - x\right)(-x\cos x + \sin x) - x\sin x - 2\cos x\right]_0^\pi$$

$$= 2\pi\left[\left(\frac{\pi}{4}(\pi) + 2\right) - (-2)\right]$$

$$= 2\pi\left[\frac{\pi}{4}(\pi) + 4\right]$$

$$= 2\pi\left(\frac{\pi^2 + 16}{4}\right)$$

$$= \frac{\pi(\pi^2 + 16)}{2}$$

22.33 According to Problem 22.23, the volume of the solid generated by rotating the region bounded by $y = \dfrac{x}{2}$ and $y = \sqrt{x}$ about the x-axis is equal to $\dfrac{8\pi}{3}$. Verify the solution using the shell method.

Rotating the region about a horizontal axis requires a horizontal radius and an integral in terms of y. Solve $y = \dfrac{x}{2}$ and $y = \sqrt{x}$ for x and construct a graph of the region that identifies the segments $d(y)$ and $h(y)$, as demonstrated in Figure 22-28.

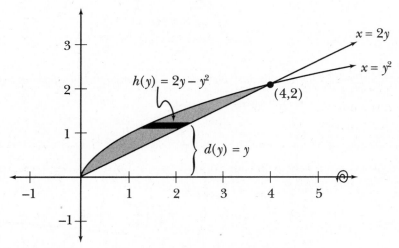

Figure 22-28   *The length of the radius $h(y)$ is the difference of its right and left boundaries; $d(y)$ is the vertical distance between the radius (y) and the axis of rotation (0). Let $a = 0$ and $b = 2$, the upper and lower boundaries of the region along the y-axis.*

Apply the shell method.

$$
\begin{aligned}
2\pi \int_a^b d(y)\,h(y)\,dy &= 2\pi \int_0^2 y\left(2y - y^2\right)dy \\
&= 2\pi \int_0^2 \left(2y^2 - y^3\right)dy \\
&= 2\pi \left(\frac{2y^3}{3} - \frac{y^4}{4}\right)\Bigg|_0^2 \\
&= 2\pi \left[\left(\frac{16}{3} - 4\right)\right] \\
&= 2\pi \left(\frac{16 - 12}{3}\right) \\
&= \frac{8\pi}{3}
\end{aligned}
$$

22.34 According to Problem 22.25, the volume of the solid generated by rotating the region bounded by $y = 3 - x^2$ and the x-axis about the line $y = -2$ is $\frac{128\pi\sqrt{3}}{5}$. Verify the solution using the shell method.

Draw a graph of the region indicating $d(y)$ and $h(y)$, as illustrated by Figure 22-29. Note that the axis of rotation is horizontal, so the radius used in the shell method must also be horizontal, and the integrand and the limits of integration must be written in terms of y. Solve $y = 3 - x^2$ for x.

> The SHELL method uses PARALLEL radii. It rhymes (sort of).

$$y = 3 - x^2$$
$$x^2 = 3 - y$$
$$x = \pm\sqrt{3 - y}$$

The portion of the parabola that is right of the x-axis has equation $x = \sqrt{3 - y}$, and the portion of the parabola left of the x-axis has equation $x = -\sqrt{3 - y}$, as indicated in Figure 22-29.

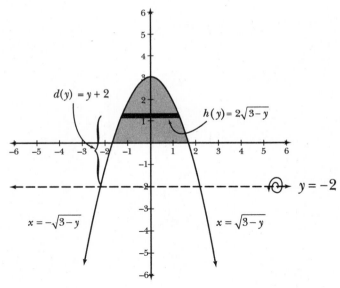

Figure 22-29 *The equation $y = 3 - x^2$ must be rewritten in terms of y in order to apply the shell method.*

Generate $h(y)$ by calculating the difference of its right and left boundaries.

$$h(y) = \sqrt{3 - y} - \left(-\sqrt{3 - y}\right) = \sqrt{3 - y} + \sqrt{3 - y} = 2\sqrt{3 - y}$$

> Subtract them in the right order—whatever's above minus whatever's underneath. In this case, $h(y)$ is above the x-axis and the rotational axis is below it.

Define $d(y)$ as the vertical distance between the radius (at position "y" along the y-axis) and the axis of rotation (at $y = -2$).

$$d(y) = y - (-2) = y + 2$$

The boundaries of the region along the y-axis are $a = 0$ and $b = 3$. Substitute a, b, $d(y)$, and $h(y)$ into the shell method formula.

$$2\pi \int_a^b d(y) h(y)\, dy = 2\pi \int_0^3 (y+2)\left(2\sqrt{3-y}\right) dy$$

$$= 4\pi \int_0^3 (y+2)\left(\sqrt{3-y}\right) dy$$

Integrate by parts: $u = y + 2$, $dv = \sqrt{3-y}\, dy$, $du = dy$, and $v = -\dfrac{2}{3}(3-y)^{3/2}$.

$$= 4\pi \left[(y+2)\left(-\frac{2}{3}\right)(3-y)^{3/2} - \frac{4}{15}(3-y)^{5/2} \right]_0^3$$

$$= 4\pi \left[0 - \left[2\left(-\frac{2}{3}\right)3^{3/2} - \frac{4}{15}\left(3^{5/2}\right) \right] \right]$$

$$= 4\pi \left(\frac{4}{3}3^{2/2} \cdot 3^{1/2} + \frac{4}{15} \cdot 3^{4/2} \cdot 3^{1/2} \right)$$

$$= 4\pi \left(\frac{4}{3} \cdot 3\sqrt{3} + \frac{4}{15} \cdot 3^2 \cdot \sqrt{3} \right)$$

$$= 4\pi \left(\frac{20\sqrt{3} + 12\sqrt{3}}{5} \right)$$

$$= 4\pi \left(\frac{32\sqrt{3}}{5} \right)$$

$$= \frac{128\pi\sqrt{3}}{5}$$

Chapter 23
ADVANCED APPLICATIONS OF DEFINITE INTEGRALS

More problems involving bounded integrals

In the preceding chapters, definite integrals have been applied to the calculation of area and volume, the identification of a function's average value, and the measurement of accumulated change. Though the majority of definite integrals are used toward one of those ends in an elementary calculus class, additional applications of the definite integral abound. This chapter discusses a small and diverse remnant of those applications: arc length, surface area of rotation, and centroids.

If you can calculate the area between curves and the volume of solids (created either by rotating something or slicing something into known cross sections), you've mastered about 85% of the definite integral problems you'll see in a typical calculus class. However, there's a small contingent of weird applications for the definite integral that textbooks like to include. These topics aren't usually covered in much depth or for very long, so some people find them very confusing. Each of the sections in this chapter is very unique, and as a result, the problems may not feel as cohesive as they did in other chapters. Don't let that bother you. Treat each section as its own island, independent of the others; also, don't spend the time you usually would figuring out WHY these formulas work—for this chapter, just worry about HOW they work.

Arc Length

How far is it from point A to point B along a curvy road?

You don't know what f(x) looks like between those points. Is it a line? a parabola? a semicircle? Who knows? That's why the generic term "arc" is used.

23.1 Given a continuous and differentiable function $f(x)$, what is the length of $f(x)$ between points $(a, f(a))$ and $(b, f(b))$?

The length of the arc formed by $f(x)$ between $x = a$ and $x = b$ is equal to $\int_a^b \sqrt{1 + \left[f'(x) \right]^2} \, dx$.

23.2 Calculate the length of the function $f(x) = \ln(\sin x)$ between $x = \dfrac{\pi}{4}$ and $x = \dfrac{3\pi}{4}$.

Differentiate $f(x)$ with respect to x.

$$f'(x) = \frac{1}{\sin x} \cdot \cos x = \frac{\cos x}{\sin x} = \cot x$$

The derivative of a natural log function is the derivative of what's inside the log divided by what's inside the log. (Same thing as 1 over what's inside times the derivative of the inside.)

Apply the arc length formula from Problem 23.1.

$$\int_a^b \sqrt{1 + \left[f'(x) \right]^2} \, dx = \int_{\pi/4}^{3\pi/4} \sqrt{1 + \cot^2 x} \, dx$$

According to a Pythagorean trigonometric identity, $1 + \cot^2 x = \csc^2 x$.

$$= \int_{\pi/4}^{3\pi/4} \sqrt{\csc^2 x} \, dx$$
$$= \int_{\pi/4}^{3\pi/4} \csc x \, dx$$

The antiderivative of $\csc x \, dx$ is $-\ln |\csc x + \cot x|$.

$$= -\ln |\csc x + \cot x| \Big\|_{\pi/4}^{3\pi/4}$$
$$= -\left[\left(\ln \left| \csc \frac{3\pi}{4} + \cot \frac{3\pi}{4} \right| \right) - \left(\ln \left| \csc \frac{\pi}{4} + \cot \frac{\pi}{4} \right| \right) \right]$$
$$= -\left[\left(\ln \left| \frac{2}{\sqrt{2}} - 1 \right| \right) - \left(\ln \left| \frac{2}{\sqrt{2}} + 1 \right| \right) \right]$$
$$= -\left[\left(\ln \left| \frac{2 - \sqrt{2}}{\sqrt{2}} \right| \right) - \left(\ln \left| \frac{2 + \sqrt{2}}{\sqrt{2}} \right| \right) \right]$$

Apply the logarithmic property that states $\log a - \log b = \log \dfrac{a}{b}$.

$$= -\ln \left| \frac{\left(2 - \sqrt{2} \right) / \sqrt{2}}{\left(2 + \sqrt{2} \right) / \sqrt{2}} \right|$$
$$= -\ln \left| \left(\frac{2 - \sqrt{2}}{\sqrt{2}} \right) \left(\frac{\sqrt{2}}{2 + \sqrt{2}} \right) \right|$$
$$= -\ln \left| \frac{2 - \sqrt{2}}{2 + \sqrt{2}} \right|$$

Rationalize the denominator by multiplying the numerator and denominator by $2-\sqrt{2}$.

$$-\ln\left|\frac{(2-\sqrt{2})(2-\sqrt{2})}{(2+\sqrt{2})(2-\sqrt{2})}\right|=-\ln\left|\frac{4-4\sqrt{2}+2}{4-2}\right|=-\ln\left|\frac{6-4\sqrt{2}}{2}\right|=-\ln\left|3-2\sqrt{2}\right|$$

The answer $\ln(3+2\sqrt{2})$ is also correct. It's the same value jumbled up using log properties.

23.3 Determine the length of $g(x)=\frac{1}{3}(4-x)^{3/2}$ between $x=0$ and $x=4$.

Differentiate $g(x)$ with respect to x.

$$g'(x)=\frac{1}{3}\cdot\frac{3}{2}(4-x)^{1/2}(-1)=-\frac{1}{2}\sqrt{4-x}$$

Don't forget to use the chain rule and take the derivative of what's inside the radical. That's where this -1 comes from.

Apply the arc length formula.

$$\int_a^b\sqrt{1+\left[g'(x)\right]^2}\,dx=\int_0^4\sqrt{1+\left(-\frac{1}{2}\sqrt{4-x}\right)^2}\,dx$$

$$=\int_0^4\sqrt{1+\frac{1}{4}(4-x)}\,dx$$

$$=\int_0^4\sqrt{1+1-\frac{x}{4}}\,dx$$

$$=\int_0^4\sqrt{2-\frac{x}{4}}\,dx$$

$$=\int_0^4\sqrt{\frac{8-x}{4}}\,dx$$

$$=\frac{1}{2}\int_0^4\sqrt{8-x}\,dx$$

Factor ¼ out of the radical to get ½, and then pull ½ out of the integral.

Integrate using variable substitution: $u=8-x$ and $-du=dx$.

$$=-\frac{1}{2}\int_8^4 u^{1/2}\,du$$

$$=-\frac{1}{2}\cdot\frac{2}{3}(u^{3/2})\Big|_8^4$$

$$=-\frac{1}{3}(4^{3/2}-8^{3/2})$$

$$=-\frac{1}{3}(8-16\sqrt{2})$$

$$=\frac{8(2\sqrt{2}-1)}{3}$$

When you change the limits of integration from 0 and 4 into 8 and 4, they feel upside down—the bigger number is suddenly on the bottom. Don't change it—everything works itself out.

23.4 Verify that the circumference of a circle is $2\pi r$.

To simplify the calculations, assume that the circle is centered at the origin—its placement in the coordinate plane will not affect its circumference. Solve the standard form equation of a circle with radius r for y.

$$x^2 + y^2 = r^2$$
$$y^2 = r^2 - x^2$$
$$y = \pm\sqrt{r^2 - x^2}$$

A circle is not a function, but it can be described by the pair of functions $y = \sqrt{r^2 - x^2}$ and $y = -\sqrt{r^2 - x^2}$, each of which is the equation for a semicircle of the same radius. Therefore, you can calculate the arc length of one semicircle, $f(x)$, and multiply the result by 2 in order to calculate the circumference of the full circle. Differentiate $f(x) = \sqrt{r^2 - x^2}$ with respect to x.

$$f'(x) = \frac{1}{2}(r^2 - x^2)^{-1/2}(-2x) = \frac{-x}{\sqrt{r^2 - x^2}}$$

Apply the arc length formula.

> The function $f(x)$ represents the top half of the circle, which will intersect the x-axis at $-r$ and r. A circle with center $(0,0)$ intersects both of the axes r units away from the center.

$$\int_a^b \sqrt{1 + \left[f'(x) \right]^2}\, dx = \int_{-r}^r \sqrt{1 + \left(\frac{-x}{\sqrt{r^2 - x^2}} \right)^2}\, dx$$

$$= \int_{-r}^r \sqrt{1 + \frac{x^2}{r^2 - x^2}}\, dx$$

$$= \int_{-r}^r \sqrt{\frac{r^2 - x^2 + x^2}{r^2 - x^2}}\, dx$$

$$= \int_{-r}^r \frac{\sqrt{r^2}}{\sqrt{r^2 - x^2}}\, dx$$

> Pull the constant r (from the numerator) outside the integral.

$$= r\int_{-r}^r \frac{dx}{\sqrt{r^2 - x^2}}$$

According to Problem 20.17, $\int \dfrac{du}{\sqrt{a^2 - u^2}} = \arcsin\left(\dfrac{u}{a}\right)$. Set $a = r$, $u = x$, and $du = dx$ and apply this antidifferentiation formula.

> According to Problem 7.28, $-\dfrac{\pi}{2} \le \arcsin x \le \dfrac{\pi}{2}$, so $\arcsin(-1) \ne \dfrac{3\pi}{2}$ even though $\sin\dfrac{3\pi}{2} = -1$.

$$= r\int_{-r}^r \frac{du}{\sqrt{a^2 - u^2}}$$

$$= r\left(\arcsin\frac{u}{a} \right)\Big|_{-r}^r$$

$$= r\left(\arcsin\frac{x}{r} \right)\Big|_{-r}^r$$

$$= r\left(\arcsin 1 - \arcsin(-1) \right)$$

$$= r\left(\frac{\pi}{2} - \left(-\frac{\pi}{2} \right) \right)$$

$$= \frac{2\pi r}{2}$$

$$= \pi r$$

If a semicircle with radius r has arc length πr, then a circle with radius r has circumference $2\pi r$.

23.5 Approximate the length of $h(x) = \sqrt{x}$ between $x = 1$ and $x = 4$ using the trapezoidal rule with $n = 6$ subdivisions and round the estimate to three decimal places.

Differentiate $h(x)$ with respect to x.

$$h'(x) = \frac{1}{2}x^{-1/2} = \frac{1}{2\sqrt{x}}$$

Apply the arc length formula.

$$\int_a^b \sqrt{1 + \left[h'(x)\right]^2}\,dx = \int_1^4 \sqrt{1 + \left(\frac{1}{2\sqrt{x}}\right)^2}\,dx$$

$$= \int_1^4 \sqrt{1 + \frac{1}{4x}}\,dx$$

$$= \int_1^4 \sqrt{\frac{4x+1}{4x}}\,dx$$

Use the trapezoidal rule to approximate the value of this integral. To clarify notation, set $k(x) = \sqrt{\frac{4x+1}{4x}}$.

If you need a quick review of the trapezoidal rule, look at Problems 17.18–17.26.

$$\int_1^4 \sqrt{\frac{4x+1}{4x}}\,dx \approx \frac{4-1}{2(6)}\left[k(1) + 2k\left(\frac{3}{2}\right) + 2k(2) + 2k\left(\frac{5}{2}\right) + 2k(3) + 2k\left(\frac{7}{2}\right) + k(4)\right]$$

$$\approx \frac{1}{4}\left[\frac{\sqrt{5}}{2} + 2\left(\frac{\sqrt{42}}{6}\right) + 2\left(\frac{3\sqrt{2}}{4}\right) + 2\left(\frac{\sqrt{110}}{10}\right) + 2\left(\frac{\sqrt{39}}{6}\right) + 2\left(\frac{\sqrt{210}}{14}\right) + \frac{\sqrt{17}}{4}\right]$$

$$\approx 3.1699645$$

You're approximating the area beneath $k(x)$ from $x = 1$ to $x = 4$, not the area beneath $\sqrt{x}$.

Surface Area

Measure the "skin" of a rotational solid

23.6 Rotate the curve defined by $f(x) = \sqrt{x}$ between $x = 0$ and $x = 3$ about the x-axis and calculate the area of the surface generated.

You're not rotating a region like most of the problems in Chapter 22. When you rotate a piece of a graph, it creates an infinitely thin three-dimensional surface, not a solid.

Differentiate $f(x)$ with respect to x.

$$f'(x) = \frac{1}{2}x^{-1/2} = \frac{1}{2\sqrt{x}}$$

The surface area generated by rotating the portion of $f(x)$ between $x = a$ and $x = b$ about the x-axis is equal to $2\pi \int_a^b f(x)\sqrt{1+\left[f(x)\right]^2}\,dx$. Substitute $f(x)$, $f'(x)$, $a = 0$, and $b = 3$ into the formula.

> **Change all the x's to y's if you're rotating around the y-axis, like in Problem 23.7.**

> **Pull ¼ out of the radical to get ½ and then pull ½ out of the integral.**

$$2\pi \int_a^b f(x)\sqrt{1+\left[f'(x)\right]^2}\,dx = 2\pi \int_0^3 \sqrt{x}\sqrt{1+\left(\frac{1}{2\sqrt{x}}\right)^2}\,dx$$

$$= 2\pi \int_0^3 \sqrt{x}\sqrt{1+\frac{1}{4x}}\,dx$$

$$= 2\pi \int_0^3 \sqrt{x}\sqrt{\frac{4x+1}{4x}}\,dx$$

$$= 2\pi \int_0^3 \sqrt{\frac{\cancel{x}(4x+1)}{4\cancel{x}}}\,dx$$

$$= 2\pi \int_0^3 \sqrt{\frac{4x+1}{4}}\,dx$$

$$= 2\pi \left(\frac{1}{2}\right)\int_0^3 \sqrt{4x+1}\,dx$$

Apply variable substitution: $u = 4x + 1$ and $du = 4dx$, so $\dfrac{du}{4} = dx$. Rewrite the limits of integration in terms of u as well: $4(0) + 1 = 1$ and $4(3) + 1 = 13$.

$$= \left(\frac{2\pi}{2}\right)\left(\frac{1}{4}\right)\int_1^{13} u^{1/2}\,du$$

$$= \frac{\pi}{4}\int_1^{13} u^{1/2}\,du$$

$$= \frac{\pi}{4}\cdot\frac{2}{3}\left(u^{3/2}\right)\Big|_1^{13}$$

$$= \frac{2\pi}{12}\left(13^{3/2} - 1^{3/2}\right)$$

$$= \frac{\pi}{6}\left(13\sqrt{13} - 1\right)$$

23.7 Find the area of the surface generated by revolving the portion of $f(y) = \dfrac{y^3}{3}$ between $y = 0$ and $y = 2$ about the y-axis.

Differentiate $f(y)$: $f'(y) = \dfrac{1}{3}\left(3y^2\right) = y^2$; apply the surface area formula.

$$2\pi \int_a^b f(y)\sqrt{1+\left[f'(y)\right]^2}\,dy = 2\pi \int_0^2 \left(\frac{1}{3}y^3\right)\sqrt{1+\left[y^2\right]^2}\,dy$$

$$= 2\pi \left(\frac{1}{3}\right)\int_0^2 y^3 \sqrt{1+y^4}\,dy$$

Apply variable substitution: $u = 1 + y^4$, $\dfrac{du}{4} = y^3 dy$, $a = 1$, and $b = 17$.

$$= \frac{2\pi}{3}\left(\frac{1}{4}\right)\int_1^{17} u^{1/2}\, du$$

$$= \frac{2\pi}{12}\left(\frac{2}{3}\right)u^{3/2}\Big|_1^{17}$$

$$= \frac{\pi}{9}\left(17^{3/2} - 1^{3/2}\right)$$

$$= \frac{\pi}{9}\left(17\sqrt{17} - 1\right)$$

23.8 Prove that the surface area of a solid right circular cylinder is $2\pi r(r + h)$, if r is the radius and h is the height of the cylinder.

As illustrated in Figure 23-1, revolving the region bounded by $x = 0$, $y = 0$, $x = r$, and $y = h$ about the y-axis generates a solid right circular cylinder.

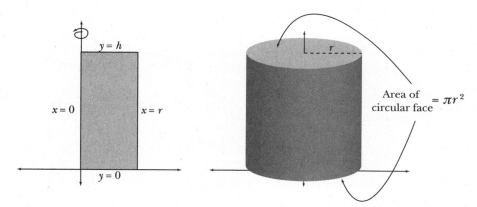

Figure 23-1   *This is a solid of rotation, not a surface of rotation. Therefore, you must account for the surface area of the cylinder's circular faces at $y = 0$ and $y = h$. Both circles have radius r, so each circle has area πr^2.*

The surface of the cylinder is (excluding the circular bases) generated by rotating the portion of $f(y) = r$ between $y = 0$ and $y = h$ about the y-axis. Apply the surface area formula.

(Note that r is a constant, so $f'(y) = 0$.)

$$2\pi\int_a^b f(y)\sqrt{1 + \left[f'(y)\right]^2}\, dy = 2\pi\int_0^h r\sqrt{1 + (0)^2}\, dy$$

$$= 2\pi\int_0^h r\, dy$$

$$= 2\pi r\int_0^h dy$$

$$= (2\pi r)\, y\Big|_0^h$$

$$= 2\pi r h$$

Even though there are no y's on the right side of the equation, this function is in terms of y because it is solved for x. It's the equation of the vertical line $x = r$ with $f(y)$ in place of x.

As noted in Figure 23-1, the total surface area of the solid is the sum of the areas of the top and bottom faces of the cylinder and the surface area of rotation.

Total surface area   = area of top face   + area of bottom face   + rotational surface area

$$= \pi r^2 \qquad\qquad +\pi r^2 \qquad\qquad +2\pi rh$$

$$= 2\pi r^2 + 2\pi rh$$

$$= 2\pi r(r+h)$$

23.9 Prove that the surface area of a sphere with radius r is equal to $4\pi r^2$.

According to Problem 22.16, you can generate a solid sphere by rotating the region bounded by $f(x) = \sqrt{r^2 - x^2}$ and the x-axis about the x-axis. Therefore, you can determine its surface area by rotating $f(x) = \sqrt{r^2 - x^2}$ between $x = -r$ and $x = r$ about the x-axis. Differentiate $f(x)$ with respect to x.

$$f'(x) = \frac{1}{2}\left(r^2 - x^2\right)^{-1/2}(-2x) = -\frac{x}{\sqrt{r^2 - x^2}}$$

Apply the surface area formula.

$$2\pi \int_a^b f(x)\sqrt{1+\left[f'(x)\right]^2}\,dx = 2\pi \int_{-r}^r \sqrt{r^2 - x^2}\sqrt{1+\left[-\frac{x}{\sqrt{r^2 - x^2}}\right]^2}\,dx$$

$$= 2\pi \int_{-r}^r \sqrt{r^2 - x^2}\sqrt{1+\frac{x^2}{r^2 - x^2}}\,dx$$

$$= 2\pi \int_{-r}^r \sqrt{r^2 - x^2}\sqrt{\frac{r^2 - x^2 + x^2}{r^2 - x^2}}\,dx$$

$$= 2\pi r \int_{-r}^r \sqrt{r^2 - x^2}\sqrt{\frac{1}{r^2 - x^2}}\,dx$$

There was an r^2 in here, but you take the square root to get r, and then move it outside the radical, because it's a constant. There's an r out front that wasn't there in the last step.

Recall that the product of two radical expressions with the same index is the root of the product: $\sqrt{a}\sqrt{b} = \sqrt{ab}$.

$$= 2\pi r \int_{-r}^r \sqrt{\frac{r^2 - x^2}{r^2 - x^2}}\,dx$$

$$= 2\pi r \int_{-r}^r 1\,dx$$

$$= 2\pi r(x)\Big|_{-r}^r$$

$$= 2\pi r[r - (-r)]$$

$$= 2\pi r(2r)$$

$$= 4\pi r^2$$

23.10 Prove that the surface area of a solid right circular cone with radius r and height h is equal to $\pi r\sqrt{h^2 + r^2}$.

As illustrated in Figure 23-2, rotating the first quadrant region bounded by $x = 0$, $y = h$, and $f(y) = \dfrac{r}{h}y$ about the y-axis results in a solid right circular cone with radius r and height h.

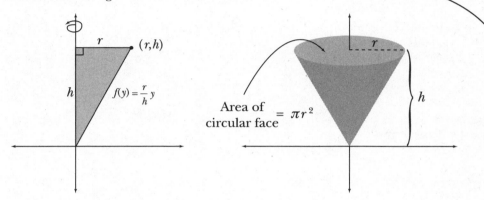

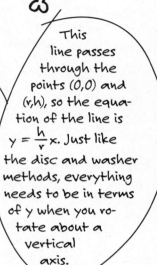

This line passes through the points (0,0) and (r,h), so the equation of the line is $y = \dfrac{h}{r}x$. Just like the disc and washer methods, everything needs to be in terms of y when you rotate about a vertical axis.

Figure 23-2 *Aside from the rotational surface area, this cone has a circular base with radius r at y = h. The area of the circle, πr^2, must be added to the rotational surface area in order to calculate the total surface area of the solid.*

Notice that $f'(y) = \dfrac{r}{h}$; plug $f(y)$, $f'(y)$, $a = 0$, and $b = h$ into the surface area formula.

$$2\pi \int_a^b f(y)\sqrt{1 + \left[f'(y)\right]^2}\,dy = 2\pi \int_0^h \left(\frac{r}{h}y\right)\sqrt{1 + \left(\frac{r}{h}\right)^2}\,dy$$

$$= 2\pi \int_0^h \left(\frac{r}{h}y\right)\sqrt{1 + \frac{r^2}{h^2}}\,dy$$

$$= 2\pi \int_0^h \left(\frac{r}{h}y\right)\frac{\sqrt{h^2 + r^2}}{\sqrt{h^2}}\,dy$$

$$= 2\pi \int_0^h \left(\frac{r}{h}y\right)\frac{\sqrt{h^2 + r^2}}{h}\,dy$$

Because $\dfrac{r}{h}$ and $\dfrac{\sqrt{h^2 + r^2}}{h}$ are constants, they can be removed from the integrand, leaving behind only $y\,dy$.

$$= 2\pi\left(\frac{r}{h}\right)\left(\frac{\sqrt{h^2 + r^2}}{h}\right)\int_0^h y\,dy \ = \ \frac{2\pi r\sqrt{h^2 + r^2}}{h^2}\left(\frac{y^2}{2}\right)\Bigg|_0^h \ = \ \frac{2\pi r\sqrt{h^2 + r^2}}{h^2}\left(\frac{h^2}{2}\right) \ = \ \pi r\sqrt{h^2 + r^2}$$

As explained in Figure 23-2, add the area of the circular face.

$$\pi r\sqrt{h^2 + r^2} + \pi r^2 = \pi r\left(r + \sqrt{h^2 + r^2}\right)$$

Centroids
Find the center of gravity for a two-dimensional shape

"Centroid" means "like a center," because it is the center of mass (or the center of gravity) for the region. If you wanted to balance the region in Figure 23-3 on your finger tip (like a geometric Harlem Globetrotter), put your finger under the centroid.

23.11 Region R is bounded by continuous functions $h(x)$ and $k(x)$, which intersect at points (x_1, y_1) and (x_2, y_2) as illustrated by Figure 23-3. Identify the centroid $(\bar{x}, \bar{y})$ of the region.

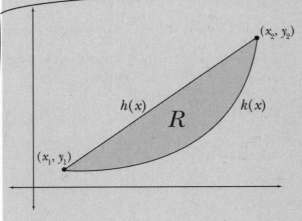

Figure 23-3

Region R is bounded above by h(x) and below by k(x).

The vertical bar is x units from the x-axis, and the horizontal bar is y units from the y-axis. Those values are marked along the axes in Figure 23-4.

Draw a horizontal and a vertical representative length across region R, as demonstrated by Figure 23-4.

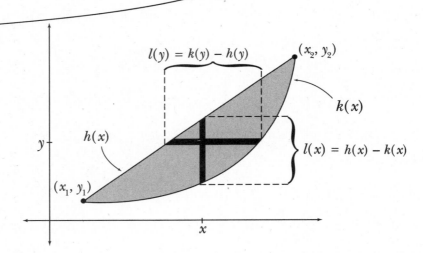

Figure 23-4 *The lengths of the representative lengths are equal to the differences of their boundaries. The vertical length equals the upper minus the lower bound (in terms of x), and the horizontal length equals the right minus the left bound (in terms of y).*

Calculate the area of the region: $A = \int_{x_1}^{x_2} \big[h(x) - k(x) \big] dx$. This value serves as the denominator in the centroid formula below.

$$\left(\bar{x}, \bar{y} \right) = \left(\frac{\int_a^b x \cdot l(x) \, dx}{A}, \frac{\int_c^d y \cdot l(y) \, dy}{A} \right)$$

$$= \left(\frac{1}{A} \int_a^b x \cdot l(x) \, dx, \frac{1}{A} \int_c^d y \cdot l(y) \, dy \right)$$

Note that $l(x)$ represents the length of the representative vertical length (in terms of x) in Figure 23-4—a length that stretches from the top of the region to the bottom; a and b are the boundaries of the region along the x-axis. Therefore, $l(x) = h(x) - k(x)$, $a = x_1$, and $b = x_2$. Similarly, $l(y)$ is the length of the representative horizontal length (in terms of y), and the boundaries of the region along the y-axis are equal to c and d. Therefore, $l(y) = k(y) - h(y)$, $c = y_1$, and $d = y_2$. Substitute these values into the centroid formula.

$$\left(\bar{x}, \bar{y} \right) = \left(\frac{1}{A} \int_{x_1}^{x_2} x \big[h(x) - k(x) \big] dx, \frac{1}{A} \int_{y_1}^{y_2} y \big[k(y) - h(y) \big] dy \right)$$

23.12 Identify the centroid of a rectangle with width w and height h.

Construct a rectangle in the coordinate plane and draw representative horizontal and vertical lengths, as illustrated in Figure 23-5.

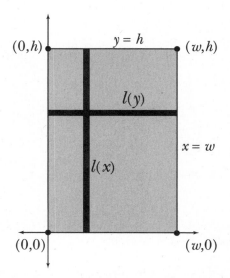

Figure 23-5   *The rectangular region is bounded by y = 0 (the x-axis), x = 0 (the y-axis), x = w, and y = h.*

A is the area of the rectangle, which is just its width times its height.

According to Figure 23-5, the length of the representative vertical length is $l(x) = h - 0$; and the x-axis boundaries of the region are $a = 0$ and $b = w$. The representative horizontal length is $l(y) = w - 0$, and it has y-axis boundaries $c = 0$ and $d = h$. Apply the centroid formula from Problem 23.11, noting that $A = wh$.

Begin by calculating $\bar{x}$, the x-coordinate of the centroid.

$$\bar{x} = \frac{1}{A} \int_{x_1}^{x_2} x \cdot l(x)\, dx$$

$$= \frac{1}{wh} \int_0^w x \cdot h \cdot dx$$

Remove the constant h from the integrand.

$$= \frac{\cancel{h}}{w\cancel{h}} \int_0^w x\, dx$$

$$= \frac{1}{w} \left(\frac{x^2}{2} \right) \Big|_0^w$$

$$= \frac{1}{w} \left(\frac{w^2}{2} \right)$$

$$= \frac{w}{2}$$

Now calculate $\bar{y}$, the y-coordinate of the centroid.

$$\bar{y} = \frac{1}{A} \int_{y_1}^{y_2} y \cdot l(y)\, dy$$

$$= \frac{1}{wh} \int_0^h y \cdot w \cdot dy$$

$$= \frac{\cancel{w}}{\cancel{w}h} \int_0^h y\, dy$$

$$= \frac{1}{h} \left(\frac{y^2}{2} \right) \Big|_0^h$$

$$= \frac{1}{h} \left(\frac{h^2}{2} \right)$$

$$= \frac{h}{2}$$

> No giant surprise here. The balancing point of a rectangle is right in the center both horizontally and vertically, where its diagonals bisect one another.

The centroid of the rectangle is $(\bar{x}, \bar{y}) = \left(\dfrac{w}{2}, \dfrac{h}{2} \right)$.

Note: Problems 23.13–23.17 refer to the region bounded by $f(x) = \sqrt{r^2 - x^2}$, a semicircle with radius r centered at the origin, and the x-axis.

23.13 Identify the x-coordinate $\bar{x}$ of the centroid for the region.

The representative vertical length is bounded above by $f(x)$ and below by $y = 0$ along the x-axis from $x = -r$ to $x = r$, as illustrated by Figure 23-6. The area of a semicircle with radius r is $A = \dfrac{\pi r^2}{2}$.

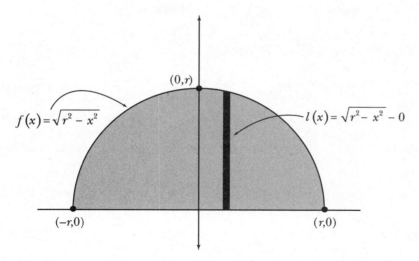

Figure 23-6 *The length l(x) equals the difference between its upper bound-ary (the semicircle) and its lower boundary (the x-axis).*

Apply the centroid formula for $\bar{x}$.

$$\bar{x} = \frac{1}{A} \int_{x_1}^{x_2} x \cdot l(x)\, dx$$

$$= \left(\frac{2}{\pi r^2}\right) \int_{-r}^{r} x\sqrt{r^2 - x^2}\, dx$$

Apply variable substitution: $u = r^2 - x^2$ and $-\dfrac{du}{2} = x\, dx$. Substitute $x = -r$ and $x = r$ into $u = r^2 - x^2$ to get the corresponding u-boundaries: $u = r^2 - (-r)^2 = 0$ and $u = r^2 - (r)^2 = 0$.

$$= \left(\frac{2}{\pi r^2}\right)\left(-\frac{1}{2}\right) \int_0^0 u^{1/2}\, du$$

$$= \left(\frac{2}{\pi r^2}\right)\left(-\frac{1}{2}\right)(0)$$

$$= 0$$

> If the limits of integration of a definite integral are equal, the definite integral equals 0. You don't cover any area if you start and stop at the same integration limit.

Note: Problems 23.13–23.17 refer to the region bounded by $f(x) = \sqrt{r^2 - x^2}$, a semicircle with radius r centered at the origin, and the x-axis.

23.14 The functions that bound the region are even. Draw a conclusion concerning the centroids of such regions and explain your answer.

If the boundary functions of a region are even, the graphs of those functions are symmetric about the y-axis. Therefore, the region is split by the y-axis into two regions of equal area. The balance point of the region, then, must lie along the line $x = 0$ (the y-axis), and $\bar{x} = 0$.

This conclusion can be further generalized: if a region is symmetric about *any* vertical line $x = c$, then $\bar{x} = c$ for that region. Similarly, if a region is symmetric about any horizontal line $y = k$, then $\bar{y} = k$.

> If f(x) is an even function, that means f(-x) = f(x). In other words, if the point (a,b) is on the graph of f(x), then so is (-a,b).

Note: Problems 23.13–23.17 refer to the region bounded by $f(x) = \sqrt{r^2 - x^2}$, a semicircle with radius r centered at the origin, and the x-axis.

23.15 Identify the y-coordinate $\bar{y}$ of the centroid for the region.

Construct a graph of the region, rewriting its boundaries in terms of y (as illustrated in Figure 23-7). To express the semicircle in terms of y, solve the standard equation of a circle (centered at the origin with radius r) for x.

$$x^2 + y^2 = r^2$$
$$x^2 = r^2 - y^2$$
$$x = \pm\sqrt{r^2 - y^2}$$

> Horizontal lengths (like horizontal radii in Chapter 22) require everything in terms of y, including functions and limits of integration.

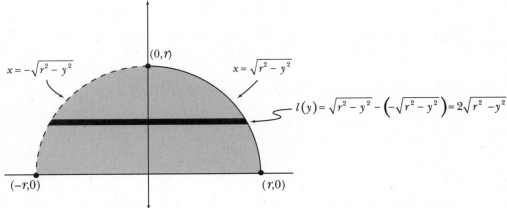

Figure 23-7   *Two functions of y are required to describe the semicircle, a positive radical expression when x > 0 and a negative radical expression when x < 0 (shown as a dotted graph). Length l(y) is defined as the difference of its right and left boundaries.*

Substitute $l(y)$ (as calculated in Figure 23-7), $A = \dfrac{\pi r^2}{2}$, $y_1 = 0$, and $y_2 = r$ into the formula for $\bar{y}$.

$$\bar{y} = \frac{1}{A}\int_{y_1}^{y_2} y \cdot l(y)\,dy = \left(\frac{2}{\pi r^2}\right)\int_0^r y \cdot 2\sqrt{r^2 - y^2}\,dy = \frac{4}{\pi r^2}\int_0^r y\sqrt{r^2 - y^2}\,dy$$

Apply variable substitution: $u = r^2 - y^2$ and $\dfrac{du}{-2} = y\,dy$. Substitute $y = 0$ and $y = r$ into $u = r^2 - y^2$ to get the corresponding u-boundaries: $u = r^2 - 0^2 = r^2$ and $u = r^2 - r^2 = 0$.

$$= \frac{4}{\pi r^2}\left(-\frac{1}{2}\right)\int_{r^2}^0 u^{1/2}\,du$$

$$= -\frac{2}{\pi r^2}\left(\frac{2}{3}\right)\left(u^{3/2}\right)\Big|_{r^2}^0$$

$$= -\frac{4}{3\pi r^2}\left(0 - \left(r^2\right)^{3/2}\right)$$

$$= -\frac{4}{3\pi r^2}\left(-r^3\right)$$

$$= \frac{4r}{3\pi}$$

Note: Problems 23.13–23.17 refer to the region bounded by $f(x) = \sqrt{r^2 - x^2}$, a semicircle with radius r centered at the origin, and the x-axis.

23.16 Verify that the alternative formula $\bar{y} = \dfrac{1}{2A} \int_{x_1}^{x_2} \left([j(x)]^2 - [k(x)]^2 \right) dx$, where $j(x)$ is the upper bound of the region and $k(x)$ is the lower bound, returns the same value for $\bar{y}$ as Problem 23.15.

> This shortcut calculates $\bar{y}$ for an infinitely thin rectangle located along $x = \bar{x}$ (the vertical line that cuts the region into two balancing halves). The thin rectangle and the region share the same $\bar{y}$-value.

The upper boundary of the region is $j(x) = f(x)$ and the lower bound is $k(x) = 0$. Therefore, $[j(x)]^2 - [k(x)]^2 = [f(x)]^2 - 0^2 = [f(x)]^2$.

$$\bar{y} = \left(\frac{2}{2 \cdot \pi r^2} \right) \int_{-r}^{r} \left[\sqrt{r^2 - x^2} \right]^2 dx$$

$$= \frac{1}{\pi r^2} \int_{-r}^{r} (r^2 - x^2) \, dx$$

$$= \frac{1}{\pi r^2} \left(r^2 x - \frac{x^3}{3} \right) \Big|_{-r}^{r}$$

$$= \frac{1}{\pi r^2} \left[\left(r^3 - \frac{r^3}{3} \right) - \left(-r^3 + \frac{r^3}{3} \right) \right]$$

$$= \frac{1}{\pi r^2} \left[\left(\frac{2r^3}{3} \right) - \left(-\frac{2r^3}{3} \right) \right]$$

$$= \frac{1}{\pi r^2} \left(\frac{4r^3}{3} \right)$$

$$= \frac{4r}{3\pi}$$

> Even though you're calculating $\bar{y}$, the shortcut is in terms of x, so use the same x integration limits as Problem 23.15: $a = -r$ and $b = r$.

Note: Problems 23.13–23.17 refer to the region bounded by $f(x) = \sqrt{r^2 - x^2}$, a semicircle with radius r centered at the origin, and the x-axis.

23.17 Explain the practical value of the alternative formula, $\bar{y} = \dfrac{1}{2A} \int_{x_1}^{x_2} \left([j(x)]^2 - [k(x)]^2 \right) dx$, as applied in Problem 23.16.

Calculating $\bar{y}$ by means of the alternative formula is usually more efficient than the formula $\bar{y} = \dfrac{1}{A} \int_{c}^{d} y \cdot l(y) \, dy$, which requires c, d, and $l(y)$ to be in terms of y.

As demonstrated in Problem 23.15, writing those values in terms of y requires you to transform one equation into a pair of equations, calculate a new representative length $l(y)$, and rewrite the limits of integration in terms of y. On the other hand, the alternative formula in Problem 23.16 used the same region boundary equations as the formula for $\bar{x}$ in Problem 23.13, the same representative length function $l(x)$, and the same limits of integration, which is both expedient and convenient.

> Here's another benefit: Sometimes rewriting everything in terms of y produces an ugly integral that's difficult, if not impossible, to calculate by hand.

Note: Problems 23.18–23.20 refer to the region bounded by the graphs of $f(x) = \sqrt{x}$, $y = 0$, and $x = 4$.

23.18 Identify $\bar{x}$, the x-coordinate of the centroid for the region.

Graph the region and $l(x)$ in order to calculate the representative vertical length, as illustrated by Figure 23-8.

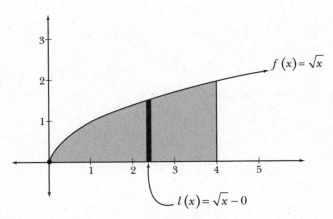

Figure 23-8   *The length of l(x) is defined as the difference of its upper and lower bounds.*

Calculate the area of the region using the fundamental theorem of calculus.

$$A = \int_0^4 x^{1/2} dx = \frac{2}{3}\left(x^{3/2}\right)\Big|_0^4 = \frac{2}{3}(8) = \frac{16}{3}$$

Evaluate the formula for $\bar{x}$ given $A = \dfrac{16}{3}$, $l(x) = \sqrt{x}$, $x_1 = 0$, and $x_2 = 4$.

$$\bar{x} = \frac{1}{A} \int_{x_1}^{x_2} x \cdot l(x)\, dx$$

$$= \frac{3}{16} \int_0^4 x\sqrt{x}\, dx$$

$$= \frac{3}{16} \int_0^4 x^{3/2}\, dx$$

$$= \frac{3}{16} \cdot \frac{2}{5}\left(x^{5/2}\right)\Big|_0^4$$

$$= \frac{3}{40}(32)$$

$$= \frac{12}{5}$$

Note: Problems 23.18–23.20 refer to the region bounded by the graphs of $f(x) = \sqrt{x}$, $y = 0$, and $x = 4$.

23.19 Calculate $\bar{y}$, the y-coordinate of the centroid for the region, using the formula from Problem 23.11.

In other words, don't use the shortcut (until Problem 23.20, that is).

Graph the region and draw $l(y)$ in order to calculate the representative horizontal length, as illustrated by Figure 23-9.

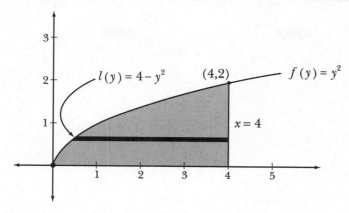

Figure 23-9 *When l(y) has nonzero boundaries, it sometimes results in a more complicated integrand. However, that is not true in this case, as you'll see when you compute the definite integral.*

According to Problem 23.18, $A = \dfrac{16}{3}$; calculate $\bar{y}$ given $l(y) = 4 - y^2$, $y_1 = 0$, and $y_2 = 2$.

> The orientation of the representative length changed (from vertical to horizontal), but the region (and its area) stayed the same.

$$\bar{y} = \frac{1}{A} \int_{y_1}^{y_2} y \cdot l(y)\,dy$$

$$= \frac{3}{16} \int_0^2 y\left(4 - y^2\right) dy$$

$$= \frac{3}{16} \int_0^2 \left(4y - y^3\right) dy$$

$$= \frac{3}{16} \left(2y^2 - \frac{y^4}{4}\right)\bigg|_0^2$$

$$= \frac{3}{16}(8 - 4)$$

$$= \frac{3}{4}$$

Note: Problems 23.18–23.20 refer to the region bounded by the graphs of $f(x) = \sqrt{x}$, $y = 0$, and $x = 4$.

23.20 Verify that the value of $\bar{y}$ generated by Problem 23.19 using the alternative formula for $\bar{y}$.

> The formula from Problem 23.16.

Evaluate the alternative formula using values written in terms of x: $j(x) = \sqrt{x}$, $k(x) = 0$, $x_1 = 0$, $x_2 = 4$, and $A = \dfrac{16}{3}$. Note that $[j(x)]^2 - [k(x)]^2 = [f(x)]^2 - 0^2 = [f(x)]^2$.

> These are the same x_1, x_2, and A values that are used to find $\bar{x}$ in Problem 23.18.

$$\bar{y} = \frac{1}{2A} \int_{x_1}^{x_2} \left(\left[j(x) \right]^2 - \left[k(x) \right]^2 \right) dx$$

$$= \frac{3}{2(16)} \int_0^4 \left[\sqrt{x} \right]^2 dx$$

$$= \frac{3}{32} \int_0^4 x \, dx$$

$$= \frac{3}{32} \left(\frac{x^2}{2} \right) \Bigg|_0^4$$

$$= \frac{3}{32}(8)$$

$$= \frac{3}{4}$$

Note: Problems 23.21–23.22 refer to the first quadrant region in Figure 23-10, bounded by the graphs of $f(x) = \cos x$, $y = \frac{1}{2}$, and $x = 0$.

23.21 Calculate $\bar{x}$, the x-coordinate of the centroid for the region.

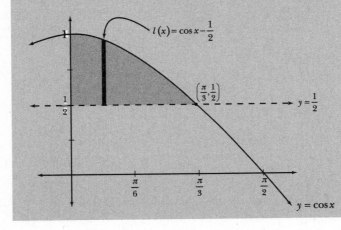

Figure 23-10

Note that the lower bound of the region, and therefore the lower bound of $l(x)$, is the horizontal line $y = \frac{1}{2}$, not the x-axis.

Calculate the area of the region using the fundamental theorem of calculus.

$$A = \int_0^{\pi/3} \left(\cos x - \frac{1}{2} \right) dx = \left(\sin x - \frac{x}{2} \right) \Bigg|_0^{\pi/3} = \frac{\sqrt{3}}{2} - \frac{\pi}{6} = \frac{3\sqrt{3} - \pi}{6}$$

> You already know that $x_1 = 0$ because the problem specifies a "first quadrant region," and you can tell from Figure 23-10 that the left edge of the region is the vertical line $x = 0$.

To determine the value of x_2, find the intersection point of $f(x) = \cos x$ and $y = \frac{1}{2}$ by setting the functions equal and solving for y.

$$\cos x = \frac{1}{2}$$

$$\arccos x = \arccos \frac{1}{2}$$

$$x = \frac{\pi}{3}$$

Evaluate the formula for $\bar{x}$, given $A = \dfrac{3\sqrt{3} - \pi}{6}$, $l(x) = \cos x - \dfrac{1}{2}$, $x_1 = 0$, and $x_2 = \dfrac{\pi}{3}$.

$$\bar{x} = \frac{1}{A}\int_{x_1}^{x_2} x \cdot l(x)\,dx$$

$$= \frac{6}{3\sqrt{3} - \pi}\int_0^{\pi/3} x\left(\cos x - \frac{1}{2}\right)dx$$

$$= \frac{6}{3\sqrt{3} - \pi}\int_0^{\pi/3}\left(x\cos x - \frac{1}{2}x\right)dx$$

According to Problem 22.32, $\int x\cos x\,dx = x\sin x + \cos x$.

$$= \frac{6}{3\sqrt{3} - \pi}\left(x\sin x + \cos x - \frac{x^2}{4}\right)\Bigg|_0^{\pi/3}$$

$$= \frac{6}{3\sqrt{3} - \pi}\left[\left(\frac{\pi}{3}\cdot\frac{\sqrt{3}}{2} + \frac{1}{2} - \frac{\pi^2}{4\cdot 9}\right) - (0 + 1 - 0)\right]$$

$$= \frac{6}{3\sqrt{3} - \pi}\left(\frac{\pi\sqrt{3}}{6} - \frac{\pi^2}{36} - \frac{1}{2}\right)$$

$$= \frac{6}{3\sqrt{3} - \pi}\left(\frac{6\pi\sqrt{3} - \pi^2 - 18}{36}\right)$$

$$= \frac{6\pi\sqrt{3} - \pi^2 - 18}{6\left(3\sqrt{3} - \pi\right)}$$

Note: Problems 23.21–23.22 refer to the first quadrant region bounded by the graphs of $f(x) = \cos x$, $y = \dfrac{1}{2}$, and $x = 0$.

23.22 Calculate $\bar{y}$, the y-coordinate of the centroid for the region.

Substitute $A = \dfrac{3\sqrt{3} - \pi}{6}$, $j(x) = \cos x$, $k(x) = \dfrac{1}{2}$, $x_1 = 0$, and $x_2 = \dfrac{\pi}{3}$ into the alternative formula for $\bar{y}$.

$$\bar{y} = \frac{1}{2A}\int_{x_1}^{x_2}\left(\left[j(x)\right]^2 - \left[k(x)\right]^2\right)dx$$

$$= \frac{6}{2(3\sqrt{3} - \pi)}\int_0^{\pi/3}\left[\left(\cos x\right)^2 - \left(\frac{1}{2}\right)^2 dx\right]$$

$$= \frac{3}{3\sqrt{3} - \pi}\int_0^{\pi/3}\left(\cos^2 x - \frac{1}{4}\right)dx$$

You can reference all of the important trig identities (like this one) in Appendix C.

Apply the power-reducing formula $\cos^2\theta = \dfrac{1+\cos 2\theta}{2} = \dfrac{1}{2} + \dfrac{\cos 2\theta}{2}$.

$$= \frac{3}{3\sqrt{3}-\pi}\int_0^{\pi/3}\left[\left(\frac{1}{2}+\frac{1}{2}\cos 2x\right)-\frac{1}{4}\right]dx$$

$$= \frac{3}{3\sqrt{3}-\pi}\int_0^{\pi/3}\left[\frac{1}{2}\cos 2x+\frac{1}{4}\right]dx$$

$$= \frac{3}{3\sqrt{3}-\pi}\left[\frac{1}{4}\sin 2x+\frac{1}{4}x\right]_0^{\pi/3}$$

$$= \frac{3}{3\sqrt{3}-\pi}\left[\frac{1}{4}\cdot\frac{\sqrt{3}}{2}+\frac{1}{4}\cdot\frac{\pi}{3}\right]$$

$$= \frac{3}{3\sqrt{3}-\pi}\left[\frac{\sqrt{3}}{8}+\frac{\pi}{12}\right]$$

$$= \frac{3}{3\sqrt{3}-\pi}\left[\frac{3\sqrt{3}+2\pi}{24}\right]$$

$$= \frac{3\sqrt{3}+2\pi}{8\left(3\sqrt{3}-\pi\right)}$$

Chapter 24
PARAMETRIC AND POLAR EQUATIONS

Writing equations using more variables than just x and y

Although the study of elementary calculus focuses primarily on equations and functions in rectangular form, a brief discussion of alternate graphical representations is in order. Though parametric and polar forms of equations are both worthy of in-depth study, this chapter will limit itself to differentiation and integration skills already discussed in terms of rectangular equations, such as calculating rates of change, measuring arc length, and calculating area.

Most of calculus (and every single chapter in this book until now) deals with "regular" functions that are defined in terms of x and y. Technically speaking, these are called "rectangular equations," because they are graphed on two axes perpendicular to one another like the sides of a rectangle. This chapter discusses parametric and polar equations, which describe equations in a fundamentally different way.

Parametric equations define the x- and y-coordinates of a graph in terms of a third variable, usually t or θ, called the "parameter." Polar equations don't use x and y coordinates at all, but describe the points on their graphs based on how far they are from the origin and what angle a line drawn through them makes with the positive x-axis.

Parametric Equations
Like revolutionaries in Boston Harbor, just add t

Note: Problems 24.1–24.2 refer to the parametric equations $x = t^2 - 1$ and $y = 1 - t^2$.

24.1 Graph the parametric curve.

> If there are trig functions in the parametric equations, plug in t values between 0 and 2π. If not, try different t-values between −10 and 10.

To visualize the shape of a parametric curve, you must substitute a sufficient number of t-values into both parametric equations. There is no single number of t-values that is appropriate for every parametric problem, but in this case, substituting integer values on the interval $[-2,2]$ is sufficient to visualize the curve. To begin, substitute $t = -2$ into both equations.

$$x = t^2 - 1 \qquad\qquad y = 1 - t^2$$
$$x = (-2)^2 - 1 \qquad\quad y = 1 - (-2)^2$$
$$x = 3 \qquad\qquad\qquad y = -3$$

> The domain of the parametric curve is the range of the x-equation, and the range of the parametric curve is the range of the y-equation.

Therefore, the point $(x,y) = (3,-3)$ is on the parametric curve. The graph in Figure 24-1 is generated by substituting additional values of t into the parametric equations and plotting the resulting coordinates. Note that the graph has domain $[-1,\infty)$ and range $(-\infty,1]$.

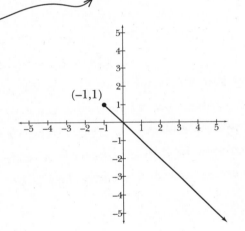

t	$x = t^2 - 1$	$y = 1 - t^2$
$t = -2$	$x = (-2)^2 - 1$ $= 3$	$y = 1 - (-2)^2$ $= -3$
$t = -1$	$x = (-1)^2 - 1$ $= 0$	$y = 1 - (-1)^2$ $= 0$
$t = 0$	$x = (0)^2 - 1$ $= -1$	$y = 1 - (0)^2$ $= 1$
$t = 1$	$x = (1)^2 - 1$ $= 0$	$y = 1 - (1)^2$ 0
$t = 2$	$x = (2)^2 - 1$ $= 3$	$y = 1 - (2)^2$ $= -3$

Figure 24-1 *The graph of the curve defined by the parametric equations $x = t^2 - 1$ and $y = 1 - t^2$.*

Note: Problems 24.1–24.2 refer to the parametric equations $x = t^2 - 1$ and $y = 1 - t^2$.

24.2 Write the equation of the parametric curve in rectangular form.

Solve the equation $x = t^2 - 1$ for t.

$$x = t^2 - 1$$
$$x + 1 = t^2$$
$$\pm\sqrt{x + 1} = t$$

Eliminate the parameter t by substituting this radical expression into $y = 1 - t^2$.

$$y = 1 - \left(\pm\sqrt{x+1}\right)^2$$
$$y = 1 - (x+1)$$
$$y = -x$$

The rectangular form of the curve is $y = -x$, a line with slope -1 and y-intercept 0. Notice that this aptly describes the graph in Figure 24-1 with one exception: that graph is only defined for $x \geq -1$. Therefore, this restriction must be applied to the rectangular form of the curve as well. ←

> Don't worry about restricting y—restricting x will take care of everything. You just want the line to abruptly cut off at the point (-1,1) like it does in Figure 24-1.

Note: Problems 24.3–24.4 refer to the parametric equations $x = \dfrac{t+1}{2}$ and $y = t - t^2$.

24.3 Graph the parametric curve.

The t-values -2, -1, 0, 1, and 2 are nearly sufficient to construct the graph. In Figure 24-2, the x- and y-equations are evaluated for those values of t, and the resulting coordinate pairs are plotted to determine the curve.

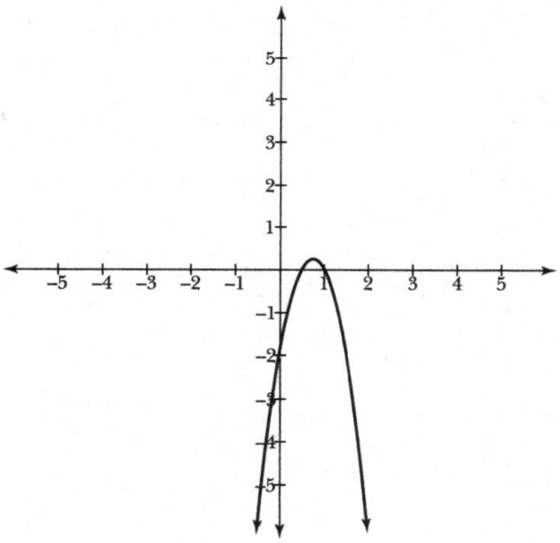

t	$x = \dfrac{t+1}{2}$	$y = t - t^2$
-2	$x = \dfrac{-2+1}{2}$ $= -\dfrac{1}{2}$	$y = -2 - (-2)^2$ $= -6$
-1	$x = \dfrac{-1+1}{2}$ $= 0$	$y = -1 - (-1)^2$ $= -2$
0	$x = \dfrac{0+1}{2}$ $= \dfrac{1}{2}$	$y = 0 - (0)^2$ $= 0$
$\dfrac{1}{2}$	$x = \dfrac{(1/2)+1}{2}$ $= \dfrac{3}{4}$	$y = \dfrac{1}{2} - \left(\dfrac{1}{2}\right)^2$ $= \dfrac{1}{4}$
1	$x = \dfrac{1+1}{2}$ $= 1$	$y = 1 - (1)^2$ $= 0$
2	$x = \dfrac{2+1}{2}$ $= \dfrac{3}{2}$	$y = 2 - (2)^2$ $= -2$

Figure 24-2 *This parametric curve is a parabola whose equation is identified in Problem 24.4. Notice that $t = \dfrac{1}{2}$ is evaluated in order to determine the vertex of the curve, thereby increasing the accuracy of the graph.*

> *Note: Problems 24.3–24.4 refer to the parametric equations $x = \dfrac{t+1}{2}$ and $y = t - t^2$.*

24.4 Write the equation of the parametric curve in rectangular form.

Solve the equation containing x for the parameter t.

$$x = \frac{t+1}{2}$$
$$t + 1 = 2x$$
$$t = 2x - 1$$

Substitute this t-value into the parametric equation containing y.

$$y = t - t^2$$
$$y = (2x - 1) - (2x - 1)^2$$
$$y = 2x - 1 - \left(4x^2 - 4x + 1\right)$$
$$y = -4x^2 + 6x - 2$$

The rectangular equation requires no restrictions, as the parametric curve and the rectangular graph are exactly equal at all points in their domains.

> The parametric curve isn't cut off anywhere like Figure 24-1 was, so it looks exactly the same as the rectangular graph of the parabola $y = -4x^2 + 6x - 2$.

> *Note: Problems 24.5–24.6 refer to the parametric equations $x = 3\cos\theta$ and $y = 4\sin\theta$.*

24.5 Graph the parametric curve.

Because the parametric equations are defined trigonometrically, substitute a range of θ-values between 0 and 2π to construct the curve, as illustrated in Figure 24-3.

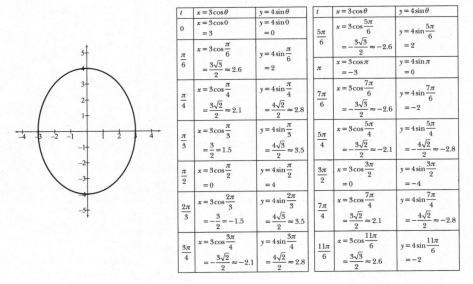

t	$x=3\cos\theta$	$y=4\sin\theta$	t	$x=3\cos\theta$	$y=4\sin\theta$
0	$x=3\cos 0$ $=3$	$y=4\sin 0$ $=0$	$\frac{5\pi}{6}$	$x=3\cos\frac{5\pi}{6}$ $=-\frac{3\sqrt{3}}{2}\approx -2.6$	$y=4\sin\frac{5\pi}{6}$ $=2$
$\frac{\pi}{6}$	$x=3\cos\frac{\pi}{6}$ $=\frac{3\sqrt{3}}{2}\approx 2.6$	$y=4\sin\frac{\pi}{6}$ $=2$	π	$x=3\cos\pi$ $=-3$	$y=4\sin\pi$ $=0$
$\frac{\pi}{4}$	$x=3\cos\frac{\pi}{4}$ $=\frac{3\sqrt{2}}{2}\approx 2.1$	$y=4\sin\frac{\pi}{4}$ $=\frac{4\sqrt{2}}{2}\approx 2.8$	$\frac{7\pi}{6}$	$x=3\cos\frac{7\pi}{6}$ $=-\frac{3\sqrt{3}}{2}\approx -2.6$	$y=4\sin\frac{7\pi}{6}$ $=-2$
$\frac{\pi}{3}$	$x=3\cos\frac{\pi}{3}$ $=\frac{3}{2}=1.5$	$y=4\sin\frac{\pi}{3}$ $=\frac{4\sqrt{3}}{2}\approx 3.5$	$\frac{5\pi}{4}$	$x=3\cos\frac{5\pi}{4}$ $=-\frac{3\sqrt{2}}{2}\approx -2.1$	$y=4\sin\frac{5\pi}{4}$ $=-\frac{4\sqrt{2}}{2}\approx -2.8$
$\frac{\pi}{2}$	$x=3\cos\frac{\pi}{2}$ $=0$	$y=4\sin\frac{\pi}{2}$ $=4$	$\frac{3\pi}{2}$	$x=3\cos\frac{3\pi}{2}$ $=0$	$y=4\sin\frac{3\pi}{2}$ $=-4$
$\frac{2\pi}{3}$	$x=3\cos\frac{2\pi}{3}$ $=-\frac{3}{2}=-1.5$	$y=4\sin\frac{2\pi}{3}$ $=\frac{4\sqrt{3}}{2}\approx 3.5$	$\frac{7\pi}{4}$	$x=3\cos\frac{7\pi}{4}$ $=\frac{3\sqrt{2}}{2}\approx 2.1$	$y=4\sin\frac{7\pi}{4}$ $=-\frac{4\sqrt{2}}{2}\approx -2.8$
$\frac{3\pi}{4}$	$x=3\cos\frac{3\pi}{4}$ $=-\frac{3\sqrt{2}}{2}\approx -2.1$	$y=4\sin\frac{3\pi}{4}$ $=\frac{4\sqrt{2}}{2}\approx 2.8$	$\frac{11\pi}{6}$	$x=3\cos\frac{11\pi}{6}$ $=\frac{3\sqrt{3}}{2}\approx 2.6$	$y=4\sin\frac{11\pi}{6}$ $=-2$

Figure 24-3 *After substituting values for θ between 0 and π, the elliptical nature of the graph is clear. However, you should substitute a few values between π and 2π to ensure that the remaining portion of the graph acts as expected (i.e., ensure that it falls below the x-axis).*

Note: Problems 24.5–24.6 refer to the parametric equations $x = 3 \cos \theta$ and $y = 4 \sin \theta$.

24.6 Write the equation of the parametric curve in rectangular form.

Solve the first parametric equation for $\cos \theta$ and the second for $\sin \theta$.

$$x = 3\cos\theta \qquad y = 4\sin\theta$$
$$\frac{x}{3} = \cos\theta \qquad \frac{y}{4} = \sin\theta$$

Substitute these expressions into the Pythagorean identity $\cos^2 \theta + \sin^2 \theta = 1$.

$$(\cos\theta)^2 + (\sin\theta)^2 = 1$$
$$\left(\frac{x}{3}\right)^2 + \left(\frac{y}{4}\right)^2 = 1$$
$$\frac{x^2}{9} + \frac{y^2}{16} = 1$$

> This ellipse has standard form $\dfrac{x^2}{b^2} + \dfrac{y^2}{a^2} = 1$, where a is half the length of the major axis and b is half the length of the minor axis.

This is the equation of an ellipse (in standard form) that is centered at the origin, has vertical major axis length $2\sqrt{16} = 8$ and horizontal minor axis of length $2\sqrt{9} = 6$.

24.7 What parametric equations define an ellipse in standard form?

The standard form of an ellipse with a horizontal major axis is $\dfrac{(x-h)^2}{a^2} + \dfrac{(y-k)^2}{b^2} = 1$, such that a is half the length of the major axis, b is half the length of the minor axis, and (h,k) is the center of ellipse. According to a Pythagorean identity, $\cos^2 \theta + \sin^2 \theta = 1$. Therefore, $\cos^2\theta = \dfrac{(x-h)^2}{a^2}$ and $\sin^2\theta = \dfrac{(y-k)^2}{b^2}$. Solve the equations for x and y, respectively.

$$\cos^2\theta = \frac{(x-h)^2}{a^2} \qquad\qquad \sin^2\theta = \frac{(y-k)^2}{b^2}$$
$$(x-h)^2 = a^2\cos^2\theta \qquad\qquad (y-k)^2 = b^2\sin^2\theta$$
$$\sqrt{(x-h)^2} = \sqrt{a^2\cos^2\theta} \qquad \sqrt{(y-k)^2} = \sqrt{b^2\sin^2\theta}$$
$$x = a\cos\theta + h \qquad\qquad y = b\sin\theta + k$$

> The identity and the equation of the ellipse both contain the sum of two things that are squared, and in both cases, the sum is 1. Set the first term in the ellipse equal to the first term in the identity, and then do the same thing for the second terms.

The above parametric equations apply only if the major axis of the ellipse is horizontal. If the major axis is vertical, reverse a and b in the parametric equations: $x = b \cos \theta + h$ and $y = a \sin \theta + k$.

> You don't have to write "±" in front of these radicals, because the cosine and sine functions handle the sign changes for you.

24.8 What parametric equations define a circle centered at the origin with radius r?

A circle is actually an ellipse with major and minor axes that are the same length. Therefore, you can use the parametric equations from Problem 24.7 and set $a = b = r$ (the radius of the circle) and $(h,k) = (0,0)$. Either set of the parametric equations defined by Problem 24.7 result in the same parametric representation of the circle.

$$x = a\cos\theta + h \qquad y = b\sin\theta + k$$
$$x = r\cos\theta \qquad y = r\sin\theta$$

Polar Coordinates

Convert from (x, y) to (r, θ) and vice versa

> The ABSCISSA is the first of the two numbers in a coordinate pair, so the abscissa of (a,b) is a.

24.9 Describe how to plot the polar coordinate pair (r, θ) in the coordinate plane.

> The ORDINATE is the second number in a coordinate pair. The ordinate of (a,b) is b.

The abscissa of a polar coordinate pair represents its distance from the pole and the ordinate is the measure of the angle formed by the polar axis and a terminal ray passing through the coordinate whose endpoint is the pole (as illustrated in Figure 24-4). Note that positive angles are measured counterclockwise and negative angles are measured clockwise.

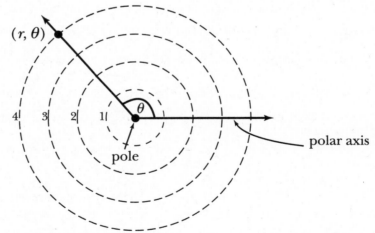

Figure 24-4 *In this diagram, θ measures approximately $135° = \dfrac{3\pi}{4}$ radians and the point (r,θ) is 4 units away from the pole. Therefore, $(r,\theta) = \left(4, \dfrac{3\pi}{4}\right)$ is a fair representation of the polar coordinate.*

Although any pole and polar axis can be used, it is customary to superimpose polar coordinates on the Cartesian plane, placing the pole at the origin and the polar axis on the positive x-axis.

24.10 Plot the polar coordinate pairs on the same plane: $A = \left(1, \frac{\pi}{4}\right)$, $B = \left(5, \frac{7\pi}{6}\right)$, $C = \left(3, -\frac{\pi}{2}\right)$, and $D = \left(-4, \frac{2\pi}{3}\right)$.

Refer to Figure 24-5 for the locations of A, B, C, and D in the coordinate plane—as stated in Problem 24.9, the pole should be placed at the origin and the positive x-axis serves as the polar axis. Note that C contains a negative angle, and D contains a negative directed distance.

> To plot D, instead of traveling up the ray in the second quadrant, extend the ray across the origin (along the dotted line in Figure 25-5), and travel 4 units into the fourth quadrant.

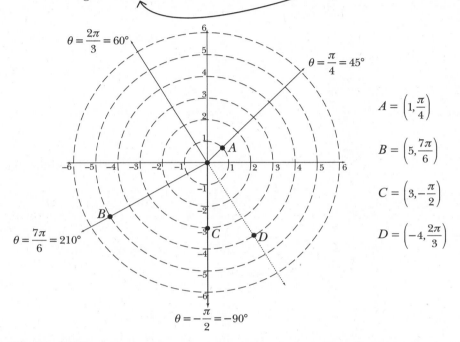

$A = \left(1, \frac{\pi}{4}\right)$

$B = \left(5, \frac{7\pi}{6}\right)$

$C = \left(3, -\frac{\pi}{2}\right)$

$D = \left(-4, \frac{2\pi}{3}\right)$

Figure 24-5 *Positive angles are rotated counterclockwise from the positive x-axis and negative angles rotate counterclockwise.*

24.11 Unlike the Cartesian coordinate system, in which each point on the plane is expressed by a unique coordinate pair, each point in the coordinate plane can be represented by an infinite number of polar coordinate pairs. Find two polar coordinate pairs that represent the same point on the plane as $\left(2, \frac{\pi}{6}\right)$.

As illustrated by Figure 24-6, replacing $\frac{\pi}{6}$ with a coterminal angle, such as $\frac{13\pi}{6}$ or $-\frac{11\pi}{6}$, will not alter the location of the point; therefore, the polar coordinate pairs $\left(2, \frac{13\pi}{6}\right)$ and $\left(2, \frac{-11\pi}{6}\right)$ represent the same point on the plane as the polar coordinate $\left(2, \frac{\pi}{6}\right)$. You can identify yet another polar coordinate pair with the same graph by adding π to θ and multiplying r by –1.

> The rays through $\theta = \frac{\pi}{6}$ and $\theta = \frac{7\pi}{6}$ form a straight line, so a directed distance of –2 along $\theta = \frac{7\pi}{6}$ is the same as a distance of 2 along $\theta = \frac{\pi}{6}$.

$$(-r, \theta + \pi) = \left(-2, \frac{\pi}{6} + \pi\right) = \left(-2, \frac{7\pi}{6}\right)$$

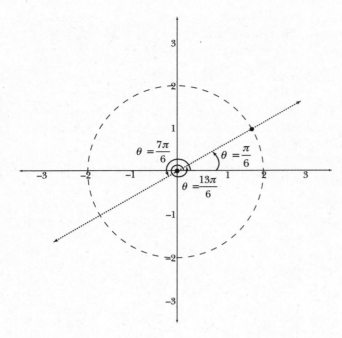

Figure 24-6

The polar coordinate pairs $\left(2,\frac{\pi}{6}\right)$, $\left(2,\frac{13\pi}{6}\right)$, *and* $\left(-2,\frac{7\pi}{6}\right)$ *all represent the same point on the coordinate plane.*

24.12 Convert the point $(-3, \pi)$ from polar to rectangular coordinates.

Given $(r, \theta) = (-3, \pi)$, apply the formulas $x = r\cos\theta$ and $y = r\sin\theta$.

$$
\begin{aligned}
x &= -3\cos\pi & y &= -3\sin\pi \\
x &= -3(-1) & y &= -3(0) \\
x &= 3 & y &= 0
\end{aligned}
$$

Therefore, the polar coordinate $(-3, \pi)$ and the rectangular coordinate $(3,0)$ represent the same point in the Cartesian plane.

24.13 Convert the point $\left(7,\frac{4\pi}{3}\right)$ from polar to rectangular coordinates.

Given $(r,\theta) = \left(7,\frac{4\pi}{3}\right)$, apply the formulas $x = r\cos\theta$ and $y = r\sin\theta$.

$$
\begin{aligned}
x &= 7\cos\frac{4\pi}{3} & y &= 7\sin\frac{4\pi}{3} \\
x &= 7\left(-\frac{1}{2}\right) & y &= 7\left(-\frac{\sqrt{3}}{2}\right) \\
x &= -\frac{7}{2} & y &= -\frac{7\sqrt{3}}{2}
\end{aligned}
$$

The rectangular coordinate pair $\left(-\frac{7}{2},-\frac{7\sqrt{3}}{2}\right)$ and the polar coordinate pair $\left(7,\frac{4\pi}{3}\right)$ represent the same point in the Cartesian plane.

24.14 Convert the point (4, −4) from rectangular to polar coordinates. ←

Given $(x,y) = (4, -4)$, apply the conversion formulas $r = \sqrt{x^2 + y^2}$ and $\tan\theta = \frac{y}{x}$.

$$r = \sqrt{4^2 + (-4)^2} \qquad \tan\theta = \frac{-4}{4}$$
$$r = \sqrt{32} \qquad\qquad \tan\theta = -1$$
$$r = 4\sqrt{2} \qquad\qquad \theta = -\frac{\pi}{4}$$

Like Problem 24.11 explains, there's more than one correct answer to this problem. Lots of polar coordinates will overlap the rectangular coordinate (4,−4).

Infinitely many angles have a tangent value of −1 and you can replace $\theta = -\frac{\pi}{4}$ with any of them, including $\theta = -\frac{5\pi}{4}, \frac{3\pi}{4}$, and $\frac{7\pi}{4}$. However, the polar coordinate must be located in the fourth quadrant to match the rectangular coordinate (4, −4). Therefore, r is positive for all angles terminating in the fourth quadrant and negative for all angles terminating in the second quadrant. Correct polar coordinate representations of the rectangular coordinate pair (4,−4) include $\left(4\sqrt{2}, -\frac{\pi}{4}\right), \left(4\sqrt{2}, \frac{7\pi}{4}\right), \left(-4\sqrt{2}, \frac{3\pi}{4}\right)$, and $\left(-4\sqrt{2}, -\frac{5\pi}{4}\right)$.

24.15 Convert the point $\left(-1, \sqrt{3}\right)$ from rectangular to polar coordinates.

Given $(x, y) = \left(-1, \sqrt{3}\right)$, apply the conversion formulas $r = \sqrt{x^2 + y^2}$ and $\tan\theta = \frac{y}{x}$. Note that $\left(-1, \sqrt{3}\right)$ is in the second quadrant, so the terminal side of θ is in the second quadrant (for $r > 0$) and in the fourth quadrant (for $r < 0$).

$$r = \sqrt{(-1)^2 + \left(\sqrt{3}\right)^2} \qquad \tan\theta = -\sqrt{3}$$
$$r = 2 \qquad\qquad\qquad \theta = \frac{2\pi}{3} ←$$

If you can't figure out how to get this angle, look at Problem 8.30.

Correct polar coordinate representations of the rectangular coordinate $\left(-1, \sqrt{3}\right)$ include $\left(2, \frac{2\pi}{3}\right), \left(2, \frac{8\pi}{3}\right), \left(-2, -\frac{\pi}{3}\right)$, and $\left(-2, \frac{5\pi}{3}\right)$.

Graphing Polar Curves
Graphing with r and θ instead of x and y

24.16 Graph the polar curve $r = 5$.

The curve consists of all the points 5 units away from the pole, regardless of the angle θ. As convention dictates that the pole be placed at the origin, then this curve is the collection of points exactly 5 units away from (0,0). In other words, its graph is a circle centered at the origin with radius 5, as illustrated in Figure 24-7.

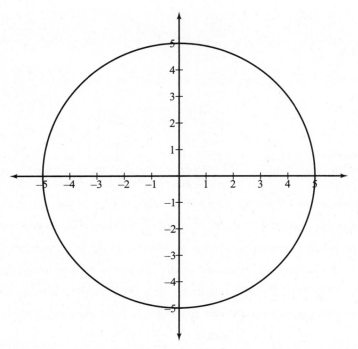

This is one of the strengths of polar equations—sometimes they're much simpler looking than the rectangular version of the same graph.

Figure 24-7 *The polar curve r = 5 has the same graph as the (comparatively more complex) rectangular curve x² + y² = 25.*

24.17 Graph the polar curve: $\theta = -\dfrac{\pi}{6}$.

Consider the line that forms the angle $-\dfrac{\pi}{6}$ with the positive *x*-axis, as illustrated in Figure 24-8.

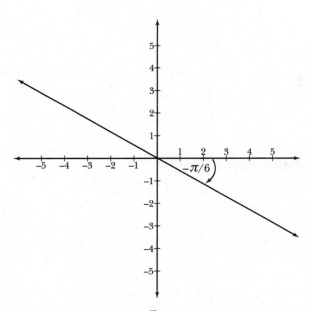

Figure 24-8 *The graph of* $\theta = -\dfrac{\pi}{6}$ *is a straight line.*

Any point along this line (regardless of its distance, r, from the origin) belongs to the curve $\theta = -\frac{\pi}{6}$, including the points in the second quadrant once the line is extended across the origin. $\longleftarrow$

You have to extend the line because r could be negative. For example, $\left(-3, -\frac{\pi}{6}\right)$ is in the second quadrant, 3 units away from the origin along the line in Figure 24-8.

24.18 Graph the polar curve: $r = 3\cos\theta$.

The most effective way to graph a polar curve is to use a table of values. The range of θ-values you should use to construct the table depends on the polar equation, but the complete graph of most polar equations containing trigonometric functions can usually be generated using the θ-interval $[0, 2\pi]$. Even fewer θ-values are needed to graph $r = 3\cos\theta$, as the entire curve is drawn between $\theta = 0$ and $\theta = \pi$ (and repeats between $\theta = \pi$ and $\theta = 2\pi$).

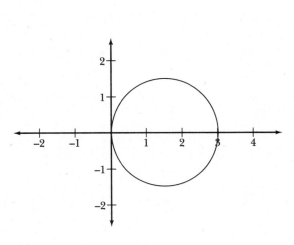

θ	$r = 3\cos\theta$
0	$r = 3\cos 0 = 3(1) = 3$
$\frac{\pi}{6}$	$r = 3\cos\frac{\pi}{6} = 3\left(\frac{\sqrt{3}}{2}\right) \approx 2.6$
$\frac{\pi}{4}$	$r = 3\cos\frac{\pi}{4} = 3\left(\frac{\sqrt{2}}{2}\right) \approx 2.1$
$\frac{\pi}{3}$	$r = 3\cos\frac{\pi}{3} = 3\left(\frac{1}{2}\right) = 1.5$
$\frac{\pi}{2}$	$r = 3\cos\frac{\pi}{2} = 3(0) = 0$
$\frac{2\pi}{3}$	$r = 3\cos\frac{2\pi}{3} = 3\left(-\frac{1}{2}\right) = -1.5$
$\frac{3\pi}{4}$	$r = 3\cos\frac{3\pi}{4} = 3\left(-\frac{\sqrt{2}}{2}\right) \approx -2.1$
$\frac{5\pi}{6}$	$r = 3\cos\frac{5\pi}{6} = 3\left(-\frac{\sqrt{3}}{2}\right) \approx -2.6$
π	$r = 3\cos\pi = 3(-1) = -3$

Figure 24-9 *The graph of $r = 3\cos\theta$ is a circle centered at $\left(\frac{3}{2}, 0\right)$ with a radius of $\frac{3}{2}$.*

24.19 Graph the polar curve: $r = 4\cos 2\theta$.

Like Problem 24.18, construct a table of values for the curve. Substituting values of θ in the interval $[0, 2\pi]$ is sufficient to construct the entire curve, as illustrated in Figure 24-10.

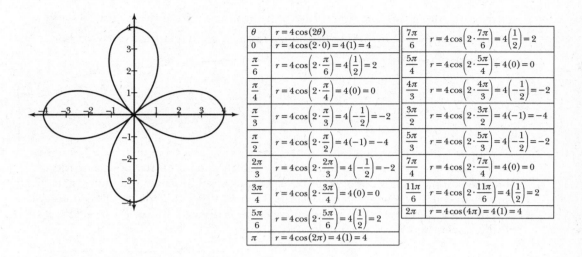

θ	$r=4\cos(2\theta)$	θ	$r=4\cos(2\theta)$
0	$r=4\cos(2\cdot 0)=4(1)=4$	$\dfrac{7\pi}{6}$	$r=4\cos\left(2\cdot\dfrac{7\pi}{6}\right)=4\left(\dfrac{1}{2}\right)=2$
$\dfrac{\pi}{6}$	$r=4\cos\left(2\cdot\dfrac{\pi}{6}\right)=4\left(\dfrac{1}{2}\right)=2$	$\dfrac{5\pi}{4}$	$r=4\cos\left(2\cdot\dfrac{5\pi}{4}\right)=4(0)=0$
$\dfrac{\pi}{4}$	$r=4\cos\left(2\cdot\dfrac{\pi}{4}\right)=4(0)=0$	$\dfrac{4\pi}{3}$	$r=4\cos\left(2\cdot\dfrac{4\pi}{3}\right)=4\left(-\dfrac{1}{2}\right)=-2$
$\dfrac{\pi}{3}$	$r=4\cos\left(2\cdot\dfrac{\pi}{3}\right)=4\left(-\dfrac{1}{2}\right)=-2$	$\dfrac{3\pi}{2}$	$r=4\cos\left(2\cdot\dfrac{3\pi}{2}\right)=4(-1)=-4$
$\dfrac{\pi}{2}$	$r=4\cos\left(2\cdot\dfrac{\pi}{2}\right)=4(-1)=-4$	$\dfrac{5\pi}{3}$	$r=4\cos\left(2\cdot\dfrac{5\pi}{3}\right)=4\left(-\dfrac{1}{2}\right)=-2$
$\dfrac{2\pi}{3}$	$r=4\cos\left(2\cdot\dfrac{2\pi}{3}\right)=4\left(-\dfrac{1}{2}\right)=-2$	$\dfrac{7\pi}{4}$	$r=4\cos\left(2\cdot\dfrac{7\pi}{4}\right)=4(0)=0$
$\dfrac{3\pi}{4}$	$r=4\cos\left(2\cdot\dfrac{3\pi}{4}\right)=4(0)=0$	$\dfrac{11\pi}{6}$	$r=4\cos\left(2\cdot\dfrac{11\pi}{6}\right)=4\left(\dfrac{1}{2}\right)=2$
$\dfrac{5\pi}{6}$	$r=4\cos\left(2\cdot\dfrac{5\pi}{6}\right)=4\left(\dfrac{1}{2}\right)=2$	2π	$r=4\cos(4\pi)=4(1)=4$
π	$r=4\cos(2\pi)=4(1)=4$		

Figure 24-10   *The graph of the polar curve r = 4 cos 2θ is described as a "rose." The number of "petals" in a rose graph (in this curve there are four) varies based on the constants in the polar equation.*

24.20 Graph the polar curve: $r = 4 \sin 2\theta$.

Use a table of values very similar to the table in Figure 24.19. All of the angles will remain the same, but rather than multiply 4 times the cosine of each angle to calculate the corresponding r, you will multiply 4 times its sine. The graph, illustrated in Figure 24-11, is very similar to the graph of $r = 4 \cos 2\theta$ as well.

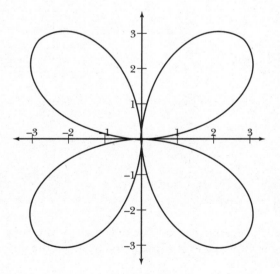

Figure 24-11   *The polar curve 4 = 4 sin 2θ is equivalent to the graph of r = 4 cos 2θ rotated $\dfrac{\pi}{4}$ radians either clockwise or counterclockwise.*

24.21 Graph the polar curve: $r = \sin\theta - \dfrac{1}{2}$.

Use a table of values to plot the curve, as illustrated in Figure 24-12.

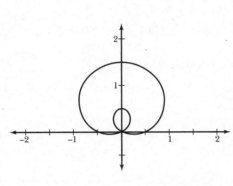

θ	$r = \sin\theta - \dfrac{1}{2}$
0	$r = \sin 0 - \dfrac{1}{2} = 0 - \dfrac{1}{2} = -\dfrac{1}{2}$
$\dfrac{\pi}{6}$	$r = \sin\dfrac{\pi}{6} - \dfrac{1}{2} = \dfrac{1}{2} - \dfrac{1}{2} = 0$
$\dfrac{\pi}{4}$	$r = \sin\dfrac{\pi}{4} - \dfrac{1}{2} = \dfrac{\sqrt{2}-1}{2} \approx 0.2$
$\dfrac{\pi}{3}$	$r = \sin\dfrac{\pi}{3} - \dfrac{1}{2} = \dfrac{\sqrt{3}-1}{2} \approx 0.4$
$\dfrac{\pi}{2}$	$r = \sin\dfrac{\pi}{2} - \dfrac{1}{2} = 1 - \dfrac{1}{2} = 0.5$
$\dfrac{2\pi}{3}$	$r = \sin\dfrac{2\pi}{3} - \dfrac{1}{2} = \dfrac{\sqrt{3}-1}{2} \approx 0.4$
$\dfrac{3\pi}{4}$	$r = \sin\dfrac{3\pi}{4} - \dfrac{1}{2} = \dfrac{\sqrt{2}-1}{2} \approx 0.2$
$\dfrac{5\pi}{6}$	$r = \sin\dfrac{5\pi}{6} - \dfrac{1}{2} = \dfrac{1}{2} - \dfrac{1}{2} = 0$

θ	$r = \sin\theta - \dfrac{1}{2}$
π	$r = \sin\pi - \dfrac{1}{2} = 0 - \dfrac{1}{2} = -\dfrac{1}{2}$
$\dfrac{7\pi}{6}$	$r = \sin\dfrac{7\pi}{6} - \dfrac{1}{2} = -\dfrac{1}{2} - \dfrac{1}{2} = -1$
$\dfrac{5\pi}{4}$	$r = \sin\dfrac{5\pi}{4} - \dfrac{1}{2} = \dfrac{-\sqrt{2}-1}{2} \approx -1.2$
$\dfrac{4\pi}{3}$	$r = \sin\dfrac{4\pi}{3} - \dfrac{1}{2} = \dfrac{-\sqrt{3}-1}{2} \approx -1.4$
$\dfrac{3\pi}{2}$	$r = \sin\dfrac{3\pi}{2} - \dfrac{1}{2} = -1 - \dfrac{1}{2} = -1.5$
$\dfrac{5\pi}{3}$	$r = \sin\dfrac{5\pi}{3} - \dfrac{1}{2} = \dfrac{-\sqrt{3}-1}{2} \approx -1.4$
$\dfrac{7\pi}{4}$	$r = \sin\dfrac{7\pi}{4} - \dfrac{1}{2} = \dfrac{-\sqrt{2}-1}{2} \approx -1.2$
$\dfrac{11\pi}{6}$	$r = \sin\dfrac{11\pi}{6} - \dfrac{1}{2} = -\dfrac{1}{2} - \dfrac{1}{2} = -1$
2π	$r = \sin(2\pi) - \dfrac{1}{2} = 0 - \dfrac{1}{2} = -\dfrac{1}{2}$

Figure 24-12 *Polar curves are classified according to their shapes. The curve* $r = \sin\theta - \dfrac{1}{2}$ *is a limaçon.*

> Rectangular graph transformations do not apply to polar equations. For example, the graph of $r = \sin\theta + 1$ is not the graph of $r = \sin\theta$ moved up one unit.

24.22 Write the polar equation $r = 3\cos\theta$, from Problem 24.18, in parametric form.

According to Problem 24.12, individual polar coordinates can be converted to rectangular coordinates using the formulas $x = r\cos\theta$ and $y = r\sin\theta$. To construct parametric equations that correspond to the polar curve, substitute $r = 3\cos\theta$ into both equations.

$$x = r\cos\theta \qquad\qquad y = r\sin\theta$$
$$x = (3\cos\theta)\cos\theta \qquad\qquad y = (3\cos\theta)\sin\theta$$
$$x = 3\cos^2\theta \qquad\qquad y = 3\cos\theta\sin\theta$$

24.23 Write the polar equation $r = \sin\theta - \dfrac{1}{2}$, from Problem 21.21, in parametric form.

Substitute the polar equation for r in the parametric equations $x = r\cos\theta$ and $y = r\sin\theta$.

$$x = r\cos\theta \qquad\qquad y = r\sin\theta$$
$$x = \left(\sin\theta - \dfrac{1}{2}\right)\cos\theta \qquad\qquad y = \left(\sin\theta - \dfrac{1}{2}\right)\sin\theta$$
$$x = \cos\theta\sin\theta - \dfrac{1}{2}\cos\theta \qquad\qquad y = \sin^2\theta - \dfrac{1}{2}\sin\theta$$

Applications of Parametric and Polar Differentiation

Teach a new dog some old differentiation tricks

24.24 Given a curve defined by the parametric equations $x = f(t)$ and $y = g(t)$, determine $\dfrac{dy}{dx}$ and $\dfrac{d^2y}{dx^2}$, the first and second derivatives.

> *Just take the derivative of the y equation (with respect to t) and divide it by the derivative of the x equation (with respect to t).*

Apply the derivative formula for parametrically-defined curves.

$$\frac{dy}{dx} = \frac{dy/dt}{dx/dt}$$

$$= \frac{g'(t)}{f'(t)}$$

Note that the second derivative of a parametrically-defined curve is not simply the derivative of $\dfrac{dy}{dx}$ with respect to t—it is the quotient of that derivative and $\dfrac{dx}{dt}$.

> *To get the second derivative, differentiate the first derivative and divide by the same denominator you divided by in the first derivative.*

$$\frac{d^2y}{dx^2} = \frac{\dfrac{d}{dt}\left(\dfrac{dy}{dx}\right)}{\dfrac{dx}{dt}}$$

Recall that $\dfrac{dy}{dx} = \dfrac{g'(t)}{f'(t)}$.

$$= \frac{\dfrac{d}{dt}\left(\dfrac{g'(t)}{f'(t)}\right)}{f'(t)}$$

24.25 Given the curve defined by the parametric equations $x = \dfrac{t+1}{2}$ and $y = t - t^2$ (from Problem 24.3), calculate $\dfrac{dy}{dx}$.

Differentiate each of the parametric equations with respect to the parameter t.

$$x = \frac{1}{2}(t+1) \qquad\qquad y = t - t^2$$

$$\frac{dx}{dt} = \frac{1}{2} \qquad\qquad \frac{dy}{dt} = 1 - 2t$$

According to Problem 24.24, the derivative of the curve is equal to the quotient of dy/dt and dx/dt.

$$\frac{dy}{dx} = \frac{dy/dt}{dx/dt}$$

$$= \frac{1-2t}{1/2}$$

$$= 2(1-2t)$$

$$= 2 - 4t$$

24.26 Verify the solution to Problem 24.25 using the rectangular form of the parametric curve and its derivative.

Problem 24.25 states that $\frac{dy}{dx} = 2 - 4t$. Rewrite the derivative in terms of x by solving the parametric equation $x = \frac{t+1}{2}$ for t.

$$x = \frac{t+1}{2}$$
$$2x = t+1$$
$$t = 2x - 1$$

Substitute t into $\frac{dy}{dx} = 2 - 4t$.

$$\frac{dy}{dx} = 2 - 4(2x - 1)$$
$$\frac{dy}{dx} = 2 - 8x + 4$$
$$\frac{dy}{dx} = -8x + 6$$

Normally, you'd stop here and leave the parametric derivative in terms of t. The rest just shows that the derivatives are equal, whether you differentiate in parametric or rectangular form.

According to Problem 24.4, the curve defined by the parametric equations $x = \frac{t+1}{2}$ and $y = t - t^2$ has rectangular form $y = -4x^2 + 6x - 2$. Differentiate with respect to x.

$$\frac{dy}{dx} = -8x + 6$$

The derivatives are equal, verifying the parametric derivative calculated in Problem 24.25.

Note: Problems 24.27–24.29 refer to the curve defined by the parametric equations $x = e^t$ and $y = \cos 3t$.

24.27 Calculate the slope of the curve when $t = 2$.

Differentiate the parametric equations with respect to t: $\frac{dx}{dt} = e^t$ and $\frac{dy}{dt} = -3\sin 3t$; apply the parametric derivative formula.

$$\frac{dy}{dx} = \frac{dy/dt}{dx/dt} = \frac{-3\sin 3t}{e^t}$$

Evaluate $\frac{dy}{dx}$ when $t = 2$.

$$\frac{-3\sin(3 \cdot 2)}{e^2} = -\frac{3\sin 6}{e^2}$$

Note: Problems 24.27–24.29 refer to the curve defined by the parametric equations $x = e^t$ and $y = \cos 3t$.

24.28 Write the equation of the tangent line to the curve when $t = 2$.

> To write the equation of a line, you need a point (the point of tangency) and a slope (the derivative at that point, calculated in the last problem).

Substitute $t = 2$ into the parametric equations to determine the point of tangency.

$$x = e^t = e^2 \qquad y = \cos 3t = \cos 6$$

When $t = 2$, the point of tangency on the curve is $(e^2, \cos 6)$. According to Problem 24.27, the slope of the tangent line at that point is $-\dfrac{3\sin 6}{e^2}$. Apply the point-slope formula to generate the equation of the tangent line.

$$y - y_1 = m\left(x - x_1\right)$$

$$y - \cos 6 = -\frac{3\sin 6}{e^2}\left(x - e^2\right)$$

$$y - \cos 6 = -\left(\frac{3\sin 6}{e^2}\right)x + 3\sin 6$$

$$y = -\left(\frac{3\sin 6}{e^2}\right)x + 3\sin 6 + \cos 6$$

Note: Problems 24.27–24.29 refer to the curve defined by the parametric equations $x = e^t$ and $y = \cos 3t$.

24.29 Determine the second derivative of the curve.

> This fraction has functions in the numerator AND the denominator (because there are t's in both), so you've got to use the quotient rule.

Apply the second derivative formula from Problem 24.24, substituting the values of $\dfrac{dy}{dx}$ and $\dfrac{dx}{dt}$ from Problem 24.27.

$$\frac{d^2 y}{dx^2} = \frac{\dfrac{d}{dt}\left(\dfrac{dy}{dx}\right)}{dx \,/\, dt}$$

$$= \frac{\dfrac{d}{dt}\left(\dfrac{-3\sin 3t}{e^t}\right)}{e^t}$$

$$= \frac{1}{e^t} \cdot \frac{d}{dt}\left(\frac{-3\sin 3t}{e^t}\right)$$

$$= \frac{1}{e^t}\left[\frac{e^t\left(-9\cos(3t)\right) - \left(-3\sin(3t)\right)\left(e^t\right)}{\left(e^t\right)^2}\right]$$

$$= \frac{1}{e^t}\left[\frac{3e^t\left(-3\cos(3t) + \sin(3t)\right)}{e^t \cdot e^t}\right]$$

$$= \frac{3\left(\sin(3t) - 3\cos(3t)\right)}{e^{2t}}$$

Note: Problems 24.30–24.32 refer to the curve defined by the parametric equations $x = 6 \cos \theta$ and $y = 2 \sin \theta$.

24.30 Find $\dfrac{dy}{dx}$, the derivative of the curve, in terms of t.

Apply the parametric derivative formula.

$$\frac{dy}{dx} = \frac{dy/d\theta}{dx/d\theta} = \frac{2\cos\theta}{-6\sin\theta} = -\frac{1}{3}\cot\theta$$

The formulas in Problem 24.24 have t's in them, but you can change t to whatever parameter is used by the problem, like θ.

Note: Problems 24.30–24.32 refer to the curve defined by the parametric equations $x = 6 \cos \theta$ and $y = 2 \sin \theta$.

24.31 Rewrite the parametrically-defined curve in rectangular form and differentiate with respect to x in order to verify the derivative in Problem 24.30.

Solve each parametric equation for the trigonometric function within.

$$x = 6\cos\theta \qquad\qquad y = 2\sin\theta$$

$$\frac{x}{6} = \cos\theta \qquad\qquad \frac{y}{2} = \sin\theta$$

Square both sides of each equation.

$$\frac{x^2}{36} = \cos^2\theta \qquad\qquad \frac{y^2}{4} = \sin^2\theta$$

Substitute into a Pythagorean trigonometric identity.

$$\cos^2\theta + \sin^2\theta = 1$$

$$\frac{x^2}{36} + \frac{y^2}{4} = 1$$

You have to use implicit differentiation. For more help, look at Problem 16.9—it has a very similar equation and derivative.

Now that the equation of the ellipse is in rectangular form, differentiate with respect to x.

$$\frac{1}{36}(2x) + \frac{1}{4}(2y)\left(\frac{dy}{dx}\right) = 0$$

$$\frac{y}{2}\left(\frac{dy}{dx}\right) = -\frac{x}{18}$$

$$\frac{dy}{dx} = -\frac{x}{18}\left(\frac{2}{y}\right)$$

$$\frac{dy}{dx} = -\frac{x}{9y}$$

Write the derivative in terms of θ, recalling that $x = 6\cos\theta$ and $y = 2\sin\theta$.

$$\frac{dy}{dx} = -\frac{6\cos\theta}{9(2\sin\theta)} = -\frac{6\cos\theta}{18\sin\theta} = -\frac{1}{3}\cot\theta$$

Note: Problems 24.30–24.32 refer to the curve defined by the parametric equations x = 6 cos θ and y = 2 sin θ.

24.32 Find $\dfrac{d^2 y}{dx^2}$, the second derivative of the curve, in terms of θ.

> The sine and cosecant functions are reciprocals, so dividing by sine is the same thing as multiplying by cosecant.

Substitute the values of $\dfrac{dy}{dx}$ and $\dfrac{dx}{d\theta}$ from Problem 24.30 into the parametric second derivative formula.

$$\frac{d^2 y}{dx^2} = \frac{\dfrac{d}{d\theta}\left(\dfrac{dy}{dx}\right)}{dx/d\theta} = \frac{\dfrac{d}{d\theta}\left(-\dfrac{1}{3}\cot\theta\right)}{-6\sin\theta} = \frac{\dfrac{1}{3}\csc^2\theta}{-6\sin\theta} = -\frac{1}{18}\csc^3\theta$$

Note: Problems 24.33–24.34 refer to the polar equation r = 2 sin θ.

24.33 Differentiate the polar equation.

> Replace r with 2 sin θ in the formulas x = r cos θ and y = r sin θ, like in Problems 24.22 and 24.23.

Express the polar equation parametrically and differentiate the parametric equations with respect to θ.

$$
\begin{array}{ll}
x = r\cos\theta & y = r\sin\theta \\
x = (2\sin\theta)\cos\theta & y = (2\sin\theta)\sin\theta \\
x = 2\sin\theta\cos\theta & y = 2\sin^2\theta
\end{array}
$$

Apply the parametric derivative formula. Note that $\dfrac{dy}{d\theta}$ requires the chain rule and $\dfrac{dx}{d\theta}$ requires the product rule.

> Check Appendix C for a list of trig identities.

$$\frac{dy}{dx} = \frac{dy/d\theta}{dx/d\theta} = \frac{4\sin\theta\cdot\cos\theta}{2[\sin\theta(-\sin\theta)+\cos\theta(\cos\theta)]} = \frac{2\sin\theta\cos\theta}{\cos^2\theta - \sin^2\theta}$$

Replace the expressions using double angle trigonometric identities.

$$= \frac{\sin 2\theta}{\cos 2\theta} = \tan 2\theta$$

Note: Problems 24.33–24.34 refer to the polar equation r = 2 sin θ.

24.34 Find the second derivative of the polar equation.

Substitute $\dfrac{dy}{dx}$ and $\dfrac{dy}{d\theta}$ from Problem 24.33 into the parametric second derivative formula.

$$\frac{d^2 y}{dx^2} = \frac{\dfrac{d}{d\theta}\left(\dfrac{dy}{dx}\right)}{dx/d\theta} = \frac{\dfrac{d}{d\theta}(\tan 2\theta)}{\cos 2\theta} = \frac{2\sec^2 2\theta}{\cos 2\theta} = 2\sec^3 2\theta$$

24.35 Identify the equation of the tangent line to the polar curve $r = \sin\theta - \cos 2\theta$ at $\theta = \dfrac{3\pi}{4}$ and write the equation in rectangular form.

Express the polar equation parametrically.

$$x = r\cos\theta \qquad\qquad y = r\sin\theta$$
$$= (\sin\theta - \cos 2\theta)\cos\theta \qquad\qquad = (\sin\theta - \cos 2\theta)\sin\theta$$
$$= \cos\theta\sin\theta - \cos\theta\cos 2\theta \qquad\qquad = \sin^2\theta - \cos 2\theta \cdot \sin\theta$$

> That means the line will contain x's and y's, not r's and θ's.

Determine the rectangular coordinates of the point of tangency by substituting $\theta = \dfrac{3\pi}{4}$ into the parametric equations.

$$x = \cos\frac{3\pi}{4}\sin\frac{3\pi}{4} - \cos\frac{3\pi}{4}\cos\frac{3\pi}{2} \qquad y = \left(\sin\frac{3\pi}{4}\right)^2 - \cos\frac{3\pi}{2}\sin\frac{3\pi}{4}$$

$$= \left(-\frac{\sqrt{2}}{2}\right)\left(\frac{\sqrt{2}}{2}\right) - \left(-\frac{\sqrt{2}}{2}\right)(0) \qquad = \left(\frac{\sqrt{2}}{2}\right)^2 - (0)\left(\frac{\sqrt{2}}{2}\right)$$

$$= -\frac{1}{2} \qquad\qquad\qquad = \frac{1}{2}$$

The point of tangency, shared by the tangent line and the polar curve, is $\left(-\dfrac{1}{2}, \dfrac{1}{2}\right)$. Differentiate the parametric equations with respect to θ, using the product and chain rules.

> A double angle trig identity tells you that $\cos^2\theta - \sin^2\theta = \cos 2\theta$.

$$\frac{dx}{d\theta} = [\cos\theta(\cos\theta) + \sin\theta(-\sin\theta)] - [\cos\theta(-2\sin 2\theta) + (\cos 2\theta)(-\sin\theta)]$$
$$= \cos^2\theta - \sin^2\theta + 2\cos\theta\sin 2\theta + \cos 2\theta\sin\theta$$
$$= \cos 2\theta + 2\cos\theta\sin 2\theta + \cos 2\theta\sin\theta$$
$$\frac{dy}{d\theta} = 2\sin\theta\cos\theta - [\cos 2\theta\cos\theta + \sin\theta(-2\sin 2\theta)]$$
$$= 2\sin\theta\cos\theta - \cos\theta\cos 2\theta + 2\sin\theta\sin 2\theta$$

Apply the parametric derivative formula.

$$\frac{dy}{dx} = \frac{dy/d\theta}{dx/d\theta} = \frac{2\sin\theta\cos\theta - \cos\theta\cos 2\theta + 2\sin\theta\sin 2\theta}{\cos 2\theta + 2\cos\theta\sin 2\theta + \cos 2\theta\sin\theta}$$

Evaluate the derivative at $\theta = \dfrac{3\pi}{4}$.

$$\frac{2\left(\sqrt{2}/2\right)\left(-\sqrt{2}/2\right) - \left(-\sqrt{2}/2\right)(0) + 2\left(\sqrt{2}/2\right)(-1)}{0 + 2\left(-\sqrt{2}/2\right)(-1) + (0)\left(\sqrt{2}/2\right)}$$

$$= \frac{-1 - \sqrt{2}}{\sqrt{2}}$$

$$= -\frac{\sqrt{2} + 2}{2}$$

Apply the point-slope formula to write the equation of the tangent line, with $(x_1, y_1) = \left(-\dfrac{1}{2}, \dfrac{1}{2}\right)$ and $m = -\dfrac{\sqrt{2}+2}{2}$.

$$y - y_1 = m(x - x_1)$$

$$y - \frac{1}{2} = -\frac{\sqrt{2}+2}{2}\left(x - \left(-\frac{1}{2}\right)\right)$$

$$y - \frac{1}{2} = -\frac{\sqrt{2}+2}{2}\left(x + \frac{1}{2}\right)$$

Applications of Parametric and Polar Integration
Maybe some integrals might also interest the new dog

24.36 Calculate the length of the curve defined by the parametric equations $x = \ln t$ and $y = \dfrac{1}{t^2}$ between $t = 1$ and $t = 3$. Use a graphing calculator to evaluate the definite integral and report the result accurate to three decimal places.

If a curve is defined by the parametric equations $x = f(t)$ and $y = g(t)$, the length of the curve between $x = a$ and $x = b$ is equal to $\displaystyle\int_a^b \sqrt{\left(\frac{dx}{dt}\right)^2 + \left(\frac{dy}{dt}\right)^2}\, dt$. Differentiate the parametric equations with respect to t.

> *This integral is too hard to calculate by hand—most parametric arc length integrals are.*

$$\frac{dx}{dt} = \frac{d}{dt}(\ln t) = \frac{1}{t} \qquad \frac{dy}{dt} = \frac{d}{dt}(t^{-2}) = -\frac{2}{t^3}$$

Substitute the derivatives into the parametric arc length formula.

$$\int_a^b \sqrt{\left(\frac{dx}{dt}\right)^2 + \left(\frac{dy}{dt}\right)^2}\, dt = \int_1^3 \sqrt{\left(\frac{1}{t}\right)^2 + \left(-\frac{2}{t^3}\right)^2}\, dt = \int_1^3 \sqrt{\frac{1}{t^2} + \frac{4}{t^6}}\, dt = \int_1^3 \sqrt{\frac{t^4 + 4}{t^6}}\, dt$$

Use a graphing calculator to determine that $\displaystyle\int_1^3 \sqrt{\frac{t^4 + 4}{t^6}}\, dt \approx 1.470$.

24.37 Prove that the circumference of a circle with radius r is equal to $2\pi r$ by calculating the arc length of a parametrically defined curve.

> *These boundaries are important—you want to draw exactly one circle. If you use boundaries of 0 and 4π, the circle gets drawn twice (even though it's hard to tell that from the graph, because the circles overlap).*

According to Problem 24.8, the parametric curve defined by $x = r\cos\theta$ and $y = r\sin\theta$ is a circle centered at the origin with radius r (for θ-values between 0 and 2π). Differentiate the parametric equations.

$$\frac{dx}{d\theta} = -r\sin\theta \qquad \frac{dy}{d\theta} = r\cos\theta$$

Apply the parametric arc length formula from Problem 24.36.

$$\int_a^b \sqrt{\left(\frac{dx}{d\theta}\right)^2 + \left(\frac{dy}{d\theta}\right)^2}\, dt = \int_0^{2\pi} \sqrt{(-r\sin\theta)^2 + (r\cos\theta)^2}\, d\theta$$

$$= \int_0^{2\pi} \sqrt{r^2\sin^2\theta + r^2\cos^2\theta}\, d\theta$$

$$= \int_0^{2\pi} \sqrt{r^2\left(\sin^2\theta + \cos^2\theta\right)}\, d\theta$$

According to a Pythagorean trigonometric identity, $\sin^2\theta + \cos^2\theta = 1$.

$$= \int_0^{2\pi}\sqrt{r^2(1)}\,d\theta = r\int_0^{2\pi} d\theta = r(\theta)\Big|_0^{2\pi} = r(2\pi - 0) = 2\pi r$$

24.38 Construct a definite integral representing the circumference of an ellipse with major axis length $2a$ and minor axis length $2b$.

The orientation of the ellipse (i.e., whether its major axis is horizontal or vertical) is irrelevant, as is the center of the ellipse. Assume the ellipse is centered at the origin and has a horizontal major axis. According to Problem 24.7, the ellipse is defined by the parametric equations $x = a\cos\theta$ and $y = b\sin\theta$ for $0 \le \theta \le 2\pi$. Differentiate the parametric equations.

$$\frac{dx}{d\theta} = -a\sin\theta \qquad \frac{dy}{d\theta} = b\cos\theta$$

Apply the parametric arc length formula.

$$\int_a^b \sqrt{\left(\frac{dx}{d\theta}\right)^2 + \left(\frac{dy}{d\theta}\right)^2}\, d\theta = \int_0^{2\pi} \sqrt{(-a\sin\theta)^2 + (b\cos\theta)^2}\, d\theta$$

$$= \int_0^{2\pi} \sqrt{a^2\sin^2\theta + b^2\cos^2\theta}\, d\theta$$

24.39 According to Problem 24.18, the graph of the polar equation $r = 3\cos\theta$ is a circle centered at $\left(\frac{3}{2}, 0\right)$ with radius $\frac{3}{2}$, which has area $\frac{9\pi}{4}$. Verify the area of the circle by calculating the area bounded by the polar curve between $\theta = 0$ and $\theta = \pi$.

Based on the formula $A = \pi r^2$ from geometry.

The area bounded by a polar curve is equal to $\frac{1}{2}\int_a^b r^2 d\theta$. Note that a and b are the bounding values of θ stated by the problem: $a = 0$ and $b = \pi$.

$$\frac{1}{2}\int_a^b r^2 d\theta = \frac{1}{2}\int_0^\pi (3\cos\theta)^2\, d\theta$$

$$= \frac{9}{2}\int_0^\pi \cos^2\theta\, d\theta$$

Polar equations are written "r =" so plug whatever r equals into the integrand r^2. Don't forget the ½ in front of the definite integral.

Apply the power-reducing formula $\cos^2\theta = \dfrac{1+\cos 2\theta}{2}$.

$$= \frac{9}{4}\int_0^\pi (1+\cos 2\theta)\,d\theta$$

$$= \frac{9}{4}\left(\theta + \frac{1}{2}\sin 2\theta\right)\Big|_0^\pi$$

$$= \frac{9\pi}{4}$$

24.40 Calculate the area bounded by one petal of the rose curve $r = \sin 3\theta$.

Each petal of the rose curve begins and ends at the origin, so set $r = 0$ and solve for θ to determine the θ-values that bound the petals.

$$\sin 3\theta = 0$$

$$3\theta = 0, \pi, 2\pi, 3\pi, \cdots$$

$$\theta = 0, \frac{\pi}{3}, \frac{2\pi}{3}, \pi, \cdots$$

Figure 24-13 illustrates the region of the polar graph bounded by $\theta = 0$ and $\theta = \dfrac{\pi}{3}$, one petal of the graph.

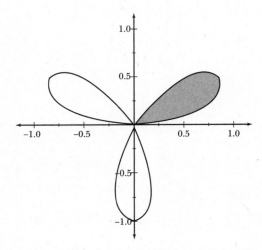

Figure 24-13   *The congruent petals of the rose curve $r = \sin 3\theta$ occur on the θ-intervals $\left[0, \dfrac{\pi}{3}\right]$, $\left[\dfrac{\pi}{3}, \dfrac{2\pi}{3}\right]$, and $\left[\dfrac{2\pi}{3}, \pi\right]$.*

Apply the polar area formula.

$$\frac{1}{2}\int_a^b r^2\,d\theta = \frac{1}{2}\int_0^{\pi/3}\sin^2 3\theta\,d\theta$$

According to a power-reducing formula, $\int \sin^2 3\theta\, d\theta = \int \dfrac{1-\cos(2\cdot 3\theta)}{2}\, d\theta$.

Replace θ in the normal power-reducing formula with 3θ.

$$= \frac{1}{2}\int_0^{\pi/3} \frac{1-\cos 6\theta}{2}\, d\theta$$

$$= \frac{1}{4}\int_0^{\pi/3} (1-\cos 6\theta)\, d\theta$$

$$= \frac{1}{4}\left(\theta - \frac{1}{6}\sin 6\theta\right)\Bigg|_0^{\pi/3}$$

$$= \frac{1}{4}\left(\frac{\pi}{3} - \frac{1}{6}\sin 2\pi\right)$$

$$= \frac{1}{4}\left(\frac{\pi}{3}\right)$$

$$= \frac{\pi}{12}$$

24.41 Calculate the area of the shaded region in Figure 24-14 bounded by the polar curve $r = 1 - 2\sin\theta$ and the axes of the Cartesian plane.

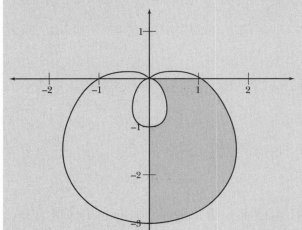

Figure 24-14

A portion of the region bounded by the polar curve $r = 1 - 2\sin\theta$.

The polar graph in Figure 24-14 is generated by plotting θ-values on the interval $[0, 2\pi]$; calculate the values of θ in that interval at which the polar curve intersects the origin (i.e., where $r = 0$).

$$1 - 2\sin\theta = 0$$

$$2\sin\theta = 1$$

$$\sin\theta = \frac{1}{2}$$

$$\theta = \frac{\pi}{6}, \frac{5\pi}{6}$$

Consider Figure 24-15, which illustrates different regions of the graph and the angles, θ, that bound those regions. According to the diagram, the shaded region in Figure 24-14 is equal to the area on the θ-interval $\left[\dfrac{3\pi}{2}, 2\pi\right]$ minus the area on the θ-interval $\left[\dfrac{\pi}{2}, \dfrac{5\pi}{6}\right]$.

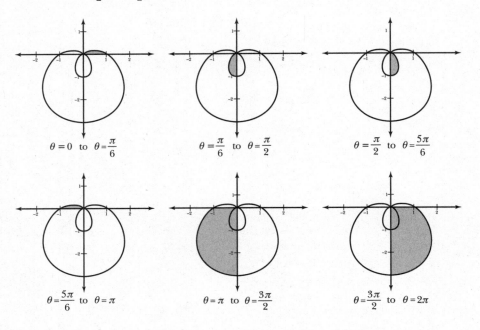

$\theta = 0$ to $\theta = \dfrac{\pi}{6}$ $\theta = \dfrac{\pi}{6}$ to $\theta = \dfrac{\pi}{2}$ $\theta = \dfrac{\pi}{2}$ to $\theta = \dfrac{5\pi}{6}$

$\theta = \dfrac{5\pi}{6}$ to $\theta = \pi$ $\theta = \pi$ to $\theta = \dfrac{3\pi}{2}$ $\theta = \dfrac{3\pi}{2}$ to $\theta = 2\pi$

Figure 24-15 *Regions bounded by $r = 1 - 2\sin\theta$ and varying θ-values on the interval $[0, 2\pi)$.*

$$\frac{1}{2}\int_{3\pi/2}^{2\pi}(1 - 2\sin\theta)^2\, d\theta - \frac{1}{2}\int_{\pi/2}^{5\pi/6}(1 - 2\sin\theta)^2\, d\theta$$

$$= \frac{1}{2}\left[\int_{3\pi/2}^{2\pi}(1 - 4\sin\theta + 4\sin^2\theta)\,d\theta - \int_{\pi/2}^{5\pi/6}(1 - 4\sin\theta + 4\sin^2\theta)\,d\theta\right]$$

Apply a power-reducing formula to determine the antiderivative of $4\sin^2\theta$.

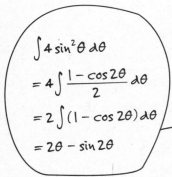

$$\int 4\sin^2\theta\, d\theta$$
$$= 4\int \frac{1 - \cos 2\theta}{2}\, d\theta$$
$$= 2\int (1 - \cos 2\theta)\, d\theta$$
$$= 2\theta - \sin 2\theta$$

$$= \frac{1}{2}\left[\left(\theta + 4\cos\theta + 2\theta - \sin 2\theta\right)\Big|_{3\pi/2}^{2\pi} - \left(\theta + 4\cos\theta + 2\theta - \sin 2\theta\right)\Big|_{\pi/2}^{5\pi/6}\right]$$

$$= \frac{1}{2}\left[\left(3\theta + 4\cos\theta - \sin 2\theta\right)\Big|_{3\pi/2}^{2\pi} - \left(3\theta + 4\cos\theta - \sin 2\theta\right)\Big|_{\pi/2}^{5\pi/6}\right]$$

$$= \frac{1}{2}\left[(6\pi + 4 - 0) - \left(\frac{9\pi}{2} + 0 - 0\right) - \left[\left(\frac{5\pi}{2} - \frac{4\sqrt{3}}{2} + \frac{\sqrt{3}}{2}\right) - \left(\frac{3\pi}{2} + 0 - 0\right)\right]\right]$$

$$= \frac{1}{2}\left[\frac{3\pi}{2} + 4 - \pi + \frac{3\sqrt{3}}{2}\right]$$

$$= \frac{1}{2}\left(\frac{\pi}{2} + \frac{8 + 3\sqrt{3}}{2}\right)$$

$$= \frac{\pi + 8 + 3\sqrt{3}}{4}$$

Chapter 25
DIFFERENTIAL EQUATIONS

Equations that contain a derivative

While the study of differential equations is sufficiently complex that it requires focused attention in a mathematician's coursework, it is both appropriate and useful to introduce them once the concepts of differentiation and integration are explored. This book limits its discussion to ordinary differential equations (equations presented in terms of a single independent variable). Equations whose solutions can be determined by means of the separation of variables are presented, as are visualization and solution approximation techniques for inseparable differential equations.

A differential equation is an equation that contains a derivative, and treats it like a variable. In the beginning of the chapter, you'll come up with equations whose derivatives are the differential equations in the problem. Through a process called "separation of variables," you'll start with $dy/dx = f(x)$ and end up with a solution $y = g(x)$, where $f(x)$ is the derivative of $g(x)$. Unfortunately, not all differential equations can be solved by separating variables, so the rest of the chapter deals with ways to visualize and estimate solutions to those kinds of equations.

Separation of Variables

Separate the y's and dy's from the x's and dx's

> **Note: Problems 25.1–25.2 refer to the differential equation $\frac{dy}{dx} = 4x$.**
>
> **25.1** Find the general solution of the differential equation.

> "Separating variables" means moving the dy and the y-terms to the left side of the equation and moving the dx and x-terms to the right side (or vice versa).

Multiply both sides of the equation by dx in order to separate the variables.

$$dy = 4x\, dx$$

Integrate both sides of the equation.

$$\int dy = 4 \int x\, dx$$

$$y = 4 \cdot \frac{x^2}{2} + C$$

$$y = 2x^2 + C$$

> The derivative of $2x^2 + C$ is $4x$. That makes $2x^2 + C$ a solution of the differential equation $4x$.

The family of curves represented by the equation $y = 2x^2 + C$ are solutions to the differential equation $\frac{dy}{dx} = 4x$.

> **Note: Problems 25.1–25.2 refer to the differential equation $\frac{dy}{dx} = 4x$.**
>
> **25.2** Find the specific solution of the differential equation that contains the point $(-3,7)$.

According to Problem 25.1, the general solution is $y = 2x^2 + C$. Substitute the given x- and y-coordinates into the solution and solve for C.

$$y = 2x^2 + C$$

$$7 = 2(-3)^2 + C$$

$$7 = 18 + C$$

$$-11 = C$$

The specific solution of the differential equation is $y = 2x^2 - 11$.

> **25.3** Describe the difference between a general solution of a differential equation (such as Problem 25.1) and a specific solution (such as Problem 25.2).

> So the general solution of $dy/dx = 4x$ contains every single equation that looks like $y = 2x^2$ plus some number. Of all those possible solutions, Problem 25.2 wants only the one, specific solution that passes through the point $(-3,7)$.

The *general solution* of a differential equation is an infinitely large collection of curves that are identical apart from the constant term in each function. Graphically, the solutions to differential equations are equivalent, except for their vertical positions in the coordinate plane. The *specific solution* of a differential equation is the unique member of the family of solutions that passes through a point identified by the problem, called the *initial value*.

Note: Problems 25.4–25.5 refer to the differential equation $\frac{dy}{dx} = \frac{3x-1}{y}$.

25.4 Find the general solution of the differential equation.

Cross multiply the proportion to separate the variables.

$$y\,dy = (3x-1)\,dx$$

Integrate both sides of the equation.

$$\frac{y^2}{2} = \frac{3x^2}{2} - x + C$$

Multiply the equation by 2 in order to eliminate the fractions. Note that the product of 2 and an arbitrary constant is another arbitrary constant.

$$y^2 = 3x^2 - 2x + C$$

There's no reason to write 2C, because you have no idea what C is. It would be weird to say "2 times larger than a number we'll never know."

Solve for y.

$$y = \pm\sqrt{3x^2 - 2x + C}$$

Note: Problems 25.4–25.5 refer to the differential equation $\frac{dy}{dx} = \frac{3x-1}{y}$.

25.5 Find the specific solution of the differential equation that has x-intercept $(-1,0)$.

According to Problem 25.4, the general solution of the differential equation is $y = \pm\sqrt{3x^2 - 2x + C}$. Substitute $x = -1$ and $y = 0$ into that equation and solve for C.

$$0 = \pm\sqrt{3(-1)^2 - 2(-1) + C}$$
$$0 = \pm\sqrt{5+C}$$
$$0 = 5+C$$
$$-5 = C$$

Mathematicians and engineers who deal with differential equations for a living might cringe at this solution, because it has a ± in it and the graph has two branches. However, this isn't a differential equations course, so don't sweat it.

Substitute C into the general solution of get the specific solution $y = \pm\sqrt{3x^2 - 2x - 5}$. Note that this solution may differ slightly from the solution reported in a differential equations course, in which the answers are typically restricted to single, continuous intervals, but it will suffice for the scope of this course.

Note: Problems 25.6–25.8 refer to the differential equation $\dfrac{dy}{dx} = \dfrac{x}{y}$.

25.6 Describe the family of solutions of the differential equation.

Cross multiply to separate the variables and antidifferentiate both sides of the resulting equation.

$$y\,dy = x\,dx$$
$$\int y\,dy = \int x\,dx$$
$$\frac{y^2}{2} = \frac{x^2}{2} + C$$

Multiply the equation by 2 to eliminate fractions.

$$y^2 = x^2 + C$$

This solution represents a family of hyperbolas centered at the origin with transverse and conjugate axes of equal length. To verify this, isolate the constant and divide each term by C—the result is a hyperbola in standard form.

$$\frac{y^2}{C} - \frac{x^2}{C} = \frac{C}{C}$$
$$\frac{y^2}{C} - \frac{x^2}{C} = 1$$

Look at Problem 6.29 if you need help understanding the standard form of a hyperbola.

Note: Problems 25.6–25.8 refer to the differential equation $\dfrac{dy}{dx} = \dfrac{x}{y}$.

25.7 Graph the specific solution of the differential equation that passes through the point $(-1,2)$.

Substitute $x = -1$ and $y = 2$ into the general solution identified by Problem 25.6.

$$\frac{y^2}{C} - \frac{x^2}{C} = 1$$
$$\frac{2^2}{C} - \frac{(-1)^2}{C} = 1$$
$$\frac{4-1}{C} = 1$$
$$C = 3$$

Substitute $C = 3$ into the general solution of determine the specific solution (illustrated in Figure 25-1) that contains the point $(-1,2)$.

$$\frac{y^2}{C} - \frac{x^2}{C} = 1$$
$$\frac{y^2}{3} - \frac{x^2}{3} = 1$$

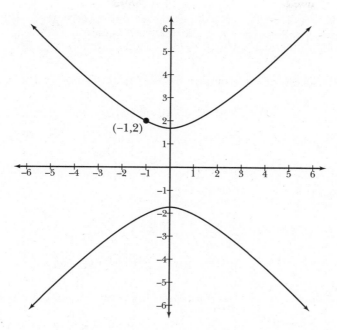

Figure 25-1 *The hyperbola* $\dfrac{y^2}{3} - \dfrac{x^2}{3} = 1$ *is centered at (0,0), has a vertical transverse axis of length* $2\sqrt{3}$*, has a horizontal conjugate axis of length* $2\sqrt{3}$*, and passes through the point (−1,2).*

Note: Problems 25.6–25.8 refer to the differential equation $\dfrac{dy}{dx} = \dfrac{x}{y}$*.*

25.8 Verify that $\dfrac{y^2}{3} - \dfrac{x^2}{3} = 1$ is a solution of the differential equation by demonstrating that the derivative of the solution is the differential equation.

Need help?
Look at Problem
16.9.

Differentiate implicitly.

$$\frac{2y}{3} \cdot \frac{dy}{dx} - \frac{2x}{3} = 0$$

$$\frac{2y}{3} \cdot \frac{dy}{dx} = \frac{2x}{3}$$

$$\left(\frac{3}{2y}\right)\left(\frac{2y}{3} \cdot \frac{dy}{dx}\right) = \frac{2x}{3}\left(\frac{3}{2y}\right)$$

$$\frac{dy}{dx} = \frac{6x}{6y}$$

$$\frac{dy}{dx} = \frac{x}{y}$$

25.9 Find the general solution of the differential equation: $\dfrac{dy}{dx} = xy(x+3)$.

Divide both sides of the equation by y and multiply both sides by dx.

$$\frac{dy}{y} = x(x+3)\,dx$$

$$\frac{dy}{y} = \left(x^2 + 3x\right)dx$$

Integrate both sides of the equation.

$$\int \frac{dy}{y} = \int \left(x^2 + 3x\right)dx$$

$$\ln|y| = \frac{x^3}{3} + \frac{3x^2}{2} + C$$

> *Drop the absolute values—e to any power is a positive number.*

Solve the equation for y by exponentiating both sides of the equation.

$$e^{\ln|y|} = e^{\left(x^3/3\right)+\left(3x^2/2\right)+C}$$

$$y = e^{\left(x^3/3\right)+\left(3x^2/2\right)+C}$$

Apply the exponential property that states $x^{a+b} = x^a x^b$. Rather than writing each term in the exponential sum separately with base e, transform only the last exponential term, C, into its own factor.

> *If C is an unknown number, then e^c is another unknown number. You might as well keep writing the unknown value as C.*

$$y = e^{\left(x^3/3\right)+\left(3x^2/2\right)}e^C$$

$$y = Ce^{\left(x^3/3\right)+\left(3x^2/2\right)}$$

25.10 Find the general solution of the differential equation: $\dfrac{dy}{dx} = \dfrac{ye^x}{e^{2x}+5}$.

Divide both sides of the equation by y and multiply both sides by dx to separate the variables.

$$\frac{dy}{y} = \frac{e^x}{e^{2x}+5}\,dx$$

Integrate both sides of the equation and isolate $\ln|y|$. Use an inverse trigonometric integration formula and variable substitution to antidifferentiate the right side of the equation.

> *The inverse tangent formula*
> $$\int \frac{du}{u^2+a^2} =$$
> $$= \frac{1}{a}\arctan\frac{u}{a} + C$$
> *with $u = e^x$ and $a = \sqrt{5}$.*

$$\int \frac{dy}{y} = \int \frac{e^x}{e^{2x}+5}\,dx$$

$$\ln|y| = \frac{1}{\sqrt{5}}\arctan\left(\frac{e^x}{\sqrt{5}}\right) + C$$

Exponentiate both sides of the equation to solve for y.

$$e^{\ln|y|} = e^{\left(1/\sqrt{5}\right)\arctan\left(e^x/\sqrt{5}\right)+C}$$

$$|y| = e^{\left(1/\sqrt{5}\right)\arctan\left(e^x/\sqrt{5}\right)} \cdot e^C$$

$$y = Ce^{\left(1/\sqrt{5}\right)\arctan\left(e^x/\sqrt{5}\right)}$$

Rationalize the exponent: $y = Ce^{\left(\sqrt{5}/5\right)\arctan\left(e^x\sqrt{5}/5\right)}$.

25.11 Identify $f(x)$, given $f''(x) = 2x - \cos x$, $f'(\pi) = 0$, and $f(0) = \dfrac{\pi}{2}$.

Integrate $f''(x)$ with respect to x to get $f'(x)$.

$$f'(x) = \int (2x - \cos x)\,dx = x^2 - \sin x + C$$

Substitute $x = \pi$ into $f'(x)$ and recall that $f'(\pi) = 0$ to determine the value of C.

$$f'(\pi) = \pi^2 - \sin \pi + C$$

$$0 = \pi^2 - 0 + C$$

$$-\pi^2 = C$$

Therefore, $f'(x) = x^2 - \sin x - \pi^2$. Integrate $f'(x)$ to find $f(x)$.

$$f(x) = \int \left(x^2 - \sin x - \pi^2\right)dx = \frac{x^3}{3} + \cos x - \pi^2 x + C$$

> Any constant b has an integral of bx. Because $-\pi^2$ is a constant, its integral is $-\pi^2 x$.

Recall that $f(0) = \dfrac{\pi}{2}$.

$$f(0) = \frac{(0)^3}{3} + \cos 0 - \pi^2(0) + C$$

$$\frac{\pi}{2} = 1 + C$$

$$\frac{\pi}{2} - 1 = C$$

Therefore, $f(x) = \dfrac{x^3}{3} + \cos x - \pi^2 x + \dfrac{\pi}{2} - 1$.

Exponential Growth and Decay
When a population's change is proportional to its size

25.12 Assume y is proportional to its rate of change $\dfrac{dy}{dt}$: $\dfrac{dy}{dt} = k \cdot y$ (where k is a nonzero real number). Find the general solution of the differential equation.

Divide both sides of the equation by y and multiply both sides by dt to separate the variables.

> The variables are y and t (according to the derivative dy/dt), even though there's no t in the equation. Keep k on the t side of the equation, because you'll eventually solve for y and would have to move it anyway.

$$\frac{dy}{y} = k\,dt$$

$$\int \frac{dy}{y} = k \int dt$$

$$\ln|y| = kt + C$$

Exponentiate both sides of the equation.

$$e^{\ln|y|} = e^{kt+C}$$

Apply an exponential property, as explained by Problem 25.9: $e^{kt+C} = e^{kt}e^{C}$.

$$e^{\ln|y|} = e^{kt}e^{C}$$

$$y = Ce^{kt}$$

Note: Problems 25.13–25.15 refer to a scientific experiment, at the start of which a scientist observes 125 bacterial colonies growing in the agar of a Petri dish. Exactly six hours later, the number of colonies has grown to 190.

25.13 Construct a mathematical model that describes the number of bacterial colonies present t hours after the start of the experiment, assuming exponential growth.

> Exponential growth means the bigger the population, the faster it grows. Exponential decay is the other side of the coin—sometimes the bigger something is, the faster it decays.

A population y experiences exponential growth if and only if y is proportional to $\frac{dy}{dt}$. According to Problem 25.12, populations experiencing exponential growth are modeled by the equation $y = Ce^{kt}$, where C is the original population, t is elapsed time, and y is the population after time t has elapsed. Determine the value of k based upon the given information.

The original bacterial population is $C = 125$; and after $t = 6$ hours, the population has grown to $y = 190$. Substitute these values into the exponential growth equation.

$$y = Ce^{kt}$$

$$190 = 125e^{k(6)}$$

Isolate e^{6k} on one side of the equation.

$$\frac{190}{125} = e^{6k}$$

$$\frac{38}{25} = e^{6k}$$

Take the natural logarithm of both sides of the equation and solve for k.

$$\ln \frac{38}{25} = \ln e^{6k}$$

$$\ln \frac{38}{25} = 6k$$

$$\frac{\ln(38/25)}{6} = k$$

It is often useful to express k as a decimal: $k \approx 0.06978505581$. Therefore, t hours after the experiment begins, there are approximately $y = 125e^{0.06978505581(t)}$ bacterial colonies in the Petri dish.

If you're going to write k as a decimal, DON'T ROUND IT—use as many decimals as your calculator can give you. Otherwise, your answers might be inaccurate.

Note: Problems 25.13–25.15 refer to a scientific experiment, at the start of which a scientist observes 125 bacterial colonies growing in the agar of a Petri dish. Exactly six hours later, the number of colonies has grown to 190.

25.14 Approximately how many bacterial colonies will be present exactly one day after the experiment begins? Round the answer to the nearest integer.

According to Problem 25.13, there are approximately $y = 125e^{0.06978505581(t)}$ colonies t hours after the experiment begins. Substitute $t = 24$ into the equation to determine how many colonies are present after 24 hours (i.e., one day).

$$y = 125e^{0.06978505581(24)} = 125e^{1.67484133943} \approx 667.244$$

There are approximately 667 bacterial colonies exactly 24 hours after the experiment begins.

Note: Problems 25.13–25.15 refer to a scientific experiment, at the start of which a scientist observes 125 bacterial colonies growing in the agar of a Petri dish. Exactly six hours later, the number of colonies has grown to 190.

25.15 Assuming that resources are not a limiting issue to growth, how many hours will it take the total bacteria population to reach 20,000? Round the answer to the nearest integer.

This is an implicit assumption in all exponential growth problems. If the bacteria eventually run out of food and space (because the population is too big for the Petri dish), the exponential growth model isn't accurate.

Apply the exponential growth model from Problem 25.13. Set $y = 20{,}000$ and isolate the natural exponential expression.

$$20{,}000 = 125e^{0.06978505581(t)}$$

$$\frac{20{,}000}{125} = e^{0.06978505581(t)}$$

$$160 = e^{0.06978505581(t)}$$

Take the natural logarithm of both sides of the equation and solve for t.

$$\ln 160 = \ln e^{0.06978505581(t)}$$

$$\ln 160 = 0.06978505581(t)$$

$$\frac{\ln 160}{0.06978505581} = t$$

$$72.726 \approx t$$

The number of bacteria colonies will reach 20,000 approximately 73 hours after the experiment begins.

Note: Problems 25.16–25.18 refer to the radioactive isotope Carbon-14, which has a half-life of 5,730 years.

25.16 Given an initial mass M of Carbon-14, construct a function that models the remaining mass of the isotope after t years have elapsed.

You can always assume that half-life problems involve exponential decay, so it's safe to use the formula $y = Ce^{kt}$ from Problem 25.12.

If an isotope has a half-life of h years, then every h years its mass is halved as its isotopes gradually decay into stable "daughter" material. In this example, the initial mass M of C-14 will decay to a mass of $\frac{M}{2}$ after 5,730 years. Apply the exponential decay formula and determine the value of k.

$$y = Ce^{kt}$$
$$\frac{M}{2} = Me^{k(5,730)}$$
$$\left(\frac{1}{M}\right)\frac{M}{2} = \left(\frac{1}{M}\right)Me^{k(5,730)}$$
$$\frac{1}{2} = e^{5,730k}$$
$$\ln\frac{1}{2} = 5,730k$$
$$\frac{\ln(1/2)}{5,730} = k$$

Exponential growth and decay both use the same formula $(y = Ce^{kt})$, but k is positive in growth problems and negative in decay problems.

Use a calculator to estimate k: $k \approx -0.000120968094$. Substitute this value and $C = M$ into the exponential decay formula.

$$y = Me^{-0.000120968094(t)}$$

Note: Problems 25.16–25.18 refer to the radioactive isotope Carbon-14, which has a half-life of 5,730 years.

25.17 Given 300 grams of C-14, what is the remaining mass of the isotope after 250 years have elapsed? Report an answer accurate to three decimal places.

Substitute $M = 300$ and $t = 250$ into the function constructed in Problem 25.16 and calculate y.

$$y = 300e^{-0.000120968094(250)}$$
$$y = 300e^{-0.030242023585}$$
$$y \approx 291.063 \text{ grams}$$

Note: Problems 25.16–25.18 refer to the radioactive isotope Carbon-14, which has a half-life of 5,730 years.

25.18 Given N grams of C-14, in approximately how many years will the mass decay to one third its original measure? Round the answer to the nearest integer.

Apply the exponential decay model from Problem 25.16, such that $M = N$ and $y = \dfrac{N}{3}$. Solve for t.

$$\frac{N}{3} = Ne^{-0.000120968094(t)}$$

$$\frac{\cancel{N}}{3\cancel{N}} = e^{-0.000120968094(t)}$$

$$\ln\frac{1}{3} = -0.000120968094(t)$$

$$t = \frac{\ln(1/3)}{-0.000120968094}$$

$$t \approx 9{,}081.835 \approx 9{,}028 \text{ years}$$

25.19 According to Newton's law of cooling, the rate at which an object cools is proportional to the difference between its temperature T and the temperature of the ambient environment T_A. Write a differential equation that expresses this relationship.

The "ambient environment" is the surrounding temperature. For example, if you're inside, the ambient temperature is room temperature.

The rate of change of the object's temperature is $\dfrac{dT}{dt}$, and the difference between its temperature and the ambient temperature is $T - T_A$. Two values are proportional if they are equal when one is multiplied by a constant of proportionality k.

$$\frac{dT}{dt} = k(T - T_A)$$

The object has to cool down for this formula to work (it is Newton's law of COOLING, after all) so T has to be greater than T_A.

25.20 Solve the differential equation in Problem 25.19 to generate a model for the temperature of the object after time t has elapsed. Let T_o be the temperature of the object when $t = 0$.

The variables in this equation are T and t; T_A is constant. Separate them by multiplying both sides of the equation by dt and dividing both sides by $T - T_A$.

$$\frac{dT}{T - T_A} = k\,dt$$

Integrate both sides of the equation and solve for T.

$$\int \frac{dT}{T - T_A} = k \int dt$$

$$\ln|T - T_A| = kt + C$$

$$e^{\ln|T - T_A|} = e^{kt + C}$$

$$T - T_A = e^{kt} e^C$$

$$T = T_A + Ce^{kt}$$

As my note on Problem 25.19 explains, $T > T_A$. Since $T - T_A$ is always positive, you don't need absolute value signs here.

Recall that $T = T_O$ when $t = 0$. Substitute these values into the equation and solve for C.

$$T_O = T_A + Ce^{k(0)}$$

$$T_O = T_A + Ce^0$$

$$T_O = T_A + C \cdot 1$$

$$T_O - T_A = C$$

Substitute C into the solution of the differential equation to generate the formula for Newton's law of cooling: $T = T_A + (T_O - T_A)e^{kt}$.

25.21 An uninsulated cup of coffee cools from 185°F to 150°F exactly four minutes after it is served. Assuming the coffee shop maintains a constant room temperature of 75°F, how much *additional* time will it take for the coffee to cool to 95°F? Report an answer accurate to three decimal places.

Apply Newton's law of cooling, substituting $T_O = 185$, $T_A = 75$, $t = 4$, and $T = 150$ into the formula from Problem 25.20. Solve for k.

$$T = T_A + \left(T_O - T_A\right)e^{kt}$$

$$150 = 75 + (185 - 75)e^{k(4)}$$

$$150 - 75 = (110)e^{4k}$$

$$\frac{75}{110} = e^{4k}$$

$$\ln\frac{15}{22} = 4k$$

$$\frac{\ln(15/22)}{4} = k$$

$$-0.095748063064 \approx k$$

Substitute k, T_A, and T_O into the formula for Newton's law of cooling to complete the model describing the temperature T of the coffee t minutes after it is served.

$$T = 75 + (185 - 75)e^{-0.095748063064(t)}$$

$$T = 75 + 110e^{-0.095748063064(t)}$$

Substitute $T = 95$ into the model and solve for t to determine when the coffee's temperature will cool to 95°.

$$95 = 75 + 110e^{-0.095748063064(t)}$$

$$\frac{20}{110} = e^{-0.095748063064(t)}$$

$$\ln\frac{2}{11} = -0.095748063064(t)$$

$$t = \frac{\ln(2/11)}{-0.095748063064}$$

$$t \approx 17.805$$

Approximately 17.805 minutes after the coffee is served, its temperature is 95°F, so the solution is $17.805 - 4 = 13.805$ minutes. ←

The question is kind of tricky. It asks how much ADDITIONAL time (AFTER the four-minute cool down from 185°F to 150°F) it takes the coffee to cool down to 95°F.

25.22 The housekeeping staff in a hotel discovers a corpse, the apparent victim of a fatal overnight heart attack, and alerts the police. Investigators arrive at 11 a.m. and note that the body's temperature is 81°F; by 12:30 p.m., the temperature has dropped to 77°F. Noting that the thermostat in the room is set to maintain a constant temperature of 67°F, and assuming that the victim's temperature was 98.6°F when he died, at what time (to the nearest minute) did the fatal heart attack occur?

Apply Newton's law of cooling, such that $T_O = 81$, $T_A = 67$, $T = 77$, and $t = 1.5$ hours.

$$T = T_A + (T_O - T_A)e^{kt}$$

$$77 = 67 + (81 - 67)e^{1.5(k)}$$

$$10 = 14e^{1.5(k)}$$

$$\ln\frac{5}{7} = 1.5k$$

$$\frac{\ln(5/7)}{1.5} = k$$

$$-0.224314824414 \approx k$$

Substitute k into the original formula to construct a function that models the temperature T of the corpse exactly t hours after 11 a.m.

$$T = 67 + (81 - 67)e^{-0.224314824414(t)}$$

$$= 67 + 14e^{-0.224314824414(t)}$$

To determine the time of death, substitute $T = 98.6$ into the mathematical model and solve for t.

$$98.6 = 67 + 14e^{-0.224314824414(t)}$$

$$31.6 = 14e^{-0.224314824414(t)}$$

$$\frac{31.6}{14} = e^{-0.224314824414(t)}$$

$$\ln\frac{31.6}{14} = -0.224314824414(t)$$

$$t = \frac{\ln(31.6/14)}{-0.224314824414}$$

$$t \approx -3.62927324622$$ ←

The person died before 11 a.m., which explains the negative answer for t. Instead of counting hours AFTER 11 a.m., count backwards from 11 a.m.

Therefore, the time of death was approximately 3.62927324622 hours before 11 a.m. Multiply the decimal portion of the number by 60 to convert to minutes.

$$60(0.62927324622) \approx 37.756 \text{ minutes}$$

The victim died roughly 3 hours and 38 minutes before investigators arrived, at approximately 7:22 a.m.

Linear Approximations

A graph and its tangent line look alike near the tangent point

25.23 Explain what is meant by "local linearity."

> Here, "similar" means "almost equal." Don't get confused with the geometric interpretation of "similar," which means "two shapes are the same but one is just a bigger version of the other."

A curve and its tangent line have very similar values near the point of tangency. In fact, if a small enough x-interval is chosen around a point of tangency, $f(x)$ resembles a straight line over that interval, as illustrated in Figure 25-2. The practical application of local linearity is the approximation of function values near a point of tangency using the equation of the tangent line rather than the function itself.

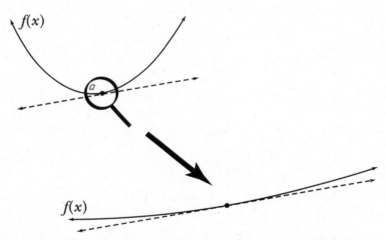

Figure 25-2 *In the region immediately surrounding a point of tangency on f(x) (indicated by the magnifying glass in the illustration), the tangent line and the graph of f(x) look remarkably similar. The higher you increase the magnification (and thus the smaller the interval around the point of tangency), the more closely f(x) will resemble the tangent line.*

25.24 Estimate $\sqrt{16.1}$ using a linear approximation.

According to Problem 25.23, the graph of $f(x) = \sqrt{x}$ and its tangent line have very similar values near a point of tangency. Therefore, $f(x)$ and the tangent line to $f(x)$ at $x = 16$ will have similar values at $x = 16.1$. Because $f'(x) = \frac{1}{2}x^{-1/2} = \frac{1}{2\sqrt{x}}$, the slope of the tangent line to $f(x)$ at $x = 16$ is $\frac{1}{2\sqrt{16}} = \frac{1}{8}$. The point of tangency is $(x, f(x)) = \left(16, \sqrt{16}\right) = (16,4)$. Plug those values into point-slope form to generate the equation of the tangent line.

$$y - 4 = \frac{1}{8}(x - 16)$$

Substitute $x = 16.1$ into the equation of the tangent line to estimate $\sqrt{16.1}$.

$$y - 4 = \frac{1}{8}(16.1 - 16)$$

$$y - 4 = \frac{1}{8}(0.1)$$

$$y - 4 = \frac{1}{8}\left(\frac{1}{10}\right)$$

$$y = 4 + \frac{1}{80}$$

$$y = \frac{321}{80}$$

Therefore, $\sqrt{16.1} \approx \frac{321}{80} \approx 4.0125$. Note that the actual value of $\sqrt{16.1}$ is 4.01248053..., so the approximation is quite accurate.

25.25 Estimate ln 1.05 using a linear approximation.

Use the equation of the tangent line to $f(x) = \ln x$ at $x = 1$ to approximate ln 1.05. Note that $f'(x) = \frac{1}{x}$, so the slope of the tangent line is 1; the point of tangency is $(1, \ln 1) = (1,0)$. Apply the point-slope formula.

$$y - 0 = 1(x - 1)$$
$$y = x - 1$$

The actual value of ln 1.05 is 0.0487901642....

Substitute $x = 1.05$ into the equation of the tangent line to approximate ln 1.05.

$$y = 1.05 - 1 = 0.05$$

Therefore, ln 1.05 ≈ 0.05.

25.26 Estimate arctan 0.85 using a linear approximation.

Because differentiable functions exhibit local linearity near a point of tangency, $f(x) = \arctan x$ and the tangent line to $f(x)$ at $x = 1$ should have similar values. Differentiate $f(x)$ and evaluate $f'(1)$.

$$f'(x) = \frac{1}{1+x^2}$$

$$f'(1) = \frac{1}{1+1^2} = \frac{1}{2}$$

The point of tangency is $(1, \arctan 1) = \left(1, \dfrac{\pi}{4}\right)$ and the slope of the tangent line is $f'(1) = \dfrac{1}{2}$. Apply the point-slope formula to write the equation of the tangent line.

$$y - \frac{\pi}{4} = \frac{1}{2}(x-1)$$

Substitute $x = 0.85$ into the equation to approximate arctan 0.85.

$$y - \frac{\pi}{4} = \frac{1}{2}(0.85-1)$$

$$y = \frac{\pi}{4} + \frac{1}{2}(-0.15)$$

$$y = \frac{\pi}{4} + \frac{1}{2}\left(-\frac{15}{100}\right)$$

$$y = \frac{50\pi - 15}{200}$$

This ugly fraction equals 0.710398... and the actual value of arctan 0.85 is 0.704494. Not a bad estimate, even though it's not pretty.

Slope Fields
They look like wind patterns on a weather map

25.27 Explain how to create a slope field.

Make it big enough to see but small enough so that it doesn't intersect any of the other segments in the slope field.

Select a coordinate (a,b) on the coordinate plane. Substitute $x = a$ and $y = b$ into the differential equation to find the slope of the tangent line to the solution curve that passes through (a,b) and draw a small line segment centered at (a,b) with that slope. Continue this process at other points on the coordinate plane until you can visualize the family of solution curves to the differential equation.

Note: Problems 25.28–25.29 refer to the differential equation $\frac{dy}{dx} = 2x$.

25.28 Draw a slope field at the coordinates indicated in Figure 25-3.

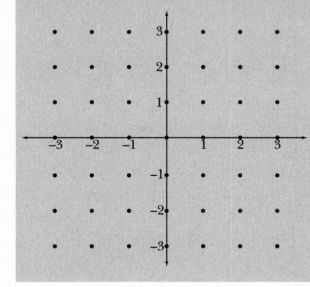

Figure 25-3

When constructing the slope field for $\frac{dy}{dx} = 2x$, include segments passing through each of these points.

All points on the y-axis will have horizontal segments associated with them—if $x = 0$, then $\frac{dy}{dx} = 2(0) = 0$, regardless of the y-value at each coordinate. In fact, every segment in this slope field has a slope equal to twice the x-value of the coordinate at which it is centered.

That's exactly what the differential equation is saying: "dy/dx" (the slope of the tangent line) "=" (is) "2x" (two times the x-value).

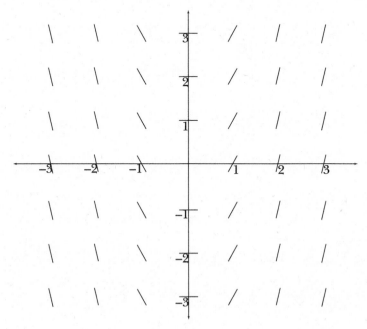

Figure 25-4 *The slope field for the differential equation $\frac{dy}{dx} = 2x$.*

Note: Problems 25.28–25.29 refer to the differential equation $\frac{dy}{dx} = 2x$.

25.29 Determine the specific solution of the differential equation that passes through (1,1) and graph it on the same coordinate axes as the slope field.

Separate the variables and integrate both sides of the equation.

$$\left(\frac{dy}{dx}\right)dx = (2x)\,dx$$

$$dy = 2x\,dx$$

$$\int dy = 2\int x\,dx$$

$$y = 2\cdot\frac{x^2}{2} + C$$

$$y = x^2 + C$$

Solve for C when $x = 1$ and $y = 1$.

$$1 = 1^2 + C$$

$$C = 0$$

The specific solution is $y = x^2$, illustrated in Figure 25-5.

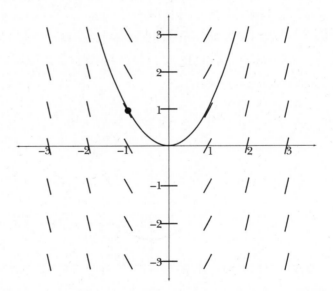

This is the same differential equation from Problems 25.7 and 25.8.

Figure 25-5 *The graph of $y = x^2$ is the specific solution of the differential equation $\frac{dy}{dx} = 2x$ that passes through the point $(-1,1)$.*

Note: Problems 25.30–25.31 refer to the differential equation $\frac{dy}{dx} = \frac{x}{y}$.

25.30 Draw the slope field for the differential equation.

The solution in Figure 25-6 is generated by a computer; hand-drawn slope fields typically contain fewer segments at more predictable coordinates. While your slope field need not mirror the number of segments in Figure 25-6 or their exact locations, it should contain a sufficient number of segments to accurately visualize the solution.

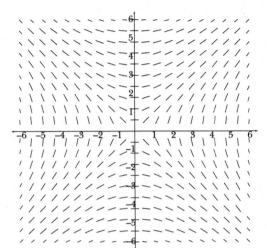

RULE OF THUMB: Keep drawing segments (in all four quadrants) until you can predict what the next segment will be without actually plugging in the numbers. Once you get that "feel" for the graph, you've probably drawn enough segments.

Figure 25-6 *The slope field for the differential equation $\dfrac{dy}{dx} = \dfrac{x}{y}$.*

Note: Problems 25.30–25.31 refer to the differential equation $\dfrac{dy}{dx} = \dfrac{x}{y}$.

25.31 Graph the specific solution $\dfrac{y^2}{3} - \dfrac{x^2}{3} = 1$ of the differential equation $\dfrac{dy}{dx} = \dfrac{x}{y}$ on the same coordinate plane as the slope field generated by Problem 25.30.

The graph of $\dfrac{y^2}{3} - \dfrac{x^2}{3} = 1$, illustrated in Figure 25-7, is a hyperbola with transverse and conjugate axes that are the same length: $2\sqrt{3}$.

If you did Problem 25.7, you already graphed this hyperbola, but it's worth graphing again to see how it's shaped by the slope field of $\dfrac{dy}{dx} = \dfrac{x}{y}$.

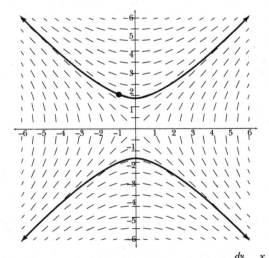

Figure 25-7 *The slope field for the differential equation $\dfrac{dy}{dx} = \dfrac{x}{y}$ and the specific solution that passes through the point (−1,2):*
$$\dfrac{y^2}{3} - \dfrac{x^2}{3} = 1.$$

25.32 Draw the slope field for the differential equation: $\dfrac{dy}{dx} = y - x$.

Notice that $\dfrac{dy}{dx} = 0$ for all points on the line $y = x$; the slope field is illustrated in Figure 25-8.

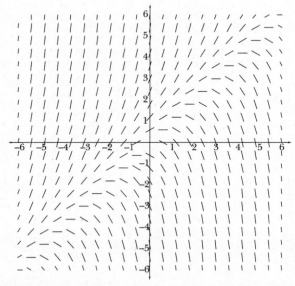

Figure 25-8 *The segments in the slope field for the differential equation*
$\dfrac{dy}{dx} = y - x$ *get steeper as the distance away from the line* $y = x$
gets larger.

25.33 Draw the slope field for the differential equation $\dfrac{dy}{dx} = \dfrac{x + y}{x}$.

The slope field is illustrated in Figure 25-9. Notice that $\dfrac{dy}{dx}$ is undefined along the y-axis (because $x = 0$ for all such points) and $\dfrac{dy}{dx} = 0$ (resulting in horizontal slopes) for all points on the line $y = -x$.

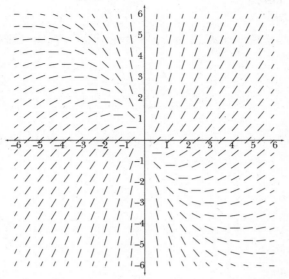

Figure 25-9 *The slope field for the differential equation $\frac{dy}{dx} = \frac{x+y}{x}$.*

25.34 Draw the slope field for the differential equation $\frac{dy}{dx} = x^2 y^2$.

The slope field illustrated in Figure 25-10 contains horizontal segments along the x- and y-axes.

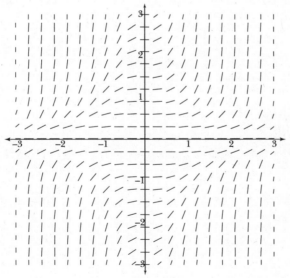

Figure 25-10 *The slope field for the differential equation $\frac{dy}{dx} = x^2 y^2$.*

Euler's Method
Take baby steps to find the differential equation's solution

25.35 Given a line with slope $\dfrac{3}{4}$ that contains points $(-1,3)$ and $\left(-\dfrac{1}{2}, d\right)$, find d.

The symbol "Δ" is called "delta" and means "change in." This is a formula from way back in beginning algebra: the slope of a line is defined as the change in y divided by the change in x.

The slope of a line is defined as the quotient of its vertical and horizontal change: $m = \dfrac{\Delta y}{\Delta x} = \dfrac{y_2 - y_1}{x_2 - x_1}$. Calculate Δx.

$$\Delta x = x_2 - x_1 = -\frac{1}{2} - (-1) = \frac{1}{2}$$

Substitute $\Delta x = \dfrac{1}{2}$ and $m = \dfrac{3}{4}$ (the slope of the line containing the points) into the slope formula.

$$m = \frac{\Delta y}{\Delta x}$$
$$\frac{3}{4} = \frac{\Delta y}{1/2}$$

Multiply both sides by $\dfrac{1}{2}$ to solve for Δy.

$$\frac{1}{2} \cdot \frac{3}{4} = \frac{\cancel{1}}{\cancel{2}}\left(\frac{\Delta y}{1/2}\right)$$
$$\frac{3}{8} = \Delta y$$

Here's what all this means. To travel from the point (-1,3) to the point $\left(-\dfrac{1}{2}, d\right)$, you have to go $\Delta x = \dfrac{1}{2}$ of a unit right and $\Delta y = \dfrac{27}{8}$ units up.

Note that $\Delta y = y_2 - y_1$; substitute $\Delta y = \dfrac{3}{8}$, $y_1 = 3$, and $y_2 = d$ into the equation and solve for d.

$$\Delta y = y_2 - y_1$$
$$\frac{3}{8} = d - 3$$
$$\frac{27}{8} = d$$

25.36 Given a line with slope $-\dfrac{1}{5}$ that contains points $(2,-9)$ and $(1,p)$, find p.

Note that $\Delta x = x_2 - x_1 = 1 - 2 = -1$. Use that value and the slope $m = -\dfrac{1}{5}$ to determine Δy.

$$m = \frac{\Delta y}{\Delta x}$$
$$-\frac{1}{5} = \frac{\Delta y}{-1}$$
$$\frac{1}{5} = \Delta y$$

Calculate p.

$$\Delta y = y_2 - y_1$$

$$\frac{1}{5} = p - (-9)$$

$$\frac{1}{5} - 9 = p$$

$$-\frac{44}{5} = p$$

"Euler" rhymes with "boiler," not "fuel-er."

25.37 Describe the role of Euler's method in the investigation of elementary differential equations.

Euler's method is used to approximate y-values on the graphs of specific solutions of differential equations, particularly when those specific solutions cannot be determined easily. The technique is based upon the principle of local linearity, and uses the equations of tangent lines near points of tangency (calculated at fixed values of Δx) to approximate the values of the function to which the lines are tangent.

If you can't actually separate variables to solve a differential equation but know that the solution contains, let's say, the point (3,4), then you can take a fixed number of "baby steps" using dy/dx to guess what other nearby points on the solution graph are.

25.38 Use Euler's method with two steps of width $\Delta x = \frac{1}{2}$ to approximate $y(4)$ if $\frac{dy}{dx} = \frac{x + y}{x}$ and the point $(3,0)$ belongs to the graph of the specific solution of the differential equation.

Apply Euler's method twice because the problem calls for two steps of width $\Delta x = \frac{1}{2}$. In order to reach $x = 4$ from $x = 3$ (the x-value of the known point on the solution graph) in two steps, those steps must have width $\Delta x = \frac{4 - 3}{2} = \frac{1}{2}$.

Step one: From point (3,0) to point (3 + Δx, 0 + Δy).

Calculate $\frac{dy}{dx}$ at the point $(3,0)$ by substituting $x = 3$ and $y = 0$ into the differential equation.

$$\frac{dy}{dx} = \frac{x + y}{x}$$

$$\frac{dy}{dx} = \frac{3 + 0}{3}$$

$$\frac{dy}{dx} = 1$$

The y-value of the point on the solution graph when x = 4.

Therefore, the slope m of the tangent line to the solution graph is 1 when $x = 3$, Use the method described in Problem 25.35 to determine Δy.

$$m = \frac{\Delta y}{\Delta x}$$

$$1 = \frac{\Delta y}{1/2}$$

$$\frac{1}{2} = \Delta y$$

Substitute $\Delta x = \Delta y = \dfrac{1}{2}$ into the destination point $(3 + \Delta x, 0 + \Delta y)$.

$$\left(3 + \frac{1}{2}, 0 + \frac{1}{2}\right) = \left(\frac{7}{2}, \frac{1}{2}\right)$$

> The whole point of step one is to figure out where you'll end up if you go Δx units right and Δy units up or down from the point (3,0) that you were given. Turns out you'll end up at $\left(\dfrac{7}{2}, \dfrac{1}{2}\right)$.

Begin at this point instead of $(3,0)$ for the second and final step of width $\Delta x = \dfrac{1}{2}$.

Step two: From point $\left(\dfrac{7}{2}, \dfrac{1}{2}\right)$ *to* $\left(\dfrac{7}{2} + \Delta x, \dfrac{1}{2} + \Delta y\right)$.

Evaluate $\dfrac{dy}{dx} = \dfrac{x + y}{x}$ when $x = \dfrac{7}{2}$ and $y = \dfrac{1}{2}$.

$$\frac{dy}{dx} = \frac{(7/2) + (1/2)}{7/2}$$

$$\frac{dy}{dx} = \frac{8/2}{7/2}$$

$$\frac{dy}{dx} = \frac{8}{7}$$

Calculate the corresponding value of Δy.

$$m = \frac{\Delta y}{\Delta x}$$

$$\frac{8}{7} = \frac{\Delta y}{1/2}$$

$$\frac{8}{14} = \Delta y$$

$$\frac{4}{7} = \Delta y$$

Substitute Δx and Δy into the coordinate $\left(\dfrac{7}{2} + \Delta x, \dfrac{1}{2} + \Delta y\right)$ to determine $y(4)$.

$$(4, y(4)) \approx \left(\frac{7}{2} + \Delta x, \frac{1}{2} + \Delta y\right)$$

$$\approx \left(\frac{7}{2} + \frac{1}{2}, \frac{1}{2} + \frac{4}{7}\right)$$

$$\approx \left(4, \frac{15}{14}\right)$$

> Start at (0,−2), and take three steps to the right (each one $\Delta x = 1/3$ units long) to reach some point (1, ?). During each step, plug the current point (x,y) into dy/dx, use that value for m, and solve $m = \dfrac{\Delta y}{\Delta x}$ for Δy.

Therefore, $y(4) \approx \dfrac{15}{14}$.

25.39 Use Euler's method with three steps of width $\Delta x = \dfrac{1}{3}$ to approximate $y(1)$ if $\dfrac{dy}{dx} = xy$ and the y-intercept of the solution of the differential equation is $(0, -2)$.

Step one: From point (0,−2) to $\left(\dfrac{1}{3}, -2 + \Delta y\right)$.

Calculate $\dfrac{dy}{dx}$ when $x = 0$ and $y = -2$.

$$\frac{dy}{dx} = xy = (0)(-2) = 0$$

Determine the value of Δy when $\Delta x = \dfrac{1}{3}$ and $m = 0$.

$$m = \frac{\Delta y}{\Delta x}$$

$$0 = \frac{\Delta y}{1/3}$$

$$0 = \Delta y$$

Therefore, step one ends at the point $\left(\dfrac{1}{3}, -2\right)$.

Step two: From point $\left(\dfrac{1}{3}, -2\right)$ *to* $\left(\dfrac{2}{3}, -2 + \Delta y\right)$.

Plug $\Delta y = 0$ into the point $\left(\dfrac{1}{3}, -2 + \Delta y\right)$ from the beginning of step one.

Calculate $\dfrac{dy}{dx}$ when $x = \dfrac{1}{3}$ and $y = -2$.

$$\frac{dy}{dx} = xy = \left(\frac{1}{3}\right)(-2) = -\frac{2}{3}$$

Determine the value of Δy when $\Delta x = \dfrac{1}{3}$ and $m = -\dfrac{2}{3}$.

$$m = \frac{\Delta y}{\Delta x}$$

$$-\frac{2}{3} = \frac{\Delta y}{1/3}$$

$$-\frac{2}{9} = \Delta y$$

Therefore, step two ends at the point $\left(\dfrac{2}{3}, -2 - \dfrac{2}{9}\right) = \left(\dfrac{2}{3}, -\dfrac{20}{9}\right)$.

Step three: From point $\left(\dfrac{2}{3}, -\dfrac{20}{9}\right)$ *to* $\left(1, -\dfrac{20}{9} + \Delta y\right)$.

Calculate $\dfrac{dy}{dx}$ when $x = \dfrac{2}{3}$ and $y = -\dfrac{20}{9}$.

$$\frac{dy}{dx} = xy = \left(\frac{2}{3}\right)\left(-\frac{20}{9}\right) = -\frac{40}{27}$$

Determine the value of Δy when $\Delta x = \dfrac{1}{3}$ and $m = -\dfrac{40}{27}$.

$$m = \frac{\Delta y}{\Delta x}$$

$$-\frac{40}{27} = \frac{\Delta y}{1/3}$$

$$-\frac{40}{81} = \Delta y$$

Therefore, step three ends at the point $\left(1, -\dfrac{20}{9} - \dfrac{40}{81}\right) = \left(1, -\dfrac{220}{81}\right)$ and $y(1) \approx -\dfrac{220}{81}$.

25.40 Use Euler's method with three steps of width $\Delta x = -\dfrac{1}{3}$ to approximate $y(-2)$ if $\dfrac{dy}{dx} = \dfrac{1}{x} - y$ and the point $(-1,4)$ belongs to the graph of the solution of the differential equation.

Step one: From point $(-1,4)$ to $\left(-\dfrac{4}{3}, 4 + \Delta y\right)$.

> Plug x and y into the differential equation: dy/dx = (1/−1) − 4 = −5.

When $x = -1$ and $y = 4$, $\dfrac{dy}{dx} = -5$, so $\Delta y = -5(\Delta x) = -5\left(-\dfrac{1}{3}\right) = \dfrac{5}{3}$. Therefore, step one ends at the point $\left(-\dfrac{4}{3}, 4 + \dfrac{5}{3}\right) = \left(-\dfrac{4}{3}, \dfrac{17}{3}\right)$.

Step two: From point $\left(-\dfrac{4}{3}, \dfrac{17}{3}\right)$ to $\left(-\dfrac{5}{3}, \dfrac{17}{3} + \Delta y\right)$.

> This is the equation $m = \dfrac{\Delta y}{\Delta x}$ already solved for Δy. If you multiply both sides of the equation by Δx, you get $\Delta y = m(\Delta x)$.

When $x = -\dfrac{4}{3}$ and $y = \dfrac{17}{3}$, $\dfrac{dy}{dx} = -\dfrac{77}{12}$, so $\Delta y = -\dfrac{77}{12}\left(-\dfrac{1}{3}\right) = \dfrac{77}{36}$. Therefore, step two ends at the point $\left(-\dfrac{5}{3}, \dfrac{17}{3} + \dfrac{77}{36}\right) = \left(-\dfrac{5}{3}, \dfrac{281}{36}\right)$.

Step three: From point $\left(-\dfrac{5}{3}, \dfrac{281}{36}\right)$ to $\left(-2, \dfrac{281}{36} + \Delta y\right)$.

When $x = -\dfrac{5}{3}$ and $y = \dfrac{281}{36}$, $\dfrac{dy}{dx} = -\dfrac{1,513}{180}$, so $\Delta y = -\dfrac{1,513}{180}\left(-\dfrac{1}{3}\right) = \dfrac{1,513}{540}$. Therefore, step three ends at the point $\left(-2, \dfrac{281}{36} + \dfrac{1,513}{540}\right) = \left(-2, \dfrac{1,432}{135}\right)$ and $y(-2) \approx \dfrac{1,432}{135}$.

25.41 Use Euler's method with three steps of width $\Delta x = \dfrac{1}{4}$ to approximate $y\left(\dfrac{7}{4}\right)$ if $\dfrac{dy}{dx} = x^y$ and the point $(1,0)$ belongs to the graph of the solution of the differential equation.

Step one: From point $(1,0)$ to $\left(\dfrac{5}{4}, 0 + \Delta y\right)$.

When $x = 1$ and $y = 0$, $\dfrac{dy}{dx} = 1^0 = 1$, so $\Delta y = \dfrac{dy}{dx}(\Delta x) = 1\left(\dfrac{1}{4}\right) = \dfrac{1}{4}$. Therefore, step one ends at the point $\left(\dfrac{5}{4}, 0 + \dfrac{1}{4}\right) = \left(\dfrac{5}{4}, \dfrac{1}{4}\right)$.

Step two: From point $\left(\dfrac{5}{4}, \dfrac{1}{4}\right)$ *to* $\left(\dfrac{3}{2}, \dfrac{1}{4} + \Delta y\right)$.

When $x = \dfrac{5}{4}$ and $y = \dfrac{1}{4}$, $\dfrac{dy}{dx} = \left(\dfrac{5}{4}\right)^{1/4} = \sqrt[4]{5/4}$, so $\Delta y = \sqrt[4]{5/4}\left(\dfrac{1}{4}\right) = \dfrac{\sqrt[4]{5/4}}{4}$.

Therefore, step two ends at the point $\left(\dfrac{3}{2}, \dfrac{1}{4} + \dfrac{\sqrt[4]{5/4}}{4}\right) = \left(\dfrac{3}{2}, \dfrac{1 + \sqrt[4]{5/4}}{4}\right)$.

Step three: From point $\left(\dfrac{3}{2}, \dfrac{1 + \sqrt[4]{5/4}}{4}\right)$ *to* $\left(\dfrac{7}{4}, \dfrac{1 + \sqrt[4]{5/4}}{4} + \Delta y\right)$.

When $x = \dfrac{3}{2}$ and $y = \dfrac{1 + \sqrt[4]{5/4}}{4}$, $\dfrac{dy}{dx} = \left(\dfrac{3}{2}\right)^{(1 + \sqrt[4]{5/4})/4}$, so $\Delta y = \left(\dfrac{3}{2}\right)^{(1 + \sqrt[4]{5/4})/4}\left(\dfrac{1}{4}\right) = \dfrac{(3/2)^{(1 + \sqrt[4]{5/4})/4}}{4}$

and $y\left(\dfrac{7}{4}\right) \approx \dfrac{1 + \sqrt[4]{5/4}}{4} + \Delta y = \dfrac{1 + \sqrt[4]{5/4} + (3/2)^{(1 + \sqrt[4]{5/4})/4}}{4}$.

(handwritten note, top right) This ugly fraction equals 0.5143428159....

(handwritten note, right) Approximately 0.82231485...

Chapter 26
BASIC SEQUENCES AND SERIES

What's uglier than one fraction? Infinitely many.

Two-semester calculus courses typically conclude with the investigation of infinite series, with a specific focus on series that approximate nontrivial function values. Since the advent of powerful, portable calculation technology, the use of interpolation via printed tables of values has waned. However, the technological tools are not merely storage vessels for said tables. Rather, they implement infinite series approximation techniques similar to those described in Chapter 28. However, before you can understand how such series are applied, you must first understand the basic concepts surrounding infinite series.

A "sequence" is a list of numbers (usually based on some defining rule), while a "series" is a list of numbers added together. You'll spend some time with sequences and finite series, but halfway through this chapter, you'll already be focusing primarily on infinite series. You'll find yourself answering one question over and over again: "Does the infinite series converge?" In other words, if you were to add up all of the numbers in the series (you have to speak hypothetically—remember, there's an infinite number of them, so you won't actually be able to count them all by hand) would you get a real number? It's weird to think that you can add an infinite list of numbers together and get a total sum of 2, but it's possible.

Sequences and Convergence

Do lists of numbers know where they're going?

26.1 List the first five terms of the sequence $\{n^2 + 1\}$.

Unless you're specifically told otherwise, assume that $n = 1$ generates the first term of the sequence (not $n = 0$). With series, it's a lot more obvious—they tell you what number to start with.

Substitute $n = 1$ into the expression to generate the first term of the sequence, $n = 2$ to generate the second term, and so forth, up to $n = 5$ to generate the fifth term.

n	1	2	3	4	5
$n^2 + 1$	$1^2 + 1 = 2$	$2^2 + 1 = 5$	$3^2 + 1 = 10$	$4^2 + 1 = 17$	$5^2 + 1 = 26$

The first five terms of the sequence are: 2, 5, 10, 17, and 26.

26.2 List the first six terms of the sequence $\left\{\dfrac{(-1)^{n-1}}{2n}\right\}$.

Substitute the integers between (and including) $n = 1$ and $n = 6$ into the expression to generate the first six terms of the sequence.

The exclamation point in here is actually a factorial sign. The factorial of an integer is the number times one less than the number times one less than that, all the way down to 1. For example: $4! = (4)(3)(2)(1) = 24$.

n	1	2	3	4	5	6
$\dfrac{(-1)^{n-1}}{2n}$	$\dfrac{(-1)^0}{2(1)} = \dfrac{1}{2}$	$\dfrac{(-1)^1}{2(2)} = -\dfrac{1}{4}$	$\dfrac{(-1)^2}{2(3)} = \dfrac{1}{6}$	$\dfrac{(-1)^3}{2(4)} = -\dfrac{1}{8}$	$\dfrac{(-1)^4}{2(5)} = \dfrac{1}{10}$	$\dfrac{(-1)^5}{2(6)} = -\dfrac{1}{12}$

The first six terms of the sequence are: $\dfrac{1}{2}, -\dfrac{1}{4}, \dfrac{1}{6}, -\dfrac{1}{8}, \dfrac{1}{10}$, and $-\dfrac{1}{12}$.

26.3 Identify a_4, the fourth term of the sequence $\{a_n\} = \left\{\dfrac{(3n)!}{n^3}\right\}$.

Note that a_4 is the fourth term in the sequence; generate it by substituting $n = 4$ into the expression.

$$a_4 = \frac{[3(4)]!}{4^3} = \frac{12!}{64} = 7,484,400$$

26.4 Determine the general term of the sequence: 0, –1, –2, –3, –4, ….

Ignore the negative signs for a second. Each term is one less than its n-value: $a_1 = 0, a_2 = 1, a_3 = 2$, etc. Basically, $a_n = n - 1$. Then multiply each term by –1, because they're all negative.

Each term in the sequence is the opposite of exactly one less than its corresponding n value: $a_n = -(n - 1)$. Therefore, the general term of the sequence is $\{1 - n\}$.

26.5 Determine the general term of the sequence: $\{a_n\} = \left\{ 1, \dfrac{5}{4}, 1, \dfrac{11}{16}, \dfrac{7}{16}, \dfrac{17}{64}, \cdots \right\}$.

Some denominators in the sequence are explicit powers of 2: $2^2 = 4$, $2^4 = 16$, and $2^6 = 64$. In fact, all of the denominators are powers of 2; this is disguised by the fact that the terms a_1, a_3, and a_5 have been reduced to lowest terms. Reinstate the original, unsimplified denominator of a_1 by multiplying its numerator and denominator by 2: $a_1 = 1\left(\dfrac{2}{2}\right) = \dfrac{2}{2}$. Similarly, reinstate the original denominators of a_3 and a_5 so that each denominator is a consecutive power of 2: $a_3 = 1\left(\dfrac{2^3}{2^3}\right) = \dfrac{8}{8}$ and $a_5 = \dfrac{7}{16}\left(\dfrac{2}{2}\right) = \dfrac{14}{32}$.

$$\{a_n\} = \left\{ \frac{2}{2}, \frac{5}{4}, \frac{8}{8}, \frac{11}{16}, \frac{14}{32}, \frac{17}{64}, \cdots \right\}$$

The numerator of the nth term is exactly one less than three times n: $3(1) - 1 = 2$, $3(2) - 1 = 5$, $3(3) - 1 = 8$, etc. Therefore, the general term of the sequence is $\left\{ \dfrac{3n-1}{2^n} \right\}$.

26.6 Does the sequence $\left\{ \dfrac{n^2}{e^n} \right\}$ converge?

> In other words, take the limit, as n approaches infinity, of the sequence expression. If a limit exists, the sequence converges to that limit.

A sequence $\{a_n\}$ converges to the real number L if and only if $\lim\limits_{n \to \infty} a_n = L$. Apply L'Hôpital's rule to evaluate the limit.

$$\lim_{n \to \infty} \frac{n^2}{e^n} = \lim_{n \to \infty} \frac{2n}{e^n} = \lim_{n \to \infty} \frac{2}{e^n}$$

According to Problem 10.28, the limit at infinity of a constant divided by an infinitely large value (e^n increases without bound as n approaches infinity) is 0: $\lim\limits_{n \to \infty} \dfrac{2}{e^n} = 0$. Therefore, the sequence $\left\{ \dfrac{n^2}{e^n} \right\}$ converges to 0.

> When a limit is indeterminate (in this case ∞/∞), take the derivative of the top and bottom of the fraction to get a simpler fraction with an equal limit value.

26.7 Does the sequence $\left\{ \dfrac{\sqrt{n}}{\ln n} \right\}$ converge?

The sequence $\left\{ \dfrac{\sqrt{n}}{\ln n} \right\}$ converges if and only if $\lim\limits_{n \to \infty} \dfrac{\sqrt{n}}{\ln n}$ exists. Apply L'Hôpital's rule to evaluate the indeterminate limit.

$$\lim_{n \to \infty} \frac{n^{1/2}}{\ln n} = \lim_{n \to \infty} \frac{\dfrac{1}{2n^{1/2}}}{\dfrac{1}{n}}$$

Multiply the numerator and denominator by n to eliminate the complex fraction.

$$\lim_{n \to \infty} \frac{\dfrac{1}{2n^{1/2}}\left(\dfrac{n}{1}\right)}{\dfrac{1}{n}\left(\dfrac{n}{1}\right)} = \lim_{n \to \infty} \frac{n}{2n^{1/2}} = \lim_{n \to \infty} \frac{1}{2} n^{1-(1/2)} = \frac{1}{2} \lim_{n \to \infty} \sqrt{n}$$

Divergent =
does not converge

As n approaches infinity, $\sqrt{n}$ increases without bound: $\dfrac{1}{2}\lim_{n \to \infty} \sqrt{n} = \infty$. Therefore, the sequence $\left\{\dfrac{\sqrt{n}}{\ln n}\right\}$ is divergent.

26.8 Determine the convergence of the sequence: $\left\{\dfrac{4n^2 - 3n + 7}{-2 + n - 5n^2}\right\}$.

The sequence converges if and only if a limit exists as n approaches infinity. Compare the degrees of the numerator and denominator to evaluate the limit, as explained in Problem 9.25.

$$\lim_{n \to \infty} \frac{4n^2 - 3n + 7}{-2 + n - 5n^2} = -\frac{4}{5}$$

The sequence converges to $-\dfrac{4}{5}$.

Series and Basic Convergence Tests
Sigma notation and the nth term divergence test

26.9 Evaluate the finite series: $\displaystyle\sum_{n=1}^{4} \frac{1}{n}$.

The series $\displaystyle\sum_{n=1}^{4} \frac{1}{n}$ is the sum of the first four terms of the sequence $\left\{\dfrac{1}{n}\right\}$.

$$\sum_{n=1}^{4} \frac{1}{n} = \frac{1}{1} + \frac{1}{2} + \frac{1}{3} + \frac{1}{4}$$

$$= \frac{12}{12} + \frac{6}{12} + \frac{4}{12} + \frac{3}{12}$$

$$= \frac{25}{12}$$

26.10 Evaluate the finite series: $\displaystyle\sum_{n=0}^{7}\cos\frac{n\pi}{2}$.

The series $\displaystyle\sum_{n=0}^{7}\cos\frac{n\pi}{2}$ is the sum of the first eight terms of the sequence $\left\{\cos\dfrac{n\pi}{2}\right\}$.

> The lower boundary is 0, so even though the upper boundary is 7, there are 8 terms in the series:
> $a_0 + a_1 + a_2 + a_3 + a_4 + a_5 + a_6 + a_7$.

$$\sum_{n=0}^{7}\cos\frac{n\pi}{2}=\cos\frac{0\pi}{2}+\cos\frac{1\pi}{2}+\cos\frac{2\pi}{2}+\cos\frac{3\pi}{2}+\cos\frac{4\pi}{2}+\cos\frac{5\pi}{2}+\cos\frac{6\pi}{2}+\cos\frac{7\pi}{2}$$

$$=1+0+(-1)+0+1+0+(-1)+0$$

$$=0$$

26.11 Evaluate the finite series: $\displaystyle\sum_{n=1}^{4}\frac{(n-1)!}{n^2}$.

The series $\displaystyle\sum_{n=1}^{4}\frac{(n-1)!}{n^2}$ is the sum of the first four terms of the sequence $\left\{\dfrac{(n-1)!}{n^2}\right\}$.

$$\sum_{n=1}^{4}\frac{(n-1)!}{n^2}=\frac{(1-1)!}{1^2}+\frac{(2-1)!}{2^2}+\frac{(3-1)!}{3^2}+\frac{(4-1)!}{4^2}$$

$$=1+\frac{1}{4}+\frac{2}{9}+\frac{6}{16}$$

$$=\frac{133}{72}$$

> $0! = 1$, so
> $a_1 = \dfrac{0!}{1^2} = \dfrac{1}{1} = 1$.

Note: Problems 26.12–26.14 refer to the arithmetic series 3 + 6 + 9 + 12 + 15 + ... + 267 + 270.

26.12 Express the series in sigma notation.

Factor the common difference out of each term.

$$3(1 + 2 + 3 + 4 + 5 + ... + 89 + 90)$$

The first term of the series is $3(1)$, the second term is $3(2)$, etc.; therefore, the nth term is $3n$. There are 90 terms, so that is the upper bound for the summation:

$$\sum_{n=1}^{90}3n.$$

> In an arithmetic series, each term is a fixed amount d (called the "common difference") larger than the previous term. In this series, each term is 3 more than the term before it, so $d = 3$.

Note: Problems 26.12–26.14 refer to the arithmetic series 3 + 6 + 9 + 12 + 15 + ... + 267 + 270.

26.13 Calculate the sum of the series.

The sum of an arithmetic series consisting of n terms is equal to $\dfrac{n}{2}(a_1 + a_n)$, where a_1 is the first term of the series and a_n is the last term. In this series of $n = 90$ terms, $a_1 = 3$ and $a_n = a_{90} = 270$.

$$\frac{n}{2}(a_1 + a_n) = \frac{90}{2}(3 + 270) = 45(273) = 12,285$$

Note: Problems 26.12–26.14 refer to the arithmetic series 3 + 6 + 9 + 12 + 15 + … + 267 + 270.

26.14 Justify the sum formula $\dfrac{n}{2}(a_1 + a_n)$ from Problem 26.13, using the arithmetic series $\displaystyle\sum_{n=1}^{90} 3n$ as an example.

The sum of the first and last terms of the series is 273: 3 + 270 = 273. Similarly, the sum of the second and second-to-last terms is 273 (6 + 267 = 273) and the sum of the third and third-to-last terms is 273 (9 + 266 = 273). In fact, the series consists of 45 pairs of numbers whose sum is 273. Therefore, the sum of the series is (45)(273), which is equivalent to the arithmetic series sum formula for this series: $\dfrac{90}{2}(3 + 270)$.

26.15 Calculate the sum of the series: 2 + 7 + 12 + 17 + … + 87 + 92.

The common difference of this arithmetic series is $d = 5$. To determine the number of terms in the series, apply the formula $n = \dfrac{a_n - a_1}{d} + 1$, where a_n is the last term of the series and a_1 is the first.

$$n = \frac{a_n - a_1}{d} + 1 = \frac{92 - 2}{5} + 1 = \frac{90}{5} + 1 = 18 + 1 = 19$$

Now apply the arithmetic series sum formula with $n = 19$, $a_1 = 2$, and $a_n = 92$.

$$\frac{n}{2}(a_1 + a_n) = \frac{19}{2}(2 + 92) = 893$$

> The sequence of partial sums is made up of the first number in the series, the sum of the first two numbers in the series, the sum of the first three numbers, the sum of the first four numbers, and so on.

26.16 Identify the first five terms of the sequence of partial sums for the infinite series $\displaystyle\sum_{n=1}^{\infty} n(n-2)$.

The *partial sum* S_n is the sum of the first n terms of the series. A sequence of partial sums, $\{S_n\}$, is the sequence $S_1, S_2, S_3, S_4, \cdots$.

$$S_1 = \sum_{n=1}^{1} n(n-2) = 1(1-2) = -1$$

$$S_2 = \sum_{n=1}^{2} n(n-2) = 1(1-2) + 2(2-2) = -1 + 2(0) = -1$$

> So S_3 is equal to S_2 plus the third term in the series, and S_4 is equal to S_3 plus the fourth term in the series.

Given the infinite series $\displaystyle\sum_{n=1}^{\infty} a_n$, notice that $S_d = S_{d-1} + a_d$.

$$S_3 = S_2 + 3(3-2) = -1 + 3(1) = 2$$
$$S_4 = S_3 + 4(4-2) = 2 + 4(2) = 10$$
$$S_5 = S_4 + 5(5-2) = 10 + 5(3) = 25$$

The first five terms of the sequence of partial sums are: –1, –1, 2, 10, and 25.

26.17 Identify the first five terms of the sequence of partial sums for the infinite series $\sum_{n=0}^{\infty} 2\left(\frac{1}{3}\right)^n$.

Note that the first term in this series corresponds to $n = 0$, so the sequence of the first five partial sums is S_0, S_1, S_2, S_3, S_4.

$$S_0 = 2\left(\frac{1}{3}\right)^0 = 2 \cdot 1 = 2$$

$$S_1 = S_0 + 2\left(\frac{1}{3}\right)^1 = 2 + 2\left(\frac{1}{3}\right) = \frac{8}{3}$$

$$S_2 = S_1 + 2\left(\frac{1}{3}\right)^2 = \frac{8}{3} + 2\left(\frac{1}{3}\right)^2 = \frac{26}{9}$$

$$S_3 = S_2 + 2\left(\frac{1}{3}\right)^3 = \frac{26}{9} + 2\left(\frac{1}{3}\right)^3 = \frac{80}{27}$$

$$S_4 = S_3 + 2\left(\frac{1}{3}\right)^4 = \frac{80}{27} + 2\left(\frac{1}{3}\right)^4 = \frac{242}{81}$$

26.18 Describe the relationship between the sequence of partial sums and the convergence of an infinite series.

An infinite series converges if and only if the sequence of its partial sums converges to a finite, real number. Consider the geometric series $\sum_{n=0}^{\infty} 2\left(\frac{1}{3}\right)^n$ from Problem 26.17, which has a sum of 3. Notice that $S_4 = \frac{242}{81} \approx 2.987654$, so the fifth term of the sequence of partial sums already approximates the actual sum relatively well. Each consecutive term of the sequence of partial sums (of which there are infinitely many) will more and more closely approach 3. For example, $S_{10} \approx 2.99998306$ and $S_{15} \approx 2.999999930$.

> Problem 26.30 explains what geometric series are and Problem 26.33 explains how to calculate their sums.

26.19 Explain why the series $\sum_{n=1}^{\infty} \frac{3n}{n-5}$ diverges, according to the nth term divergence test.

According to the nth term divergence test, if $\lim_{n \to \infty} a_n \neq 0$, then the infinite series $\sum a_n$ is divergent.

Apply L'Hôpital's rule to evaluate the limit as n approaches infinity.

$$\lim_{n \to \infty} \frac{3n}{n-5} = \lim_{n \to \infty} \frac{3}{1} = 3$$

The terms of the series approach 3; essentially, this series adds the number 3 infinitely many times, resulting in an infinitely large sum. Unless the nth term of a series approaches 0, a series cannot converge, because only 0 has a finite sum when added to itself infinitely many times.

> The nth term convergence test can NEVER be used to prove that a series converges, and in many cases, it won't be able to help at all. However, it is an efficient and simple test that picks out obviously divergent series.

26.20 Why does the series $\displaystyle\sum_{n=1}^{\infty} \frac{(n+2)!}{n!}$ diverge, according to the nth term divergence test?

> This is like writing 5! as (5)(4)(3!), which still equals 120.

Rewrite the numerator of the expression and reduce the fraction.

$$\frac{(n+2)!}{n!} = \frac{(n+2)(n+1)\,\cancel{n!}}{\cancel{n!}} = (n+2)(n+1) = n^2 + 3n + 2$$

Because $\displaystyle\lim_{n\to\infty}\left(n^2 + 3n + 2\right) \neq 0$, the series diverges according to the nth term divergence test.

26.21 Why does the series $2 + \dfrac{\sqrt{5}}{2} + \dfrac{\sqrt{6}}{3} + \dfrac{\sqrt{7}}{4} + \dfrac{2\sqrt{2}}{5} + \dfrac{1}{2} + \dfrac{\sqrt{10}}{7} + \cdots$ diverge, according to the nth term divergence test?

> The numerator stays fixed and the denominator gets infinitely large. 1 divided by a gigantic number is basically 0.

Identify the nth term of the series (as demonstrated in Problems 26.4 and 26.5): $\displaystyle\sum_{n=1}^{\infty} \frac{\sqrt{n+3}}{n}$. Apply the nth term divergence test and calculate the limit at infinity using L'Hopital's rule.

$$\lim_{n\to\infty} \frac{\sqrt{n+3}}{n} = \lim_{n\to\infty} \frac{(1/2)(n+3)^{-1/2}}{1} = \lim_{n\to\infty} \frac{1}{2\sqrt{n+3}} = 0$$

The nth term divergence test can conclude only that the series $\{a_n\}$ diverges if $\displaystyle\lim_{n\to\infty} a_n \neq 0$. However, that limit does equal 0 for this series, so the nth term divergence test does not allow you to draw any conclusion whatsoever. In order to determine the convergence of this series, you will have to apply one of the convergence tests presented in Chapter 27.

Telescoping Series and p-Series
How to handle these easy-to-spot series

26.22 Describe the behavior of a telescoping series and explain how to calculate its sum.

A telescoping series contains infinitely many pairs of opposite values, which have a sum of 0. Although the series contains an infinite number of terms, all but a finite number of those terms are eliminated by their corresponding opposites. The sum of the series, therefore, is the sum of the remaining terms.

26.23 Calculate the sum of the telescoping series: $\sum_{n=1}^{\infty}\left(\dfrac{1}{n}-\dfrac{1}{n+1}\right)$.

Expand the first five terms of the infinite series.

$$\sum_{n=1}^{\infty}\left(\frac{1}{n}-\frac{1}{n+1}\right)=\left(1-\frac{1}{2}\right)+\left(\frac{1}{2}-\frac{1}{3}\right)+\left(\frac{1}{3}-\frac{1}{4}\right)+\left(\frac{1}{4}-\frac{1}{5}\right)+\left(\frac{1}{5}-\frac{1}{6}\right)+\cdots$$

Notice that the series contains pairs of opposite real numbers such as $-\dfrac{1}{2}$ and $\dfrac{1}{2}$, $-\dfrac{1}{3}$ and $\dfrac{1}{3}$, etc. More specifically, every rational number in the series (except for 1) has a corresponding opposite.

$$\sum_{n=1}^{\infty}\left(\frac{1}{n}-\frac{1}{n+1}\right)=1+\left(\frac{1}{2}-\frac{1}{2}\right)+\left(\frac{1}{3}-\frac{1}{3}\right)+\left(\frac{1}{4}-\frac{1}{4}\right)+\left(\frac{1}{5}-\frac{1}{5}\right)+\cdots$$
$$=1+0+0+0+0+\cdots$$

Therefore, $\sum_{n=1}^{\infty}\left(\dfrac{1}{n}-\dfrac{1}{n+1}\right)=1$.

26.24 Calculate the sum of the telescoping series: $\sum_{n=3}^{\infty}\left(\dfrac{1}{n-1}-\dfrac{1}{n+2}\right)$.

Expand the first seven terms of the series. Note that $n = 3$ represents the first term in the series, not $n = 1$.

$$\sum_{n=3}^{\infty}\left(\frac{1}{n-1}-\frac{1}{n+2}\right)=\left(\frac{1}{2}-\frac{1}{5}\right)+\left(\frac{1}{3}-\frac{1}{6}\right)+\left(\frac{1}{4}-\frac{1}{7}\right)+\left(\frac{1}{5}-\frac{1}{8}\right)+\left(\frac{1}{6}-\frac{1}{9}\right)+\left(\frac{1}{7}-\frac{1}{10}\right)+\left(\frac{1}{8}-\frac{1}{11}\right)+\cdots$$

Except for $\dfrac{1}{2}$, $\dfrac{1}{3}$, and $\dfrac{1}{4}$, all of the rational numbers in the series have a corresponding opposite.

$$\sum_{n=3}^{\infty}\left(\frac{1}{n-1}-\frac{1}{n+2}\right)=\frac{1}{2}+\frac{1}{3}+\frac{1}{4}+\left(\frac{1}{5}-\frac{1}{5}\right)+\left(\frac{1}{6}-\frac{1}{6}\right)+\left(\frac{1}{7}-\frac{1}{7}\right)+\left(\frac{1}{8}-\frac{1}{8}\right)+\cdots$$
$$=\frac{1}{2}+\frac{1}{3}+\frac{1}{4}+0+0+0+0+\cdots$$
$$=\frac{13}{12}$$

26.25 Define p-series and describe the conditions under which a p-series converges.

> Most of the convergence tests in Chapters 26-28 apply only to positive series—series that contain only positive terms.

Positive infinite series of the form $\sum\dfrac{1}{n^p}$, where p is a positive real number, are called p-series; they converge when $p > 1$ but diverge when $0 < p \le 1$.

26.26 Determine the convergence of the series: $\sum_{n=1}^{\infty} \dfrac{1}{n^2}$.

Because $\sum_{n=1}^{\infty} \dfrac{1}{n^2}$ has form $\sum \dfrac{1}{n^p}$, it is a p-series with $p = 2$. According to Problem 26.25, a p-series converges when $p > 1$, so $\sum_{n=1}^{\infty} \dfrac{1}{n^2}$ is convergent.

> You can pull constants out of summation notation just like you can pull them out of integrals—every term in this series will be multiplied by 3. A constant affects the sum of a series, but doesn't affect its convergence.

26.27 Determine the convergence of the series: $\sum_{n=1}^{\infty} 3n^{-6/5}$.

Rewrite the series, eliminating the negative exponent.

$$3\sum_{n=1}^{\infty} \dfrac{1}{n^{6/5}}$$

Note that $\sum_{n=1}^{\infty} \dfrac{1}{n^{6/5}}$ is a p-series with $p = \dfrac{6}{5}$; the series is convergent because $\dfrac{6}{5} > 1$ (as explained in Problem 26.25).

26.28 Determine the convergence of the series: $\sum_{n=1}^{\infty} \dfrac{\sqrt[3]{n^2}}{5 \cdot \sqrt[4]{n^3}}$.

Rewrite the series using rational exponents.

$$\frac{1}{5}\sum_{n=1}^{\infty} \frac{n^{2/3}}{n^{3/4}} = \frac{1}{5}\sum_{n=1}^{\infty} n^{(2/3)-(3/4)} = \frac{1}{5}\sum_{n=1}^{\infty} n^{-1/12} = \frac{1}{5}\sum_{n=1}^{\infty} \frac{1}{n^{1/12}}$$

Note that $\sum_{n=1}^{\infty} \dfrac{1}{n^{1/12}}$ is a p-series with $p = \dfrac{1}{12}$; the series is divergent because $0 < \dfrac{1}{12} \le 1$ (as explained in Problem 26.25).

26.29 For what values of a does the series $\sum_{n=1}^{\infty} \dfrac{1}{n^{3a+2}}$ converge?

Because $\sum_{n=1}^{\infty} \dfrac{1}{n^{3a+2}}$ is a p-series with $p = 3a + 2$, it converges when $p > 1$.

$$p > 1$$
$$3a + 2 > 1$$
$$3a > -1$$
$$a > -\frac{1}{3}$$

The series $\sum_{n=1}^{\infty} \dfrac{1}{n^{3a+2}}$ converges when $a > -\dfrac{1}{3}$.

Geometric Series

Do they converge, and if so, what's the sum?

26.30 Determine the common ratio of the geometric series: $3 + \dfrac{3}{4} + \dfrac{3}{16} + \dfrac{3}{64} + \cdots$.

A geometric series has the form $\displaystyle\sum_{n=0}^{\infty} ar^n$, where a and r are real numbers; r is as the *ratio* of the geometric series, and a is the *scale factor* of the series. The first term of the geometric series $\displaystyle\sum_{n=0}^{\infty} ar^n$ is $a \times r^0 = a \times 1 = a$, so in this series $a = 3$. Factor 3 out of each term.

$$3\left(1 + \frac{1}{4} + \frac{1}{16} + \frac{1}{64} + \cdots\right)$$

Each term in the resulting series is $\left(\dfrac{1}{4}\right)^n$, i.e. the $n = 1$ term contains $\dfrac{1}{4}$, the $n = 2$ term contains $\left(\dfrac{1}{4}\right)^2$, etc. Therefore, the common ratio is $r = \dfrac{1}{4}$.

> A geometric series starts with a and every term is r times the term before it. For example, if $a = 5$ and $r = 2$, the geometric series $\displaystyle\sum_{n=0}^{\infty} 5(2)^n$ is $5 + 10 + 20 + 40 + \ldots$.

26.31 Write the geometric series using summation notation: $\dfrac{1}{2} + \dfrac{1}{3} + \dfrac{2}{9} + \dfrac{4}{27} + \cdots$.

The first term of the geometric series $\displaystyle\sum_{n=0}^{\infty} ar^n$ is a; therefore, $a = \dfrac{1}{2}$. Factor $\dfrac{1}{2}$ out of each term.

$$\frac{1}{2}\left(1 + \frac{2}{3} + \frac{4}{9} + \frac{8}{27} + \cdots\right)$$

> Factoring ½ out of a number is like dividing it by ½, which is the same as multiplying it by 2.

Notice that the nth term of the series within the parentheses is $\left(\dfrac{2}{3}\right)^n$, so $r = \dfrac{2}{3}$.

Substitute a and r into the geometric series formula $\displaystyle\sum_{n=0}^{\infty} ar^n$ to get $\displaystyle\sum_{n=0}^{\infty} \frac{1}{2}\left(\frac{2}{3}\right)^n$.

Note: Problems 26.32–26.33 are based on the infinite geometric series $\displaystyle\sum_{n=0}^{\infty} 6\left(\frac{4}{7}\right)^n$.

26.32 Determine the convergence of the series.

Notice that $\displaystyle\sum_{n=0}^{\infty} 6\left(\frac{4}{7}\right)^n$ has the form $\displaystyle\sum_{n=0}^{\infty} ar^n$, so it is a geometric series with $a = 6$ and $r = \dfrac{4}{7}$. Geometric series converge if $0 < |r| < 1$; because $0 < \dfrac{4}{7} < 1$, the geometric series $\displaystyle\sum_{n=0}^{\infty} 6\left(\frac{4}{7}\right)^n$ converges.

> Geometric series diverge if $|r| \geq 1$.

26.33 Calculate the sum of the series.

A convergent geometric series $\sum_{n=0}^{\infty} ar^n$ has sum $\dfrac{a}{1-r}$. Substitute $a = 6$ and $r = \dfrac{4}{7}$ into the formula to calculate the sum of the series $\sum_{n=0}^{\infty} 6\left(\dfrac{4}{7}\right)^n$.

$$\sum_{n=0}^{\infty} 6\left(\frac{4}{7}\right)^n = \frac{a}{1-r} = \frac{6}{1-(4/7)} = \frac{6}{(7/7)-(4/7)} = \frac{6}{3/7} = 6 \cdot \frac{7}{3} = \frac{42}{3} = 14$$

26.34 Determine the convergence of the series $\sum_{n=0}^{\infty} \left(\dfrac{8}{5}\right)\left(\dfrac{3}{2}\right)^n$; if the series converges, calculate its sum.

The geometric series $\sum_{n=0}^{\infty} \left(\dfrac{8}{5}\right)\left(\dfrac{3}{2}\right)^n$ is divergent because $|r| > 1 : \left|\dfrac{3}{2}\right| > 1$.

26.35 Determine the convergence of the series $\sum_{n=0}^{\infty} 2\left(-\dfrac{3}{5}\right)^n$; if the series converges, calculate its sum.

$\dfrac{3}{5}$ (the absolute value of r) is between 0 and 1.

Notice that $\sum_{n=0}^{\infty} 2\left(-\dfrac{3}{5}\right)^n$ is a geometric series with $a = 2$ and $r = -\dfrac{3}{5}$. Because $0 < |r| < 1$, the series converges to sum of $\dfrac{a}{1-r}$.

$$\sum_{n=0}^{\infty} 2\left(-\frac{3}{5}\right)^n = \frac{a}{1-r} = \frac{2}{1-(-3/5)} = \frac{2}{(5/5)+(3/5)} = \frac{2}{8/5} = 2 \cdot \frac{5}{8} = \frac{5}{4}$$

26.36 Determine the convergence of the series $\sum_{n=6}^{\infty} \left(-\dfrac{5}{3}\right)\left(-\dfrac{2}{3}\right)^n$; if the series converges, calculate its sum.

Although the lower bound of the sum is $n = 6$ and not $n = 0$, subtracting a finite number of terms will not affect the convergence of the series. Because $0 < \left|-\dfrac{2}{3}\right| < 1$, the geometric series converges. However, when the lower bound does not equal 0, an alternative formula must be used to calculate the sum of the series:

$$\sum_{n=k}^{\infty} ar^n = \frac{ar^k}{1-r}.$$

$$\sum_{n=6}^{\infty} \left(-\frac{5}{3}\right)\left(-\frac{2}{3}\right)^n = \frac{\left(-\frac{5}{3}\right)\left(-\frac{2}{3}\right)^6}{1-\left(-\frac{2}{3}\right)} = \frac{-\left(\frac{5}{3}\right)\left(\frac{64}{729}\right)}{\frac{5}{3}} = -\frac{64}{729}$$

The Integral Test

Infinite series and improper integrals are related

26.37 Assume that $\sum\limits_{n=1}^{\infty} a_n$ is a positive, decreasing series. Describe how to determine the convergence of $\sum\limits_{n=1}^{\infty} a_n$ using the integral test.

Consider the function $f(n) = a_n$. Assuming $f(n)$ is continuous, evaluate the improper integral $\int_1^{\infty} f(n)\,dn$. If the integral converges (i.e., equals a finite, real number), then the series converges. Similarly, a divergent integral indicates a divergent series.

> If you don't know how to deal with improper integrals, look at Problems 21.30– 21.37.

26.38 According to Problem 26.26, $\sum\limits_{n=1}^{\infty} \dfrac{1}{n^2}$ is a convergent p-series. Use the integral test to verify the convergence of the series.

According to the integral test, the series converges if and only if the improper integral $\int_1^{\infty} \dfrac{1}{n^2}\,dn$ converges.

$$\int_1^{\infty} \frac{1}{n^2}\,dn = \lim_{a \to \infty} \int_1^{a} \frac{dn}{n^2}$$

$$= \lim_{a \to \infty} \int_1^{a} n^{-2}\,dn$$

$$= \lim_{a \to \infty} \left(-\frac{1}{n}\bigg|_1^{a} \right)$$

$$= \lim_{a \to \infty} \left(-\frac{1}{a} + \frac{1}{1} \right)$$

$$= \lim_{a \to \infty} \left(-\frac{1}{a} \right) + \lim_{a \to \infty} 1$$

$$= 0 + 1$$

$$= 1$$

> The integral converges to 1, but that doesn't mean $\sum\limits_{n=1}^{\infty} \dfrac{1}{n^2} = 1$. The definite integral value does not equal the sum of the series.

Because the improper integral $\int_1^{\infty} \dfrac{1}{n^2}\,dn$ converges, the series $\sum\limits_{n=1}^{\infty} \dfrac{1}{n^2}$ also converges.

26.39 Determine the convergence of the series $\sum\limits_{n=1}^{\infty} \dfrac{(\ln n)^2}{n}$.

Apply the integral test by determining the convergence of the corresponding improper integral.

> *Use variable substitution to integrate: $u = \ln n$ and $du = (1/n)dn$.*

$$\int_1^{\infty} \frac{(\ln n)^2}{n}\, dn = \lim_{a \to \infty} \int_1^{a} \frac{(\ln n)^2}{n}\, dn$$

$$= \lim_{a \to \infty} \int_0^{\ln a} u^2\, du$$

$$= \lim_{a \to \infty} \left(\frac{u^3}{3} \Big|_0^{\ln a} \right)$$

$$= \lim_{a \to \infty} \left[\frac{(\ln a)^3}{3} \right]$$

$$= \infty$$

Because the improper integral $\int_1^{\infty} \dfrac{(\ln n)^2}{n}\, dn$ diverges, the infinite series $\sum\limits_{n=1}^{\infty} \dfrac{(\ln n)^2}{n}$ diverges as well.

26.40 Determine the convergence of the series: $\sum\limits_{n=0}^{\infty} \dfrac{1}{8+n^2}$.

Apply the integral test by evaluating $\int_0^{\infty} \dfrac{dn}{8+n^2}$.

> *Use the inverse trig antiderivative formula $\int \dfrac{du}{a^2+u^2} = \dfrac{1}{a} \arctan \dfrac{u}{a} + C$, setting $a = \sqrt{8} = 2\sqrt{2}$ and $u = x$.*

$$\int_0^{\infty} \frac{dn}{8+n^2} = \lim_{a \to \infty} \int_0^{a} \frac{dn}{8+n^2}$$

$$= \lim_{a \to \infty} \left[\frac{1}{\sqrt{8}} \left(\arctan \frac{n}{\sqrt{8}} \Big|_0^{a} \right) \right]$$

$$= \lim_{a \to \infty} \left[\frac{1}{2\sqrt{2}} \left(\arctan \frac{a}{2\sqrt{2}} - \arctan 0 \right) \right]$$

$$= \lim_{a \to \infty} \left[\frac{1}{2\sqrt{2}} \arctan \frac{a}{2\sqrt{2}} \right]$$

Note that $\lim\limits_{n \to \infty} \left(\arctan \dfrac{a}{2\sqrt{2}} \right) = \dfrac{\pi}{2}$.

> *Because tangent gets infinitely big as x approaches $\dfrac{\pi}{2}$, arctangent approaches $\dfrac{\pi}{2}$ when x gets infinitely big.*

$$= \frac{1}{2\sqrt{2}} \cdot \frac{\pi}{2} = \frac{\pi}{4\sqrt{2}} = \frac{\pi\sqrt{2}}{8}$$

Because the improper integral $\int_0^{\infty} \dfrac{dn}{8+n^2}$ converges, the infinite series $\sum\limits_{n=0}^{\infty} \dfrac{1}{8+n^2}$ converges as well.

26.41 According to the nth term divergence test, the series $\sum\limits_{n=1}^{\infty} \dfrac{n^3+1}{n^3}$ diverges. Verify that conclusion using the integral test.

Apply the integral test by evaluating $\int_{1}^{\infty} \dfrac{n^3+1}{n^3}\,dn$.

$$\int_{1}^{\infty} \frac{n^3+1}{n^3}\,dn = \lim_{a\to\infty} \int_{1}^{a} \frac{n^3+1}{n^3}\,dn$$

$$= \lim_{a\to\infty}\left[\int_{1}^{a}\left(\frac{n^3}{n^3}+\frac{1}{n^3}\right)dn\right]$$

$$= \lim_{a\to\infty}\left[\int_{1}^{a}\left(1+n^{-3}\right)dn\right]$$

$$= \lim_{a\to\infty}\left[\left(n-\frac{1}{2n^2}\right)\bigg|_{1}^{a}\right]$$

$$= \lim_{a\to\infty}\left[\left(a-\frac{1}{2a^2}\right)-\left(1-\frac{1}{2}\right)\right]$$

$$= \lim_{a\to\infty}\left[\left(\frac{2a^3-1}{2a^2}\right)-\frac{1}{2}\right]$$

$$= \infty$$

$\lim\limits_{a\to\infty}\dfrac{2a^3-1}{2a^2}=\infty$ because the degree of the numerator is greater than the degree in the denominator.

Because the improper definite integral $\int_{1}^{\infty} \dfrac{n^3+1}{n^3}\,dn$ diverges, the series $\sum\limits_{n=1}^{\infty} \dfrac{n^3+1}{n^3}$ diverges as well.

Chapter 27
ADDITIONAL INFINITE SERIES CONVERGENCE TESTS

For use with uglier infinite series

Chapter 26 presented examples of simple infinite series that exhibited definitive characteristics; however, the majority of series cannot be classified so easily. This chapter presents an assortment of tests to supplement and enhance those discussed in Chapter 26, significantly increasing the number and variety of series for which you can determine convergence. Note that the majority of the tests apply only to series consisting of positive terms, although the treatment of negative terms is discussed in the last section of the chapter.

In Chapter 26, you learned how to handle specific kinds of series: geometric series, which look like $\sum_{n=0}^{\infty} ar^n$; p-series, which look like $\sum_{n=1}^{\infty} \frac{1}{n^p}$; telescoping series, where just about all of the terms cancel one another out; and the integral test, which is useful only if you can integrate the generic nth term of the series. This chapter will broaden your horizons and introduce new convergence tests to use on series you come across that don't fit cleanly into any of the categories covered so far.

Comparison Test
Proving series are bigger than big and smaller than small

27.1 Describe how the comparison test determines the convergence or divergence of an infinite series.

> *If every a term is smaller than (or equal to) the corresponding b term ($a_1 \leq b_1$, $a_2 \leq b_2$, etc.) and the b's add up to a finite number, the a's will have a smaller (or equal) finite sum.*

Given the positive, infinite series $\sum a_n$ and $\sum b_n$ such that $a_n \leq b_n$, the convergence of $\sum b_n$ guarantees the convergence of $\sum a_n$. Similarly, the divergence of $\sum a_n$ guarantees the divergence of $\sum b_n$.

27.2 Determine the convergence of the series using the comparison test: $\sum\limits_{n=1}^{\infty} \dfrac{n}{n+1}$.

Consider the comparison series $\sum\limits_{n=1}^{\infty} \dfrac{1}{n+1}$. Because the denominators of the two series are equal, comparing their corresponding values is a simple matter. Because the numerator of $\dfrac{n}{n+1}$ is greater than (or equal to) the numerator of $\dfrac{1}{n+1}$, each

> *The first terms of both series are the same: ½. After that (for all $n > 1$), the terms of $\sum\limits_{n=1}^{\infty} \dfrac{n}{n+1}$ are larger.*

term of $\sum\limits_{n=1}^{\infty} \dfrac{n}{n+1}$ is greater than or equal to the corresponding term of $\sum\limits_{n=1}^{\infty} \dfrac{1}{n+1}$:

$\sum\limits_{n=1}^{\infty} \dfrac{1}{n+1} \leq \sum\limits_{n=1}^{\infty} \dfrac{n}{n+1}$. Because $\lim\limits_{a\to\infty} \int_1^a \dfrac{dn}{n+1} = \infty$, $\sum\limits_{n=1}^{\infty} \dfrac{1}{n+1}$ is divergent according to

the integral test. By comparison, $\sum\limits_{n=1}^{\infty} \dfrac{n}{n+1}$ must diverge as well, as it is greater than a divergent series. Note that the nth term divergence test gives the same result.

27.3 Determine the convergence of the series: $\sum\limits_{n=0}^{\infty} \dfrac{1}{\sqrt{n^5+7}}$.

Consider the comparison series $\sum\limits_{n=0}^{\infty} \dfrac{1}{\sqrt{n^5}} = \sum\limits_{n=0}^{\infty} \dfrac{1}{n^{5/2}}$. The denominator of $\dfrac{1}{\sqrt{n^5+7}}$

> *A bigger denominator means a smaller fraction, but a bigger numerator means a bigger fraction.*

is greater than the denominator of $\dfrac{1}{\sqrt{n^5}}$, so each term of the series $\sum\limits_{n=0}^{\infty} \dfrac{1}{\sqrt{n^5+7}}$ is

less than the corresponding term of the series $\sum\limits_{n=0}^{\infty} n^{-5/2}$: $\sum\limits_{n=0}^{\infty} \dfrac{1}{\sqrt{n^5+7}} \leq \sum\limits_{n=0}^{\infty} \dfrac{1}{n^{5/2}}$.

Notice that $\sum\limits_{n=0}^{\infty} \dfrac{1}{n^{5/2}}$ is a convergent p-series (since $\dfrac{5}{2} > 1$), so according to the

comparison test, $\sum\limits_{n=0}^{\infty} \dfrac{1}{\sqrt{n^5+7}}$ also converges because its sum is less than or equal

to the sum of a convergent series.

27.4 Determine the convergence of the series: $\displaystyle\sum_{n=2}^{\infty}\frac{\ln n}{\sqrt{n}}$.

Consider the comparison series $\displaystyle\sum_{n=2}^{\infty}\frac{1}{\sqrt{n}}=\sum_{n=2}^{\infty}\frac{1}{n^{1/2}}$. Because $\ln n \geq 1$ for all $n \geq 3$,

$\displaystyle\sum_{n=2}^{\infty}\frac{\ln n}{\sqrt{n}}\geq\sum_{n=2}^{\infty}\frac{1}{n^{1/2}}$. Note that $\displaystyle\sum_{n=2}^{\infty}\frac{1}{n^{1/2}}$ is a divergent p-series, so $\displaystyle\sum_{n=2}^{\infty}\frac{\ln n}{\sqrt{n}}$ must also be divergent according to the comparison test.

> The numerator on the left is bigger than the numerator on the right (and the denominators are the same), so the left sum is bigger than the right sum.

27.5 Determine the convergence of the series: $\displaystyle\sum_{n=1}^{\infty}\frac{1}{n^{\sqrt{n}}}$.

Consider the geometric comparison series $\displaystyle\sum a_n=\sum_{n=1}^{\infty}\left(\frac{1}{2}\right)^n$. Although the first few terms of $\displaystyle\sum_{n=1}^{\infty}\left(\frac{1}{2}\right)^n$ are smaller than the corresponding terms of $\displaystyle\sum b_n=\sum_{n=1}^{\infty}\frac{1}{n^{\sqrt{n}}}$, the opposite is true for all $n \geq 5$, as illustrated by the table of values below.

n	1	2	3	4	5	6
a_n	$\frac{1}{2}=0.5$	$\frac{1}{4}=0.25$	$\frac{1}{8}=0.125$	$\frac{1}{16}=0.0625$	$\frac{1}{32}=0.03125$	$\frac{1}{64}=0.01563$
b_n	1	$\frac{1}{2^{\sqrt{2}}}\approx 0.375$	$\frac{1}{3^{\sqrt{3}}}\approx 0.149$	$\frac{1}{4^{\sqrt{4}}}=0.0625$	$\frac{1}{5^{\sqrt{5}}}\approx 0.02736$	$\frac{1}{6^{\sqrt{6}}}\approx 0.01241$

Note that $\displaystyle\sum_{n=1}^{\infty}\left(\frac{1}{2}\right)^n$ is a convergent geometric series (because $0<\left|\frac{1}{2}\right|<1$). Because

$\displaystyle\sum_{n=1}^{\infty}\frac{1}{n^{\sqrt{n}}}\leq\sum_{n=1}^{\infty}\left(\frac{1}{2}\right)^n$, $\displaystyle\sum_{n=1}^{\infty}\frac{1}{n^{\sqrt{n}}}$ also converges, according to the comparison test.

> Terms at the beginning of $\sum b_n$ are allowed to be smaller than the corresponding terms of $\sum a_n$, as long as the terms of $\sum b_n$ eventually get bigger and STAY bigger. A few bad apples don't spoil the whole bushel.

27.6 Determine the convergence of the series: $\displaystyle\sum_{n=1}^{\infty}\frac{3n+1}{n^2+4}$.

Apply the integral test to determine the convergence of the comparison series $\displaystyle\sum_{n=1}^{\infty}\frac{n}{n^2+4}$.

$$\int_1^{\infty}\frac{n\,dn}{n^2+4}=\lim_{a\to\infty}\int_1^{a}\frac{n\,dn}{n^2+4}$$

$$=\lim_{a\to\infty}\left[\frac{1}{2}\int_5^{a^2+4}\frac{du}{u}\right]$$

$$=\lim_{a\to\infty}\left[\frac{1}{2}\left(\ln|u|\right)\Big|_5^{a^2+4}\right]$$

$$=\lim_{a\to\infty}\left(\frac{1}{2}\left[\ln\left(a^2+4\right)-\ln 5\right]\right)$$

$$=\infty$$

> Use variable substitution to integrate: $u = n^2 + 4$ and $du/2 = n\,dn$.

Because the numerator $3x + 1$ is greater than the numerator x in the other series.

Therefore, $\sum_{n=1}^{\infty} \dfrac{3x+1}{x^2+4}$ is greater than or equal to the divergent series $\sum_{n=1}^{\infty} \dfrac{x}{x^2+4}$, so $\sum_{n=1}^{\infty} \dfrac{3x+1}{x^2+4}$ diverges according to the comparison test.

27.7 Determine the convergence of the series: $\sum_{n=0}^{\infty} \dfrac{7^n+2}{6^n}$.

Consider the geometric comparison series $\sum_{n=0}^{\infty} \dfrac{7^n}{6^n} = \sum_{n=0}^{\infty} \left(\dfrac{7}{6}\right)^n$, which diverges because $\left|\dfrac{7}{6}\right| > 1$. Because $7^n + 2 > 7^n$, it follows that $\sum_{n=0}^{\infty} \dfrac{7^n+2}{6^n} \geq \sum_{n=0}^{\infty} \left(\dfrac{7}{6}\right)^n$, and $\sum_{n=0}^{\infty} \dfrac{7^n+2}{6^n}$ diverges according to the comparison test.

Limit Comparison Test
Series that converge or diverge by association

The comparison series is usually made up of the "biggest parts" of a_n, the things that would matter most if n were really big. This includes the highest powers of n and things raised to the n power.

27.8 Describe how to determine the convergence of a positive infinite series using the limit comparison test.

Given the series $\sum a_n$, create a comparison series $\sum b_n$ for which you can determine the convergence and compute $\lim\limits_{n\to\infty} \dfrac{a_n}{b_n}$. If the limit is equal to a positive, real number, then both series either converge or diverge. If the limit either equals 0 or does not exist, the limit comparison test is inconclusive.

27.9 Determine the convergence of the series $\sum_{n=1}^{\infty} \dfrac{5n^2+9}{7n^3-2n+8}$.

As n approaches infinity, the highest exponents of n in the numerator and denominator (2 and 3 respectively) are more influential than the surrounding terms. The comparison series $\sum b_n$ need not include the leading coefficients of the numerator and denominator (5 and 7 respectively).

Because they get much bigger than the n terms and the constants when n gets huge.

$$\sum_{n=1}^{\infty} b_n = \sum_{n=1}^{\infty} \dfrac{n^2}{n^3} = \sum_{n=1}^{\infty} n^{-1} = \sum_{n=1}^{\infty} \dfrac{1}{n}$$

Now apply the limit comparison test by evaluating the limit, as $n \to \infty$, of the series' quotient.

$$\lim_{n\to\infty} \dfrac{a_n}{b_n} = \lim_{n\to\infty} \dfrac{\dfrac{5n^2+9}{7n^3-2n+8}}{\dfrac{1}{n}}$$

Multiply the numerator and denominator by the reciprocal of the denominator to eliminate the complex fraction.

$$= \lim_{n \to \infty} \frac{5n^2 + 9}{7n^3 - 2n + 8} \cdot \frac{n}{1} = \lim_{n \to \infty} \frac{5n^3 + 9n}{7n^3 - 2n + 8} = \frac{5}{7}$$

The limit exists, so both series either converge or diverge according to the limit comparison test. Because $\sum_{n=1}^{\infty} b_n = \sum_{n=1}^{\infty} \frac{1}{n}$ is a divergent p-series, $\sum_{n=1}^{\infty} \frac{5n^2 + 9}{7n^3 - 2n + 8}$ diverges as well.

> When the highest powers of the numerator and denominator are equal, divide their coefficients to get the limit as n approaches ∞.

27.10 Determine the convergence of $\sum_{n=1}^{\infty} \frac{n+2}{\sqrt[3]{n^2 - 3n + 5}}$.

Construct the comparison series $\sum b_n$ using the degrees of the numerator and the denominator.

$$\sum_{n=1}^{\infty} b_n = \sum_{n=1}^{\infty} \frac{n}{n^{2/3}} = \sum_{n=1}^{\infty} n^{1-(2/3)} = \sum_{n=1}^{\infty} n^{1/3} = \sum_{n=1}^{\infty} \sqrt[3]{n}$$

Apply the limit comparison test.

$$\lim_{n \to \infty} \frac{\frac{n+2}{\sqrt[3]{n^2 - 3n + 5}}}{\frac{\sqrt[3]{n}}{1}} = \lim_{n \to \infty} \frac{n+2}{\sqrt[3]{n^2 - 3n + 5}} \cdot \frac{1}{\sqrt[3]{n}} = \lim_{n \to \infty} \frac{n+2}{\sqrt[3]{n^3 - 3n^2 + 5n}}$$

The degrees of the numerator and denominator are equal $\left(\sqrt[3]{n^3} = n^{3/3} = 1 \right)$, so the limit equals the quotient of the leading coefficients of the numerator and the denominator.

$$\lim_{n \to \infty} \frac{n+2}{\sqrt[3]{n^3 - 3n^2 + 5n}} = \frac{1}{1} = 1$$

The limit exists, so both series either converge or diverge according to the limit comparison test; because $\sum_{n=1}^{\infty} \sqrt[3]{n}$ diverges according to the nth term divergence test, $\sum_{n=1}^{\infty} \frac{n+2}{\sqrt[3]{n^2 - 3n + 5}}$ must diverge as well.

27.11 Determine the convergence of the series: $\sum_{n=1}^{\infty} \frac{3^n + 6}{5^n + 1}$.

Apply the limit comparison test using the geometric comparison series $\sum_{n=1}^{\infty} b_n = \sum_{n=1}^{\infty} \left(\frac{3}{5} \right)^n$.

$$\lim_{n \to \infty} \frac{\frac{3^n + 6}{5^n + 1}}{\frac{3^n}{5^n}} = \lim_{n \to \infty} \frac{3^n + 6}{5^n + 1} \cdot \frac{5^n}{3^n} = \lim_{n \to \infty} \frac{(3 \cdot 5)^n + 6(5^n)}{(3 \cdot 5)^n + 3^n} = \lim_{n \to \infty} \frac{15^n + 6(5^n)}{15^n + 3^n}$$

> This time, you're not using the highest powers of n to make the comparison series (like in Problems 27.9 and 27.10), but the things raised to the n power.

15ⁿ is much larger than 5ⁿ when n gets infinitely large. If multiple terms are raised to the n power, only the ones with the biggest bases in the numerator and denominator matter.

As n approaches infinity, $\lim\limits_{n\to\infty}\dfrac{15^n+6(5^n)}{15^n+3^n}=\lim\limits_{n\to\infty}\dfrac{15^n}{15^n}=1$. Therefore, according to the limit comparison test, $\sum\limits_{n=1}^{\infty}\dfrac{3^n+6}{5^n+1}$ converges because $\sum\limits_{n=1}^{\infty}\left(\dfrac{3}{5}\right)^n$ is a convergent geometric series.

27.12 Determine the convergence of the series: $\sum\limits_{n=1}^{\infty}\dfrac{1}{\sqrt{n+1}+\sqrt{n+2}}$.

Apply the limit comparison test using the divergent p-series $\sum\limits_{n=1}^{\infty}\dfrac{1}{\sqrt{n}}=\sum\limits_{n=1}^{\infty}\dfrac{1}{n^{1/2}}$.

$$\lim_{n\to\infty}\dfrac{\dfrac{1}{\sqrt{n+1}+\sqrt{n+2}}}{\dfrac{1}{\sqrt{n}}}=\lim_{n\to\infty}\dfrac{1}{\sqrt{n+1}+\sqrt{n+2}}\cdot\dfrac{\sqrt{n}}{1}=\lim_{n\to\infty}\dfrac{\sqrt{n}}{\sqrt{n+1}+\sqrt{n+2}}$$

Multiply the numerator and denominator by $\dfrac{1}{\sqrt{n}}$.

As n approaches infinity, 1/n and 2/n approach 0, because dividing any real number by an infinitely big number gives you a microscopically small result that's basically 0.

$$=\lim_{n\to\infty}\dfrac{\dfrac{\sqrt{n}}{\sqrt{n}}}{\sqrt{\dfrac{n}{n}+\dfrac{1}{n}}+\sqrt{\dfrac{n}{n}+\dfrac{2}{n}}}=\lim_{n\to\infty}\dfrac{1}{\sqrt{1+\dfrac{1}{n}}+\sqrt{1+\dfrac{2}{n}}}=\dfrac{1}{\sqrt{1+0}+\sqrt{1+0}}=\dfrac{1}{2}$$

The limit exists, so $\sum\limits_{n=1}^{\infty}\dfrac{1}{\sqrt{n+1}+\sqrt{n+2}}$ must diverge according to the limit comparison test because $\sum\limits_{n=1}^{\infty}\dfrac{1}{\sqrt{n}}$ diverges.

27.13 Determine the convergence of the series $\sum\limits_{n=1}^{\infty}\dfrac{\sqrt{n+n^5}}{\sqrt[3]{n^8+n^6+1}}$.

Construct a comparison series using the highest powers of the numerator and denominator.

$$\sum_{n=1}^{\infty}\dfrac{n^{5/2}}{n^{8/3}}=\sum_{n=1}^{\infty}n^{(5/2)-(8/3)}=\sum_{n=1}^{\infty}n^{-1/6}=\sum_{n=1}^{\infty}\dfrac{1}{n^{1/6}}$$

Apply the limit comparison test.

You have to rewrite $n^{1/6}$ as a square root if you want to multiply it by another square root.

$$\lim_{n\to\infty}\dfrac{\dfrac{\sqrt{n+n^5}}{\sqrt[3]{n^8+n^6+1}}}{\dfrac{1}{n^{1/6}}}=\lim_{n\to\infty}\dfrac{\sqrt{n+n^5}}{\sqrt[3]{n^8+n^6+1}}\cdot\dfrac{n^{1/6}}{1}$$

Note that $n^{1/6}=\left(n^{1/3}\right)^{1/2}=\sqrt{n^{1/3}}$.

$$=\lim_{n\to\infty}\dfrac{\sqrt{n+n^5}}{\sqrt[3]{n^8+n^6+1}}\cdot\dfrac{\sqrt{n^{1/3}}}{1}=\lim_{n\to\infty}\dfrac{\sqrt{n^{4/3}+n^{16/3}}}{\sqrt[3]{n^8+n^6+1}}$$

The degrees of the numerator and denominator are equal ($\sqrt{n^{16/3}} = (n^{16/3})^{1/2} = n^{8/3}$ and $\sqrt[3]{n^8} = n^{8/3}$), so the limit equals the quotient of the leading coefficients.

$$= \lim_{n \to \infty} \frac{\sqrt[3]{n^8}}{\sqrt[3]{n^8}} = 1$$

Because the limit exists and the comparison series is a divergent p-series,

$\sum_{n=1}^{\infty} \dfrac{\sqrt{n+n^5}}{\sqrt[3]{n^8+n^6+1}}$ diverges according to the limit comparison test.

Ratio Test
Compare neighboring terms of a series

27.14 Explain how to determine the convergence of the positive series $\sum_{n=1}^{\infty} a_n$ using the ratio test.

If $\lim_{n \to \infty} \dfrac{a_{n+1}}{a_n} = L$, and $L < 1$, then $\sum_{n=1}^{\infty} a_n$ converges. If $L > 1$ or $L = \infty$, then the series diverges. The ratio test cannot determine the convergence of a series if $L = 1$.

27.15 Demonstrate that the ratio test cannot be used to determine the convergence of the series: $\sum_{n=1}^{\infty} \dfrac{3}{n^4}$.

> The denominator a_n is just the expression $3/n^4$. To get the numerator a_{n+1}, replace the n inside $3/n^4$ with $n+1$.

Evaluate $\lim_{n \to \infty} \dfrac{a_{n+1}}{a_n}$.

$$\lim_{n \to \infty} \frac{a_{n+1}}{a_n} = \lim_{n \to \infty} \frac{\dfrac{3}{(n+1)^4}}{\dfrac{3}{n^4}}$$

Multiply the numerator and denominator by the reciprocal of the denominator to eliminate the complex fraction.

> You have to do this almost every time, so save yourself a step by writing $a_{n+1} \cdot \dfrac{1}{a_n}$ instead of $\dfrac{a_{n+1}}{a_n}$.

$$= \lim_{n \to \infty} \frac{\dfrac{3}{(n+1)^4}}{\dfrac{3}{n^4}} \cdot \frac{n^4}{3}$$

$$= \lim_{n \to \infty} \frac{\cancel{3}n^4}{\cancel{3}\left(n^4 + 4n^3 + 6n^2 + 4n + 1\right)}$$

$$= \lim_{n \to \infty} \frac{n^4}{n^4 + 4n^3 + 6n^2 + 4n + 1}$$

$$= 1$$

> The numerator and denominator have the same degree, so divide their coefficients to get the limit: $1/1 = 1$.

According to Problem 27.14, the ratio test cannot be used to determine the convergence of the series $\sum_{n=1}^{\infty} a_n$ when $\lim_{n \to \infty} \dfrac{a_{n+1}}{a_n} = 1$.

RULE OF THUMB: The ratio test is good for series containing terms that will get humongous as n approaches infinity. Series containing factorials (like in this problem), n's raised to powers, and things raised to n powers are good candidates for the ratio test.

27.16 Determine the convergence of the series: $\sum_{n=1}^{\infty} \dfrac{n+3}{n!}$.

Apply the ratio test.

$$\lim_{n \to \infty} \frac{a_{n+1}}{a_n} = \lim_{n \to \infty} \frac{\dfrac{(n+1)+3}{(n+1)!}}{\dfrac{n+3}{n!}}$$

$$= \lim_{n \to \infty} \frac{n+4}{(n+1)!} \cdot \frac{n!}{n+3}$$

$$= \lim_{n \to \infty} \frac{(n+4)\,n!}{(n+1)\,(n!)\,(n+3)}$$

$$= \lim_{n \to \infty} \frac{n+4}{n^2 + 4n + 3}$$

$$= 0$$

The denominator's degree is larger, so the limit as $n \to \infty$ is automatically 0.

Because $\lim_{n \to \infty} \dfrac{a_{n+1}}{a_n} = 0$ and $0 < 1$, the series $\sum_{n=1}^{\infty} \dfrac{n+3}{n!}$ converges according to the ratio test.

27.17 Determine the convergence of the series: $\sum_{n=1}^{\infty} \dfrac{3^n}{n!}$.

Apply the ratio test.

$$\lim_{n \to \infty} \frac{a_{n+1}}{a_n} = \lim_{n \to \infty} \frac{\dfrac{3^{n+1}}{(n+1)!}}{\dfrac{3^n}{n!}}$$

$$= \lim_{n \to \infty} \frac{3^{n+1}}{(n+1)!} \cdot \frac{n!}{3^n}$$

This is the property $(x^a)(x^b) = x^{a+b}$ in action. Instead of adding the exponents of a product to make one expression, you're taking that one expression and writing it as a product: $x^{a+b} = (x^a)(x^b)$.

Note that $3^{n+1} = 3^n \cdot 3^1 = 3 \cdot 3^n$.

$$= \lim_{n \to \infty} \frac{\left(3 \cdot 3^n\right)(n!)}{(n+1)\,(n!)\,(3^n)} = \lim_{n \to \infty} \frac{3}{n+1} = 0$$

Because $\lim_{n \to \infty} \dfrac{a_{n+1}}{a_n} = 0$ and $0 < 1$, $\sum_{n=1}^{\infty} \dfrac{3^n}{n!}$ converges according to the ratio test.

27.18 Determine the convergence of the series: $\displaystyle\sum_{n=1}^{\infty} n^{-2}e^{n}$.

Rewrite the series by eliminating the negative exponent: $\displaystyle\sum_{n=1}^{\infty} \frac{e^{n}}{n^{2}}$. Apply the ratio test.

$$\lim_{n\to\infty} \frac{a_{n+1}}{a_{n}} = \lim_{n\to\infty} \frac{\dfrac{e^{n+1}}{(n+1)^{2}}}{\dfrac{e^{n}}{n^{2}}}$$

$$= \lim_{n\to\infty} \frac{e^{n+1}}{(n+1)^{2}} \cdot \frac{n^{2}}{e^{n}}$$

$$= \lim_{n\to\infty} \frac{e \cdot \cancel{e^{n}} \cdot n^{2}}{\cancel{e^{n}}\,(n+1)^{2}}$$

$$= \lim_{n\to\infty} \frac{e \cdot n^{2}}{n^{2}+2n+1}$$

$$= e$$

> The numerator and denominator have degree 2, so take the coefficients of the n^2's to find the limit: $\frac{e}{1} = e$.

> $e \approx 2.71828182846$

Because $\displaystyle\lim_{n\to\infty} \frac{a_{n+1}}{a_{n}} = e$ and $e > 1$, $\displaystyle\sum_{n=1}^{\infty} n^{-2}e^{n}$ diverges according to the ratio test.

27.19 Determine the convergence of the series: $\displaystyle\sum_{n=1}^{\infty} \frac{(n+1)^{2}}{n\cdot 2^{n}}$.

Apply the ratio test.

$$\lim_{n\to\infty} \frac{a_{n+1}}{a_{n}} = \lim_{n\to\infty} \frac{\dfrac{[(n+1)+1]^{2}}{(n+1)\cdot 2^{n+1}}}{\dfrac{(n+1)^{2}}{n\cdot 2^{n}}}$$

$$= \lim_{n\to\infty} \frac{(n+2)^{2}}{(n+1)\cdot 2^{n+1}} \cdot \frac{n\cdot 2^{n}}{(n+1)^{2}}$$

$$= \lim_{n\to\infty} \frac{(n+2)^{2}\,(n)\,\cancel{(2^{n})}}{(n+1)^{3}\,\cancel{(2^{n})}\,(2)}$$

$$= \lim_{n\to\infty} \frac{n\left(n^{2}+4n+4\right)}{2\left(n^{3}+3n^{2}+3n+1\right)}$$

$$= \lim_{n\to\infty} \frac{n^{3}+4n^{2}+4n}{2n^{3}+6n^{2}+6n+2}$$

$$= \frac{1}{2}$$

> If you're not sure where $(2^{n})(2)$ comes from, look at my note on Problem 27.17.

Because $\displaystyle\lim_{n\to\infty} \frac{a_{n+1}}{a_{n}} = \frac{1}{2}$ and $\frac{1}{2} < 1$, $\displaystyle\sum_{n=1}^{\infty} \frac{(n+1)^{2}}{n\cdot 2^{n}}$ converges according to the ratio test.

27.20 Determine the convergence of the series: $\displaystyle\sum_{n=5}^{\infty} \frac{4^n \cdot n!}{n \cdot (n+4)!}$.

Apply the ratio test.

$$\lim_{n\to\infty} \frac{a_{n+1}}{a_n} = \lim_{n\to\infty} \frac{\dfrac{4^{n+1}(n+1)!}{(n+1)\left[(n+1)+4\right]!}}{\dfrac{4^n \cdot n!}{n \cdot (n+4)!}}$$

$$= \lim_{n\to\infty} \frac{4^{n+1}(n+1)!}{(n+1)(n+5)!} \cdot \frac{n \cdot (n+4)!}{4^n \cdot n!}$$

Notice that $4^{n+1} = 4(4^n)$, $(n+1)! = (n+1)(n!)$, and $(n+5)! = (n+5)(n+4)!$.

$$= \lim_{n\to\infty} \frac{4\,(\cancel{4^n})\,(\cancel{n+1})\,(\cancel{n!})\,(n)\,(\cancel{n+4)!}}{(\cancel{n+1})\,(n+5)\,(\cancel{n+4)!}\,(\cancel{4^n})\,(\cancel{n!})}$$

$$= \lim_{n\to\infty} \frac{4n}{n+5}$$

$$= 4$$

Because $\displaystyle\lim_{n\to\infty} \frac{a_{n+1}}{a_n} = 4$ and $4 > 1$, $\displaystyle\sum_{n=5}^{\infty} \frac{4^n \cdot n!}{n \cdot (n+4)!}$ diverges according to the ratio test.

Root Test
Helpful for terms inside radical signs

These are the same three possible conclusions about L that the ratio test uses.

27.21 Explain how to determine the convergence of the positive infinite series $\displaystyle\sum_{n=1}^{\infty} a_n$ using the root test.

If $\displaystyle\lim_{n\to\infty} \sqrt[n]{a_n} = L$, and $L < 1$, then $\displaystyle\sum_{n=1}^{\infty} a_n$ converges. If $L > 1$ or $L = \infty$, then the series diverges. The root test cannot be used to determine the convergence of a series if $L = 1$.

27.22 Applying the root test occasionally results in the expression $\displaystyle\lim_{n\to\infty} \sqrt[n]{n}$. Evaluate the limit so that you may reference it in the problems that follow.

Begin by setting $y = \sqrt[n]{n}$ and rewriting the radical using a rational exponent: $y = n^{1/n}$. Take the natural logarithm of both sides of the equation.

The log property $\log x^a = a \log x$ allows you to pull exponents out of logs and write them out front like coefficients.

$$\ln y = \ln n^{1/n}$$

$$= \frac{1}{n} \ln n$$

$$= \frac{\ln n}{n}$$

Evaluate $\lim\limits_{n \to \infty} \dfrac{\ln n}{n}$ using L'Hôpital's rule.

$$\lim_{n \to \infty} \frac{\ln n}{n} = \lim_{n \to \infty} \frac{1/n}{1} = \lim_{n \to \infty} \frac{1}{n} = 0$$

Recall that $\ln y = \dfrac{\ln n}{n}$. Therefore, $\lim\limits_{n \to \infty} \ln y = \lim\limits_{n \to \infty} \dfrac{\ln n}{n} = 0$. Now that you have established that $\lim\limits_{n \to \infty} \ln y = 0$, evaluate the limit specified by the original problem.

$$\lim_{n \to \infty} \sqrt[n]{n} = \lim_{n \to \infty} n^{1/n}$$

Substitute $y = n^{1/n}$.

$$= \lim_{n \to \infty} y$$

Note that $y = e^{\ln y}$. ⟵

$$= \lim_{n \to \infty} e^{\ln y}$$

Recall that $\lim\limits_{n \to \infty} \ln y = 0$.

$$= \lim_{n \to \infty} e^{0}$$
$$= 1$$

Therefore, $\lim\limits_{n \to \infty} \sqrt[n]{n} = 1$.

> Because e^x and $\ln x$ are inverse functions, you know that $\ln e^x = e^{\ln x} = x$. Once you rewrite y as $e^{\ln y}$, you can use all of the work you did a few steps back with ln y.

27.23 Determine the convergence of the series: $\sum\limits_{n=1}^{\infty} \left(\dfrac{1}{n} \right)^n$.

Note that the general term of the series is raised to the n power, an indication that the root test is likely the best candidate to determine convergence.

$$\lim_{n \to \infty} \sqrt[n]{a_n} = \lim_{n \to \infty} \sqrt[n]{\left(\frac{1}{n} \right)^n}$$
$$= \lim_{n \to \infty} \frac{1}{n}$$
$$= 0$$

> The nth root and the nth power cancel each other out: $\sqrt[n]{x^n} = (x^n)^{1/n} = x^{n/n} = x^1 = x$.

Because $\lim\limits_{n \to \infty} \sqrt[n]{a_n} = 0$ and $0 < 1$, $\sum\limits_{n=1}^{\infty} \left(\dfrac{1}{n} \right)^n$ converges according to the root test.

27.24 Determine the convergence of the series: $\sum\limits_{n=1}^{\infty} \left(\dfrac{6n + 5n^3}{2n^3 + 3n^2 - n + 1} \right)^n$.

Apply the root test.

$$\lim_{n \to \infty} \sqrt[n]{a_n} = \lim_{n \to \infty} \sqrt[n]{\left(\frac{6n + 5n^3}{2n^3 + 3n^2 - n + 1} \right)^n}$$
$$= \lim_{n \to \infty} \frac{6n + 5n^3}{2n^3 + 3n^2 - n + 1}$$
$$= \frac{5}{2}$$

> The numerator and denominator both have degree 3 so the limit is the n^3 coefficient of the numerator over the n^3 coefficient of the denominator.

Because $\lim_{n\to\infty}\sqrt[n]{a_n}=\frac{5}{2}$ and $\frac{5}{2}>1$, $\sum_{n=1}^{\infty}\left(\frac{6n+5n^3}{2n^3+3n^2-n+1}\right)^n$ diverges according to the root test.

27.25 Determine the convergence of the series: $\sum_{n=1}^{\infty}\frac{2}{(\ln n)^n}$.

Constants can be factored out of a series without affecting the series' convergence.

$$2\sum_{n=1}^{\infty}\frac{1}{(\ln n)^n}=2\sum_{n=1}^{\infty}\left(\frac{1}{\ln n}\right)^n$$

Apply the root test to determine the convergence of $\sum_{n=1}^{\infty}\left(\frac{1}{\ln n}\right)^n$.

$$\lim_{n\to\infty}\sqrt[n]{a_n}=\lim_{n\to\infty}\sqrt[n]{\left(\frac{1}{\ln n}\right)^n}=\lim_{n\to\infty}\frac{1}{\ln n}=0$$

Because $\lim_{n\to\infty}\sqrt[n]{a_n}=0$ and $0<1$, $\sum_{n=1}^{\infty}\left(\frac{1}{\ln n}\right)^n$ converges according to the root test.

Therefore, $2\sum_{n=1}^{\infty}\left(\frac{1}{\ln n}\right)^n=\sum_{n=1}^{\infty}\frac{2}{(\ln n)^n}$ converges as well.

No matter what n is, $1^n=1$, so you can also raise the numerator to the n.

If $\sum_{n=1}^{\infty}\left(\frac{1}{\ln n}\right)^n$ converges, then it has a finite sum (even though you can't figure out what it is). That means $\sum_{n=1}^{\infty}\frac{2}{(\ln n)^n}$ also converges, because its sum is twice as big.

27.26 Determine the convergence of the series: $\sum_{n=1}^{\infty}\frac{n}{e^n}$.

Although only the denominator is raised to the n power, the root test is still applicable.

$$\lim_{n\to\infty}\sqrt[n]{a_n}=\lim_{n\to\infty}\sqrt[n]{\frac{n}{e^n}}$$
$$=\lim_{n\to\infty}\left(\frac{n}{e^n}\right)^{1/n}$$
$$=\lim_{n\to\infty}\frac{n^{1/n}}{e^{n/n}}$$
$$=\lim_{n\to\infty}\frac{n^{1/n}}{e}$$

According to Problem 27.22, $\lim_{n\to\infty}n^{1/n}=1$, so $\lim_{n\to\infty}\frac{n^{1/n}}{e}=\frac{1}{e}$. Because $\lim_{n\to\infty}\sqrt[n]{a_n}=\frac{1}{e}$ and $\frac{1}{e}<1$, $\sum_{n=1}^{\infty}\frac{n}{e^n}$ converges according to the root test.

27.27 Determine the convergence of the series: $\displaystyle\sum_{n=1}^{\infty}\left(2+3\sqrt[n]{n}\right)^{n}$.

Apply the root test.

$$\lim_{n\to\infty}\sqrt[n]{a_n}=\lim_{n\to\infty}\sqrt[n]{\left(2+3\sqrt[n]{n}\right)^{n}}$$

$$=\lim_{n\to\infty}\left(2+3\sqrt[n]{n}\right)$$

$$=\lim_{n\to\infty}2+3\lim_{n\to\infty}\sqrt[n]{n}$$

According to Problem 27.22, $\displaystyle\lim_{n\to\infty}\sqrt[n]{n}=1$.

$$=2+3(1)$$

$$=5$$

Because $\displaystyle\lim_{n\to\infty}\sqrt[n]{a_n}=5$ and $5>1$, $\displaystyle\sum_{n=1}^{\infty}\left(2+3\sqrt[n]{n}\right)^{n}$ diverges according to the root test.

27.28 Determine the convergence of the series: $\displaystyle\sum_{n=1}^{\infty}\frac{(4n)^{n}}{3^{2n+1}}$.

Apply the root test.

$$\lim_{n\to\infty}\sqrt[n]{a_n}=\lim_{n\to\infty}\sqrt[n]{\frac{(4n)^{n}}{3^{2n+1}}}$$

$$=\lim_{n\to\infty}\frac{\sqrt[n]{(4n)^{n}}}{\sqrt[n]{3^{2n+1}}}$$

$$=\lim_{n\to\infty}\frac{\left[(4n)^{n}\right]^{1/n}}{\left(3^{2n+1}\right)^{1/n}}$$

$$=\lim_{n\to\infty}\frac{4n}{3^{(2n/n)+(1/n)}}$$

$$=\lim_{n\to\infty}\frac{4n}{3^{2+(1/n)}}$$

$$=\infty$$

To raise 3^{2n+1} to the $1/n$ power, multiply the powers together: $3^{(2n+1)(1/n)}$.

Because $\displaystyle\lim_{n\to\infty}\sqrt[n]{a_n}=\infty$, $\displaystyle\sum_{n=1}^{\infty}\frac{(4n)^{n}}{3^{2n+1}}$ diverges according to the root test.

Alternating Series Test and Absolute Convergence
What if series have negative terms?

27.29 What is the defining characteristic of an alternating series, and how is the alternating series test used to determine convergence?

The two conditions, ignoring the negative signs, are: (1) the general term of the series has to approach 0 when n is really huge, and (2) each term in the series has to be smaller than the term before it.

An alternating series contains both positive and negative terms; consecutive terms have opposite signs. An alternating series usually contains -1 raised to a power of n to generate the positive and negative terms, for example: $\sum_{n=1}^{\infty}(-1)^n n^2 = -1+4-9+16-25+\cdots$. However, other functions, such as sine and cosine, can generate terms of alternating sign as well: $\sum_{n=1}^{\infty}\frac{\cos(n\pi)}{n+1} = -\frac{1}{2}+\frac{1}{3}-\frac{1}{4}+\frac{1}{5}-\frac{1}{6}+\cdots$. According to the alternating series test, the alternating series $\sum_{n=1}^{\infty}a_n$ converges if two conditions are satisfied: $\lim_{n\to\infty}|a_n| = 0$ and $|a_{n+1}| \le |a_n|$.

Note: Problems 27.30–27.31 refer to the alternating series $1 - \frac{3}{4} + \frac{9}{9} - \frac{27}{16} + \frac{81}{25} - \cdots$.

27.30 Write the series using summation notation.

There are other ways you can write this series, like $\sum_{n=0}^{\infty}\frac{(-1)^{n+2}\cdot 3^n}{(n+1)^2}$, $\sum_{n=0}^{\infty}\frac{\cos(n\pi)\cdot 3^n}{(n+1)^2}$, and $\sum_{n=1}^{\infty}\frac{(-1)^{n+1}\cdot 3^{n-1}}{n^2}$.

Let $a_1 = 1$, $a_2 = -\frac{3}{4}$, $a_3 = \frac{9}{9}$, etc. Notice that the numerator of each term is a power of 3, and each denominator is a perfect square: $\sum_{n=0}^{\infty}\frac{(-1)^n 3^n}{(n+1)^2}$.

Note: Problems 27.30–27.31 refer to the alternating series $1 - \frac{3}{4} + \frac{9}{9} - \frac{27}{16} + \frac{81}{25} - \cdots$.

27.31 Determine the convergence of the series.

These absolute values tell you to leave off the part of the alternating series that changes the sign, in this case $(-1)^{n+1}$, when you're testing for convergence.

In order to satisfy the alternating series test, two conditions must be satisfied. First, $\lim_{n\to\infty}|a_n|$ must equal 0. Use L'Hôpital's rule to evaluate $\lim_{n\to\infty}\frac{3^n}{(n+1)^2}$, from Problem 27.30.

$$\lim_{n\to\infty}\frac{3^n}{n^2+2n+1} = \lim_{n\to\infty}\frac{(\ln 3)3^n}{2n+1} = \lim_{n\to\infty}\frac{(\ln 3)^2 3^n}{2} = \infty$$

The nth term does not approach 0 as n approaches infinity, so this series fails the first condition of the alternating series test and diverges according to the nth term divergence test.

27.32 Determine the convergence of the series: $\displaystyle\sum_{n=3}^{\infty}\frac{(-1)^n \ln(n-2)}{n-2}$.

Apply the alternating series test. First, ensure that $\displaystyle\lim_{n\to\infty}|a_n| = 0$ using L'Hôpital's rule.

$$\lim_{n\to\infty}\frac{\ln(n-2)}{n-2} = \lim_{n\to\infty}\frac{1/(n-2)}{1} = \lim_{n\to\infty}\frac{1}{n-2} = 0$$

Now expand the series to verify visually that $|a_{n+1}| \le |a_n|$ (i.e., each term is less than the term that precedes it).

$$\sum_{n=3}^{\infty}\frac{\ln(n-2)}{n-2} = \frac{\ln 1}{1} + \frac{\ln 2}{2} + \frac{\ln 3}{3} + \frac{\ln 4}{4} + \frac{\ln 5}{5} + \frac{\ln 6}{6} + \frac{\ln 7}{7} + \frac{\ln 8}{8} + \frac{\ln 9}{9}\cdots$$

$$\approx 0 + 0.3466 + 0.3662 + 0.3466 + 0.3219 + 0.2986 + 0.2780 + 0.2599 + 0.2441$$

Although $a_4 > a_3$ ($0.3466 > 0$) and $a_5 > a_4$ ($0.3662 > 0.3466$), once $n \ge 5$, each term is less than or equal to the term that precedes it. To more rigorously prove that the terms of the series decrease, differentiate $f(x) = \dfrac{\ln(x-2)}{x-2}$.

$$f'(x) = \frac{(x-2)\left(\dfrac{1}{x-2}\right) - \ln(x-2)}{(x-2)^2} = \frac{1 - \ln(x-2)}{(x-2)^2}$$

Note that $f'(x) < 0$ for all $x > 3$, so the function (and the terms of the series it generates) is decreasing on that interval.

Because both conditions of the alternating series test are met, $\displaystyle\sum_{n=3}^{\infty}\frac{(-1)^n \ln(n-2)}{n-2}$ converges.

> **Leave off $(-1)^n$ again, just like you did when checking the first condition of the alternating series test.**

> **So not EVERY term has to be less than the term before it, but there has to be some cut-off point (like $n = 5$) where that's true and stays true for the rest of the series.**

27.33 The remainder R_n of an alternating series describes how accurately the partial sum S_n reflects the actual sum of the series S: $|R_n| = |S - S_n| \le |a_{n+1}|$. Use R_6 to identify an interval of values within which the sum of the series $\displaystyle\sum_{n=3}^{\infty}\frac{(-1)^n \ln(n-2)}{n-2}$ is contained. Report the boundaries of the interval accurate to three decimal places.

According to the given information, $|R_6| \le |a_{6+1}|$, so the partial sum S_6 approximates the actual sum of the series with an error of a_7. Begin by calculating S_6, the sum of the terms up to and including the $n = 6$ term. (Note that the series begins with $n = 3$.) Use a calculator to approximate the sum.

$$S_6 = \frac{(-1)^3 \ln(3-2)}{3-2} + \frac{(-1)^4 \ln(4-2)}{4-2} + \frac{(-1)^5 \ln(5-2)}{5-2} + \frac{(-1)^6 \ln(6-2)}{6-2}$$

$$= -\frac{\ln 1}{1} + \frac{\ln 2}{2} - \frac{\ln 3}{3} + \frac{\ln 4}{4}$$

$$\approx 0.32694308433724$$

Now calculate a_7.

$$a_7 = \frac{(-1)^7 \ln(7-2)}{7-2} = -\frac{\ln 5}{5} \approx -0.32188758248682$$

Therefore, the actual sum of $\sum_{n=3}^{\infty} \frac{(-1)^n \ln(n-2)}{n-2}$ is no more than

$a_7 \approx 0.32188758248682$ units greater or less than $S_6 \approx 0.32694308433724$. Add a_7

to and subtract a_7 from S_6 to generate an interval that contains the actual sum

of $\sum_{n=3}^{\infty} \frac{(-1)^n \ln(n-2)}{n-2}$.

$$\left(S_6 - |a_7|, S_6 + |a_7|\right) \approx (0.32694308433724 - 0.32188758248682, 0.32694308433724 + 0.32188758248682)$$
$$\approx (0.005, 0.649)$$

27.34 What conclusions can be drawn if $\sum_{n=1}^{\infty} a_n$ exhibits *absolute convergence*?

> Absolute convergence is used only if some of the terms in the series are negative, because it changes everything into positive terms. That means you can use the ratio, root, and integral tests (which apply only to positive series).

If $\sum_{n=1}^{\infty} |a_n|$ converges, then $\sum_{n=1}^{\infty} a_n$ converges *absolutely*; there is no need to actually

test the convergence of $\sum_{n=1}^{\infty} a_n$ in such cases—it is guaranteed. However, if

$\sum_{n=1}^{\infty} |a_n|$ diverges, $\sum_{n=1}^{\infty} a_n$ may still converge; if it does, $\sum_{n=1}^{\infty} a_n$ exhibits *conditional*

convergence.

27.35 Determine whether $\sum_{n=3}^{\infty} \frac{(-1)^n \ln(n-2)}{n-2}$, the series defined in Problem 27.32,

converges absolutely or conditionally.

> The whole point of testing for absolute convergence is to see if the series converges when you take the absolute value of every term. That's why you drop the negatives.

Discard the factor $(-1)^n$ that causes the terms to alternate signs and apply the

integral test by evaluating the improper integral $\int_3^{\infty} \frac{\ln(n-2)}{n-2} dn$.

$$\int_3^{\infty} \frac{\ln(n-2)}{n-2} dn = \lim_{a \to \infty} \int_3^a \frac{\ln(n-2)}{n-2} dn$$

Integrate using variable substitution: $u = \ln(n-2)$ and $du = \frac{dn}{n-2}$.

$$= \lim_{a \to \infty} \int_0^{\ln(a-2)} u \, du$$
$$= \lim_{a \to \infty} \left(\frac{u^2}{2} \Big|_0^{\ln(a-2)} \right)$$
$$= \lim_{a \to \infty} \frac{[\ln(a-2)]^2}{2}$$
$$= \infty$$

The series $\displaystyle\sum_{n=3}^{\infty} \frac{\ln(n-2)}{n-2}$ diverges according to the integral test because

the corresponding improper integral, $\displaystyle\int_{3}^{\infty} \frac{\ln(n-2)}{n-2}\, dn$, diverges. Therefore,

$\displaystyle\sum_{n=3}^{\infty} \frac{(-1)^n \ln(n-2)}{n-2}$ does not converge absolutely. However, it does converge

conditionally, according to Problem 27.32.

> So it converges when you leave the negatives in (and use the alternating series test), but without the negative terms, the series diverges.

27.36 Determine whether the series $\displaystyle\sum_{n=3}^{\infty} \frac{(-1)^{n-1} 4^n}{(n+4)!}$ converges absolutely or conditionally.

Discard the factor that causes the terms of the series to alternate sign, $(-1)^{n-1}$, and test the resulting series for convergence using the ratio test.

$$\lim_{n\to\infty} \frac{a_{n+1}}{a_n} = \lim_{n\to\infty} \frac{\dfrac{4^{n+1}}{[(n+1)+4]!}}{\dfrac{4^n}{(n+4)!}}$$

$$= \lim_{n\to\infty} \frac{4^{n+1}}{(n+5)!} \cdot \frac{(n+4)!}{4^n}$$

$$= \lim_{n\to\infty} \frac{\left(4^n\right)(4)(n+4)!}{\left(4^n\right)(n+5)(n+4)!}$$

$$= \lim_{n\to\infty} \frac{4}{n+5}$$

$$= 0$$

Because $\displaystyle\lim_{n\to\infty} \frac{a_{n+1}}{a_n} = 0$ and $0 < 1$, $\displaystyle\sum_{n=3}^{\infty} \frac{4^n}{(n+4)!}$ converges according to the ratio test.

Therefore, $\displaystyle\sum_{n=3}^{\infty} \frac{(-1)^{n-1} 4^n}{(n+4)!}$ converges absolutely.

Chapter 28
ADVANCED INFINITE SERIES

Series that contain x's

The final chapter of the book concerns infinite series that represent functions; hence, the series that follow will contain a variable (usually *x*) in addition to *n*. The exercises begin with power series; although a brief discussion about the representation of known functions via power series will ensue, the majority of the focus is paid to the determination of a radius and an interval of convergence. Following that, Taylor and Maclaurin series are used to estimate function values, methods more elaborate than linear approximations and capable of providing more accurate approximations further from the *x*-values about which they are centered.

This chapter deals with three specific types of series. Actually, it's two types of series, because Maclaurin series are a specific kind of Taylor series, but more on that later. You'll start with power series, and your job will be to figure out where the series converge. That means you'll have to figure out which x-values make the series converge when you plug them in and which make the series diverge. After a hefty dose of generic power series, you'll deal with two specific kinds of power series: Taylor and Maclaurin series (when there are an infinite number of terms) and polynomials (when there are a finite number of terms).

Power Series

Finding intervals of convergence

28.1 Write the power series using summation notation and identify the x-value about which it is centered: $0 + \frac{1}{3}(x+3) + \frac{2}{4}(x+3)^2 + \frac{3}{5}(x+3)^3 + \cdots$.

> A power series always converges where it is centered, so $\sum_{n=0}^{\infty} \frac{n}{n+2}(x+3)^n$ converges at $x = -3$. It's proven in Problem 28.3.

Let the first term be a_0, the second term be a_1, etc. Notice that the numerator of each coefficient is n, and each denominator is $n + 2$. Furthermore, each term contains the quantity $(x + 3)$ raised to the n power: $\sum_{n=0}^{\infty} \frac{n}{n+2}(x+3)^n$. When compared to the general form of a power series centered about $x = c$, $\sum_{n=0}^{\infty} a_n (x - c)^n$,

$a_n = \dfrac{n}{n+2}$ and $c = -3$; therefore, the power series is centered about $x = -3$.

28.2 Write the power series in summation notation and identify the x-value about which it is centered: $x + \frac{x^2}{4} + \frac{x^3}{8} + \frac{x^4}{16} + \cdots$.

> So the first term will be $(x - c)^1 = x - c$ instead of $(x - c)^0 = 1$.

Each term in this series contains x, so it is beneficial to begin the series with $n = 1$ instead of $n = 0$: $\sum_{n=1}^{\infty} \frac{x^n}{n^2}$. Note that the power series is centered at $c = 0$, as $(x - c)^n = (x - 0)^n = x^n$.

28.3 Prove that the power series $\sum_{n=1}^{\infty} a_n (x - c)^n$ converges at $x = c$.

> The series starts at $n = 1$ because it would contain 0^0 if it began at $n = 0$, which is indeterminate.

Expand the power series.

$$\sum_{n=1}^{\infty} a_n (x-c)^n = a_1 (x-c)^1 + a_2 (x-c)^2 + a_3 (x-c)^3 + a_4 (x-c)^4 + \cdots$$

To determine the convergence of a power series at a specific x-value, substitute it into the series.

$$\sum_{n=1}^{\infty} a_n (c-c)^n = a_1 (c-c)^1 + a_2 (c-c)^2 + a_3 (c-c)^3 + a_4 (c-c)^3 + \cdots$$

$$= a_1 (0)^1 + a_2 (0)^2 + a_3 (0)^3 + a_4 (0)^4 + \cdots$$

$$= 0 + 0 + 0 + 0 + \cdots$$

$$= 0$$

The series has finite sum 0, so it converges when $x = c$.

28.4 According to Problem 28.3, the power series $\sum\limits_{n=0}^{\infty} a_n (x-c)^n$ converges at $x = c$. On what other intervals might the power series converge?

A power series converges on exactly one of the following sets of x-values: (1) $x = c$ only; (2) all real numbers: $(-\infty,\infty)$; or (3) on the interval $(c-r, c+r)$, where r is the radius of convergence (i.e., $|x - c| < r$).

For example, a power series centered at $c = 2$ with a radius of convergence $r = 5$ converges for all x-values up to 5 units away from $x = 2$: $(2 - 5, 2 + 5) = (-3,7)$.

28.5 What is the difference between the *radius* of convergence and the *interval* of convergence of a power series?

If the power series $\sum\limits_{n=0}^{\infty} a_n (x-c)^n$ has radius of convergence r, then the series converges for x between $c - r$ and $c + r$. However, you must test the endpoints $x = c - r$ and $x = c + r$ individually to determine whether or not the series converges at each.

A series might converge at both of its endpoints, neither endpoint, or only one of them.

Note: Problems 28.6–28.9 refer to the power series $\sum\limits_{n=0}^{\infty} \left(-\dfrac{x}{4} \right)^n$.

28.6 Determine the radius of convergence for the series.

Apply the ratio test to determine where the series converges absolutely.

$$\lim_{n \to \infty} \left| \frac{a_{n+1}}{a_n} \right| = \lim_{n \to \infty} \left| \frac{\left(-\dfrac{x}{4} \right)^{n+1}}{\left(-\dfrac{x}{4} \right)^n} \right|$$

$$= \lim_{n \to \infty} \left| \left(\frac{x}{4} \right)^{n+1} \cdot \left(\frac{4}{x} \right)^n \right|$$

$$= \lim_{n \to \infty} \left| \frac{x^{n+1}}{4^{n+1}} \cdot \frac{4^n}{x^n} \right|$$

$$= \lim_{n \to \infty} \left| \frac{(x^n)(x)(4^n)}{(4^n)(4)(x^n)} \right|$$

$$= \lim_{n \to \infty} \left| \frac{x}{4} \right|$$

The value of the expression $\left| \dfrac{x}{4} \right|$ is unaffected as n approaches infinity, so

Because $\left| \dfrac{x}{4} \right|$ doesn't have any n's in it.

$\lim\limits_{n \to \infty} \left| \dfrac{x}{4} \right| = \left| \dfrac{x}{4} \right|$. According to the ratio test, the series converges only if the limit is

less than 1: $\left| \dfrac{x}{4} \right| < 1$. Multiply both sides of the inequality to solve for x: $|x| < 4$. Note

that this expression has form $|x - c| < r$ if $c = 0$ and $r = 4$. Therefore, the series

$\sum\limits_{n=0}^{\infty} \left(-\dfrac{x}{4} \right)^n$ has radius of convergence 4.

Note: Problems 28.6–28.9 refer to the power series $\sum_{n=0}^{\infty}\left(-\frac{x}{4}\right)^n$.

28.7 Does the series converge at the left endpoint of its interval of convergence?

According to Problem 28.6, the left endpoint of the interval of convergence is $x=-4$. Substitute that value into the power series.

$$\sum_{n=0}^{\infty}\left(-\frac{(-4)}{4}\right)^n = \sum_{n=0}^{\infty}\left(\frac{4}{4}\right)^n = \sum_{n=0}^{\infty}1^n$$

The series diverges according to the nth term divergence test.

Note: Problems 28.6–28.9 refer to the power series $\sum_{n=0}^{\infty}\left(-\frac{x}{4}\right)^n$.

28.8 Does the series converge at the right endpoint of its interval of convergence?

Ignoring $(-1)^n$, (which is normal during the alternating series test), all that's left in the series is 1, because technically $(-1)^n = 1(-1)^n$. That means the nth term is 1, not 0, so the series diverges.

Substitute the right endpoint, $x=4$, into the series.

$$\sum_{n=0}^{\infty}\left(-\frac{4}{4}\right)^n = \sum_{n=0}^{\infty}(-1)^n$$

The alternating series diverges according to the nth term divergence test.

Note: Problems 28.6–28.9 refer to the power series $\sum_{n=0}^{\infty}\left(-\frac{x}{4}\right)^n$.

28.9 Identify the interval of convergence of the series and verify the answer using the root test.

According to Problems 28.7 and 28.8, the series diverges at both of its endpoints, so the interval of convergence does not include either: $(-4,4)$. Apply the root test to determine whether the series converges absolutely.

You can plug positive and negative x's into this power series, but the ratio and root tests work only with positive terms, so you need absolute values.

$$\lim_{n\to\infty}\sqrt[n]{\left|\left(-\frac{x}{4}\right)^n\right|} = \lim_{n\to\infty}\left|\frac{x}{4}\right|$$

According to the root test, the series converges only if the limit is less than 1.

$$\left|\frac{x}{4}\right| < 1$$
$$|x| < 4$$

Therefore, the radius of convergence is $r=4$, which matches the radius of convergence determined by Problem 28.6. You already know that the series diverges at $x=-4$ and $x=4$ (according to Problems 28.7 and 28.8), so the interval of convergence is $(-4,4)$.

Note: Problems 28.10–28.12 refer to the power series $\displaystyle\sum_{n=0}^{\infty} \frac{(2n)x^n}{3^n}$.

28.10 Determine the radius of convergence of the series.

Apply the ratio test to determine where the series converges absolutely.

$$\lim_{n\to\infty}\left|\frac{a_{n+1}}{a_n}\right| = \lim_{n\to\infty}\left|\frac{\dfrac{2(n+1)x^{n+1}}{3^{n+1}}}{\dfrac{(2n)x^n}{3^n}}\right|$$

$$= \lim_{n\to\infty}\left|\frac{2(n+1)x^{n+1}}{3^{n+1}}\cdot\frac{3^n}{(2n)x^n}\right|$$

$$= \lim_{n\to\infty}\left|\frac{2(n+1)\left(x^n\right)(x)\left(3^n\right)}{\left(3^n\right)(3)(2n)\left(x^n\right)}\right|$$

$$= \lim_{n\to\infty}\left|\frac{n+1}{3n}\cdot x\right|$$

Note that $\displaystyle\lim_{n\to\infty}\frac{n+1}{3n} = \frac{1}{3}$.

$$= \left|\frac{1}{3}\cdot x\right|$$

$$= \left|\frac{x}{3}\right|$$

> The numerator and denominator have the same degree (1) so divide the coefficients of the n terms to get 1/3.

According to the ratio test, the series converges only if the limit is less than 1.

$$\left|\frac{x}{3}\right| < 1$$

Multiply each side by 3 to solve for x.

$$|x| < 3$$

A power series has radius of convergence r about $x = c$ given $|x - c| < r$. In this problem, $c = 0$ and $r = 3$, so the radius of convergence is 3, and the series converges on the interval $(c - r, c + r) = (0 - 3, 0 + 3) = (-3,3)$.

Note: Problems 28.10–28.12 refer to the power series $\displaystyle\sum_{n=0}^{\infty} \frac{(2n)x^n}{3^n}$.

28.11 Does the series converge at the left endpoint of its interval of convergence?

According to Problem 28.10, the left endpoint of the interval of convergence is $x = -3$. Substitute $x = -3$ into the power series to determine whether or not the series converges.

$$\sum_{n=0}^{\infty} \frac{(2n)(-3)^n}{3^n}$$

Note that $(-3)^n = (-1)^n \cdot 3^n$.

$$\sum_{n=0}^{\infty} \frac{(2n)(-1)^n \, \cancel{3^n}}{\cancel{3^n}} = \sum_{n=0}^{\infty} (-1)^n (2n)$$

This alternating series diverges according to the nth term divergence test: $\lim_{n \to \infty} 2n \neq 0$. Therefore, $\sum_{n=0}^{\infty} \frac{(2n)x^n}{3^n}$ diverges at $x = -3$.

Note: Problems 28.10–28.12 refer to the power series $\sum_{n=0}^{\infty} \frac{(2n)x^n}{3^n}$.

28.12 Does the series converge at the right endpoint of its interval of convergence? Indicate the interval of convergence for the series.

Substitute $x = 3$ into the power series.

$$\sum_{n=0}^{\infty} \frac{(2n)\cancel{3^n}}{\cancel{3^n}} = \sum_{n=0}^{\infty} 2n$$

The series diverges by the nth term divergence test. Because the series $\sum_{n=0}^{\infty} \frac{(2n)x^n}{3^n}$ diverges at $x = -3$ and $x = 3$, its interval of convergence is $(-3,3)$.

> Problem 28.10 says that the interval of convergence is AT LEAST (-3,3). The actual interval could have been [-3,3), (-3,3], or [-3,3], if either or both endpoints had worked.

Note: Problems 28.13–28.14 refer to the power series $\sum_{n=0}^{\infty} \frac{(-1)^n (x+2)^n}{n(2^n)}$.

28.13 Identify the radius of convergence for the series.

Use the ratio test to determine where the series converges absolutely.

$$\lim_{n \to \infty} \left| \frac{a_{n+1}}{a_n} \right| = \lim_{n \to \infty} \left| \frac{\dfrac{(x+2)^{n+1}}{(n+1)(2^{n+1})}}{\dfrac{(x+2)^n}{n(2^n)}} \right|$$

$$= \lim_{n \to \infty} \left| \frac{(x+2)^{n+1}}{(n+1)(2^{n+1})} \cdot \frac{n(2^n)}{(x+2)^n} \right|$$

$$= \lim_{n \to \infty} \left| \frac{\cancel{(x+2)^n}(x+2)(n)\cancel{(2^n)}}{(n+1)\cancel{(2^n)}(2)\cancel{(x+2)^n}} \right|$$

$$= \lim_{n \to \infty} \left| \frac{n(x+2)}{2(n+1)} \right|$$

$$= \lim_{n \to \infty} \left| \frac{n}{2n+2}(x+2) \right|$$

$$= \left| \frac{1}{2}(x+2) \right|$$

According to the ratio test, the series converges if $\left|\dfrac{1}{2}(x+2)\right| < 1$. Multiply both sides of the inequality by 2 to reach the form $|x-c| \le r$:

$$|x+2| < 2$$

The power series is centered at $c = -2$ and has radius of convergence $r = 2$.

> Take the opposite of the number in parentheses to get c, since the formula $|x-c| \le r$ contains $-c$.

Note: Problems 28.13–28.14 refer to the power series $\displaystyle\sum_{n=0}^{\infty} \dfrac{(-1)^n (x+2)^n}{n(2^n)}$.

28.14 Identify the interval of convergence for the series.

The series converges on the open interval $(c-r,\, c+r) = (-4, 0)$ and may converge at either (or both) of the endpoints as well. Substitute $x = -4$ and $x = 0$ into the series.

$$\sum_{n=0}^{\infty} \frac{(-1)^n (-4+2)^n}{n(2^n)} = \sum_{n=0}^{\infty} \frac{(-1)^n (-2)^n}{n(2^n)} = \sum_{n=0}^{\infty} \frac{(-1)^n (-1)^n\, 2^n}{n(2^n)} = \sum_{n=0}^{\infty} \frac{2^n}{n(2^n)} = \sum_{n=0}^{\infty} \frac{1}{n}$$

$$\sum_{n=0}^{\infty} \frac{(-1)^n (0+2)^n}{n(2^n)} = \sum_{n=0}^{\infty} \frac{(-1)^n\, 2^n}{n(2^n)} = \sum_{n=0}^{\infty} \frac{(-1)^n}{n}$$

> $(-1)^n(-1)^n = [(-1)^n]^2$. No matter what n is, you'll end up squaring -1, so $(-1)^n(-1)^n = 1$.

Substituting $x = -4$ into the power series results in $\displaystyle\sum_{n=0}^{\infty} \dfrac{1}{n}$, a divergent p-series; substituting $x = 0$ results in a convergent alternating series (according to the alternating series test). Therefore, the interval of convergence for $\displaystyle\sum_{n=0}^{\infty} \dfrac{(-1)^n (x+2)^n}{n(2^n)}$ is $(-4,0]$.

28.15 Identify the interval of convergence for the power series: $\displaystyle\sum_{n=0}^{\infty} \dfrac{(x+4)^n}{(n+1)(n+5)}$.

Apply the ratio test to determine the radius of convergence.

$$\lim_{n\to\infty} \left| \frac{\dfrac{(x+4)^{n+1}}{[(n+1)+1][(n+1)+5]}}{\dfrac{(x+4)^n}{(n+1)(n+5)}} \right| = \lim_{n\to\infty} \left| \frac{(x+4)^{n+1}}{(n+2)(n+6)} \cdot \frac{(n+1)(n+5)}{(x+4)^n} \right| = \lim_{n\to\infty} \left| \frac{n^2+6n+5}{n^2+8n+12}(x+4) \right| = |x+4|$$

According to the ratio test, $\displaystyle\sum_{n=0}^{\infty} \dfrac{(x+4)^n}{(n+1)(n+5)}$ converges when $|x+4| < 1$. The series is centered at $c = -4$, has radius of convergence $r = 1$, so it converges on the interval $(-5, -3)$. Substituting the endpoints of the interval into the series produces two convergent series, as demonstrated on the next page.

Use the comparison test: $\frac{1}{n^2+6n+5} < \frac{1}{n^2}$. All the terms in the series are less than the corresponding terms of the convergent p-series, so $\sum_{n=0}^{\infty} \frac{1}{n^2+6n+5}$ converges.

$$\sum_{n=0}^{\infty} \frac{(-5+4)^n}{(n+1)(n+5)} = \sum_{n=0}^{\infty} \frac{(-1)^n}{n^2+6n+5}$$

$$\sum_{n=0}^{\infty} \frac{(-3+4)^n}{(n+1)(n+5)} = \sum_{n=0}^{\infty} \frac{1^n}{n^2+6n+5} = \sum_{n=0}^{\infty} \frac{1}{n^2+6n+5}$$

Substituting $x = -5$ results in a convergent alternating series, and substituting $x = -3$ produces another convergent series. Therefore, both endpoints should be included in the interval of convergence: $[-5,-3]$.

28.16 Identify the interval of convergence for the power series: $\sum_{n=0}^{\infty} \frac{3^n(x-1)^n}{n!}$.

Apply the ratio test to determine the radius of convergence.

$$\lim_{n\to\infty} \left| \frac{\frac{3^{n+1}(x-1)^{n+1}}{(n+1)!}}{\frac{3^n(x-1)^n}{n!}} \right| = \lim_{n\to\infty} \left| \frac{3^{n+1}(x-1)^{n+1}}{(n+1)!} \cdot \frac{n!}{3^n(x-1)^n} \right| = \lim_{n\to\infty} \left| \frac{3^n \cdot 3 \cdot (x-1)^n (x-1) \cdot n!}{(n+1)(n!)(3^n)(x-1)^n} \right| = \lim_{n\to\infty} \left| \frac{3(x-1)}{n+1} \right| = 0$$

According to the ratio test, the series converges when this limit is less than 1. Because the limit equals 0 for any real number x, $\sum_{n=0}^{\infty} \frac{(x+4)^n}{(n+1)(n+5)}$ converges for all real numbers: $(-\infty, \infty)$.

28.17 Identify the radius of convergence for the power series: $\sum_{n=2}^{\infty} \frac{n!(x-3)^n}{(n+5)^2}$.

Apply the ratio test to determine where the series converges absolutely.

The degree of the numerator is 3 and the degree of the denominator is 2. Since $3 > 2$, the limit is ∞.

$$\lim_{n\to\infty} \frac{\frac{(n+1)!(x-3)^{n+1}}{[(n+1)+5]^2}}{\frac{n!(x-3)^n}{(n+5)^2}} = \lim_{n\to\infty} \left| \frac{(n+1)!(x-3)^{n+1}}{(n+6)^2} \cdot \frac{(n+5)^2}{n!(x-3)^n} \right| = \lim_{n\to\infty} \left| \frac{(n+1)(n+5)^2}{(n+6)^2}(x-3) \right| = \infty$$

The ratio test stipulates that the limit must be less than 1 in order for the series to converge. Clearly, $\infty > 1$, so $\sum_{n=2}^{\infty} \frac{n!(x-3)^n}{(n+5)^2}$ diverges for all values of x except for the center $x = 3$. The radius of convergence is $r = 0$.

Power series ALWAYS converge where they're centered, no matter what.

28.18 The power series $\sum\limits_{n=0}^{\infty} \dfrac{x^n}{n!} = 1 + x + \dfrac{x^2}{2!} + \dfrac{x^3}{3!} + \dfrac{x^4}{4!} + \cdots$ generates the exact values of the function $f(x) = e^x$. Verify that the power series, like e^x, is its own derivative.

> In other words, you'll get the same thing if you plug some number x into e^x and $\sum\limits_{n=0}^{\infty} \dfrac{x^n}{n!}$.

Differentiate each term of the convergent power series with respect to x.

$$\frac{d}{dx}\left[1 + x + \frac{x^2}{2!} + \frac{x^3}{3!} + \frac{x^4}{4!} + \cdots \right] = 0 + 1 + \frac{2x}{2!} + \frac{3x^2}{3!} + \frac{4x^3}{4!} + \cdots$$

$$= 1 + x + \frac{\cancel{2}x^2}{\cancel{2}\cdot 2!} + \frac{\cancel{4}x^3}{\cancel{4}\cdot 3!} + \cdots$$

$$= 1 + x + \frac{x^2}{2!} + \frac{x^3}{3!} + \cdots$$

Notice that the derivative is equivalent to the original power series.

> Problem 28.23 proves that this power series is convergent for all real numbers.

28.19 The power series $\sum\limits_{n=0}^{\infty} \dfrac{(-1)^n x^{2n+1}}{(2n+1)!}$ generates the exact values of $g(x) = \sin x$. What power series generates the values for the function $h(x) = \cos x$?

Because $\dfrac{d}{dx}(\sin x) = \cos x$, you can differentiate the convergent power series representing $\sin x$ to create the convergent power series representing $\cos x$.

$$\frac{d}{dx}\left(\frac{(-1)^n x^{2n+1}}{(2n+1)!} \right) = \frac{(-1)^n (2n+1) x^{2n+1-1}}{(2n+1)!}$$

> $\dfrac{(-1)^n}{(2n+1)!}$ is a constant, so ignore it while you take the derivative of x^{2n+1} with the power rule.

Write $(2n+1)!$ as $(2n+1)[(2n+1)-1]! = (2n+1)(2n)!$.

$$= \frac{(-1)^n (2n+1) x^{2n}}{(2n+1)(2n)!}$$

$$= \frac{(-1)^n x^{2n}}{(2n)!}$$

Therefore, $\cos x = \sum\limits_{n=0}^{\infty} \dfrac{(-1)^n x^{2n}}{(2n)!}$.

Taylor and Maclaurin Series
Series that approximate function values

28.20 Define the pth Maclaurin polynomial of a function $f(x)$, assuming $f(x)$ is differentiable at least p times.

> $f^{(n)}$ means the nth derivative of $f(x)$. You can use primes for the first through third derivatives (f', f'', and f'''), but use parentheses for fourth derivatives and higher ($f^{(4)}$, $f^{(5)}$, and $f^{(6)}$).

The Maclaurin polynomial is defined as the first p terms of an infinite series.

$$\sum_{n=0}^{p} \frac{f^{(n)}(0)}{n!} x^n = f(0) + \frac{f'(0)}{1!}x + \frac{f''(0)}{2!}x^2 + \frac{f'''(0)}{3!}x^3 + \frac{f^{(4)}(0)}{4!} + \cdots + \frac{f^{(p)}(0)}{p!}x^p$$

28.21 What is the difference between the pth Maclaurin polynomial of a function $f(x)$ and the pth Taylor polynomial?

A Taylor polynomial is also defined as the first p terms of an infinite series.

$$\sum_{n=0}^{p} \frac{f^{(n)}(c)}{n!} (x-c)^n = f(c) + \frac{f'(c)}{1!}(x-c) + \frac{f''(c)}{2!}(x-c)^2 + \frac{f'''(c)}{3!}(x-c)^3 + \cdots + \frac{f^{(p)}(c)}{p!}(x-c)^p$$

Both the Maclaurin and Taylor series are power series of form $\sum_{n=0}^{\infty} a_n (x-c)^n$, where $a_n = \dfrac{f^{(n)}(c)}{n!}$. However, Maclaurin series are always centered about $c = 0$, whereas Taylor series can be centered about any real number c.

Note: Problems 28.22–28.23 refer to the power series $\sum_{n=0}^{\infty} \dfrac{x^n}{n!} = 1 + x + \dfrac{x^2}{2!} + \dfrac{x^3}{3!} + \dfrac{x^4}{4!} + \cdots$, *originally introduced in Problem 28.18.*

28.22 Demonstrate that the power series is actually the Maclaurin series for $f(x) = e^x$.

> This is not usually true—only a few functions are equal to their derivatives, and e^x is one of them.

The terms of a Maclaurin series have the form $\sum_{n=0}^{\infty} \dfrac{f^{(n)}(0)}{n!} x^n$; the nth term contains the nth derivative of $f(x)$. Note that the nth derivative of e^x (with respect to x) is e^x.

$$e^x = f(x) = f'(x) = f''(x) = f'''(x) = f^{(4)}(x) = \cdots$$

Expand the first few terms of the Maclaurin series to verify that its terms correspond with the terms of the series $1 + x + \dfrac{x^2}{2!} + \dfrac{x^3}{3!} + \dfrac{x^4}{4!} + \cdots$.

> $f^{(0)}$ is the "zeroth" derivatives of $f(x)$, which is just $f(x)$.

$$\sum_{n=0}^{\infty} \frac{f^{(n)}(0)}{n!} x^n = \frac{f^{(0)}(0)}{0!}x^0 + \frac{f'(0)}{1!}x^1 + \frac{f''(0)}{2!}x^2 + \frac{f'''(0)}{3!}x^3 + \frac{f^{(4)}(0)}{4!}x^4 + \cdots$$

$$= \frac{e^0}{0!}x^0 + \frac{e^0}{1!}x^1 + \frac{e^0}{2!}x^2 + \frac{e^0}{3!}x^3 + \frac{e^0}{4!}x^4 + \cdots$$

$$= 1 + \frac{x}{1} + \frac{x^2}{2!} + \frac{x^3}{3!} + \frac{x^4}{4!} + \cdots$$

Note: Problems 28.22–28.23 refer to the power series $\sum_{n=0}^{\infty} \dfrac{x^n}{n!} = 1 + x + \dfrac{x^2}{2!} + \dfrac{x^3}{3!} + \dfrac{x^4}{4!} + \cdots$, originally introduced in Problem 28.18.

28.23 Prove that the series converges for all real numbers.

Apply the ratio test to determine the absolute convergence of the power series.

$$\lim_{n \to \infty} \left| \frac{a_{n+1}}{a_n} \right| = \lim_{n \to \infty} \left| \frac{\dfrac{x^{n+1}}{(n+1)!}}{\dfrac{x^n}{n!}} \right| = \lim_{n \to \infty} \left| \frac{x^{n+1}}{(n+1)!} \cdot \frac{n!}{x^n} \right| = \lim_{n \to \infty} \left| \frac{(x^n)(x)(n!)}{(n+1)(n!)(x^n)} \right| = \lim_{n \to \infty} \left| \frac{1}{n+1} \cdot x \right| = |0 \cdot x| = 0$$

According to the ratio test, the series converges when this limit is less than 1. Because $0 < 1$ for all x, the series converges for all real numbers.

> No matter what x you plug in, $|0 \times| = 0$.

Note: Problems 28.24–28.25 refer to the power series $\sum_{n=0}^{\infty} \dfrac{(-1)^n x^{2n+1}}{(2n+1)!}$, originally introduced in Problem 28.19.

28.24 Demonstrate that the power series is actually the Maclaurin series for $g(x) = \sin x$.

Expand the first five terms of the series ($n = 0$ through $n = 4$) in order to discern a pattern.

$$\sum_{n=0}^{\infty} \frac{(-1)^n x^{2n+1}}{(2n+1)!} = \frac{x^1}{1!} - \frac{x^3}{3!} + \frac{x^5}{5!} - \frac{x^7}{7!} + \frac{x^9}{9!} - \cdots$$

Calculate derivatives of $g(x) = \sin x$ and evaluate each at $x = 0$.

$g(x) = \sin x$	$g'(x) = \cos x$	$g''(x) = -\sin x$	$g'''(x) = -\cos x$	$g^{(4)}(x) = \sin x$	$g^{(5)}(x) = \cos x$
$g(0) = \sin 0$	$g'(0) = \cos 0$	$g''(0) = -\sin 0$	$g'''(0) = -\cos 0$	$g^{(4)}(0) = \sin 0$	$g^{(5)}(0) = \cos 0$
$= 0$	$= 1$	$= 0$	$= -1$	$= 0$	$= 1$

Note that the derivative values repeat: 0, 1, 0, –1, 0, 1, 0, –1, ... Substitute the values of $g(x)$ and its derivatives into the Maclaurin series formula.

$$\sum_{n=0}^{\infty} \frac{g^{(n)}(0)}{n!} x^n = \frac{g(0)}{0!} x^0 + \frac{g'(0)}{1!} x^1 + \frac{g''(0)}{2!} x^2 + \frac{g'''(0)}{3!} x^3 + \frac{g^{(4)}(0)}{4!} x^4 + \cdots$$

$$= \frac{0}{1} \cdot 1 + \frac{1}{1!} \cdot x + \frac{0}{2!} \cdot x^2 + \frac{-1}{3!} \cdot x^3 + \frac{0}{4!} x^4 + \cdots$$

$$= \frac{x}{1!} - \frac{x^3}{3!} + \cdots$$

When n is even, the corresponding term of $\sum_{n=0}^{\infty} \dfrac{g^{(n)}(0)}{n!} x^n$ equals 0. Expand the series through the $n = 9$ term to verify that the terms match the power series expanded at the outset of the problem.

$$\sum_{n=0}^{\infty} \frac{g^{(n)}(0)}{n!} x^n = \frac{x^1}{1} - \frac{x^3}{3!} + \frac{x^5}{5!} - \frac{x^7}{7!} + \frac{x^9}{9} - \cdots$$

Note: Problems 28.24–28.25 refer to the power series $\sum\limits_{n=0}^{\infty} \dfrac{(-1)^n \, x^{2n+1}}{(2n+1)!}$, *originally introduced in Problem 28.19.*

28.25 Prove that the series converges for all real numbers.

x^{2n+1} and x^2 have the same bases, so multiplying them together means adding the exponents: $x^{2n+1+2} = x^{2n+3}$. It's a good idea to write x^{2n+3} this way because it lets you cancel the factor out with the x^{2n+1} in the denominator.

Apply the ratio test to determine the absolute convergence of the power series.

$$\lim_{n\to\infty}\left|\frac{a_{n+1}}{a_n}\right| = \lim_{n\to\infty}\left|\frac{\dfrac{x^{2(n+1)+1}}{[2(n+1)+1]!}}{\dfrac{x^{2n+1}}{(2n+1)!}}\right|$$

$$= \lim_{n\to\infty}\left|\frac{x^{2n+3}}{(2n+3)!} \cdot \frac{(2n+1)!}{x^{2n+1}}\right|$$

Note that $x^{2n+3} = (x^{2n+1})(x^2)$ because $(2n+1)+2 = 2n+3$.

Writing $(2n+3)!$ as $(2n+3)(2n+2)(2n+1)!$ is the same as writing $12!$ as $(12)(11)(10!)$. Subtract 1 from each factor as you go.

$$= \lim_{n\to\infty}\left|\frac{(x^{2n+1})(x^2)(2n+1)!}{(2n+3)(2n+2)(2n+1)!(x^{2n+1})}\right|$$

$$= \lim_{n\to\infty}\left|\frac{1}{4n^2+10n+6} \cdot x^2\right|$$

Note that $\lim\limits_{n\to\infty}\left|\dfrac{1}{4n^2+10n+6}\right| = 0$, so $\lim\limits_{n\to\infty}\left|\dfrac{1}{4n^2+10n+6} \cdot x^2\right| = 0 \cdot x^2 = 0$. According to the ratio test, the series converges when this limit is less than 1. Because $0 < 1$ for all x, $\sum\limits_{n=0}^{\infty} \dfrac{(-1)^n \, x^{2n+1}}{(2n+1)!}$ converges for all real numbers.

28.26 Write the sixth-degree Maclaurin polynomial for $h(x) = \cos x$ and verify that its terms are generated by the series identified by Problem 28.19:

$$\cos x = \sum_{n=0}^{\infty} \frac{(-1)^n \, x^{2n}}{(2n)!}.$$

Differentiate $h(x) = \cos x$ six times and evaluate each derivative at $x = 0$.

When n is odd, $h^{(n)} = 0$, so the series will contain only x raised to even powers.

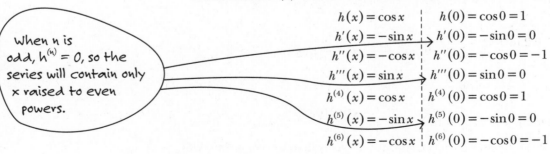

$$\begin{aligned}
h(x) &= \cos x & h(0) &= \cos 0 = 1 \\
h'(x) &= -\sin x & h'(0) &= -\sin 0 = 0 \\
h''(x) &= -\cos x & h''(0) &= -\cos 0 = -1 \\
h'''(x) &= \sin x & h'''(0) &= \sin 0 = 0 \\
h^{(4)}(x) &= \cos x & h^{(4)}(0) &= \cos 0 = 1 \\
h^{(5)}(x) &= -\sin x & h^{(5)}(0) &= -\sin 0 = 0 \\
h^{(6)}(x) &= -\cos x & h^{(6)}(0) &= -\cos 0 = -1
\end{aligned}$$

Substitute $h(0)$, $h'(0)$, $h''(0)$, $\cdots$, $h^{(6)}(0)$ into the Maclaurin polynomial formula.

$$\sum_{n=0}^{6} \frac{h^{(n)}(0)}{n!} x^n = \frac{h(0)}{0!} x^0 + \frac{h'(0)}{1!} x^1 + \frac{h''(0)}{2!} x^2 + \frac{h'''(0)}{3!} x^3 + \frac{h^{(4)}(0)}{4!} x^4 + \frac{h^{(5)}(0)}{5!} x^5 + \frac{h^{(6)}(0)}{6!} x^6$$

$$= \frac{1}{1} \cdot 1 + \frac{0}{1!} \cdot x + \frac{-1}{2!} \cdot x^2 + \frac{0}{3!} \cdot x^3 + \frac{1}{4!} \cdot x^4 + \frac{0}{5!} \cdot x^5 + \frac{-1}{6!} \cdot x^6$$

$$= 1 - \frac{x^2}{2!} + \frac{x^4}{4!} - \frac{x^6}{6!}$$

Expand the series $\sum_{n=0}^{3} \frac{(-1)^n x^{2n}}{(2n!)}$ to verify that its terms match those of the Maclaurin series.

$$\sum_{n=0}^{3} \frac{(-1)^n x^{2n}}{(2n)!} = \frac{(-1)^0 x^0}{0!} + \frac{(-1)^1 x^{2(1)}}{(2 \cdot 1)!} + \frac{(-1)^2 x^{2(2)}}{(2 \cdot 2)!} + \frac{(-1)^3 x^{2(3)}}{(2 \cdot 3)!}$$

$$= 1 - \frac{x^2}{2!} + \frac{x^4}{4!} - \frac{x^6}{6!}$$

28.27 Given $\cos 0.2 \approx 0.98006657784124$, estimate $\cos 0.2$ using the second, fourth, and sixth degree Maclaurin polynomials for $h(x) = \cos x$ to demonstrate that the larger the degree of the approximating polynomial, the more accurate the approximation.

According to Problem 28.26, the Maclaurin series for $y = \cos x$ is $\sum_{n=0}^{\infty} \frac{(-1)^n x^{2n}}{(2n!)}$.

The first two terms of the series constitute the second degree Maclaurin polynomial. Add an additional term to create the fourth degree Maclaurin polynomial and then another to create the sixth degree polynomial.

> Don't get hung up on the degree of the Maclaurin polynomial versus the number of terms it has versus the upper limit of the summation notation. Just remember this: a second degree Maclaurin polynomial ends with x^2.

$$\sum_{n=0}^{1} \frac{(-1)^n x^{2n}}{(2n!)} = 1 - \frac{x^2}{2!}$$

$$\sum_{n=0}^{2} \frac{(-1)^n x^{2n}}{(2n!)} = 1 - \frac{x^2}{2!} + \frac{x^4}{4!}$$

$$\sum_{n=0}^{3} \frac{(-1)^n x^{2n}}{(2n!)} = 1 - \frac{x^2}{2!} + \frac{x^4}{4!} - \frac{x^6}{6!}$$

Substitute $n = 0.2$ into each polynomial to approximate $\cos (0.2)$.

$$1 - \frac{(0.2)^2}{2!} = 0.98$$

$$1 - \frac{(0.2)^2}{2!} + \frac{(0.2)^4}{4!} \approx 0.98006666666667$$

$$1 - \frac{(0.2)^2}{2!} + \frac{(0.2)^4}{4!} - \frac{(0.2)^6}{6!} \approx 0.98006657777778$$

Although the second degree Maclaurin polynomial produces a fairly accurate approximation of $\cos (0.2)$, the sixth degree polynomial generates an approximation accurate to nine decimal places.

28.28 Approximate ln (2.95) using a fourth degree Maclaurin polynomial, given ln 3 ≈ 1.09861228867.

Maclaurin polynomials only provide accurate function approximations for x-values very close to 0. Begin by constructing the function $f(x) = \ln(3 - x)$. Calculate the first four derivatives of $\ln(3 - x)$ and evaluate each at $x = 0$.

> Use a function that almost equals ln 2.95 when x = 0. The simplest one is f(x) = ln (3 − x) because 3 is an integer and f(0) = ln 3, which is close to ln 2.95.

$$f(x) = \ln(3 - x)$$

$$f'(x) = -(3 - x)^{-1} = -\frac{1}{3 - x}$$

$$f''(x) = -(3 - x)^{-2} = -\frac{1}{(3 - x)^2}$$

$$f'''(x) = -2(3 - x)^{-3} = -\frac{2}{(3 - x)^3}$$

$$f^{(4)}(x) = -6(3 - x)^{-4} = -\frac{6}{(3 - x)^4}$$

$$f(0) = \ln(3 - 0) = \ln 3$$

$$f'(0) = -\frac{1}{3 - 0} = -\frac{1}{3}$$

$$f''(0) = -\frac{1}{(3 - 0)^2} = -\frac{1}{9}$$

$$f'''(0) = -\frac{2}{(3 - 0)^3} = -\frac{2}{27}$$

$$f^{(4)}(0) = -\frac{6}{(3 - 0)^4} = -\frac{2}{27}$$

Apply the Maclaurin series formula.

$$f(x) \approx \sum_{n=0}^{4} \frac{f^{(n)}(0)}{n!} x^n$$

$$\approx \frac{f(0)}{0!} x^0 + \frac{f'(0)}{1!} x^1 + \frac{f''(0)}{2!} x^2 + \frac{f'''(0)}{3!} x^3 + \frac{f^{(4)}(0)}{4!} x^4$$

$$\approx \left(\frac{\ln 3}{1}\right) \cdot 1 + \left(\frac{-1/3}{1}\right) \cdot x + \left(\frac{-1/9}{2!}\right) x^2 + \left(\frac{-2/27}{3!}\right) x^3 + \left(\frac{-2/27}{4!}\right) x^4$$

$$\approx \ln 3 - \frac{x}{3} - \frac{x^2}{18} - \frac{x^3}{81} - \frac{x^4}{324}$$

Approximate ln (2.95) by substituting $x = 0.05$ and the given estimate of ln 3 into the Maclaurin polynomial.

> The actual value of ln 2.95 is 1.0818051703517....

$$f(x) \approx \ln 3 - \frac{x}{3} - \frac{x^2}{18} - \frac{x^3}{162} - \frac{x^4}{324}$$

$$f(0.05) \approx \ln 3 - \frac{0.05}{3} - \frac{(0.05)^2}{18} - \frac{(0.05)^3}{81} - \frac{(0.05)^4}{324}$$

$$\approx 1.08180517061$$

Note: Problems 28.29–28.30 refer to the function f(x) = sin x.

28.29 Identify the fourth degree Taylor polynomial for $f(x)$ centered at $c = \dfrac{3\pi}{2}$.

Differentiate $f(x) = \sin x$ four times and evaluate each derivative at $c = \dfrac{3\pi}{2}$.

When you're building Maclaurin polynomials, you plug 0 into the derivatives, but when you're working with Taylor polynomials, plug in x = c.

$$f(x) = \sin x \qquad f\left(\frac{3\pi}{2}\right) = \sin\frac{3\pi}{2} = -1$$

$$f'(x) = \cos x \qquad f'\left(\frac{3\pi}{2}\right) = \cos\frac{3\pi}{2} = 0$$

$$f''(x) = -\sin x \qquad f''\left(\frac{3\pi}{2}\right) = -\sin\frac{3\pi}{2} = 1$$

$$f'''(x) = -\cos x \qquad f'''\left(\frac{3\pi}{2}\right) = -\cos\frac{3\pi}{2} = 0$$

$$f^{(4)}(x) = \sin x \qquad f^{(4)}\left(\frac{3\pi}{2}\right) = \sin\frac{3\pi}{2} = -1$$

Apply the Taylor series formula.

$$\sin x \approx \sum_{n=0}^{4} \frac{f^{(n)}(c)}{n!}(x-c)^n$$

$$\approx \left(\frac{f(3\pi/2)}{0!}\right)\left(x-\frac{3\pi}{2}\right)^0 + \left(\frac{f'(3\pi/2)}{1!}\right)\left(x-\frac{3\pi}{2}\right)^1 + \left(\frac{f''(3\pi/2)}{2!}\right)\left(x-\frac{3\pi}{2}\right)^2 +$$

$$\left(\frac{f'''(3\pi/2)}{3!}\right)\left(x-\frac{3\pi}{2}\right)^3 + \left(\frac{f^{(4)}(3\pi/2)}{4!}\right)\left(x-\frac{3\pi}{2}\right)^4$$

$$\approx -1 + \frac{0}{1!}\left(x-\frac{3\pi}{2}\right) + \frac{1}{2!}\left(x-\frac{3\pi}{2}\right)^2 + \frac{0}{3!}\left(x-\frac{3\pi}{2}\right)^3 + \frac{-1}{4!}\left(x-\frac{3\pi}{2}\right)^4$$

$$\approx -1 + \frac{1}{2}\left(x-\frac{3\pi}{2}\right)^2 - \frac{1}{24}\left(x-\frac{3\pi}{2}\right)^4$$

Note: Problems 28.29–28.30 refer to the function f(x) = sin x.

28.30 Estimate $\sin\dfrac{5\pi}{3}$ using the Taylor polynomial generated in Problem 28.29. Compare the approximation to the actual value and identify two ways you could better approximate $\sin\dfrac{5\pi}{3}$ using a Taylor polynomial.

Substitute $x = \dfrac{5\pi}{3}$ into the polynomial generated in Problem 28.29.

$$\sin\frac{5\pi}{3} \approx -1 + \frac{1}{2}\left(\frac{5\pi}{3} - \frac{3\pi}{2}\right)^2 - \frac{1}{24}\left(\frac{5\pi}{3} - \frac{3\pi}{2}\right)^4$$

$$\approx -1 + \frac{1}{2}\left(\frac{10\pi - 9\pi}{6}\right)^2 - \frac{1}{24}\left(\frac{10\pi - 9\pi}{6}\right)^4$$

$$\approx -1 + \frac{1}{2}\left(\frac{\pi}{6}\right)^2 - \frac{1}{24}\left(\frac{\pi}{6}\right)^4$$

$$\approx -0.86605388341574$$

Recall, from the unit circle, that $\sin\dfrac{5\pi}{3} = -\dfrac{\sqrt{3}}{2} \approx -0.86602540378443$, so the approximation is accurate to four decimal places. Increase the accuracy of the approximation by increasing the degree of the Taylor polynomial or centering the polynomial about a c-value closer to $\dfrac{5\pi}{3}$ than $\dfrac{3\pi}{2}$.

28.31 Approximate ln 2.95 using the fourth degree Taylor polynomial for $f(x) = \ln x$ centered about $c = 3$. Compare the results to the approximation generated by the Maclaurin polynomial of equal degree calculated by Problem 28.28.

> If you don't feel like rewriting a function to use a Maclaurin polynomial (like you had to do in Problem 28.28), use a Taylor polynomial instead, since you can center it at any x-value, not just x = 0.

Differentiate $f(x) = \ln x$ four times and evaluate each derivative at $c = 3$.

$$f(x) = \ln x \qquad\qquad f(3) = \ln 3$$

$$f'(x) = x^{-1} = \frac{1}{x} \qquad f'(3) = \frac{1}{3}$$

$$f''(x) = -x^{-2} = -\frac{1}{x^2} \qquad f''(3) = -\frac{1}{9}$$

$$f'''(x) = 2x^{-3} = \frac{2}{x^3} \qquad f'''(3) = \frac{2}{27}$$

$$f^{(4)}(x) = -6x^{-4} = -\frac{6}{x^4} \qquad f^{(4)}(3) = -\frac{6}{81} = -\frac{2}{27}$$

Apply the Taylor series formula.

$$\ln x \approx \sum_{n=0}^{4} \frac{f^{(n)}(3)}{n!}(x-3)^n$$

$$\approx \frac{f(3)}{0!}\cdot 1 + \left(\frac{f'(3)}{1!}\right)(x-3) + \left(\frac{f''(3)}{2!}\right)(x-3)^2 + \left(\frac{f'''(3)}{3!}\right)(x-3)^3 + \left(\frac{f^{(4)}(3)}{4!}\right)(x-3)^4$$

$$\approx \ln 3 + \frac{1}{3}(x-3) - \frac{1}{9\cdot 2!}(x-3)^2 + \frac{2}{27\cdot 3!}(x-3)^3 - \frac{2}{27\cdot 4!}(x-3)^4$$

$$\approx \ln 3 + \frac{1}{3}(x-3) - \frac{1}{18}(x-3)^2 + \frac{2}{162}(x-3)^3 - \frac{2}{648}(x-3)^4$$

Approximate ln (2.95) by substituting $x = 2.95$ into the polynomial.

$$\ln 2.95 \approx \ln 3 + \frac{1}{3}(2.95-3) - \frac{1}{18}(2.95-3)^2 + \frac{2}{162}(2.95-3)^3 - \frac{2}{648}(2.95-3)^4$$

$$\approx \ln 3 - \frac{1}{3}(0.05) - \frac{1}{18}(0.05)^2 - \frac{1}{81}(0.05)^3 - \frac{1}{324}(0.05)^4$$

This approximation is exactly equal to the approximation in Problem 28.28, so the Taylor and Maclaurin polynomial estimates of ln 2.95 are identical.

Appendix A

Important graphs to memorize and graph transformations

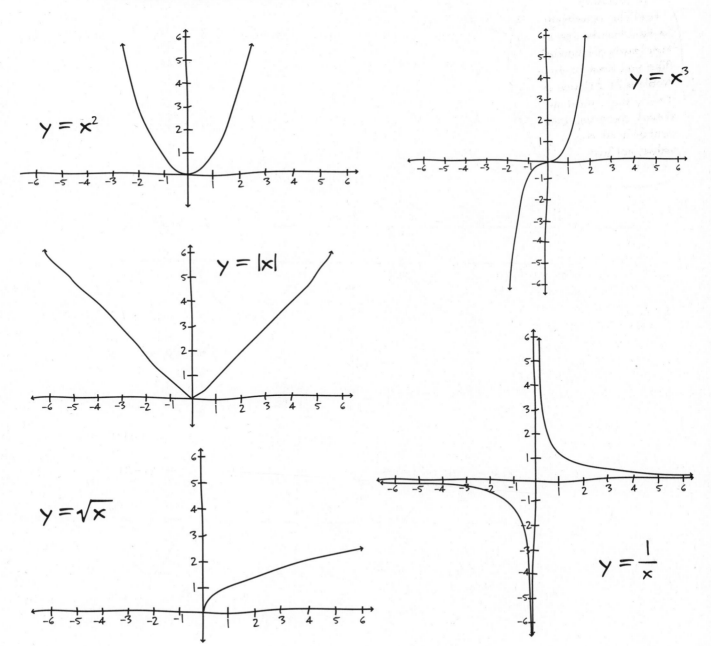

$y = x^2$

$y = x^3$

$y = |x|$

$y = \sqrt{x}$

$y = \dfrac{1}{x}$

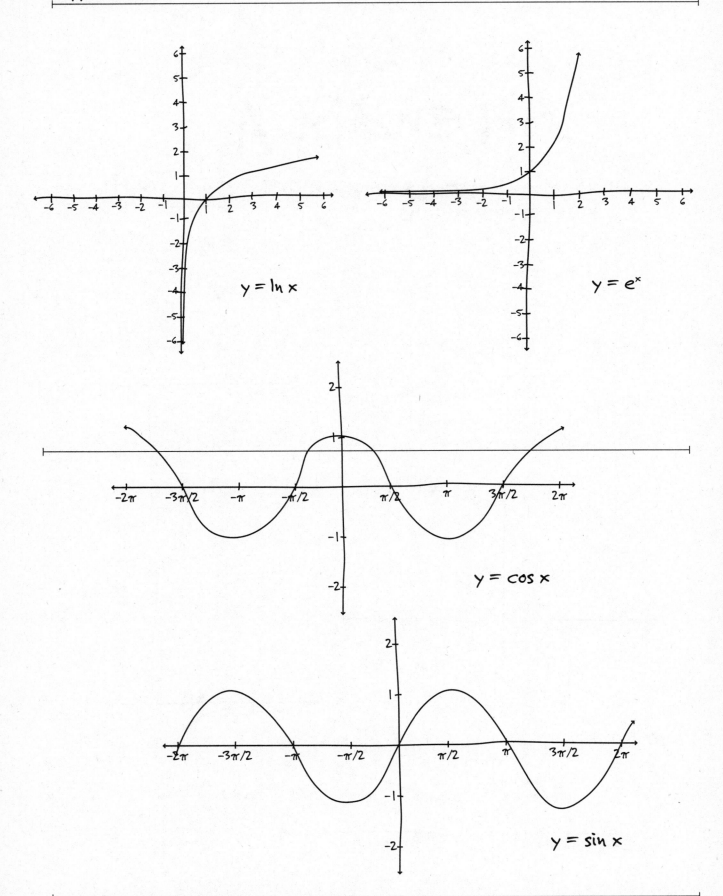

$y = \ln x$

$y = e^x$

$y = \cos x$

$y = \sin x$

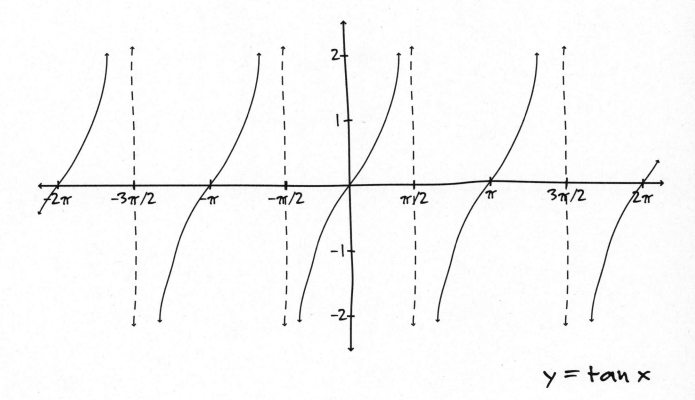

$y = \tan x$

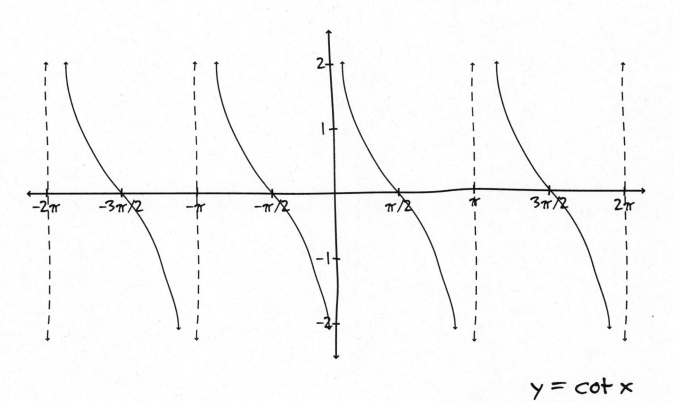

$y = \cot x$

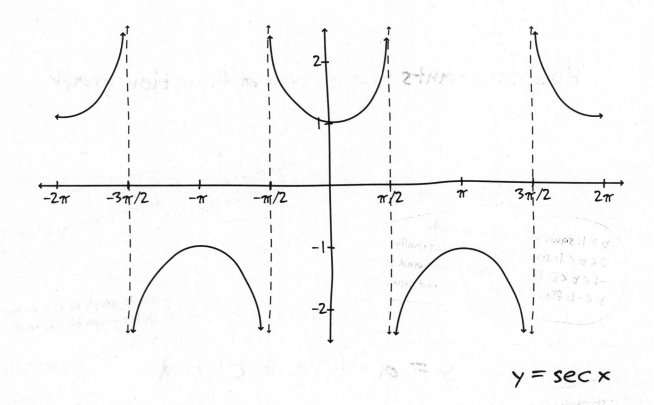

$y = \sec x$

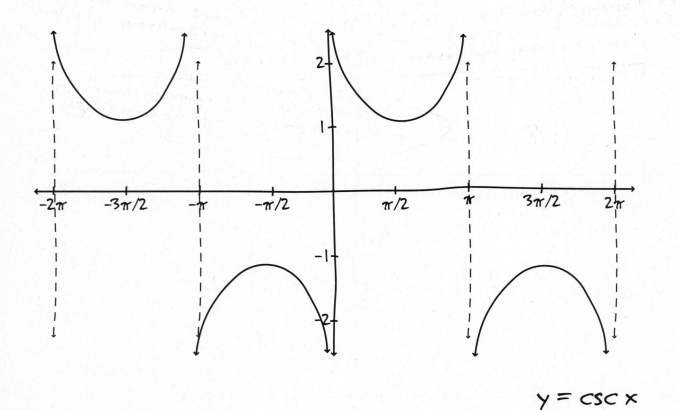

$y = \csc x$

How constants transform a function graph

Absolute value of $a \cdot f(bx + c) + d$ reflects portions of $f(x)$ below the x-axis across the x-axis

$b > 1$: squishes $f(x)$ horizontally
$0 < b < 1$: stretches $f(x)$ horizontally
$-1 < b < 0$: flips $f(x)$ over y-axis and stretches it horizontally
$b < -1$: flips $f(x)$ over y-axis and squishes it horizontally

$c > 0$: shifts graph of $f(x)$ to the left
$c < 0$: shifts graph of $f(x)$ to the right

$$y = a \cdot f(bx + c) + d$$

$a > 1$: stretches $f(x)$ vertically
$0 < a < 1$: squishes $f(x)$ vertically
$-1 < a < 0$: flips $f(x)$ over x-axis and squishes it vertically
$a < -1$: flips $f(x)$ over x-axis and stretches it vertically

$d > 0$: shifts graph of $f(x)$ up
$d < 0$: shifts graph of $f(x)$ down

Absolute value of $(bx + c)$ erases graph left of y-axis and replaces it with reflection of $f(x)$ across the y-axis

Appendix B

The unit circle

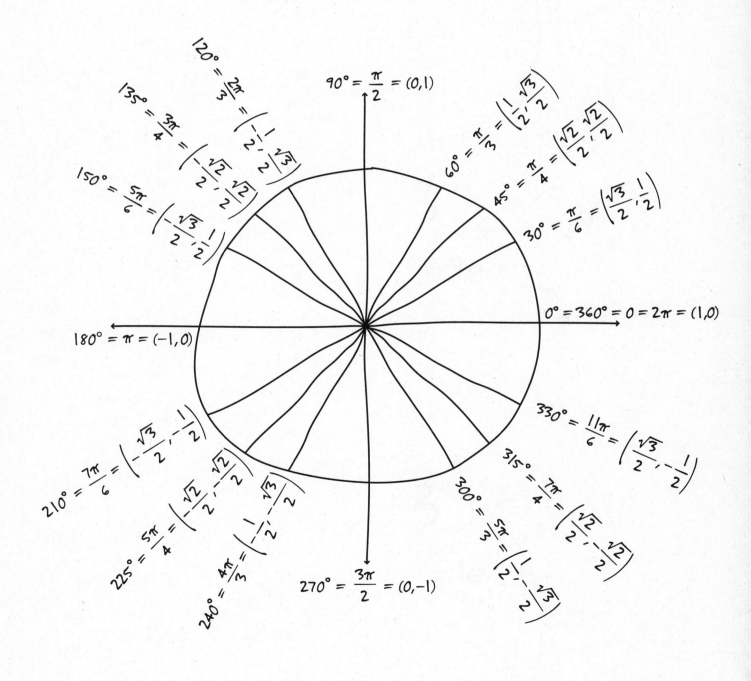

Appendix C
Trigonometric identities

Reciprocal Identities

$$\cos x = \frac{1}{\sec x}$$

$$\sin x = \frac{1}{\csc x}$$

$$\tan x = \frac{\sin x}{\cos x} = \frac{1}{\cot x}$$

$$\cot x = \frac{\cos x}{\sin x} = \frac{1}{\tan x}$$

$$\sec x = \frac{1}{\cos x}$$

$$\csc x = \frac{1}{\sin x}$$

Cofunction Identities

$$\cos\left(\frac{\pi}{2} - x\right) = \sin x \qquad \sin\left(\frac{\pi}{2} - x\right) = \cos x$$

$$\tan\left(\frac{\pi}{2} - x\right) = \cot x \qquad \cot\left(\frac{\pi}{2} - x\right) = \tan x$$

$$\sec\left(\frac{\pi}{2} - x\right) = \csc x \qquad \csc\left(\frac{\pi}{2} - x\right) = \sec x$$

Pythagorean Identities

$$\cos^2 x + \sin^2 x = 1 \qquad 1 + \tan^2 x = \sec^2 x$$

$$1 + \cot^2 x = \csc^2 x$$

Double Angle Identities

$$\sin 2x = 2\sin x \cos x$$

$$\cos 2x = \cos^2 x - \sin^2 x = 2\cos^2 x - 1 = 1 - 2\sin^2 x$$

$$\tan 2x = \frac{2\tan x}{1 - \tan^2 x}$$

Sum and Difference Identities

$$\cos(A \pm B) = \cos A \cos B \mp \sin A \sin B$$

$$\sin(A \pm B) = \sin A \cos B \pm \cos A \sin B$$

$$\tan(A \pm B) = \frac{\tan A \pm \tan B}{1 \mp \tan A \tan B}$$

Product-to-Sum Identities

$$\cos A \cos B = \frac{\cos(A - B) + \cos(A + B)}{2}$$

$$\sin A \sin B = \frac{\cos(A - B) - \cos(A + B)}{2}$$

$$\sin A \cos B = \frac{\sin(A + B) + \sin(A - B)}{2}$$

Power-Reduction Formulas

$$\cos^2 x = \frac{1 + \cos 2x}{2} \qquad \sin^2 x = \frac{1 - \cos 2x}{2}$$

Appendix D
Derivative Formulas

Trig Derivatives

$$\frac{d}{dx}(\cos x) = -\sin x$$

$$\frac{d}{dx}(\sin x) = \cos x$$

$$\frac{d}{dx}(\tan x) = \sec^2 x$$

$$\frac{d}{dx}(\cot x) = -\csc^2 x$$

$$\frac{d}{dx}(\sec x) = \sec x \tan x$$

$$\frac{d}{dx}(\csc x) = -\csc x \cot x$$

Exponential/Log Derivatives

$$\frac{d}{dx}(e^x) = e^x$$

$$\frac{d}{dx}(\ln x) = \frac{1}{x}$$

$$\frac{d}{dx}(a^x) = (\ln a)\, a^x$$

$$\frac{d}{dx}(\log_a x) = \frac{1}{x\,(\ln a)}$$

Inverse Trig Derivatives

$$\frac{d}{dx}(\arcsin x) = \frac{1}{\sqrt{1-x^2}}$$

$$\frac{d}{dx}(\arccos x) = -\frac{1}{\sqrt{1-x^2}}$$

$$\frac{d}{dx}(\arctan x) = \frac{1}{1+x^2}$$

$$\frac{d}{dx}(\text{arccot}\, x) = -\frac{1}{1+x^2}$$

$$\frac{d}{dx}(\text{arcsec}\, x) = \frac{1}{|x|\sqrt{x^2-1}}$$

$$\frac{d}{dx}(\text{arccsc}\, x) = -\frac{1}{|x|\sqrt{x^2-1}}$$

Appendix E

Antiderivative Formulas

Trig Antiderivatives

$\int \cos x \, dx = \sin x + C$

$\int \sin x \, dx = -\cos x + C$

$\int \tan x \, dx = -\ln |\cos x| + C$

$\int \cot x \, dx = \ln |\sin x| + C$

$\int \sec x \, dx = \ln |\sec x + \tan x| + C$

$\int \csc x \, dx = -\ln |\csc x + \cot x| + C$

Exponential/Log Antiderivatives

$\int e^x \, dx = e^x + C$

$\int \ln x \, dx = x \ln x - x + C$

Inverse Trig Antiderivatives

$\int \dfrac{du}{\sqrt{a^2 - u^2}} = \arcsin \dfrac{u}{a} + C$

$\int \dfrac{du}{a^2 + u^2} = \dfrac{1}{a} \arctan \dfrac{u}{a} + C$

$\int \dfrac{du}{u\sqrt{u^2 - a^2}} = \dfrac{1}{a} \operatorname{arcsec} \dfrac{|u|}{a} + C$

Index

ALPHABETICAL LIST OF CONCEPTS WITH PROBLEM NUMBERS

This comprehensive index organizes the concepts and skills discussed within the book alphabetically. Each entry is accompanied by one or more problem numbers in which the topics are most prominently featured.

All of these numbers refer to problems, not pages, in the book. For example, 8.2 is the second problem in Chapter 8.

S

T

U–V

washer method: 22.19–22.28, 22.30
wiggle graph: see *sign graph*